ENCYCLOPEDIA OF MATHEMATICS AND ITS APPLICATIONS

EDITED BY G.-C. ROTA

Volume 28

Rational approximation of real functions

ENCYCLOPEDIA OF MATHEMATICS AND ITS APPLICATIONS

Rational approximation of real functions

P.P. PETRUSHEV, V.A. POPOV
Bulgarian Academy of Sciences

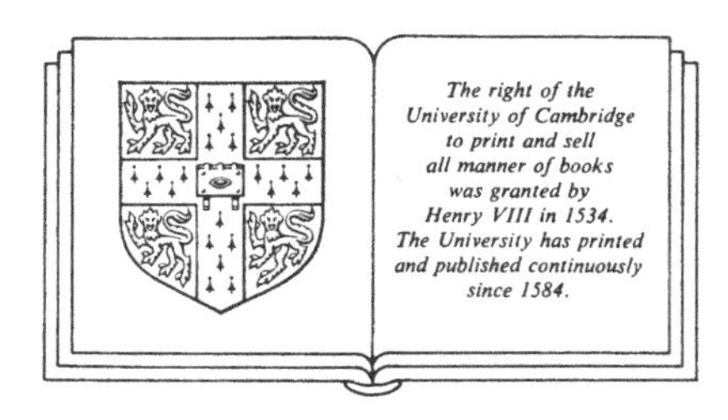

CAMBRIDGE UNIVERSITY PRESS
Cambridge
New York New Rochelle Melbourne Sydney

CAMBRIDGE UNIVERSITY PRESS
Cambridge, New York, Melbourne, Madrid, Cape Town, Singapore,
São Paulo, Delhi, Dubai, Tokyo, Mexico City

Cambridge University Press
The Edinburgh Building, Cambridge CB2 8RU, UK

Published in the United States of America by Cambridge University Press, New York

www.cambridge.org
Information on this title: www.cambridge.org/9780521331074

First published 1987

A catalogue record for this publication is available from the British Library

Library of Congress Cataloguing in Publication data

Petrushev, P.P. (Penčo Petrov), 1949–
Rational approximation of real functions.

(Encyclopedia of mathematics and its applications; v. 28)
Bibliography
Includes index.
1. Functions of real variables. 2. Approximation theory.
I. Popov, Vasil A. (Vasil Atanasov) II. Title.
III. Series
QA331.5.P48 1987 515.8 86-34331

ISBN 978-0-521-33107-4 Hardback

CONTENTS

PREFACE

Rational functions are a classical tool for approximation. They turn out to be a more convenient tool for approximation in many cases than polynomials which explains the constant increase of interest in them. On the other hand rational functions are a nonlinear approximation tool and they possess some intrinsic peculiarities creating a lot of difficulties in their investigation. After the classcial results of Zolotarjov from the end of the last century substantial progress was achieved in 1964 when D. Newman showed that $|x|$ is uniformly approximated by rational functions much better than by algebraic polynomials. Newman's result stimulated the appearance of many substantial results in the field of rational real approximations.

Our aim in this book is to present the basic achievements in rational real approximations. Nevertheless, for the sake of completeness we have included some results referring to the field of complex rational approximations in Chapters 6 and 12. Also, in order to stress some peculiarities of rational approximations we have included for comparison some classical and more recent results from the linear theory of approximation. On the other hand, since rational approximations are closely connected with spline approximations, we have included as well some results concerning spline approximations.

As usual the specific topics selected reflect the authors' interests and preferences.

We now sketch briefly the contents of the book. Chapters 1 and 3 contain some basic facts concerning linear approximation theory. A basic problem in approximation theory is to find complete direct and converse theorems. In our opinion the most natural way to obtain such theorems in linear and nonlinear approximations is to prove pairs of adjusted inequalities of Jackson and Bernstein type and then to characterize the corresponding approximations by the K-functional of Peetre. This main viewpoint is given and illustrated at the end of Chapter 3 and next applied to the spline approximation in Chapter 7.

Chapter 2 is devoted to the study of the qualitative theory of rational approximation such as the existence, the uniqueness and the characterization problems, the problem of continuity of metric projection and numerical methods.

The heart of the book is contained in Chapters 4 to 11. Chapter 4 presents the uniform rational approximation of some important functions such as $|x|$, $\sqrt{x}$, e^x. In Chapter 5 the uniform rational approximation of a number of classes is considered. The exact orders of approximation are established. The basic methods for rational approximation are given. In Chapter 6 some converse theorems for rational uniform approximation are proved. In Chapter 7 complete direct and converse theorems for the spline approximation in L_p, C, BMO are proved using Besov spaces. Chapter 8 investigates the relations between the rational and spline approximations. Chapter 9 deals with rational approximation in Hausdorff metric. A characteristic particularity of rational approximation is the appearance of the so-called 'o small' effect in the order of rational approximation of each individual function of some function classes. This phenomenon is investigated and characterized for some function classes in Chapter 10. The exactness of the proved estimates is established and discussed in Chapter 11.

Chapter 12 considers some special problems, connected with Padé approximants – some of the so-called direct and converse problems for convergence of the rows and diagonal of the Padé-table. Finally some numerical results and graphs are presented in the Appendix.

ACKNOWLEDGEMENTS

First of all we owe thanks to Professor G.G. Lorentz for the suggestion to write this book. We are grateful to Academician Bl. Sendov for constant attention and support. In the course of preparing the manuscript we enjoyed the helpful attention of many colleagues. We would especially like to mention A. Andreev and P. Marinov for their aid in numerical methods and calculations, K. Ivanov for the algebraical polynomial approximation in Chapter 3, R. Kovacheva for Padé approximations and E. Moskona for help in writing Chapter 6. Our colleagues A. Andreev, D. Drjanov, K. Ivanov, R. Kovacheva, R. Maleev, Sv.Markov, E. Sendova, S. Tashev read parts of the manuscript and we appreciate their remarks.

We owe special thanks to Professor D. Braess who submitted to us in manuscript the chapter on the rational approximation of his unpublished book on nonlinear approximation. We would like to express our gratitude to all our colleagues who were so kind as to give us additional information for the manuscript: J.-E. Anderson, B. Brosovski, Yu.A. Brudnyi, A.P. Bulanov, R. DeVore, E.P. Dolženko, T. Ganelius, A.A. Gonchar, J. Karlsson, G. López, A.A. Pekarskii, J. Peetre, V.N. Russak, E.A. Sevastijanov, K. Scherer, J. Szabados.

1

Qualitative theory of linear approximation

We shall begin with a short survey of the basic results related to linear approximations (i.e. approximation by means of linear subspaces) so that one can feel better the peculiarities, the advantages as well as some shortcomings of the rational approximation. In this chapter we shall consider the problems of existence, uniqueness and characterization of the best approximation (best polynomial approximation). At the end of the chapter we shall consider also numerical algorithms for finding the best uniform polynomial approximation.

1.1 Approximation in normed linear spaces

Let X be a normed linear space. Recall that X is said to be a normed linear space if:

(i) *X is a linear space, i.e. for its elements sum, and product with real numbers, are defined so that the standard axioms of commutativity and associativity are satisfied;*

(ii) *X is a normed space, i.e. to each $x \in X$ there corresponds a nonnegative real number $\|x\|$ satisfying the axioms*
   (a) $\|x\| \geqslant 0$, $\|x\| = 0$ *iff* $x = 0$,
   (b) $\|\lambda x\| = |\lambda| \, \|x\|$, *$\lambda$ a real number,*
   (c) $\|x + y\| \leqslant \|x\| + \|y\|$ (the triangle inequality).

Let $\{\varphi_i\}_{i=1}^n$ be a system of n linearly independent elements of X. Let us consider the linear subspace of X: $G = \{\varphi : \varphi = \sum_{i=1}^n a_i \varphi_i,\ a_i$ real numbers$\}$, generated by the system $\{\varphi_i\}_{i=1}^n$. For each element $f \in X$ we denote by $E_G(f)$ the best approximation to f by means of elements of G:

$$E_G(f) = \inf\{\|f - \varphi\| : \varphi \in G\}. \tag{1}$$

The following basic problems (basic not only for linear approximation theory, but for the theory of approximation in general) arise.

(i) *Existence problem: does an element $\varphi \in G$ of best approximation for $f \in X$ exist, i.e. is there $\varphi = \varphi(f) \in G$ such that*

$$\| f - \varphi(f) \| = E_G(f)?$$

(ii) *Uniqueness problem: if there exists an element of best approximation for $f \in X$, is it unique?*

(iii) *Characterization problem: in the case where the element of best approximation for $f \in X$ exists and is unique, can we characterize it in some way?*

(iv) *Can we estimate how big $E_G(f)$ is?*

(v) *Numerical methods: assuming that we know that the answer to the first two (or three) problems is positive, how can we find $\varphi(f)$ in practice?*

The whole theory of approximation represents full or partial (for the present, unfortunately) answers to the above problems when we approximate different classes of functions in different normed linear spaces (or, more generally, in metric spaces) with respect to different approximation tools (e.g. algebraic polynomials, trigonometric polynomials, rational functions, spline functions, linear combinations of exponential functions).

In the case of approximation in a normed linear space by a finite dimensional subspace we can give a positive answer to the first question. More precisely the following theorem holds.

Theorem 1.1 (Existence theorem). *Let G be a finite dimensional subspace of the normed linear space X. For every $f \in X$ there is an element of best approximation in G.*

Proof. The proof of this theorem is based on the following well-known fundamental property of finite dimensional normed spaces: every bounded closed subset in a finite dimensional normed linear space is compact. The idea of the proof is to show that the inf in (1) may be taken over a compact subset of G.

Let $\varphi_0 \in G$ be arbitrary. Then the set $A \subset G$:

$$A = \{\varphi : \varphi \in G, \| f - \varphi \| \leqslant \| f - \varphi_0 \| \}$$

is nonempty ($\varphi_0 \in A$), closed and bounded (since if $\varphi \in A$ then $\| \varphi \| \leqslant \| \varphi - f \| + \| f \| \leqslant \| \varphi_0 - f \| + \| f \|$). Therefore A is compact and obviously

$$E_G(f) = \inf \{ \| f - \varphi \| : \varphi \in G \} = \inf \{ \| f - \varphi \| : \varphi \in A \}.$$

The norm $\| f - \varphi \|$ is a continuous function of φ (by the triangle inequality $| \| f - \varphi \| - \| f - \psi \| | \leqslant \| \varphi - \psi \|$), therefore $\| f - \varphi \|$ attains its inf on the compact set A at some point $\varphi(f) \in A \subset G$. □

If the set $G \subset X$ has the property that every $f \in X$ has an element of best approximation in G, we shall call G an existence set. Obviously every existence set must be closed (every boundary point of G must belong to G). Theorem 1.1 gives us that every finite dimensional subspace of a linear normed space is an existence set.

Unfortunately the element of best approximation in an existence set G is not always unique. Let us denote by $P_G(f)$ the set

$$P_G(f) = \{\varphi : \varphi \in G, \| f - \varphi \| = E_G(f)\}$$

of all elements of best approximation of f.

Theorem 1.2. *Let X be a normed linear space and G a subspace of X, G an existence set. Then for every $f \in X$ the set $P_G(f)$ is convex and closed.*

Proof. Indeed, if $\varphi \in P_G(f)$ and $\psi \in P_G(f)$ then for every $\alpha \in [0, 1]$ we have

$$E_G(f) \leqslant \| f - (\alpha\varphi + (1-\alpha)\psi) \| \leqslant \alpha \| f - \varphi \| + (1-\alpha) \| f - \psi \| = E_G(f).$$

From this it follows that

$$E_G(f) = \| f - (\alpha\varphi + (1-\alpha)\psi) \|,$$

i.e. $\alpha\varphi + (1-\alpha)\psi \in P_G(f)$, therefore $P_G(f)$ is convex.

If $\| \varphi_m - \varphi \| \underset{m\to\infty}{\to} 0$, $\varphi_m \in G$, then φ also $\in G$, since G is closed. If $\varphi_m \in P_G(f)$ then

$$E_G(f) \leqslant \| f - \varphi \| \leqslant \| f - \varphi_m \| + \| \varphi_m - \varphi \| \underset{m\to\infty}{\to} E_G(f),$$

i.e.

$$E_G(f) = \| f - \varphi \|,$$

therefore $\varphi \in P_G(f)$. □

We shall see now that, when the normed linear space is strictly normed, there exists a unique element of best approximation in every subspace of X, which is an existence set (in particular in every finite dimensional subspace). Let us recall that a normed linear space X is said to be strictly normed if the equality $\| x + y \| = \| x \| + \| y \|$ implies that $x = \alpha y$, α a real number.

Theorem 1.3 (Uniqueness theorem). *Let X be a strictly normed linear space and G a subspace of X, G an existence set. Then for every $f \in X$ there exists a unique element of best approximation in G, i.e. $P_G(f)$ consists of exactly one element.*

Proof. Let $\varphi \in P_G(f)$ and $\psi \in P_G(f)$. In virtue of theorem 1.2 $(\varphi + \psi)/2 \in P_G(f)$ and therefore

$$E_G(f) = \| f - (\varphi + \psi)/2 \| \leqslant \tfrac{1}{2} \| f - \varphi \| + \tfrac{1}{2} \| f - \psi \| = E_G(f).$$

From this it follows that

$$\| f-(\varphi+\psi)/2 \| = \left\| \frac{f-\varphi}{2} \right\| + \left\| \frac{f-\psi}{2} \right\|.$$

Since X is strictly normed, the last equality implies $f-\varphi=\alpha(f-\psi)$. If $\alpha \neq 1$ it follows that $f \in G$ and in this case $P_G(f)=\{f\}$, i.e. $\varphi=\psi$. If $\alpha=1$ we obtain $\varphi=\psi$. □

Corollary 1.1. *Let G be a finite dimensional subspace of a linear strictly normed space X. Then for every $f \in X$ there exists a unique element of best approximation in G.*

In this book we shall use mostly the following function spaces.

(i) The space $C[a,b]$ of all functions which are continuous in the closed finite interval $[a,b]$. This space becomes a normed one (even a Banach space, i.e. a complete one) if we introduce the so-called uniform or Chebyshev norm,

$$\| f \|_{C[a,b]} = \| f \|_C = \max \{ | f(x) | : x \in [a,b] \}.$$

The approximations in $C[a,b]$ are usually called uniform or Chebyshev approximations.

(ii) The space $L_p(a,b)$, $1 \leqslant p < \infty$, (a,b) a finite or infinite interval,[†] consisting of all functions f such that $|f|^p$ is Lebesgue-integrable in the interval (a,b). If we consider all equivalent (in the sense of Lebesgue) functions as one, $L_p(a,b)$ becomes a normed (even Banach) space with respect to the so-called L_p-norm

$$\| f \|_{L_p(a,b)} = \| f \|_{L_p} = \| f \|_p = \left\{ \int_a^b | f(x) |^p \mathrm{d}x \right\}^{1/p}. \tag{2}$$

The approximations in $L_p(a,b)$ will be called L_p-approximations.

(iii) We shall use the notation (2) also in the case $0<p<1$ when $\| f \|_p$ is not a norm (since the triangle inequality does not hold), but only a quasinorm

$$\| f+g \|_p \leqslant c(p)(\| f \|_p + \| g \|_p).$$

(iv) The space $L_\infty[a,b]$ consisting of all essentially bounded functions in the interval $[a,b]$ supplied with the norm

$$\| f \|_{L_\infty[a,b]} = \| f \|_{L_\infty} = \| f \|_\infty = \text{ess sup.} | f(x) | = \inf \{ \lambda : \text{mes} \{ x : | f(x) | > \lambda \} = 0 \}$$

where mes$\{A\}$ denotes the Lebesgue measure of the set $\{A\}$.

If $f \in L_\infty$ then $\| f \|_p \to \| f \|_\infty$ when $p \to \infty$. Furthermore it is clear that if $f \in C[a,b]$ then $\| f \|_C = \| f \|_\infty$. Sometimes we shall use the notation $\| f \|_C$ also for bounded functions and we shall interpret it as $\sup \{ | f(x) | : x \in [a,b] \}$.

[†] We shall use also the notation $L_p[a,b]$.

Beside these spaces we shall use in some paragraphs Orlich spaces, Besov spaces, Hardy spaces and BMO spaces.

The spaces $C[a,b]$, $L_p(a,b)$, $1 \leqslant p < \infty$, $L_\infty[a,b]$ are normed linear ones. Therefore, in virtue of theorem 1.1 for each of their elements there exists an element of best approximation with respect to an arbitrary finite dimensional subspace of theirs. The main subspace used is that of algebraic polynomials of nth degree, denoted by P_n. It is the $(n+1)$-dimensional subspace generated by the functions $1, x, \dots, x^n$. Applying theorem 1.1 in this case we obtain the following.

Theorem 1.4 (E. Borel). *Let $f \in C[a,b]$ (or $L_p[a,b]$, $1 \leqslant p < \infty$). Then for every natural number n there exists an algebraic polynomial $p \in P_n$ of best uniform (or L_p) approximation in P_n.*

It is often necessary to approximate 2π-periodic functions. Without pointing it out explicitly every time, we shall use the notations we introduced in the case of an interval also for linear spaces of 2π-periodic functions, namely $C[0,2\pi]$, $L_p[0,2\pi]$, $1 \leqslant p \leqslant \infty$. The tools used most often in this case are the trigonometric polynomials. We shall denote by T_n the set of all trigonometric polynomials of nth order, i.e. T_n is the $(2n+1)$-dimensional subspace generated by the functions $1, \cos x, \sin x, \dots, \cos nx, \sin nx$. In the periodic case theorem 1.1 implies the following.

Theorem 1.4'. *Let f be a 2π-periodic function and $f \in C[0,2\pi]$ ($f \in L_p[0,2\pi]$). For every natural number n there exists a trigonometric polynomial $q \in T_n$ of best uniform (L_p) approximation in T_n.*

Let us consider now the question of uniqueness. One can show that the spaces L_p, $1 < p < \infty$, are strictly normed (see for example S.M. Nikol'skij (1969)). Then theorem 1.3 implies the following.

Theorem 1.5. *Let $f \in L_p(a,b)$ (let f be 2π-periodic and $f \in L_p[0,2\pi]$), $1 < p < \infty$. Then for every natural number n there exists a unique algebraic (trigonometric) polynomial of nth degree of best L_p-approximation in P_n (in T_n).*

However, the spaces C, L_∞, $L = L_1$ are not strictly normed. Let us show this for instance for $C[0,1]$. If we consider the functions 1 and x, we have

$$\| 1 + x \|_{C[0,1]} = \| 1 \|_{C[0,1]} + \| x \|_{C[0,1]} = 2$$

but the functions 1 and x are not linearly dependent.

It is easy to see by examples that in the general case in L_1 we do not have uniqueness. Let us consider the function

$$\sigma(x) = \begin{cases} -1, & -1 \leqslant x \leqslant 0, \\ 1, & 0 < x \leqslant 1. \end{cases}$$

In $L_1(-1,1)$ every constant c, $-1 \leqslant c \leqslant 1$, is a polynomial of degree zero of best approximation to σ.

Fortunately enough it turns out that the algebraic polynomial of nth degree of best uniform approximation is unique. This follows from the Chebyshev theorem, which gives a characterization of the algebraic polynomial of best uniform approximation by alternation. This theorem as well as its proof can be modified for the best rational uniform approximation. That is why it will be of special interest of us.

1.2 Characterization of the algebraic polynomial of best uniform approximation

Now we are going to solve the third basic problem of the theory of approximation in the case of uniform approximation by means of algebraic polynomials – characterization of the algebraic polynomial of best uniform approximation. This problem was solved by P.L. Chebyshev in the last century with his famous alternation theorem.

Let $f \in C[a,b]$. We shall denote by $E_n(f)_C$ the best uniform approximation of the function f by means of algebraic polynomials of nth degree:

$$E_n(f)_{C[a,b]} = E_n(f)_C = \inf\{\|f-p\|_{C[a,b]}: p \in P_n\}.$$

In what follows in this section we shall write $E_n(f)$ instead of $E_n(f)_C$ and $\|f\|$, $\|f-p\|$ instead of $\|f\|_C$, $\|f-p\|_C$.

Definition 1.1. *Let $f \in C[a,b]$. The polynomial $p \in P_n$ is said to realize Chebyshev alternation (or simply alternation) for f in $[a,b]$ if there exist $n+2$ points x_i, $i=1,\dots,n+2$, $a \leqslant x_1 < \cdots < x_{n+2} \leqslant b$, such that*

$$f(x_i) - p(x_i) = \varepsilon(-1)^i \|f-p\|, \quad i=1,\dots,n+2,$$

where the number ε is $+1$ or -1.

The Chebyshev alternation has the following geometric interpretation: let $p \in P_n$ realize Chebyshev alternation for $f \in C[a,b]$ in $[a,b]$. Let us consider the functions $\varphi(x) = f(x) + \|f-p\|$ and $\psi(x) = f(x) - \|f-p\|$. Then the graph of the polynomial lies in the strip between φ and ψ, touching alternately the upper function φ and the lower function ψ at least $n+2$ times.

Theorem 1.6 (Chebyshev alternation theorem). *Let $f \in C[a,b]$. The necessary and sufficient condition for the algebraic polynomial $p \in P_n$ to be a polynomial of best uniform approximation for f in P_n is that p realizes Chebyshev alternation for f in $[a,b]$.*

Proof. Let $p \in P_n$ realize Chebyshev alternation for f in $[a,b]$. Assume that p is not a polynomial of best uniform approximation, but $q \in P_n$ is. Then

$$E_n(f) = \|f-q\| < \|f-p\|.$$

The above inequality implies that the polynomial $s = p - q \in P_n$ has the sign of $p-f$ in the points x_i, $i=1,\dots,n+2$, since $|p(x_i)-f(x_i)| = \|f-p\| > \|f-q\|$, $p(x_i)-q(x_i) = p(x_i)-f(x_i)-(q(x_i)-f(x_i))$. Therefore $s \in P_n$ will change its sign at least $n+1$ times, i.e. s must have at least $n+1$ zeros in $[a,b]$. Since $s \in P_n$, it follows that $s \equiv 0$, i.e. $p=q$, which is a contradiction with the assumption.

Let now $p \in P_n$ be an algebraic polynomial of best uniform approximation for f in P_n. We shall show that p realizes Chebyshev alternation for f. Let us assume, contrary to this, that $m+2$ is the highest number of points $x_1 < x_2 < \cdots < x_{m+2}$ in $[a,b]$ such that

$$f(x_i) - p(x_i) = \varepsilon(-1)^i \|f-p\| = \varepsilon(-1)^i E_n(f), \quad i=1,\dots,m+2, \tag{1}$$

where $\varepsilon = 1$ or -1 and $m < n$. Then there exist $m+3$ points $\xi_0, \xi_1, \dots, \xi_{m+2}$ which satisfy the inequalities

$$a = \xi_0 \leqslant x_1 < \xi_1 < x_2 < \xi_2 < \cdots < \xi_{m+1} < x_{m+2} \leqslant \xi_{m+2} = b$$

and are such that for every $x \in [\xi_{i-1}, \xi_i]$ we have

$$\varepsilon(-1)^i (f(x) - p(x)) > -E_n(f), \quad i=1,\dots,m+2. \tag{2}$$

From (1) it follows that the continuous function $f-p$ changes its sign in the interval $[x_i, x_{i+1}]$, therefore the points $\xi_1, \xi_2, \dots, \xi_{m+1}$ can be chosen so that

$$f(\xi_i) = p(\xi_i), \quad i=1,\dots,m+1. \tag{3}$$

Since $[\xi_{i-1}, \xi_i]$, $i=1,\dots,m+2$, are a finite number of closed intervals and $f-p$ is a continuous function in each of them, from (2) it follows that there exists $\delta > 0$ such that for every $x \in [\xi_{i-1}, \xi_i]$, $i=1,\dots,m+2$, we have the inequality

$$\varepsilon(-1)^i (f(x) - p(x)) > \delta - E_n(f). \tag{4}$$

Let us set

$$Q(x) = (-1)^{m+1} \lambda (x-\xi_1) \cdots (x-\xi_{m+1}),$$

where

$$\lambda = \frac{\delta}{2\|(x-\xi_1)\cdots(x-\xi_{m+1})\|_{C[a,b]}}.$$

Since $m < n$, we have $Q \in P_n$.

From this definition of Q it follows also that

$$|Q(x)| \leqslant \delta/2 \quad \text{for } x \in [a,b], \tag{5}$$

$$(-1)^i Q(x) > 0 \quad \text{for } x \in (\xi_i, \xi_{i+1}), \quad i=0,\dots,m+1, \tag{6}$$

$$Q(\xi_0) > 0, \quad (-1)^{m+1} Q(\xi_{m+2}) > 0, \tag{7}$$

$$Q(\xi_i) = 0, \quad i = 1,\dots,m+1. \tag{8}$$

Since p is an algebraic polynomial of best approximation to f in P_n, we have

$$-E_n(f) \leqslant f(x) - p(x) \leqslant E_n(f) \quad \text{for } x \in [a,b]. \tag{9}$$

Let us consider the difference

$$f(x) - p(x) - \varepsilon Q(x).$$

In view of (4), (5) and (6), for every $x \in [\xi_i, \xi_{i+1}]$, we have, for $i = 0,\dots,m+1$,

$$\begin{aligned}\varepsilon(-1)^i(f(x) - p(x) - \varepsilon Q(x)) &= \varepsilon(-1)^i(f(x) - p(x)) - (-1)^i Q(x) \\ &> \delta - E_n(f) - \delta/2 = \delta/2 - E_n(f).\end{aligned} \tag{10}$$

From (5)–(9) we also obtain that, for every $x \in (\xi_i, \xi_{i+1})$ and $x = \xi_0, \xi_{m+2}$, we have

$$\begin{aligned}\varepsilon(-1)^i(f(x) - p(x) - \varepsilon Q(x)) = (-1)^i \varepsilon(f(x) - p(x)) - (-1)^i Q(x) \leqslant E_n(f) \\ -(-1)^i Q(x) < E_n(f).\end{aligned} \tag{11}$$

For $x = \xi_i$, $i = 1,\dots,m+1$, we have, from (8),

$$f(\xi_i) - p(\xi_i) - \varepsilon Q(\xi_i) = 0. \tag{12}$$

Consequently the inequalities (10)–(12) give us that, for every $x \in [a,b]$, we have

$$|f(x) - p(x) - \varepsilon Q(x)| < E_n(f). \tag{13}$$

Since $f - p - \varepsilon Q$ is a continuous function in $[a,b]$, from (13) it follows that

$$\| f - p - \varepsilon Q \| < E_n(f),$$

i.e. a contradiction, since $p + \varepsilon Q \in P_n$. □

From theorem 1.6 there follows easily the uniqueness of the algebraical polynomial of best uniform approximation as follows.

Theorem 1.7. *Let $f \in C[a,b]$. For every natural number n there exists a unique algebraic polynomial $p \in P_n$ of best uniform approximation to f in P_n.*

Proof. Let $p \in P_n$ and $q \in P_n$ be two algebraic polynomials of best uniform approximation to f:

$$\| f - p \| = \| f - q \| = E_n(f). \tag{14}$$

From theorem 1.2 the polynomial $g = (p+q)/2 \in P_n$ is also a polynomial of best uniform approximation to f. By theorem 1.6 g realizes Chebyshev alternation for f, i.e. there exist $n+2$ points x_i, $i = 1,\dots,n+2$, $a \leqslant x_1 <$

$x_2 < \cdots < x_{n+2} \leqslant b$, such that

$$f(x_i) - \frac{p(x_i) + q(x_i)}{2} = \varepsilon(-1)^i E_n(f), \quad i = 1, \ldots, n+2, \tag{15}$$

where $\varepsilon = 1$ or -1.

From (14) it follows that

$$\left.\begin{aligned} |f(x_i) - p(x_i)| \leqslant E_n(f), \\ |f(x_i) - q(x_i)| \leqslant E_n(f). \end{aligned}\right\} \tag{16}$$

Therefore the equality (15) can be fulfilled only if we have

$$f(x_i) - p(x_i) = f(x_i) - q(x_i),$$

i.e. if $p(x_i) = q(x_i)$ for $i = 1, \ldots, n+2$.

We thus have that the algebraical polynomials $p \in P_n$ and $q \in P_n$ coincide in $n+2$ different points. Consequently $p = q$. □

The following theorem of de la Vallée-Poussin is very useful in the numerical methods for obtaining the polynomial of best uniform approximation.

Theorem 1.8. *Let $f \in C[a,b]$, $p \in P_n$ and x_i, $i = 1, \ldots, n+2$, $a \leqslant x_1 < x_2 < \cdots < x_{n+2} \leqslant b$, be $n+2$ different points in $[a,b]$. If the difference $f - p$ has alternate signs at the points x_i, $i = 1, \ldots, n+2$, then*

$$E_n(f) \geqslant \mu = \min\{|f(x_i) - p(x_i)| : i = 1, \ldots, n+2\}.$$

Proof. Let us assume that $E_n(f) < \mu$. Let $q \in P_n$ be the algebraic polynomial of best uniform approximation to f, i.e. $\| f - q \| = E_n(f) < \mu$.

From this it follows that the difference $p - q$ must have the sign of $p(x_i) - f(x_i)$ at the points x_i, $i = 1, \ldots, n+2$. By the conditions of the theorem therefore $p - q$ must have alternate signs at $n+2$ points x_i, $i = 1, 2, \ldots, n+2$, i.e. the algebraic polynomial $p - q \in P_n$ must have at least $n+1$ different zeros in $[a,b]$; consequently $p - q \equiv 0$ which contradicts

$$\| f - q \| = E_n(f) < \mu \leqslant \| f - p \|.$$ □

1.3 Numerical methods

We shall describe in this section the so-called Remez algorithms for numerical solution of basic problem (v) from section 1.1 – finding the polynomial of best uniform approximation. The algorithms are more general and can be used for best uniform approximation by means of arbitrary Haar subspaces of $C[a,b]$.

Definition 1.2. *The system $\{\varphi_i\}_{i=1}^n$ of functions $\varphi_i \in C[a,b]$, $i = 1, \ldots, n$, is said to be a Chebyshev system on the interval $[a,b]$ if every generalized polynomial*

$\varphi = \sum_{i=1}^{n} a_i \varphi_i$ *can have at most* $n-1$ *zeros in* $[a, b]$ *(every zero calculated with its multiplicity).*

Let $C^{(k)}[a, b]$ denote the space of all functions in the interval $[a, b]$ which have kth derivative $f^{(k)}$ in $[a, b]$, which belongs to $C[a, b]$.

We shall say that $x_0 \in [a, b]$ is a zero of $f \in C^{(k)}[a, b]$ of order k (or multiplicity k) if

$$f(x_0) = f'(x_0) = \cdots = f^{(k-1)}(x_0) = 0, \quad f^{(k)}(x_0) \neq 0.$$

Definition 1.3. *A subspace* $G \subset C[a, b]$, $G = \{\varphi \colon \varphi = \sum_{i=1}^{n} a_i \varphi_i\}$, *generated by the Chebyshev system* $\{\varphi_i\}_{i=1}^{n}$, *is said to be a Haar subspace.*

Let $\{\varphi_i\}_{i=1}^{n}$ be a Chebyshev system. In this section we shall use the following notations. Let $f \in C[a, b]$. Then

$$E_n(f) = \inf\left\{ \| f - \varphi \| \colon \varphi = \sum_{i=1}^{n} a_i \varphi_i \right\},$$

$$\Delta(a) = \left\| \sum_{i=1}^{n} a_i \varphi_i - f \right\|_{C[a,b]},$$

$$\mathbf{a} = (a_1, \ldots, a_n) \in \mathbb{R}^n, \quad r(\mathbf{a}, x) = \sum_{i=1}^{n} a_i \varphi_i(x) - f(x),$$

$\mathbb{R}^n$ the n-dimensional Euclidean space.

Our aim is to find real numbers $\{c_i^*\}_{i=1}^{n}$ such that

$$\left\| f - \sum_{i=1}^{n} c_i^* \varphi_i \right\|_{C[a,b]} = E_n(f).$$

First Remez algorithm

The algorithm consists of the following recursive procedure.

(i) *Select* $n+1$ *points* $X^{(0)} = \{x_i\}_0^n$, *where* $a \leqslant x_0 < x_1 < \cdots < x_n \leqslant b$;

(ii) *Set* $k = 0$;

(iii) *Given the set* $X^{(k)}$ *find a vector* $\mathbf{c}^{(k)} \in \mathbb{R}^n$ *such that if we denote* $\Delta^{(k)}(\mathbf{c}) = \max\{|r(\mathbf{c}, x)| \colon x \in X^{(k)}\}$ *then*

$$\Delta^{(k)}(\mathbf{c}^{(k)}) = \inf\{\Delta^{(k)}(\mathbf{c}) \colon \mathbf{c} \in \mathbb{R}^n\};$$

(iv) *Find a point* $x_{n+k+1} \in [a, b]$ *such that* $\Delta(\mathbf{c}^{(k)}) = |r(\mathbf{c}^{(k)}, x_{n+k+1})|$;

(v) *Form the set* $X^{(k+1)} = X^{(k)} \cup \{x_{n+k+1}\}$;

(vi) *Set* $k = k+1$;

(vii) *Go to* (iii).

The choice of the initial set X^0 can be done in different ways (equidistant point in the trigonometrical case, the roots of the $(n+1)$-th polynomial of Chebyshev in the algebraic case and so on) and there exist no strong rules for this.

At step (iii) we have to find in fact the polynomial of the best uniform approximation on a set which consists of a finite number of points.

Step (iv) is usually the most laborious point in the algorithm.

The execution of the algorithm stops when the polynomial obtained at the kth iteration satisfies some demands.

The method (i)–(vii) generates a sequence of vectors $\{\mathbf{c}^{(k)}\}_{k=0}^{\infty}$ for which we have the following.

Theorem 1.9. *Let* $\mathbf{c}^*$ *be a cluster point of the sequence* $\{\mathbf{c}^{(k)}\}_{k=0}^{\infty}$. *Then* $E_n(f) = \Delta(\mathbf{c}^*)$.

Proof. Let us set $|\mathbf{c}| = \sum_{i=1}^{n} |c_i|$ and

$$\theta = \min_{|\mathbf{c}|=1} \max_{x \in X^{(0)}} \left| \sum_{i=1}^{n} c_i \varphi_i(x) \right|.$$

Since $X^{(0)}$ contains $n+1$ different points and $\{\varphi_i\}_{i=1}^{n}$ is a Chebyshev system on the interval $[a,b]$ we have $\theta > 0$. From $X^{(k)} \subset X^{(k+1)} \subset [a,b]$ we get for every $\mathbf{c} \in \mathbb{R}^n$ that

$$\Delta^{(k)}(\mathbf{c}) \leqslant \Delta^{(k+1)}(\mathbf{c}) \leqslant \Delta(\mathbf{c})$$

and consequently ($\bar{\mathbf{c}}$: $\Delta(\bar{\mathbf{c}}) = E_n(f)$)

$$\Delta^{(k)}(\mathbf{c}^{(k)}) \leqslant \Delta^{(k)}(\mathbf{c}^{(k+1)}) \leqslant \Delta^{(k+1)}(\mathbf{c}^{(k+1)}) \leqslant \Delta^{(k+1)}(\bar{\mathbf{c}}) \leqslant \Delta(\bar{\mathbf{c}}) = E_n(f).$$

The last inequalities show that the sequence $\{\Delta^{(k)}(\mathbf{c}^{(k)})\}_0^{\infty}$ is monotone nondecreasing and bounded from above. This means that there exists $\varepsilon \geqslant 0$ such that $\lim_{k \to \infty} \Delta^{(k)}(\mathbf{c}^{(k)}) = E_n(f) - \varepsilon$. We shall show that $\varepsilon = 0$.

First we prove that the sequence $\{\mathbf{c}^{(k)}\}_{k=0}^{\infty}$ is bounded. Indeed,

$$\Delta^0(\mathbf{c}) = \max_{x \in X^0} \left| \sum_{i=1}^{n} c_i \varphi_i(x) - f(x) \right| \geqslant \max_{x \in X^{(0)}} \left| \sum_{i=1}^{n} c_i \varphi_i(x) \right| - \| f \|_C \geqslant \theta |\mathbf{c}| - \| f \|$$

and if $|\mathbf{c}| > 2\| f \|/\theta$ then $\Delta^{(k)}(\mathbf{c}) \geqslant \Delta^0(\mathbf{c}) > \| f \| = \Delta^{(k)}(0)$, i.e. $\mathbf{c}$ can not minimize any of the functions $\Delta^{(k)}$. So the sequence $\{\mathbf{c}^{(k)}\}_{k=1}^{\infty}$ generated by the algorithm is bounded.

Further let us set $M = \max_{1 \leqslant i \leqslant n} \| \varphi_i \|_{C[a,b]}$. Then for an arbitrary vector $\mathbf{b}$

$$|r(\mathbf{b}, x) - r(\mathbf{c}, x)| = \left| \sum_{i=1}^{n} (b_i - c_i)\varphi_i(x) \right| \leqslant M|\mathbf{b} - \mathbf{c}|$$

and therefore $|r(\mathbf{b}, x)| \leqslant |r(\mathbf{c}, x)| + M|\mathbf{b} - \mathbf{c}|$, i.e.

$$\Delta(\mathbf{b}) = \| r(\mathbf{b}, \cdot) \|_{C[a,b]} = |r(\mathbf{b}, \bar{x})| \leqslant |r(\mathbf{c}, \bar{x})| + M|\mathbf{b} - \mathbf{c}| \leqslant \Delta(\mathbf{c}) + M|\mathbf{b} - \mathbf{c}|. \tag{1}$$

Let us suppose now that $\varepsilon > 0$ and $\mathbf{c}^* \in \mathbb{R}^n$ is a cluster point of the sequence $\{\mathbf{c}^{(k)}\}_{k=0}^{\infty}$. For every $\delta > 0$ there exists an index k such that $|\mathbf{c}^* - \mathbf{c}^{(k)}| < \delta$ and an index $i > k$ such that $|\mathbf{c}^* - \mathbf{c}^{(i)}| < \delta$. Then $|\mathbf{c}^{(k)} - \mathbf{c}^{(i)}| < 2\delta$ and, using (1), setting $\mathbf{c}^*$ in place of $\mathbf{b}$, we obtain

$$\begin{aligned} E_n(f) &\leqslant \Delta(\mathbf{c}^*) \leqslant \Delta(\mathbf{c}^{(k)}) + M\delta = |r(\mathbf{c}^{(k)}, x^{(k+1)})| + M\delta \\ &\leqslant |r(\mathbf{c}^{(i)}, x^{(k+1)})| + 3M\delta \leqslant \Delta^{(k+1)}(\mathbf{c}^{(i)}) + 3M\delta \\ &\leqslant \Delta^{(i)}(\mathbf{c}^{(i)}) + 3M\delta \leqslant E_n(f) - \varepsilon + 3M\delta. \end{aligned}$$

The number $\delta > 0$ was arbitrary, so for $\varepsilon > 3M\delta$ this leads to a contradiction. Therefore $\varepsilon = 0$ and $\Delta(\mathbf{c}^*) = E_n(f)$. □

Corollary 1.2. *Let* $\{\varphi_i(x)\}_1^n = \{x^i\}_0^{n-1}$. *Then there exists* $\lim_{k\to\infty} \mathbf{c}^{(k)} = \mathbf{c}^*$.

This follows from the uniqueness of the best uniform algebraic approximation (theorem 1.7).

Corollary 1.2 gives that the first Remez algorithm is convergent for the case of approximation by means of algebraic polynomials.

Remark. The uniqueness theorem is also valid for approximation in the uniform metric by means of a Chebyshev system. So we have convergence of the first Remez algorithm also in the general case of a Chebyshev system.

Second Remez algorithm

We shall describe the second Remez algorithm again for an arbitrary Chebyshev system and we shall prove the order of convergence for the case of uniform approximation by means of algebraic polynomials.

(i) *Take* $n+1$ *different points* x_i, $i = 0, \ldots, n$, $a \leqslant x_0 < x_1 < \cdots < x_n \leqslant b$;

(ii) *Solve the linear system*

$$f(x_j) - \sum_{i=1}^{n} c_i \varphi_i(x_j) = (-1)^j \lambda, \quad j = 0, 1, \ldots, n,$$

with respect to the unknowns $c_1, \ldots, c_n$ *and* λ;

(iii) *Find the points* $\{z_i\}_{i=0}^{n+1}$ *such that* $z_0 = a$, $z_{n+1} = b$ *and* $r(z_i) = 0$ *for* $i = 1, \ldots, n$;†

(iv) *Select the points* $y_i \in [z_i, z_{i+1}]$, $i = 0, 1, \ldots, n$, *such that*

$$(\operatorname{sign} r(x_i)) r(y_i) = \max\{r(x) \operatorname{sign} r(x_i) : x \in [z_i, z_{i+1}]\},$$

(v) *If* $\|r(\mathbf{c}; \cdot)\|_{C[a,b]} > \max\{|r(\mathbf{c}; y_i)| : 0 \leqslant i \leqslant n\}$ *then there exists a point* $y \in [a, b]$ *such that* $|r(\mathbf{c}; y)| = \|r(\mathbf{c}; \cdot)\|_{C[a,b]}$ – *we put the point* y *in place of some point among* $y_0, y_1, \ldots, y_n$ *so that the function* $r(\mathbf{c}; x)$ *would preserve the alternating signs on the newly obtained points which we denote again by* $y_0, y_1, \ldots, y_n$;

(vi) *Go to* (ii) *and instead of the points* $\{x_i\}_{i=0}^n$ *consider the points* $\{y_i\}_{i=0}^n$.

† $r(x) \equiv r(\mathbf{c}; x)$

This procedure can be easily carried out using computers and numerical experiments show that it is not very sensitive to the choice of the initial points.

Usually we go out of the iterative process and stop the calculation when on the kth step $\|r(\mathbf{c};\cdot)\|$ differs negligibly from $|\lambda|$. This stop-condition comes from the Chebyshev theorem of alternation.

The second Remez algorithm has quadratic convergence under some restrictions on the smoothness of the function f (see L. Veidinger (1960)). We shall prove here the linear convergence of the algorithm for every $f\in C[a,b]$ in the case of polynomial approximations.

Theorem 1.10. *Let $\{\varphi_i\}_{i=1}^{n}=\{x^i\}_{i=0}^{n-1}$ and let $f\in C[a,b]$. The polynomial $p^{(k)}=\sum_{i=0}^{n-1}c_ix^i$ generated on the kth step by the second Remez algorithm satisfies the condition $\|p^{(k)}-p\|_{C[a,b]}\leqslant c\theta^k$, where p is the algebraic polynomial of best uniform approximation for f of $(n-1)$-th degree, $0<\theta<1$ and c is a constant, independent of k.*

Proof. We again use the abbreviation $r(x)\equiv r(\mathbf{c};x)$. Since we described a single cycle of the second Remez algorithm let us denote $\alpha=|r(x_0)|=\cdots=|r(x_n)|=|\lambda|$, $\beta=\max\{|r(y_i)|:i=0,\ldots,n\}=\|r(\mathbf{c};\cdot)\|_C$, $\gamma=\min\{|r(y_i)|:i=0,\ldots,n\}$, $\bar{\beta}=\|f-p\|_C$.

From de la Vallée-Poussin's theorem (theorem 1.8) we get $\alpha\leqslant\gamma\leqslant\bar{\beta}\leqslant\beta$. Let us agree that on the next cycle of the algorithm the constants corresponding to α, β, γ, λ and the coefficient vector $\mathbf{c}$ will be denoted by α', β', γ', λ' and $\mathbf{c}'$. According to this convention it is clear that the vector $\mathbf{c}'$ is selected by the system

$$(-1)^i\lambda'+\sum_{j=0}^{n-1}c_j'y_i^j=f(y_i),\quad i=0,\ldots,n,$$

and

$$\lambda'=\frac{\begin{vmatrix} f(y_0) & 1 & y_0\cdots y_0^{n-1}\\ & \cdots & \\ f(y_n) & 1 & y_n\cdots y_n^{n-1}\end{vmatrix}}{\begin{vmatrix} 1 & 1 & y_0\cdots y_0^{n-1}\\ -1 & 1 & y_1\cdots y_1^{n-1}\\ & \cdots & \\ (-1)^n & 1 & y_n\cdots y_n^{n-1}\end{vmatrix}}=\frac{\sum_{i=0}^{n}(-1)^if(y_i)M_i}{\sum_{i=0}^{n}M_i},$$

where M_i are the minors corresponding to the first column of the matrix in the denominator.

If f has the form $f=\sum_{j=0}^{n-1}a_jx^j$ then the approximation has to be exact and $\lambda'=0$, i.e.

$$\sum_{i=0}^{n}(-1)^i\sum_{j=0}^{n-1}a_jy_i^jM_i=0.$$

Thus we may replace $f(y_i)$ by

$$r(y_i) = f(y_i) - \sum_{j=0}^{n-1} c_j y_i^j$$

in the expression for λ'. Taking into account that $\operatorname{sign} r(y_i) = -\operatorname{sign} r(y_{i+1})$ we obtain

$$\alpha' = |\lambda'| = \left(\sum_{i=0}^{n} M_i |r(y_i)| \right) \Big/ \sum_{i=0}^{n} M_i.$$

Since $M_i > 0$,

$$M_i = \prod (y_k - y_j), \tag{2}$$

where the product is taken over all k, j such that $k > j$, $k, j = 0, 1, \ldots, i-1, i+1, \ldots, n$ and $y_k > y_j$ for $k > j$.

Now let $\theta_i = M_i / \sum_{i=0}^{n} M_i$. Then

$$\alpha' = \sum_{i=0}^{n} \theta_i |r(y_i)| \geqslant \gamma \sum_{i=0}^{n} \theta_i = \gamma \geqslant \alpha. \tag{3}$$

We shall show that there exists θ, $0 < \theta < 1$, such that for all numbers θ_i generated at the kth iteration of the algorithm we have

$$1 - \theta < \theta_i < 1, \quad i = 0, 1, \ldots, n. \tag{4}$$

From (2) it follows that this will be true if there exists $\delta > 0$ independent of k such that

$$y_{i+1}^{(k)} - y_i^{(k)} \geqslant \delta > 0, \quad i = 0, \ldots, n-1, \quad k = 1, 2, \ldots. \tag{5}$$

Let us assume that this inequality is not true. Then the sequence $\{y_0^{(k)}, \ldots, y_n^{(k)}\}_{k=1}^{\infty}$ will have a cluster point $(\bar{y}_0, \bar{y}_1, \ldots, \bar{y}_n)$, where at least two points $\bar{y}_i$ coincide. Consequently there exists an algebraic polynomial $q(x) = \sum_{i=0}^{n-1} a_i x^i$ which interpolates f at the points $\bar{y}_0, \bar{y}_1, \ldots, \bar{y}_n$ (the number of the different points is at most n). By definition $\alpha^{(k+1)}$ is the best approximation of f at the points $y_0^{(k)}, y_1^{(k)}, \ldots, y_n^{(k)}$ at the kth iteration and

$$\begin{aligned} \alpha^{(k+1)} &\leqslant \max \{ |f(y_i^{(k)}) - q(y_i^{(k)})| : i = 0, \ldots, n \} \\ &= \max \{ |f(y_i^{(k)}) - q(y_i^{(k)}) - f(\bar{y}_i) + q(\bar{y}_i)| : i = 0, \ldots, n \}, \end{aligned} \tag{6}$$

since $f(\bar{y}_i) = q(\bar{y}_i)$, $i = 0, \ldots, n$. This inequality contradicts the fact that $\alpha' \geqslant \alpha$ (see (3)), i.e. $\alpha^{(1)} \leqslant \alpha^{(2)} \leqslant \cdots \leqslant \alpha^{(k+1)} \leqslant \cdots$. Really, if k is such that $\max \{ |y_i^{(k)} - \bar{y}_i| : i = 0, \ldots, n \}$ is small enough, then $(\alpha^{(1)} > 0)$

$$\max \{ |f(y_i^{(k)}) - f(\bar{y}_i) - (q(y_i^{(k)}) - q(\bar{y}_i))| : i = 0, \ldots, n \} < \alpha^{(1)} \tag{7}$$

since q and f are continuous functions. From (6) and (7) we get the contradiction $\alpha^{(k+1)} < \alpha^{(1)}$.

Therefore (5), and consequently (4), hold true. Using (4) we obtain

$$\gamma' - \gamma \geqslant \alpha' - \gamma = \sum_{i=0}^{n} \theta_i(|r(y_i)| - \gamma) \geqslant (1-\theta)(\beta - \gamma) \geqslant (1-\theta)(\bar{\beta} - \gamma)$$

and

$$\bar{\beta} - \gamma' = (\bar{\beta} - \gamma) - (\gamma' - \gamma) \leqslant (\bar{\beta} - \gamma) - (1-\theta)(\bar{\beta} - \gamma) = \theta(\bar{\beta} - \gamma),$$

i.e. $\bar{\beta} - \gamma^{(k)} \leqslant \theta^k(\bar{\beta} - \gamma^{(0)})$ and

$$\beta^{(k)} - \beta \leqslant \beta^{(k)} - \gamma^{(k)} \leqslant \frac{\gamma^{(k+1)} - \gamma^{(k)}}{1-\theta} \leqslant \frac{\bar{\beta} - \gamma^{(k)}}{1-\theta} \leqslant \frac{\theta^k(\bar{\beta} - \gamma^{(0)})^{\dagger}}{1-\theta}. \tag{8}$$

Finally we shall apply the strong uniqueness theorem 2.5 from Chapter 2 (obviously the theorem remains true for P_{n-1}, i.e. when $m = 0$). By this theorem if p is the polynomial of best uniform approximation for f of $(n-1)$-th degree, then there exists a constant $c(f) > 0$, depending only on f, such that for every polynomial $q \in P_{n-1}$ we have:

$$\| f - q \| \geqslant \| f - p \| + c(f) \| q - p \|. \tag{9}$$

Denoting by $p^{(k)}$ the algebraic polynomial generated at the kth step of the algorithm (at the kth iteration), we obtain, from (8) and (9),

$$\begin{aligned} \| p^{(k)} - p \| &\leqslant \frac{1}{c(f)}(\| f - p^{(k)} \| - \| f - p \|) = \frac{1}{c(f)}(\beta^{(k)} - E_n(f)) \\ &= \frac{1}{c(f)}(\beta^{(k)} - \bar{\beta}) \leqslant \frac{\bar{\beta} - \gamma^{(0)}}{c(f)(1-\theta)}\theta^k \end{aligned}$$

which completes the proof. □

Remark. Theorem 1.10 remains valid also for an arbitrary Chebyshev system.

1.4 Notes

The classical theorems for characterization and uniqueness of the best polynomial uniform approximation are given by P.L. Chebyshev (see P.L. Tchebycheff (1899), see also Ch.de la Vallée-Poussin (1910)).

The abstract theory of linear approximations is a very developed domain. We recommend the following books, which contain some more details than given here: I. Singer (1970), E.W. Cheney (1966), J. Rice (1964), (1969), Collatz, Krabs (1973).

Usually uniform approximation by means of a Chebyshev system is considered. We shall give only the formulations of some theorems.

Let K be compact and let $C(K)$ be the set of all continuous functions on

† $\gamma^{(k)}$, $\beta^{(k)}$ are γ, β at the kth iteration.

K (real- or complex-valued). The following characterization theorem is known as the Kolmogorov criterion (A.N. Kolmogorov, 1948).

Let $f \in C(K)$ and let G be a linear subspace of $C(K)$. A function $\varphi_0 \in G$ is a best approximation of f with respect to G if and only if the inequality

$$\min_{x \in A} \operatorname{Re} \overline{(f(x) - \varphi_0(x))}\, \varphi(x) \leqslant 0$$

holds for every $\varphi \in G$, where A is the set of the extremal points of $f - \varphi_0$, i.e.

$$A = \{x : x \in K, |f(x) - \varphi_0(x)| = \| f - \varphi_0 \|_{C(K)}\},$$

and $\bar{\alpha}$ is the conjugate of α.

The uniqueness theorem 1.7 has the following form.

Let G be a Haar subspace of $C(K)$ (see section 1.3). *Then for every $f \in C(K)$ there is exactly one best uniform approximation of f with respect to G* (A. Haar, 1918, A.N. Kolmogorov, 1948).

The theorem (1.6) of Chebyshev also is true for Chebyshev systems (Haar subspaces), as follows.

Let G be a Haar subspace of $C[a, b]$ with dimension n. Let $\varphi \in G$ be the best uniform approximation to $f \in C[a, b]$ with respect to G. Then there exist $n + 1$ points x_i, $i = 1, \ldots, n + 1$, $a \leqslant x_1 < \cdots < x_{n+1} \leqslant b$, such that

$$f(x_i) - \varphi(x_i) = \varepsilon(-1)^i \| f - \varphi \|_{C[a,b]}, \quad i = 1, \ldots, n + 1, \varepsilon = \pm 1.$$

For the first and second Remez algorithms see Remez (1969). There are many modifications of these algorithms, see the books of Cheney (1966), Rice (1964, 1969), Meinardus (1967). We have used in section 1.3 the book of Cheney (1966).

2

Qualitative theory of the best rational approximation

The most essential problems in the qualitative theory of the best approximation are the problems of existence, uniqueness and characterization of the best approximation. Finally the problems connected with the continuity of the operator of the best approximation, or, as is mainly used, the continuity of the metric projection, are considered. In this chapter we shall consider these questions for the best rational approximation. The difficulties arise from the fact that the set R_{nm} of all rational functions of order (n, m) (see the exact definition in section 2.1) is not a finite dimensional linear space and the bounded sets in R_{nm} are not compact in $C[a, b]$ or in $L_p(a, b)$. Nevertheless we shall see that there exists an element of best approximation in $C[a, b]$ and $L_p(a, b)$ (section 2.1). Moreover in $C[a, b]$ we have uniqueness and characterization of the best approximation by means of an alternation, as in the linear case (see section 2.2). Unfortunately in $L_p(a, b)$, $1 \leqslant p < \infty$, we do not have uniqueness (section 2.3). In section 2.4 we consider the problem of continuity of the metric projection in $C[a, b]$ – the metric projection is continuous only in the so-called 'normal points' (see section 2.4). In section 2.5 we consider numerical methods for obtaining the rational function of best uniform approximation. We should like to remark that we examine only the usual rational approximation. Some references for the qualitative theory of generalized rational approximations are given in the notes at the end of the chapter.

2.1 Existence

We shall denote by R_{nm} the set of all real-valued rational functions with numerator an algebraic polynomial of degree at most n and denominator an algebraic polynomial of degree at most m, i.e. $r \in R_{nm}$ if r has the form

$$r(x) = \frac{a_n x^n + a_{n-1} x^{n-1} + \cdots + a_0}{b_m x^m + b_{m-1} x^{m-1} + \cdots + b_0} \tag{1}$$

where a_i, $i = 0, \ldots, n$, b_i, $i = 0, \ldots, m$, are real numbers.

If $r \in R_{nm}$ has the form (1) with $a_n \neq 0$, or $b_m \neq 0$, we say that r is nondegenerate.

If $r = p/q$, p and q algebraic polynomials without common zeros, we say that r is a reduced rational function, or r has a reduced form, or r is irreducible.

Since the set R_{nm} is nonlinear when $m \geqslant 1$, we cannot apply the general theory of linear approximation to obtain the existence of the best rational approximation in the spaces $C[a, b]$ and $L_p(a, b)$, $1 \leqslant p < \infty$. So we shall prove its existence directly.

We define the best rational approximation in $C[a, b]$ and $L_p(a, b)$, $1 \leqslant p < \infty$, of order (n, m) as usual:

$$R_{nm}(f)_{C[a,b]} = \inf\{\|f - r\|_{C[a,b]} : r \in R_{nm}\},$$

$$R_{nm}(f)_{L_p(a,b)} = \inf\{\|f - r\|_{L_p(a,b)} : r \in R_{nm}\}.$$

When it is clear we shall write briefly $R_{nm}(f)_C$ or $R_{nm}(f)$ and $R_{nm}(f)_{L_p}$ or $R_{nm}(f)_p$. When $m = n$ we shall use the notations $R_n(f)_{C[a,b]}$, $R_n(f)_C$ or $R_n(f)$ and $R_n(f)_{L_p(a,b)}$, $R_n(f)_{L_p}$ or $R_n(f)_p$.

Theorem 2.1 (Existence theorem). *Let $f \in C[a, b]$ (or $f \in L_p(a, b)$, $1 \leqslant p < \infty$). Then there exists a rational function $r \in R_{nm}$ (respectively $r_p \in R_{nm}$) such that*

$$\|f - r\|_{C[a,b]} = R_{nm}(f)_{C[a,b]}$$

(respectively

$$\|f - r_p\|_{L_p[a,b]} = R_{nm}(f)_{L_p[a,b]}).$$

Remark. The rational function r, respectively r_p, is called a rational function of best approximation to f in $C[a, b]$, or of best uniform approximation to f, respectively a rational function of the best L_p-approximation to f, of order (n, m).

Proof of theorem 2.1. Let X denote the space $C[a, b]$ or $L_p(a, b)$, $1 \leqslant p < \infty$. Let $f \in X$ and $r_N \in R_{nm}$ be such that

$$\|f - r_N\|_X \leqslant R_{nm}(f)_X + 1/N, \quad N = 1, 2, \ldots. \tag{2}$$

Then it follows that

$$\|r_N\|_X \leqslant R_{nm}(f)_X + \|f\|_X + 1 = A, \quad N = 1, 2, \ldots. \tag{3}$$

Let $r_N = p_N/q_N$, where $p_N \in P_n$, $q_N \in P_m$. We can assume that r_N is normalized so that

$$\| q_N \|_{C[a,b]} = 1, \quad N = 1, 2, \ldots. \tag{4}$$

Now (3) and (4) give us

$$\| p_N \|_X = \| q_N r_N \|_X \leqslant \| r_N \|_X \| q_N \|_{C[a,b]} \leqslant A. \tag{5}$$

From (4) and (5) it follows that the sets $\{p_N : N = 1, 2, \ldots\} \subset P_n$ and $\{q_N : N = 1, 2, \ldots\} \subset P_m$ are sequences in compact sets (P_n, P_m are finite dimensional spaces), so there exists a subsequence N_i, $i = 1, 2, \ldots, \infty$, and $p \in P_n, q \in P_m$ such that

$$\left.\begin{aligned} &\| p - p_{N_i} \|_X \underset{N_i \to \infty}{\longrightarrow} 0; \quad \| p - p_{N_i} \|_{C[a,b]} \underset{N_i \to \infty}{\longrightarrow} 0; \\ &\| q - q_{N_i} \|_{C[a,b]} \underset{N_i \to \infty}{\longrightarrow} 0. \end{aligned}\right\} \tag{6}$$

(all norms in a finite dimensional linear normed space are equivalent).

From (4) and (6) we obtain

$$\| q \|_{C[a,b]} = 1. \tag{7}$$

If x is not a zero of q, in view of (6) $q_{N_i}(x) \to q(x)$ and therefore $q_{N_i}(x) \neq 0$ for sufficiently large N_i. Using (6) we obtain ($r = p/q$):

$$|r(x) - r_{N_i}(x)| \leqslant \frac{1}{|q(x) q_{N_i}(x)|} \{ \| p \|_C \| q - q_{N_i} \|_C + \| q \|_C \| p - p_{N_i} \|_C \} \underset{N_i \to \infty}{\longrightarrow} 0. \tag{8}$$

Therefore, for every $x \in [a, b]$, x not a zero of q, we get from (2) and (8)

$$|r(x) - f(x)| \leqslant |r(x) - r_{N_i}(x)| + |r_{N_i}(x) - f(x)| \underset{N_i \to \infty}{\longrightarrow} R_{nm}(f)_C \tag{9}$$

or

$$|r(x) - f(x)| \leqslant R_{nm}(f)_{C[a,b]}. \tag{10}$$

On the other hand we have from (3) for every $x \in [a, b]$

$$\left| \frac{p_{N_i}(x)}{q_{N_i}(x)} \right| \leqslant A \quad \text{or} \quad |p_{N_i}(x)| \leqslant A |q_{N_i}(x)|.$$

The last inequality together with (6) gives us

$$|p(x)| \leqslant A |q(x)|, \quad x \in [a, b]. \tag{11}$$

The inequality (11) shows that every zero of q in $[a, b]$ is also a zero of p with at least the same multiplicity. Therefore $r = p/q$ is a continuous function in $[a, b]$. Then, since (10) is valid for $x \in [a, b]$ which are not zeros of q, (10) is valid for all $x \in [a, b]$, so (10) gives us

$$\| f - r \|_C = R_{nm}(f)_C.$$

Now let $X = L_p[a, b]$. Let K be some collection of intervals $\Delta_i = [\alpha_i, \beta_i] \subset [a, b]$ such that Δ_i does not contain a zero of q. Then, by (10), (6), (2), we have

$$\begin{aligned}
&\left\{\int_K |f(x) - r(x)|^p \mathrm{d}x\right\}^{1/p} \\
&\leqslant \|f - r_{N_i}\|_p + \left\{\int_K |r(x) - r_{N_i}(x)|^p \mathrm{d}x\right\}^{1/p} \\
&\leqslant \|f - r_{N_i}\|_p + (\mathrm{mes}(K))^{1/p}\{\|p\|_C \|q - q_{N_i}\|_C + \|q\|_C \|p - p_{N_i}\|_C\} \\
&\xrightarrow[N_i \to \infty]{} R_{nm}(f)_p,
\end{aligned}$$

i.e.

$$\left\{\int_K |f(x) - r(x)|^p \mathrm{d}x\right\}^{1/p} \leqslant R_{nm}(f)_p$$

for every such compact K. Since the number of the zeros of q is finite, it follows from the definition of the Lebesgue integral that $\|f - r\|_p \leqslant R_{nm}(f)_p$, and since $r \in R_{nm}$ we must have

$$\|f - r\|_p = R_{nm}(f)_p. \qquad \square$$

The proof of this existence theorem shows the difficulties which arise when we work with rational functions. Roughly speaking, we must think in terms of the poles of the rational function – the proof of theorem 2.1 is so long because we have to consider the poles of r. Indeed it follows from the proof that in the uniform case it is not possible that r has poles on $[a, b]$, because, if q has a zero, on $[a, b]$, p should have the same zero at least with the same multiplicity. But from here follows the possibility for the best rational approximation r to be degenerate: this means that $p \in P_{n-1}$, $q \in P_{m-1}$ if $r = p/q \in R_{nm}$.

We shall see that in questions connected with the continuity of the metric projection in $C[a, b]$ on R_{nm} this possibility of degeneracy will be the main problem.

2.2 Uniqueness and characterization of the best uniform approximation

We have seen that if $f \in C[a, b]$ then there exists a rational function $r \in R_{nm}$ of best uniform approximation. The set of rational functions R_{nm} is a nonlinear one; nevertheless it still has uniqueness of the rational function of best uniform approximation and also characterization of this best approximation by means of alternation. In order to formulate this theorem we shall need the notion of the defect of a rational function.

Let $r \in R_{nm}$ and the reduced form of r be $r = p/q$, i.e. p and q have no common zeros. The defect $d(r)$ of r is given by

$$d(r) = \begin{cases} \min\{n - \deg p, m - \deg q\}, & r \not\equiv 0 \\ m, & r \equiv 0, \end{cases}$$

where $\deg p$ denotes the exact degree of the algebraic polynomial p ($\deg p = k$ if $p \in P_k$ and $p \notin P_{k-1}$).

It follows directly from the definition that:

(a) *r is degenerate if and only if $d(r) > 0$;*
(b) *$d(r)$ is the greatest number s for which $r \in R_{(n-s)(m-s)}$.*

Theorem 2.2 *Let $f \in C[a,b]$. For all natural numbers n and m the rational function $r \in R_{nm}$ is a rational function of best uniform approximation to f of order (n,m) if and only if there exist $N = n + m + 2 - d(r)$ points x_i, $i = 1, \ldots, N$, $a \leqslant x_1 < x_2 < \cdots < x_N \leqslant b$, such that*

$$f(x_i) - r(x_i) = \varepsilon(-1)^i \| f - r \|_{C[a,b]}, \quad i = 1, \ldots, N, \ \varepsilon = \pm 1.$$

Moreover the rational function of order (n,m) of best uniform approximation to f is unique.

In other words r is the rational function of order (n,m) of best uniform approximation to f if and only if $f - r$ alternates at least $n + m + 2 - d(r)$ times in the interval $[a,b]$.

Before proving theorem 2.2 we shall give some lemmas.

Lemma 2.1. *Let $\varphi \in C^1[a,b]$ and let x_i, $i = 1, \ldots, k+1$, $a \leqslant x_1 < x_2 < \cdots < x_{k+1} \leqslant b$, be $k+1$ different points in the interval $[a,b]$ such that*

$$\begin{aligned} &\varphi(x_1) \neq 0, \varphi(x_2) = \cdots = \varphi(x_k) = 0, \quad \varphi(x_{k+1}) \neq 0, \\ &\operatorname{sign} \varphi(x_1) = (-1)^k \operatorname{sign} \varphi(x_{k+1}). \end{aligned} \tag{1}$$

Then φ has at least k zeros on (x_1, x_{k+1}), if we compute every zero with its multiplicity.

Proof. The function φ has $k-1$ zeros on (x_1, x_{k+1}) $x_2, x_3, \ldots, x_k$. We must show that there exists in (x_1, x_{k+1}) a zero z of φ, different from $x_2, x_3, \ldots, x_k$, or that one of the zeros $x_2, \ldots, x_k$ has multiplicity at least 2.

If there does not exist a zero of φ in (x_1, x_{k+1}) different from $x_2, \ldots, x_k$, then in each interval (x_i, x_{i+1}), $i = 1, \ldots, k$, the function φ has constant sign. If the sign of φ is the same in two adjacent intervals (x_i, x_{i+1}), (x_{i+1}, x_{i+2}), then x_{i+1} must be at least a double zero of φ, since $\varphi \in C^1[a,b]$. If we assume that in all adjacent intervals (x_i, x_{i+1}), (x_{i+1}, x_{i+2}), $i = 1, \ldots, k-1$, φ has a different sign, we obtain that $\operatorname{sign} \varphi(x_1) = (-1)^{k+1} \operatorname{sign} \varphi(x_{k+1})$ and we come to contradiction with the condition (1) of the lemma. □

Lemma 2.2. *Let $\{\varphi_i\}_{i=1}^n$ be a Chebyshev system on the interval $[a,b]$,*

$\varphi_i \in C^1[a,b]$, $i=1,\dots,n$, and $G=\{\varphi: \varphi\sum_{i=1}^n a_i\varphi_i\}$ be the Haar subspace, generated by $\varphi_1,\dots,\varphi_n$. Let x_i, $i=0,\dots,n$, $a\leqslant x_0<x_1<\cdots<x_n\leqslant b$, be $n+1$ different points on $[a,b]$. If for $\varphi\in G$ we have

$$(-1)^i\varphi(x_i)\geqslant 0, \quad i=0,\dots,n,$$

or

$$(-1)^i\varphi(x_i)\leqslant 0, \quad i=0,\dots,n,$$

then $\varphi\equiv 0$.

Proof. Let us assume that $\varphi\not\equiv 0$. Let us have for example

$$(-1)^i\varphi(x_i)\geqslant 0, \quad i=0,\dots,n. \tag{2}$$

We shall prove that φ has at least n zeros in the interval $[a,b]$, every zero counted with its multiplicity, which contradicts the assumption of the lemma, that $\{\varphi_i\}_{i=1}^n$ is a Chebyshev system on $[a,b]$.

If $\varphi(x_i)\neq 0$, $i=0,\dots,n$, from (2) and the continuity of the function φ it follows at once that φ has at least n zeros in $[a,b]$. Let now $\varphi(x_i)=0$ for some i. If $\varphi(x_i)\neq 0$ for only one value of i, then the same result follows. There remains the case when $\varphi(x_i)\neq 0$ for at least two values of i. Let the first two be j and $j+k$, i.e.

$$\left.\begin{aligned} \varphi(x_0)=\cdots=\varphi(x_{j-1})=0, \quad \varphi(x_j)\neq 0,\\ \varphi(x_{j+1})=\cdots=\varphi(x_{j+k-1})=0, \quad \varphi(x_{j+k})\neq 0. \end{aligned}\right\} \tag{3}$$

From the hypothesis (2) it follows that

$$\operatorname{sign}\varphi(x_j)=(-1)^k\operatorname{sign}\varphi(x_{j+k}). \tag{4}$$

Since $\varphi\in C^1[a,b]$, from (3), (4) and lemma 2.1 it follows that φ has at least k zeros in the interval (x_j,x_{j+k}) and therefore φ has at least $j+k$ zeros in the interval $[a,x_{j+k}]$. Going on in this way, we obtain that there exist n zeros of φ in $[a,x_n]$, every zero counted with its multiplicity. □

In the proof of theorem 2.2 we shall use also the following modification of the well-known Vallée-Poussin theorem for polynomials.

Theorem 2.3. *Let $f\in C[a,b]$. Let $p\in P_n$, $q\in P_m$ and let q have no zeros on $[a,b]$. Let there exist $N=n+m+2-d(p/q)$ points $\{x_i\}_{i=1}^N$, $a\leqslant x_1<x_2<\cdots<x_N\leqslant b$, in $[a,b]$ such that*

$$f(x_i)-\frac{p(x_i)}{q(x_i)}=\varepsilon(-1)^i\lambda_i, \quad \varepsilon=\pm 1,\ \lambda_i>0,\ i=1,\dots,N. \tag{5}$$

Then

$$R_{nm}(f)_{C[a,b]}\geqslant \min\{\lambda_i: i=1,\dots,N\}.$$

Proof. Let us assume that there exists a rational function $r_1 = p_1/q_1 \in R_{nm}$, (p_1/q_1)-irreducible, such that

$$\| f - r_1 \|_{C[a,b]} < \min \{\lambda_i : i = 1, \dots, N\}. \tag{6}$$

Let us consider the values of the difference $s = p/q - r_1$ at the points x_i, $i = 1, \dots, N$. We obtain from (5) and (6) that

$$\operatorname{sign} s(x_i) = \operatorname{sign} \left\{ \left(\frac{p(x_i)}{q(x_i)} - f(x_i) \right) - (r_1(x_i) - f(x_i)) \right\}$$

$$= \operatorname{sign} \left(\frac{p(x_i)}{q(x_i)} - f(x_i) \right) = \varepsilon(-1)^{i+1}, \quad i = 1, \dots, N.$$

Hence s has at least $N - 1$ different zeros y_i, $i = 1, \dots, N - 1$, in the interval $[a, b]$, i.e.

$$s(y_i) = 0, \quad i = 1, \dots, N - 1.$$

Let us note now that $r_1 = p_1/q_1$ has a reduced form and $\| r_1 \|_{C[a,b]} < \infty$, and consequently q_1 has no zeros on $[a, b]$. So from

$$s(y_i) = \frac{p(y_i)}{q(y_i)} - \frac{p_1(y_i)}{q_1(y_i)} = 0, \quad i = 1, \dots, N - 1,$$

it follows that

$$p(y_i)q_1(y_i) - p_1(y_i)q(y_i) = 0, \quad i = 1, \dots, N - 1,$$

i.e. the algebraic polynomial $pq_1 - p_1 q \in P_M$, $M \leqslant n + m - d(p/q) = N - 2$, has at least $N - 1 > M$ different zeros in the interval $[a, b]$. This contradiction proves the theorem. □

Let us mention that later on we shall use theorem 2.3 in the numerical method of Remez for finding the rational function of best approximation (see section 2.5).

Proof of theorem 2.2. First we shall prove that if $r \in R_{nm}$ realizes an alternation, then r is a rational function of best uniform approximation to f of order (n, m). If we apply theorem 2.3 to f and r with

$$\lambda_i = \lambda = \| f - r \|_{C[a,b]},$$

we obtain that $\lambda \leqslant R_{nm}(f)_{C[a,b]}$, and since $r \in R_{nm}$ we must really have $\lambda = \| f - r \|_{C[a,b]} = R_{nm}(f)$, i.e. r is a rational function of best uniform approximation to f of order (n, m).

Now let r be a rational function of best uniform approximation to f of order (n, m). We shall prove that $f - r$ must alternate at least $N = n + m +$

$2-d(r)$ times in $[a,b]$. Let us assume the opposite, that $M \leqslant N-1$ is the highest number of points $x_1 < x_2 < \cdots < x_M$ in $[a,b]$ such that

$$f(x_i)-r(x_i)=\varepsilon(-1)^i \|f-r\|_{C[a,b]}=\varepsilon(-1)^i R_{nm}(f)_{C[a,b]}, \quad i=1,\ldots,M, \varepsilon=\pm 1. \tag{7}$$

Then there exist $M+1$ points $\xi_i, i=0,\ldots,M$, $a=\xi_0<\xi_1<\cdots<\xi_M=b$ such that for every $x\in[\xi_{i-1},\xi_i]$ we have

$$\varepsilon(-1)^i(f(x)-r(x)) > -R_{nm}(f)_{C[a,b]}, \quad i=1,\ldots,M. \tag{8}$$

In view of (7) the continuous function $f-r$ changes its sign in $[x_i,x_{i+1}]$, therefore we can assume, as in section 1.2, that the points $\xi_i, i=1,\ldots,M-1$, are such that

$$f(\xi_i)-r(\xi_i)=0, \quad i=1,\ldots,M-1. \tag{9}$$

Let us consider the algebraic polynomial

$$s(x)=(-1)^M(x-\xi_1)\cdots(x-\xi_{M-1})\in P_{M-1}. \tag{10}$$

Let $r=p/q$ and p and q have no common zeros. Since $s\in P_{N-2}$, $p\in P_{n-d(r)}$, $q\in P_{m-d(r)}$, there exist two algebraic polynomials $p_1\in P_m, q_1\in P_n$, such that

$$s=pp_1-qq_1.$$

Let us consider the rational function

$$\tilde{r}=\frac{p-\varepsilon\delta q_1}{q-\varepsilon\delta p_1}\in R_{nm}, \tag{11}$$

where ε ($\varepsilon=1$ or -1) is the same as in (7), and δ, $\delta>0$, will be chosen later.

Since $\|f-r\|_{C[a,b]}<\infty$, p and q have no common zeros, and q has no zeros in $[a,b]$, we can find δ_1 so that for $0<\delta<\delta_1$ the polynomial $q-\varepsilon\delta p_1$ has the same sign as q in $[a,b]$.

Let us consider the difference $f-\tilde{r}$. We have

$$f-\tilde{r}=f-r+\frac{p}{q}-\frac{p-\varepsilon\delta q_1}{q-\varepsilon\delta p_1}=f-r-\frac{\varepsilon\delta(pp_1-qq_1)}{q(q-\varepsilon\delta p_1)}=f-r-\frac{\varepsilon\delta s}{q(q-\varepsilon\delta p_1)}.$$

Let $\delta_2\leqslant\delta_1$ be such that for δ, $0<\delta<\delta_2$, we have for $x\in[\xi_{i-1},\xi_i]$

$$\varepsilon(-1)^i(f(x)-\tilde{r}(x))=\varepsilon(-1)^i(f(x)-r(x)) \\ +(-1)^{i+1}\frac{\delta s(x)}{q(x)(q(x)-\varepsilon\delta p_1(x))}>-R_{nm}(f)_{C[a,b]}. \tag{12}$$

This is possible in view of (8), since $f-r$ is a continuous function in $[a,b]$. On the other hand for $x\in(\xi_{i-1},\xi_i)$, $i=1,\ldots,M$, $x=\xi_0$, $x=\xi_M$, we have for $0 < \delta < \delta_2$

$$\varepsilon(-1)^i(f(x) - \tilde{r}(x)) = \varepsilon(-1)^i(f(x) - r(x)) - \varepsilon(-1)^i \frac{\varepsilon\delta s(x)}{q(x)(q(x) - \varepsilon\delta p_1(x))}$$

$$\leqslant R_{mn}(f)_{C[a,b]} - \frac{\delta(-1)^i s(x)}{q(x)(q(x) - \varepsilon\delta p_1(x))}$$

$$< R_{nm}(f)_{C[a,b]}, \tag{13}$$

since by (10) we have $(-1)^i s(x) > 0$ for $x \in (\xi_{i-1}, \xi_i)$, $i = 1, \ldots, M$, $x = \xi_0$, $x = \xi_M$.

In view of (9) we have also (13) for $x = \xi_i$, $i = 1, \ldots, M-1$, i.e. for all $x \in [a, b]$. Since $f - \tilde{r}$ is a continuous function on $[a, b]$, the inequalities (12) and (13) give us

$$\| f - \tilde{r} \|_{C[a,b]} < R_{nm}(f)_{C[a,b]}. \tag{14}$$

By (11) $\tilde{r} \in R_{nm}$, and therefore (14) is a contradiction. Consequently $f - r$ must alternate at least N times.

Now let us prove the uniqueness of the best rational approximation of order (n, m).

Let us assume that there exist two different rational functions $r_1 = p_1/q_1 \in R_{nm}$ and $r_2 = p_2/q_2 \in R_{nm}$ such that

$$\| f - r_1 \|_{C[a,b]} = \| f - r_2 \|_{C[a,b]} = R_{nm}(f)_{C[a,b]}.$$

We can assume that $r_1 = p_1/q_1$ and $r_2 = p_2/q_2$ have a reduced form and q_1, q_2 have no zeros in $[a, b]$.

Let $N_1 = n + m + 2 - d(r_1)$, $N_2 = n + m + 2 - d(r_2)$ and let us assume for definiteness that $N_1 \geqslant N_2$, or, which is the same, $d(r_1) \leqslant d(r_2)$. Let x_i, $i = 1, \ldots, N_1$, $a \leqslant x_1 < \cdots < x_{N_1} \leqslant b$, be the points of alternation for r_1, i.e.

$$f(x_i) - r_1(x_i) = \varepsilon(-1)^i R_{nm}(f)_{C[a,b]}, \quad i = 1, \ldots, N_1, \varepsilon = \pm 1. \tag{15}$$

Let us consider the difference $s = r_1 - r_2$ at the points x_i, $i = 1, \ldots, N_1$. There are two possibilities:

(a) $s(x_i) = 0$, $i = 1, \ldots, N_1$,
(b) $s(x_i) \neq 0$ for some i.

In case (b), since $| f(x_i) - r_1(x_i) | = R_{nm}(f)$, we must have

$$\tilde{\varepsilon}(r_2(x_i) - f(x_i)) < \tilde{\varepsilon}(r_1(x_i) - f(x_i)), \quad \tilde{\varepsilon} = \operatorname{sign}(r_1(x_i) - f(x_i)),$$

and therefore

$$\operatorname{sign} s(x_i) = \operatorname{sign}(r_1(x_i) - f(x_i)). \tag{16}$$

From (15) and (16) it follows that

$$\varepsilon(-1)^{i+1} s(x_i) \geqslant 0, \quad i = 1, 2, \ldots, N_1. \tag{17}$$

On the other hand s can change its sign on $[a, b]$ at most $N_1 - 1$ times, since

$$s = r_1 - r_2 = \frac{p_1}{q_1} - \frac{p_2}{q_2} = \frac{p_1 q_2 - p_2 q_1}{q_1 q_2}$$

and $p_1 q_2 - p_2 q_1 \in P_M$, where $M = n - d(r_1) + m - d(r_2) \leqslant n + m - d(r_2) = N_2 - 2$ and therefore s belongs to the Haar subspace

$$\left\{ \frac{\varphi}{q_1 q_2} : \varphi \in P_{N_2 - 2} \right\}.$$

From here, (17) and lemma 2.2 it follows that $s \equiv 0$, i.e. $r_1 \equiv r_2$. □

2.3 Nonuniqueness in $L_p(a, b)$, $1 \leqslant p < \infty$

One of the unpleasant facts in the theory of the rational approximation is the nonuniqueness of rational functions of best approximation in L_p, $1 \leqslant p < \infty$. More precisely there exist functions $f \in L_p(a, b)$, $1 \leqslant p < \infty$, which have more than one best approximating rational function of order (n, m) in L_p.

Next, we get a function which has at least two best approximating rational functions of order $(0, 2)$ in $L_p(-1, 1)$, $1 \leqslant p < \infty$.

Fix $A > 1$. Choose ε such that $0 < \varepsilon \leqslant \frac{1}{6}$ and when $p > 1$

$$1/(1 + (6\varepsilon)^{1/(p-1)})^{p-1} > 1 - \frac{1}{A}\left(\arctan A - \frac{\pi}{4}\right). \tag{1}$$

Obviously such a choice of ε is possible.

Consider the function

$$f(x) = \begin{cases} 0, & x \in [-1, -\frac{1}{2} - \varepsilon] \cup [-\frac{1}{2} + \varepsilon, \frac{1}{2} - \varepsilon] \cup [\frac{1}{2} + \varepsilon, 1] \\ 1, & x \in (-\frac{1}{2} - \varepsilon, -\frac{1}{2} + \varepsilon) \cup (\frac{1}{2} - \varepsilon, \frac{1}{2} + \varepsilon). \end{cases}$$

Clearly f is an even function. We shall show that there is no even best approximating rational function of order $(0, 2)$ to f in L_p. This fact, obviously, implies the required nonuniqueness.

Consider the rational function $r^*(x) = B^2/((x - \frac{1}{2})^2 + B^2)$, $B = \varepsilon/A$, $r^* \in R_{0,2}$. Clearly $0 \leqslant r^*(x) \leqslant 1$, $x \in (-\infty, \infty)$, and hence

$$\begin{aligned} \| f - r^* \|_{L_p(-1,1)} &\leqslant \left(\int_{-1}^{1} | f(x) - r^*(x) | \, \mathrm{d}x \right)^{1/p} \\ &\leqslant \left(4\varepsilon - \int_{-\varepsilon}^{\varepsilon} \frac{B^2 \mathrm{d}x}{x^2 + B^2} + 2 \int_{\varepsilon}^{\infty} \frac{B^2 \mathrm{d}x}{x^2 + B^2} \right)^{1/p} \\ &= (4\varepsilon)^{1/p} \left(1 - \frac{1}{A}\left(\arctan A - \frac{\pi}{4} \right) \right)^{1/p}. \end{aligned} \tag{2}$$

Now, let $r \in R_{0,2}$ and r be an even function. It is readily seen that r is monotone on $[0, 1]$. Suppose that r is nondecreasing on $[0, 1]$ (the case when r is decreasing is considered similarly). Consider the case when $0 \leqslant r(x) \leqslant 1$, $x \in (\frac{1}{2} - \varepsilon, 1)$ (the other possible cases are trivial). Let $p > 1$. Then we get

$$\| f - r \|_{L_p(-1,1)} \geqslant \left(2 \int_{1/2-\varepsilon}^{1/2+\varepsilon} (1 - r(\tfrac{1}{2} + \varepsilon))^p \mathrm{d}x + 2 \int_{1/2+\varepsilon}^{1} r(\tfrac{1}{2} + \varepsilon)^p \mathrm{d}x \right)^{1/p}$$

$$\geqslant \inf_{0 \leqslant c \leqslant 1} \left(2 \int_{1/2-\varepsilon}^{1/2+\varepsilon} (1 - c)^p \mathrm{d}x + 2 \int_{1/2+\varepsilon}^{1} c^p \mathrm{d}x \right)^{1/p}$$

$$\geqslant (4\varepsilon)^{1/p} \inf_{0 \leqslant c \leqslant 1} \left((1 - c)^p + \frac{1}{6\varepsilon} c^p \right)^{1/p}$$

$$\geqslant (4\varepsilon)^{1/p} / (1 + (6\varepsilon)^{1/(p-1)})^{(p-1)/p}.$$

When $p = 1$ we have immediately $\| f - r \|_{L_1(-1,1)} \geqslant 4\varepsilon$. The last estimates, (1) and (2) imply that any even rational function $r \in R_{0,2}$ is not best approximating to f in $L_p(-1, 1)$. Consequently, there exist at least two best approximating functions.

2.4 Properties of the rational metric projection in $C[a, b]$

Let X be a metric space with a distance d. The set $G \subset X$ is called a Chebyshev set if for every $f \in X$ there exists a unique $g \in G$ such that

$$d(f, g) = \inf \{ d(f, \varphi) : \varphi \in G \}.$$

This unique element is usually denoted by $P_G f$ or Pf and is called a metric projection of f on G.

Here we shall discuss the problem of continuity of the metric projection from $C[a, b]$ on the set R_{nm}. We define the operator P of the metric projection in $C[a, b]$ on R_{nm} in the following way:

If $f \notin R_{nm}$, then $Pf = r$ where r is the rational function of best uniform approximation of order (n, m).

If $f \in R_{nm}$ we set $Pf = f$.

The metric projection is said to be continuous at the point f if from $f_k \to f$ (in the corresponding metric) it follows that $Pf_k \to Pf$ (in the same metric).

The interesting fact about the problem of continuity of the metric projection in $C[a, b]$ on R_{nm} is that the continuity is connected with the degeneracy of the rational function of best approximation to f, i.e. with the degeneracy of Pf. We repeat that the rational function $r \in R_{nm}$ is degenerate if its defect is >0, i.e. $d(r) > 0$ (see section 2.2).

In this section we shall write $\| f \|$ instead of $\| f \|_{C[a,b]}$.

The following theorem holds.

Theorem 2.4. *The metric projection in $C[a,b]$ on R_{nm} is continuous at $f \in C[a,b]$ if and only if $d(Pf)=0$.*

The points $f \in C[a,b]$ for which $d(Pf)=0$ are often called 'normal points'.

Theorem 2.4 is connected with the names of Maehly and Witzgall (1960), Cheney and Loeb (1964), Werner (1964) (see the notes at the end of this chapter).

We shall obtain the part 'if' in theorem 2.4 as a consequence of another property of Pf at points where $d(Pf)=0$ – the strong uniqueness of the best approximation.

Let X be a metric space with a distance d and let Pf be the metric projection of $f \in X$ on the Chebyshev set $G \subset X$.

The metric projection is called strongly unique at the point $f \in X$ if for every $\varphi \in G$ we have

$$d(f,\varphi) \geqslant d(f,Pf) + \gamma d(\varphi, Pf)$$

where $\gamma > 0$ is a constant, possibly depending on f.

Let us remark that the last statement of theorem 2.2 can be reformulated, using the notion of Chebyshev sets, as follows: the set R_{nm} is a Chebyshev set in $C[a,b]$.

The following theorem holds.

Theorem 2.5. *The metric projection Pf in $C[a,b]$ on R_{nm} is strongly unique if $d(Pf)=0$.*

In the proof of theorem 2.5 we shall use the following lemma.

Lemma 2.3. *Let $R = S/Q \in R_{nm}$, $Q(x) > 0$ for $x \in [a,b]$. The set $A = \{p + Rq: p \in P_n, q \in P_m\}$ in $C[a,b]$ has dimension $k = n + m + 1 - d(R)$. Moreover A is a Haar subspace on the interval $[a,b]$.*

Proof. Let $\dim D$ denote the dimension of the set $D \subset C[a,b]$. Evidently we have

$$\begin{aligned} \dim A &= \dim P_n + \dim RP_m - \dim(P_n \cap RP_m) \\ &= n + 1 + m + 1 - \dim(P_n \cap RP_m), \end{aligned} \tag{1}$$

where RP_m denotes the set $RP_m = \{\varphi: \varphi = Rq: q \in P_m\}$. If $S \equiv 0$ then $d(R) = m$ and $\dim A = n+1$. Let $S \not\equiv 0$. An element $\varphi \in RP_m$ belongs to P_n if and only if $\varphi = R\psi$ where $\psi = Qq_1$. Let us estimate the degree of q_1. Since $\psi \in P_m$, $Q \in P_{m-\operatorname{def} Q}$, q_1 must belong to $P_{\operatorname{def} Q}$, where $\operatorname{def} Q$ is the defect of the algebraical polynomial Q in P_m, i.e. $\operatorname{def} Q = m - \deg Q$.

On the other hand, $\varphi = R\psi = Sq_1$ must belong to P_n, therefore q_1 must belong to $P_{\operatorname{def} S}$. Consequently $q_1 \in P_{d(R)}$, $d(R) = \min(\operatorname{def} S, \operatorname{def} Q)$, and the dimension of $P_n \cap RP_m$ is exactly $d(R)+1$. From (1) we obtain that $\dim A = k$.

Now let us prove that A is a Haar subspace. Suppose that $\varphi = p + Rq$,

$p \in P_n$, $q \in P_m$, has at least k zeros in $[a,b]$. Then the algebraic polynomial $pQ + Sq \in P_{n+m-d(R)}$ has $k = n+m+1-d(R)$ zeros in $[a,b]$, which is impossible. □

Proof of theorem 2.5 (Cheney, 1966). Let us set for $r \neq Pf$

$$\gamma(r) = \frac{\|f-r\| - \|f-Pf\|}{\|r-Pf\|} = \frac{\|f-r\| - R_{nm}(f)_{C[a,b]}}{\|r-Pf\|}. \tag{2}$$

We must prove that

$$\inf\{\gamma(r): r \in R_{nm}, r \neq Pf\} = \gamma > 0.$$

Let us suppose the contrary: $\gamma = 0$. Then there exists a sequence $\{r_k\}_{k=1}^{\infty}$ such that $r_k = p_k/q_k$, $r_k \neq Pf$, $p_k \in P_n$, $q_k \in P_m$ and $\gamma(r_k) \underset{k\to\infty}{\longrightarrow} 0$. We may assume that

$$\|p_k\| + \|q_k\| = 1 \tag{3}$$

(if we multiply p_k and q_k by a constant r_k does not change).

Let $Pf = p/q, p \in P_n, q \in P_m$. From condition (3) it follows (passing to sub-sequences if necessary) that there exist $p^* \in P_n$, $q^* \in P_m$ such that

$$\|p_k - p^*\| \underset{k\to\infty}{\longrightarrow} 0, \quad \|q_k - q^*\| \underset{k\to\infty}{\longrightarrow} 0.$$

Let us first remark that $\|p^*/q^*\|_{C[a,b]} < \infty$. Indeed, in the opposite case from (2) it follows that $\gamma = 1$.

We shall show that

$$p^*/q^* = r^* = p/q = Pf. \tag{4}$$

In fact if $r^* \neq p/q = Pf$ then from (2) the contradiction follows:

$$0 = \gamma = \lim_{k\to\infty} \gamma(r_k) = \lim_{k\to\infty} \frac{\|f-r_k\| - \|f-Pf\|}{\|r_k - Pf\|}$$

$$= \frac{\|f-r^*\| - \|f-Pf\|}{\|r^*-Pf\|} > 0,$$

since $\|f-r^*\| > \|f-Pf\|$ if $r^* \neq Pf$ by the uniqueness of the best uniform approximation (theorem 2.2).

Since $d(Pf) = 0$, $Pf = p/q$ is irreducible and p has degree exactly n or q has degree exactly m. So from (4) it follows that $p^* = cp$, $q^* = cq$, c a constant. Since we can set $\|p\| + \|q\| = 1$ by (3), it follows that we can set $c = 1$.

Let

$$Y = \{x: |f(x) - (Pf)(x)| = \|f-Pf\|\},$$

$$\sigma(x) = \operatorname{sign}(f(x) - (Pf)(x)).$$

For every $y \in Y$ we have

$$\begin{aligned}\gamma(r_k)\|r_k - Pf\| &= \|f - r_k\| - \|f - Pf\| \\ &\geqslant \sigma(y)(f(y) - r_k(y)) - \sigma(y)(f(y) - (Pf)(y)) \\ &= \sigma(y)((Pf)(y) - r_k(y)) = \frac{\sigma(y)(q_k(y)(Pf)(y) - p_k(y))}{q_k(y)}. \end{aligned} \quad (5)$$

Now since $\|q_k - q\| \underset{k\to\infty}{\longrightarrow} 0$ and $|q(x)| > 0$ for $x \in [a, b]$ it follows that there exist $\varepsilon > 0$ and $N > 0$ such that for $k > N$ we have $|q_k(x)| \geqslant \varepsilon > 0$ for $x \in [a, b]$. We have

$$\inf\{\max_{y\in Y} \sigma(y)(\tilde{q}Pf - \tilde{p})(y): \tilde{q} \in P_m, \tilde{p} \in P_n, \|\tilde{q}Pf - \tilde{p}\| = 1\} = c > 0. \quad (6)$$

Let us assume the contrary, that $c = 0$. Since the set $\|\tilde{q}Pf - \tilde{p}\| = 1, \tilde{q} \in P_m$, $\tilde{p} \in P_n$ is compact, there exist $\bar{q} \in P_m$, $\bar{p} \in P_n$, $\|\bar{q}Pf - \bar{p}\| = 1$, such that

$$\max_{y\in Y} \sigma(y)(\bar{q}Pf - \bar{p})(y) = 0. \quad (7)$$

By theorem 2.2 the set Y contains $N = n + m + 2$ points x_i, $i = 1, \ldots, N$, $a \leqslant x_1 < x_2 < \cdots < x_N \leqslant b$ (remember that $d(Pf) = 0$), such that

$$\sigma(x_i) = \operatorname{sign}(f(x_i) - (Pf)(x_i)) = \tilde{\varepsilon}(-1)^i, \quad i = 1, \ldots, N, \tilde{\varepsilon} = \pm 1. \quad (8)$$

From (7) and (8) we obtain

$$\varepsilon(-1)^i(\bar{q}(x_i)(Pf)(x_i) - \bar{p}(x_i)) \leqslant 0, \quad i = 1, \ldots, N. \quad (9)$$

But $\bar{q}Pf - \bar{p} \in A = \{\varphi: \varphi = p + qPf, p \in P_n, q \in P_m\}$. By lemma 2.3 the set A is a Haar subspace with dimension $k = n + m + 1 - d(Pf) = n + m + 1$. So (9) and lemma 2.2 give us $\bar{q}Pf - \bar{p} \equiv 0$, which contradicts $\|\bar{q}Pf - \bar{p}\| = 1$. Therefore $c > 0$.

From (5)–(7), using that $\|q_k\| \leqslant 1$ (see (3)), $|q_k(x)| > \varepsilon$ for $k > N, x \in [a, b]$, we have

$$\begin{aligned}\gamma(r_k)\|r_k - Pf\| &\geqslant \max_{y\in Y} \frac{\sigma(y)(q_kPf - p_k)(y)}{q_k(y)} \\ &\geqslant \max_{y\in Y} \sigma(y)(q_kPf - p_k)(y) \\ &\geqslant c\|q_kPf - p_k\| \geqslant c\varepsilon\|Pf - r_k\|, \end{aligned}$$

i.e. $\gamma(r_k) \geqslant c\varepsilon$ which contradicts $\gamma(r_k) \underset{k\to\infty}{\longrightarrow} 0$. □

Theorem 2.6. *Let $f \in C[a, b]$ and $d(Pf) = 0$, where P is the operator of the metric projection in $C[a, b]$ on R_{nm}. Then there exists a constant $c(f)$ such that* *for every $g \in C[a,b]$ we have*

$$\|Pf - Pg\| \leqslant c(f)\|g - f\|.$$

Proof. Using the strong uniqueness of the metric projection in $C[a,b]$ on R_{nm} (theorem 2.5) at the normal point f $(d(Pf) = 0)$ we obtain that there exists a constant $\gamma(f) > 0$ depending possibly on f such that for every $r \in R_{nm}$ we have

$$\|f - r\| \geqslant \|f - Pf\| + \gamma(f)\|Pf - r\|.$$

Setting $r = Pg$ we obtain

$$\begin{aligned}\gamma(f)\|Pf - Pg\| &\leqslant \|f - Pg\| - \|f - Pf\| \\ &\leqslant \|f - g\| + \|g - Pg\| - \|f - Pf\| \\ &\leqslant \|f - g\| + \|g - Pf\| - \|f - Pf\| \\ &\leqslant \|f - g\| + \|g - f\| + \|f - Pf\| - \|f - Pf\| \\ &= 2\|f - g\|,\end{aligned}$$

i.e.

$$\|Pf - Pg\| \leqslant \frac{2}{\gamma(f)}\|f - g\|. \qquad \square$$

The part 'if' of theorem 2.4 immediately follows from theorem 2.6. Now we shall prove the part 'only if'. This part follows immediately from the following theorem.

Theorem 2.7. *Let $f \in C[0,1]$ and let*

$$R_{nm}(f)_C = \|f - r\|_{C[0,1]}, \quad r \in R_{nm},$$

realize an alternation of $n + m + 2 - d$ points, $d(r) = d > 0$.

There exists $\delta > 0$ such that for every $\varepsilon, 0 < \varepsilon < \delta$, there is a function $f_\varepsilon \in C[0,1]$ such that

$$\|f - f_\varepsilon\|_{C[0,1]} \leqslant 2\varepsilon,$$

$$R_{nm}(f_\varepsilon) = \|f_\varepsilon - r_\varepsilon\|_{C[0,1]}, r_\varepsilon \in R_{nm}, d(r_\varepsilon) = d - 1$$

and

$$\|r - r_\varepsilon\| \geqslant \delta.$$

In the proof of theorem 2.7 we shall use the following lemma.

Lemma 2.4. *Let $\tilde{x} > 0$ be given. There exist α, β, γ such that the function $\varphi(x) = \gamma(x - \alpha)/(x - \beta)$ has the following properties:*

$$\varphi(0) = a, \quad \varphi(\tilde{x}) = b, \quad 0 < b < a; \quad \varphi'(x) < 0, \quad x \in [0, \infty).$$

Proof. From the conditions on φ we obtain

$$\gamma\frac{\alpha}{\beta}=a,\quad \gamma\frac{\tilde{x}-\alpha}{\tilde{x}-\beta}=b.$$

We get

$$\gamma\alpha=\beta a,$$

$$\gamma\tilde{x}-\gamma\alpha=\tilde{x}b-\beta b,$$

$$\gamma\tilde{x}-\beta a=\tilde{x}b-\beta b,$$

$$\beta(b-a)=\tilde{x}(b-\gamma),$$

$$\beta=\frac{\tilde{x}(b-\gamma)}{b-a},\quad \alpha=\frac{a\tilde{x}(b-\gamma)}{\gamma(b-a)}.$$

Since $a>b$, for $0<\gamma<b$ we have $\alpha<\beta<0$ and

$$\varphi'(x)=\gamma\frac{\alpha-\beta}{(x-\beta)^2}<0\quad\text{for } x\in[0,\infty).$$ □

Proof of theorem 2.7. Let the points of alternation of $f-r$ be x_i, $i=1,\ldots,k$, $0\leqslant x_1<x_2<\cdots<x_k\leqslant 1$, $k=n+m+2-d$. Without loss of generality we can assume that the first extremum of $f-r$ is positive, that x_2 is the first point in $[0,1]$ for which $f(x_2)-r(x_2)=-R_{nm}(f)$ and that

$$f(x_1)-r(x_1)=R_{nm}(f), x_1=\max\{x\colon f(x)-r(x)=R_{nm}(f), x<x_2\}.$$

Let $z=\min\{x\colon f(x)-r(x)=0, x_1<x<x_2\}$. Let us denote

$$I_1=\{x\colon 0\leqslant x\leqslant z\},\quad I_2=\{x\colon z\leqslant x\leqslant 1\}.$$

From the assumption given above it follows that

$$f(x)-r(x)>-R_{nm}(f)\quad\text{for } x\in I_1$$

and therefore there exists $\delta>0$ such that

$$f(x)-(r(x)+\delta)>-R_{nm}(f)\quad\text{for } x\in I_1.$$

We can assume also that $\delta<R_{nm}(f)$. Let $0<\varepsilon<\delta$. Let $\tilde{x}_1$ be such that

$$\tilde{x}_1=\max\{x\colon x\in[x_1,z], f(x)-r(x)=R_{nm}(f)-\varepsilon\}.$$

Let $r=p/q$, $q(x)\geqslant\mu>0$ for $x\in[0,1]$.

From lemma 2.4 it follows that there exists a function $\varphi_a(x)=\gamma(x-\alpha)/(x-\beta)$ such that

$$\varphi_a(0)=a, \varphi(\tilde{x}_1)=\min\left(\varepsilon\mu,\frac{a}{2}\right)$$

and

$$\varphi_a'(x)<0\quad\text{for } x\in[0,1].$$

Set $\psi_a=\varphi_a/q$.

From these properties of φ_a it follows that

$$f(x)-(r(x)+\psi_a(x)) \geqslant -R_{nm}(f) \quad \text{for } x\in I_1, \quad 0<a\leqslant \delta\mu.$$

Since φ_a is a continuous function of a, when a increases we shall have an $\bar{a}$ for which there is a point $x_0\in[0,\tilde{x}_1)$ such that

$$f(x_0)-(r(x_0)+\psi_{\bar{a}}(x_0)) = -R_{nm}(f)$$

and for every $x\in I_1$ we have

$$|f(x)-(r(x)+\psi_{\bar{a}}(x))| \leqslant R_{nm}(f).$$

Obviously we must have $\bar{a}\geqslant\delta$.

In the interval $[\tilde{x}_1, z)$ there is a point $\bar{x}_1$ for which

$$f(\bar{x}_1)-(r(\bar{x}_1)+\psi_{\bar{a}}(\bar{x}_1)) = \max\{f(x)-(r(x)+\psi_{\bar{a}}(x)): x\in[\tilde{x}_1, z)\} = R_{nm}(f)-\xi,$$

and, since $\varphi'_{\bar{a}}(x)<0, \varphi_{\bar{a}}(\tilde{x}_1)=\min(\varepsilon\mu, \bar{a}/2)$, it follows that $0<\xi\leqslant 2\varepsilon$.

Let us define

$$r_\varepsilon(x) = r(x)+\psi_{\bar{a}}(x).$$

Then $r_\varepsilon\in R_{n-d+1,m-d+1}\subset R_{nm}$.

In the interval I_2 we define $f_\varepsilon(x)=f(x)+\psi_{\bar{a}}(x)$. In the interval I_1 we define $f_\varepsilon(x)=f(x)$ for $x\in[0,x_0]$, in $[x_0,z]$ we define f_ε so that $f_\varepsilon(\bar{x}_1)=f(\bar{x}_1)+\xi$, $\|f-f_\varepsilon\|_C\leqslant 2\varepsilon$, $f_\varepsilon\in C[0,1]$, $\|f_\varepsilon-r_\varepsilon\|_{C[\tilde{x}_1,z]}\leqslant R_{nm}(f)_{C[0,1]}$.

From the construction given above it follows that the points $x_0, \bar{x}_1, x_2, \ldots, x_k$ are points of alternation for $f_\varepsilon - r_\varepsilon$, and, since $r_\varepsilon\in R_{n-d+1,m-d+1}$, from theorem 2.2 it follows that

$$R_{nm}(f_\varepsilon)_C = \|f_\varepsilon - r_\varepsilon\|_C = \|f-r\|_C = R_{nm}(f)_C$$

Since $\|r-r_\varepsilon\|_C = \|\varphi_{\bar{a}}\|\geqslant\delta$ and $\|f-f_\varepsilon\|\leqslant 2\varepsilon$, the theorem is proved. □

2.5 Numerical methods for best uniform approximation

Best rational approximations with respect to uniform distance are often used for representation of functions because of their perfect approximation properties and the possibilities which we have with the advent of high-speed digital computers. Two algorithms have turned out to be suitable – the differential correction algorithm and the second Remez algorithm.

Differential correction algorithm

This algorithm for finding the rational function of order (n, m) of best uniform approximation to $f\in C[a,b]$ is due to Cheney and Loeb (1961).

In this paragraph we set

$$R_{nm} = \{r: r=p/q, p\in P_n, q\in P_m, q(x)>0, x\in[a,b]\},$$

$$\Delta(r) = \|f-r\|_{C[a,b]}.$$

The algorithm can be described by the following iterative process:

(i) *Choose an initial approximation* $r_0 = p_0/q_0$ *such that* $r_0 \in R_{nm}$;

(ii) *Set* $k = 0$;

(iii) *Compute the number* $\Delta_k = \Delta(r_k) = \| f - r_k \|_{C[a,b]}$;

(iv) *Form the following function of* $r = p/q$, $r \in R_{nm}$ –

$$\delta_k(r) = \max_{x \in [a,b]} \{ |f(x)q(x) - p(x)| - \Delta_k q(x) \};$$

(v) *Select* $r_{k+1} = p_{k+1}/q_{k+1}$, $r_{k+1} \in R_{nm}$, *so as to minimize the function* $\delta_k(r)$ *subject to the constraint* $\| q_{k+1} \|_{C[a,b]} = 1$ –

$$\delta_k(r_{k+1}) = \min \{ \delta_k(r) : r \in R_{nm}, \| q \|_C = 1 \};$$

(vi) *If* $\delta_k(r_{k+1}) \geqslant 0$ *go to* (viii), *if* $\delta_k(r_{k+1}) < 0$ *go to* (vii);

(vii) *Set* $k = k + 1$; *go to* (iii);

(viii) *Stop*; r_k *is the rational function of order* (n, m) *of best uniform approximation to* f.

Remarks (1) $r_0 \in R_{nm}$ may be arbitrary (for example $r_0 = 0/1$).

(2) The minimization of $\delta_k(r)$ is a problem of convex programming and there exist effective methods for its solution. Evidently this step of the algorithm is the most difficult to carry out.

(3) In the differential correction algorithm originally given by Cheney and Loeb (1961), in step (iv) the function $\delta_k(r)$ is defined as

$$\delta_k(r) = \max_{x \in [a,b]} \left\{ \frac{|f(x)q(x) - p(x)| - \Delta_k q(x)}{q_k(x)} \right\}.$$

In Barrodale, Powell and Roberts (1972) it is shown that this algorithm is quadratically convergent if f is (n, m) normal (see section 2.4) in contrast to the algorithm described which exhibits a linear convergence rate – theorem 2.8 below.

(4) In practical applications the constraint $\| q \|_{C[a,b]} = 1$ is less convenient than $|b_i| \leqslant 1$, $i = 0, \ldots, m$, where $q(x) = b_0 + b_1 x + \cdots + b_m x^m$, and this condition does not change the proof of convergence.

(5) In practical calculations step (vi) is changed to

(vi) *If* $\delta_k(r_{k+1}) \geqslant -\varepsilon$ *go to* (viii), *where* ε *is a sufficiently small positive number.*

The following theorem shows the effectiveness of the suggested method.

Theorem 2.8. *Let* $f \in C[a, b]$. *Then the sequence* $\{\Delta_k\}_{k=0}^{\infty}$ *generated by the differential correction algorithm satisfies*

$$0 \leqslant \Delta_{k+1} - R_{nm}(f)_{C[a,b]} \leqslant \theta^{k+1}(\Delta_0 - R_{nm}(f)_{C[a,b]}),$$

where $0 < \theta < 1$.

Proof. First we shall prove that if $q_0(x) > 0$ for $x \in [a, b]$ then $q_k(x) > 0$ for $x \in [a, b]$ if r_{k-1} is not the rational function of order (n, m) of best uniform approximation to f. Let us assume that k_0 is the smallest index for which there exists a point $x_0 \in [a, b]$ such that $q_{k_0+1}(x_0) \leqslant 0$. If $r_{k_0} \in R_{nm}$ is not the rational function of best uniform approximation to f in R_{nm}, then there exists $r = p/q, r \in R_{nm}, \| q \|_C = 1$, such that $\Delta(r) < \Delta(r_{k_0})$ and

$$\delta_{k_0}(r_{k_0+1}) \leqslant \delta_{k_0}(r) = \max_{x \in [a,b]} \{(|f(x) - r(x)| - \Delta_{k_0}) q(x)\} < 0.$$

But from $q_{k_0+1}(x_0) \leqslant 0$ we get

$$\delta_{k_0}(r_{k_0+1}) \geqslant |f(x_0) q_{k_0+1}(x_0) - p_{k_0+1}(x_0)| - \Delta_{k_0} q_{k_0+1}(x_0) \geqslant 0,$$

which contradicts the previous inequality. Thus we have proved that $q_k(x) > 0$ for $x \in [a, b]$ if r_{k-1} is not the best rational approximation to f.

Now we shall prove that $\delta_k(r_{k+1}) \leqslant 0$, and $\delta_k(r_{k+1}) = 0$ if r_k is the best approximation to f. In fact

$$\delta_k(r_{k+1}) \leqslant \delta_k(r_k) = 0$$

and if r_k is not the best approximation to f then as above there exists $r \in R_{nm}$, $r = p/q$, $\| q \|_C = 1$, such that $\Delta(r) < \Delta(r_k)$ and consequently $\delta_k(r_{k+1}) \leqslant \delta_k(r) < 0$. From

$$\begin{aligned} 0 > \delta_k(r_{k+1}) &= \max \{(|f(x) - r_{k+1}(x)| - \Delta_k) q_{k+1}(x) : x \in [a, b]\} \\ &\geqslant \max \{|f(x) - r_{k+1}(x)| - \Delta_k : x \in [a, b]\} = \Delta_{k+1} - \Delta_k \end{aligned}$$

we find that the sequence $\Delta_0, \Delta_1, \Delta_2, \ldots$ converges downward to a limit L. If we suppose that $L > R_{nm}(f)_C$ then there exists $r \in R_{nm}$, $r = p/q$, $\| q \|_C = 1$, such that $\Delta(r) < L$ and

$$|f(x) - r(x)| \leqslant \Delta(r) < L \leqslant \Delta_k, \quad x \in [a, b].$$

On the other hand

$$\begin{aligned} \delta_k(r_{k+1}) \leqslant \delta_k(r) &= \max \{(|f(x) - r(x)| - \Delta_k) q(x) : x \in [a, b]\} \\ &\leqslant \min \{q(x) : x \in [a, b]\} \max \{|f(x) - r(x)| - \Delta_k : x \in [a, b]\} \\ &= \alpha(\Delta(r) - \Delta_k), \end{aligned}$$

where $\alpha = \min \{q(x) : x \in [a, b]\}$. Therefore

$$\Delta_{k+1} \leqslant \delta_k(r_{k+1}) + \Delta_k \leqslant \alpha(\Delta(r) - \Delta_k) + \Delta_k$$

and setting $k \to \infty$ we get $L \leqslant \alpha(\Delta(r) - L) + L$, i.e. the contradiction $\Delta(r) \geqslant L$. Thus we obtain that

$$\lim_{k \to \infty} \Delta_k = R_{nm}(f)_{C[a,b]} \cdot$$

Let $r \in R_{nm}$, $r = p/q$, $\| q \|_C = 1$, be the best approximation to f. From above we have $\Delta_{k+1} - \Delta_k \leqslant \delta_k(r_{k+1})$ and $\delta_k(r_{k+1}) \leqslant \alpha(R_{nm}(f)_{C[a,b]} - \Delta_k)$, hence

$$\begin{aligned}\Delta_{k+1} - R_{nm}(f)_C &= (\Delta_k - R_{nm}(f)_C) + \Delta_{k+1} - \Delta_k \\ &\leqslant \Delta_k - R_{nm}(f)_C + \alpha(R_{nm} - \Delta_k) = (1-\alpha)(\Delta_k - R_{mn}(f)_C) \\ &= \theta(\Delta_k - R_{nm}(f)_C)\end{aligned}$$

and recursively

$$\Delta_{k+1} - R_{nm}(f)_C \leqslant \theta^{k+1}(\Delta_0 - R_{nm}(f)_C).$$

Since $0 < q(x) \leqslant 1$, it follows that $0 < \alpha \leqslant 1$ and $0 \leqslant \theta < 1$. Thus the theorem is proved. □

Theorem 2.9. *Let $f \in C[a,b]$ and $\tilde{r}$ be the rational function of order (n,m) of best uniform approximation to f. Suppose we have $d(\tilde{r}) = 0$ (i.e., f is a normal point with respect to R_{nm}). Then for the sequence $\{r_k\}_{k=0}^{\infty}$ of rational functions $r_k \in R_{nm}$ generated by means of the differential correction algorithm we have:*

$$\| r_k - \tilde{r} \|_{C[a,b]} \leqslant \theta^k \gamma^{-1}(f) | \Delta_0 - R_{nm}(f)_{C[a,b]} |,$$

where $0 < \theta < 1$ and the constant $\gamma^{-1}(f)$ depends only on f.

Proof. Since $d(\tilde{r}) = 0$, using theorem 2.5 we obtain from the strong uniqueness of $\tilde{r}$

$$\begin{aligned}\gamma(f) \| r_k - \tilde{r} \|_C &\leqslant \| f - r_k \|_C + \| f - \tilde{r} \|_C = \Delta_k - R_{nm}(f)_C \\ &\leqslant \theta^k(\Delta_0 - R_{nm}(f)_C), \\ \| r_k - \tilde{r} \|_C &\leqslant \theta^k \gamma^{-1}(f)(\Delta_0 - R_{nm}(f)_C)\end{aligned}$$

□

Remez algorithm

The idea of the Remez algorithm is the determination of the alternation according to the Chebyshev theorem (E. Remez, 1934a, b).

We shall describe the Remez algorithm in the case when $f \in C[a,b]$ is a normal point with respect to R_{nm} (see section 2.4).

(i) *Set $k = 0$ and $N = m + n + 2$;*
(ii) *Choose a point set $a \leqslant x_1^{(k)} < x_2^{(k)} \cdots < x_N^{(k)} \leqslant b$;*
(iii) *Solve the nonlinear system of equations*

$$f(x_i^{(k)}) - \frac{p_k(x_i^{(k)})}{q_k(x_i^{(k)})} = (-1)^i \lambda_k, \quad i = 1, \ldots, N,$$

for the unknowns $a_0^k, \ldots, a_n^k, b_0^k, \ldots, b_m^k, \lambda_k$, where

$$p_k(x) = \sum_{i=0}^{n} a_i^k x^i, \quad q_k(x) = \sum_{i=0}^{m} b_i^k x^i;$$

(iv) *If* $\| f - r_k \|_{C[a,b]} = |\lambda_k|$ *go to* (vii), *else go to* (v);†

(v) *Select the points* $a \leqslant x_1^{(k+1)} < \cdots < x_N^{(k+1)} \leqslant b$ *such that*

$$\operatorname{sign}(f(x_i^{(k+1)})) - r_k(x_i^{(k+1)})) = -\operatorname{sign}(f(x_{i+1}^{(k+1)}) - r_k(x_{i+1}^{(k+1)})), i = 1, \ldots, N-1,$$
$$|f(x_i^{(k+1)}) - r_k(x_i^{(k+1)})| \geqslant |\lambda_k|, \qquad i = 1, \ldots, N,$$

and for some s, $1 \leqslant s \leqslant N$, *we have*

$$|f(x_s^{(k+1)}) - r_k(x_s^{(k+1)})| = \| f - r_k \|_{C[a,b]};$$

(vi) *Set* $k = k + 1$; *go to* (iii);

(vii) *Stop; the function* r_k *is the rational function of best uniform approximation in* R_{nm} *to* f.

Remarks. (1) The initial points $\{x_i^{(0)}\}_{i=1}^N$ can be chosen to be the extremal points of the Nth Chebyshev polynomials translated to $[a, b]$.

(2) at step (iii) for solving the system one may write the system in the form

$$p_k(x_i^{(k)}) - (f(x_i^{(k)}) - (-1)^i \lambda_k^s)(q_k(x_i^{(k)}) - 1)$$
$$= f(x_i^{(k)}) - (-1)^i \lambda_k^{s+1}, \quad i = 1, \ldots, N, \ s = 0, 1, \ldots,$$

and, setting $\lambda_k^0 = 0$, we solve the linear system for $a_0^k, \ldots, a_n^k$, $b_0^k, \ldots, b_m^k$, λ_k^1. Then substituting the λ_k^1 obtained in place of λ_k^s we find λ_k^2 and so on. Usually the sequence $\{\lambda_k^s\}_{s=0}^\infty$ converges. The other approach is to use Newton's method to solve the nonlinear system at step (iii).

(3) The convergence theorems for the Remez algorithm are given by H. Werner (1962) and A. Ralston (1965).

(4) In the practical applications step (iv) is usually changed to

(iv) If $|\| f - r_k \|_{C[a,b]} - |\lambda_k|| \leqslant \varepsilon$ go to (vii), else go to (v), where ε is a sufficiently small positive number.

(5) The numerical experiments show that if the Remez algorithm works without failure it converges faster than the differential correction algorithm.

2.6 Notes

Usually the qualitative theory of rational approximation considers the so-called generalized rational approximation, i.e. approximation by means of expressions of the type

$$\frac{a_n \varphi_n + \cdots a_0 \varphi_0}{b_m \psi_m + \cdots + b_0 \psi_0},$$

where $\{\varphi_i\}_{i=0}^n$, $\{\psi_i\}_{i=0}^m$ are Chebyshev systems. Almost all results of Chapter 2

† $r_k = p_k / q_k$.

can be generalized for approximations of this type, see the books of Cheney (1966), Rice (1969), Collatz and Krabs (1973), Meinardus (1967), Braess (1986).

The existence theorem 2.1 and the characterization and uniqueness theorem 2.2 go back to Ch.de la Vallée-Poussin (1910).

The generalizations of theorems 2.2 and 2.3 for approximations of the type

$$\inf\{\|f-sr\|_{C[a,b]}: r\in R_{nm}, s(x)>0, x\in[a,b]\}$$

are given by Ahiezer (1930); see also Ahiezer (1970) and section 4.3.

For approximations by means of generalized rational functions, concerning existence, characterization and uniqueness, besides the books given above we want to mention the papers of B. Boehm (1965), Cheney and Loeb (1961), Newman and Shapiro (1964), L. Collatz (1960). There exist also generalizations of the Kolmogorov criteria, see for example Meinardus and Schwedt (1964), B. Brosovski (1965a, b, 1969), Brosovski and Guerreiro (1984).

There are many investigations concerning the properties of uniform metric projection with respect to a given generalized rational system of functions. Besides the books given above and the works of Cheney and Loeb (1964) and H. Werner (1964) we want to mention the work of Goldstein (1963); see also Chalmers and Taylor (1983).

For other details on numerical methods for obtaining the best uniform rational approximation see the books given above and Veidinger (1960), J. Maehly (1963), L. Collatz (1960), Cheney and Loeb (1962), Wetterling (1963), Werner (1963), Ralston (1965), Barrodale, Powell, Roberts (1972), Kaufman, Leeming, Taylor (1978), Dunham (1967a, b).

Let us mention finally some interesting results regarding real and complex uniform approximation by rational functions on an interval and on the disk.

We have considered till now real uniform rational approximation on the interval $[a,b]$:

$$R_{nm}(f)_{C[a,b]}=\inf\{\|f-r\|_{C[a,b]}: r\in R_{nm}\},$$

where f is a real-valued function on $[a,b]$, and $r\in R_{nm}$ has real coefficients.

We can consider also complex uniform rational approximation:

$$R_{nm}^{\mathbb{C}}(f)_{C[a,b]}=\{\|f-r\|_{C[a,b]}: r\in R_{nm}^{\mathbb{C}}\}$$

where f is also a real-valued function, but $R_{nm}^{\mathbb{C}}$ is the set of rational functions of the type

$$\frac{\alpha_n x^n+\cdots+\alpha_0}{\beta_m x^m+\cdots+\beta_0},$$

where α_i, $i=0,\ldots,n,\beta_i$, $i=0,\ldots,m$, are complex numbers.

Lungu (1971) and Saff and Varga (1977, 1978) found that for all n and m

there is a function f with

$$R^{\mathbb{C}}_{nm}(f)_{C[a,b]} < R_{nm}(f)_{C[a,b]}. \tag{1}$$

From here, using symmetry arguments, it is easy to see that the best uniform complex rational approximation of a real valued function is not always unique; see also Ruttan (1977).

Trefethen and Gutknecht (1983a, b) have many results in this topic. For example let us set

$$\gamma_{nm} = \inf\{R^{\mathbb{C}}_{nm}(f)/R_{nm}(f): f \in C[a,b],\ f \text{ real-valued}\}.$$

Then

$$\gamma_{nm} = 0 \quad \text{for } n \geqslant 0, \quad m \geqslant n+3.$$

Trefethen and Gutknecht have shown also that for the complex rational uniform approximation on the disk there are analogs of (1) and of non-uniqueness.

3

Some classical results in the linear theory

The most essential question in the quantitative theory of approximation is the connection between the degree of the best approximation to a given function f by means of some tool for approximation (algebraic polynomials, trigonometric polynomials, rational functions, spline functions and others) with respect to a given metric (uniform, L_p and others) and the smoothness properties of f (differentiability, Lipschitz conditions etc.).

The solutions of these questions in linear approximations usually use the moduli of continuity and smoothness. So we shall begin in section 3.1 with some definitions and properties of the moduli of smoothness in $C[a,b]$ and in $L_p[a,b]$. In section 3.2 and 3.3 we give the classical theorems of Jackson and Bernstein for best trigonometrical L_p approximation. In section 3.4 we consider briefly the best approximation by means of algebraical polynomials in $[-1,1]$ and the singularities connected with them. Finally in section 3.5 we consider the K-functional of J. Peetre, which is the abstract version of the moduli of smoothness, and its application for the characterization of the degree of the best approximation in the abstract case, using abstract Jackson type and Bernstein type theorems.

3.1 Moduli of continuity and smoothness in C and L_p

Let the function f be bounded on the interval $[a,b]$.

Definition 3.1. *The modulus of continuity (in $C[a,b]$) of the function f is the following function of $\delta\in[0,\infty)$:*

$$\omega(f;\delta)_{C[a,b]}=\sup\{|f(x')-f(x'')|:|x'-x''|\leqslant\delta, x',x''\in[a,b]\}. \qquad (1)$$

We shall write, when it is clear, $\omega(f;\delta)$ or $\omega(f;\delta)_C$ instead of $\omega(f;\delta)_{C[a,b]}$.

Obviously the necessary and sufficient condition for the function f to be continuous in the finite closed interval $[a,b]$ is $\omega(f;\delta)\underset{\delta\to 0}{\longrightarrow} 0$.

A natural generalization of the modulus of continuity is the moduli of smoothness.

For every bounded function and every natural number k we define the kth difference with step h in the point x by

$$\triangle_h^k f(x)=\sum_{m=0}^{k}(-1)^{m+k}\binom{k}{m}f(x+mh),\quad \triangle_h^1 f(x)=\triangle_h f(x); \tag{2}$$

where $\binom{k}{m}=k!/(m!(k-m)!)$ are the Newton binomial coefficients.

Let f be defined and bounded in the interval $[a,b]$ and k be a natural number.

Definition 3.2. *The modulus of smoothness in $C[a,b]$ of kth order of the function f is the following function of $\delta\in[0,\infty)$:*

$$\omega_k(f;\delta)_{C[a,b]}=\sup\{|\triangle_h^k f(x)|:|h|\leqslant\delta, x, x+kh\in[a,b]\}. \tag{3}$$

We shall write, when it is clear, $\omega_k(f;\delta)$ or $\omega_k(f;\delta)_C$ instead of $\omega_k(f;\delta)_{C[a,b]}$. In some cases we shall also use the notations:

$$\omega_k(f;\Delta)=\omega_k(f;[a,b])=\sup\{|\triangle_h^k f(x)|:x,x+kh\in[a,b]=\Delta\}. \tag{4}$$

From definition 3.2 it follows immediately that the modulus of smoothness of first order is exactly the modulus of continuity: $\omega_1(f;\delta)_{C[a,b]}=\omega(f;\delta)_{C[a,b]}$.

The second modulus of smoothness $\omega_2(f;\delta)_C$ is often called the modulus of smoothness or Zygmund's modulus.

The moduli of smoothness have the following basic properties

(i) $\omega_k(f;\delta')\leqslant\omega_k(f;\delta'')$ *if* $\delta'\leqslant\delta''$.
(ii) $\omega_k(f+g;\delta)\leqslant\omega_k(f;\delta)+\omega_k(g;\delta)$.
(iii) $\omega_k(f;\delta)\leqslant 2\omega_{k-1}(f;\delta)$, $k\geqslant 2$.
(iv) *If f' exists and is bounded in $[a,b]$, then*

$$\omega_k(f;\delta)\leqslant\delta\omega_{k-1}(f';\delta),\quad k\geqslant 2.$$

(iv′) $\omega_1(f;\delta)\leqslant\delta\|f'\|_{C[a,b]}$.
(v) *If n is a natural number, then*

$$\omega_k(f;n\delta)\leqslant n^k\omega_k(f;\delta).$$

(v′) $\omega_k(f;\lambda\delta)\leqslant([\lambda]+1)^k\omega_k(f;\delta)\leqslant(\lambda+1)^k\omega_k(f;\delta)$, $\lambda>0$.†

The proofs of properties (i) and (ii) follow directly from definition 3.2. The

† $[\lambda]$ denotes the integer part of λ.

proof of property (iii) follows from the equality

$$\triangle_h^k f(x) = \triangle_h^{k-1} f(x+h) - \triangle_h^{k-1} f(x).$$

Properties (iv), (iv′) make sense for functions with bounded first derivative. For every $h, |h| \leqslant \delta$, we have

$$\begin{aligned}
|\triangle_h^k f(x)| &= |\triangle_h^{k-1}(f(x+h) - f(x))| \\
&= \left| \triangle_h^{k-1} \int_0^h f'(x+t)\mathrm{d}t \right| = \left| \int_0^h \triangle_h^{k-1} f'(x+t)\mathrm{d}t \right| \\
&\leqslant \int_{\min(0,h)}^{\max(0,h)} |\triangle_h^{k-1} f'(x+t)|\mathrm{d}t \leqslant \int_{\min(0,h)}^{\max(0,h)} \omega_{k-1}(f';|h|)\mathrm{d}t \\
&= |h|\omega_{k-1}(f';h) \leqslant \delta\omega_{k-1}(f';\delta).
\end{aligned}$$

Consequently

$$\sup\{|\triangle_h^k f(x)| : |h| \leqslant \delta, x, x+k\delta \in [a,b]\} = \omega_k(f;\delta) \leqslant \delta\omega_{k-1}(f';\delta).$$

For $k=1$ we obtain, for $|h| \leqslant \delta$,

$$|\Delta_h f(x)| = \left| \int_0^h f'(x+t)\mathrm{d}t \right| \leqslant |h| \, \| f' \|_{C[a,b]} \leqslant \delta \| f' \|_{C[a,b]}.$$

In the proof of property (v) we shall use the equality

$$\triangle_{nh}^k f(x) = \sum_{i_1=0}^{n-1} \sum_{i_2=0}^{n-1} \cdots \sum_{i_k=0}^{n-1} \triangle_h^k f(x + i_1 h + i_2 h + \cdots + i_k h). \tag{5}$$

It is easy to prove this equality by induction with respect to k. For $k=1$ we have

$$\begin{aligned}
\triangle_{nh} f(x) &= f(x+nh) - f(x) = \sum_{i=0}^{n-1} (f(x+ih+h) - f(x+ih)) \\
&= \sum_{i=0}^{n-1} \triangle_h f(x+ih).
\end{aligned}$$

Let us assume that (5) is true for a given natural number k. Then

$$\begin{aligned}
\triangle_{nh}^{k+1} f(x) &= \triangle_{nh}^k (f(x+nh) - f(x)) \\
&= \triangle_{nh}^k(\triangle_{nh} f(x)) = \sum_{i_1=0}^{n-1} \cdots \sum_{i_k=0}^{n-1} \triangle_h^k(\triangle_{nh} f(x + i_1 h + \cdots + i_k h)) \\
&= \sum_{i_1=0}^{n-1} \cdots \sum_{i_k=0}^{n-1} \triangle_h^k \left(\sum_{i_{k+1}=0}^{n-1} \triangle_h f(x + i_1 h + \cdots + i_k h + i_{k+1} h) \right) \\
&= \sum_{i_1=0}^{n-1} \cdots \sum_{i_{k+1}=0}^{n-1} \triangle_h^{k+1} f(x + i_1 h + \cdots i_{k+1} h).
\end{aligned}$$

From (5) for $|h| \leqslant \delta$ it follows that

$$|\Delta_{nh}^k f(x)| \leqslant = \sum_{i_1=0}^{n-1} \cdots \sum_{i_k=0}^{n-1} |\Delta_h^k f(x + i_1 h + \cdots + i_k h)| \leqslant n^k \omega_k(f;\delta),$$

hence property (v).

Property (v′) follows immediately from properties (i) and (v):

$$\omega_k(f;\lambda\delta) \leqslant \omega_k(f;([\lambda]+1)\delta) \leqslant ([\lambda]+1)^k \omega_k(f;\delta) \leqslant (\lambda+1)^k \omega_k(f;\delta).$$

From properties (iv) and (iv′) it follows that

(vi) *If* $f^{(k)} \in C[a,b]$ *(or* $f^{(k)}$ *exists and is bounded on* $[a,b]$*), then*

$$\omega_k(f;\delta) \leqslant \delta^k \| f^{(k)} \|_{C[a,b]}.$$

Later on (in section 7.1) we shall prove the following more complicated property of ω_k.

(vii) Theorem of Marchaud. *For every* $m < k$ *we have*

$$\omega_m(f;\delta) \leqslant c(k)\delta^m \left\{ \int_\delta^{(b-a)/k} t^{-m-1} \omega_k(f;t)\mathrm{d}t + (b-a)^{-m} \| f \|_{C[a,b]} \right\},$$

where the constant $c(k)$ *depends only on* k.

Using the integral norms L_p, $1 \leqslant p < \infty$, instead of the uniform norm, it is possible to obtain analogues of the moduli of smoothness, which are usually called integral moduli of continuity or smoothness, L_p-moduli or p-moduli.

Let the function f belong to $L_p(a,b)$.

Definition 3.3. *The integral modulus (L_p-modulus, p-modulus) of order k of the function f is the following function of* $\delta \in [0,\infty)$:

$$\omega_k(f;\delta)_{L_p(a,b)} \equiv \omega_k(f;\delta)_{L_p} \equiv \omega_k(f;\delta)_p = \sup_{0<h\leqslant\delta} \left\{ \int_a^{b-kh} |\Delta_h^k f(x)|^p \mathrm{d}x \right\}^{1/p}. \quad (6)$$

It is easy to see that $\omega_k(f;\delta)_p$ has the following properties.

(i) $\omega_k(f;\delta')_p \leqslant \omega_k(f;\delta'')_p$, $\delta' \leqslant \delta''$.
(ii) $\omega_k(f+g;\delta)_p \leqslant \omega_k(f;\delta)_p + \omega_k(g;\delta)_p$.
(iii) $\omega_k(f;\delta)_p \leqslant 2\omega_{k-1}(f;\delta)_p$.
(iv) *If* $f' \in L_p(a,b)$ *then*

$$\omega_k(f;\delta)_p \leqslant \delta\omega_{k-1}(f';\delta)_p, \quad k \geqslant 2.$$

(iv′) *If* $f' \in L_p(a,b)$ *then*

$$\omega_1(f;\delta)_p \leqslant \delta \| f' \|_{L_p(a,b)}.$$

(v) $\omega_k(f;n\delta)_p \leqslant n^k \omega_k(f;\delta)_p$.

(v′) $\omega_k(f;\lambda\delta)_p \leqslant ([\lambda]+1)^k\omega_k(f;\delta)_p \leqslant (\lambda+1)^k\omega_k(f;\delta)_p$.

(vi) *If $f^{(k)}$ exists and $f^{(k)}\in L_p(a,b)$ then*

$$\omega_k(f;\delta)_p \leqslant \delta^k \| f^{(k)} \|_{L_p(a,b)}.$$

Property (vii) has the form

(vii) *If $f\in L_p(a,b)$, then for every $m<k$ we have*

$$\omega_m(f;\delta)_p \leqslant c(k)\delta^m\left\{\int_\delta^{(b-a)/k} t^{-m-1}\omega_k(f;t)_p\,\mathrm{d}t + (b-a)^{-m}\|f\|_{L_p}\right\}.$$

The proofs of properties (i)–(vi) are similar to the uniform case.

We shall give one specific property of $\omega_1(f;\delta)_1 = \omega(f;\delta)_1$.

(viii) *Let the function f have bounded variation on the interval $[a,b]$. Then*

$$\omega(f;\delta)_1 \leqslant \delta V_a^b f, \tag{7}$$

where $V_a^b f$ denotes the variation of the function f in the interval $[a,b]$.

Proof. In fact, we have

$$\int_a^{b-h} |f(x+h)-f(x)|\,\mathrm{d}x \leqslant \int_a^{b-h} V_x^{x+h} f\,\mathrm{d}x = \int_a^{b-h} (V_a^{x+h} f - V_a^x f)\,\mathrm{d}x$$

$$= \int_{a+h}^b V_a^x\,\mathrm{d}x - \int_a^{b-h} V_a^x\,\mathrm{d}x \leqslant \int_{b-h}^b V_a^x f\,\mathrm{d}x \leqslant hV_a^b f. \qquad \square$$

It is very essential that it is possible to have a converse of the inequality (7) in the following sense.

Theorem 3.1** **(Hardy, Littlewood, 1928). *If $\omega(f;\delta)_1 = O(\delta)$ then the function f is equivalent to a function of bounded variation.*

Proof. Let $f\in L(a,b)$ and $\omega(f;\delta)_1 \leqslant M\delta$. Let us set $c=(a+b)/2$ and

$$f_h(x) = \begin{cases} \dfrac{1}{h}\displaystyle\int_0^h f(x+t)\,\mathrm{d}t, & x\in[a,c], \\ \dfrac{1}{h}\displaystyle\int_{-h}^0 f(x+t)\,\mathrm{d}t, & x\in(c,b] \end{cases}$$

where $h\in(0,(b-a)/2)$. Since $f\in L(a,b)$ then for every $x\in[a,b]\backslash Q$, where $Q\subset[a,b]$ is a set with Lebesgue measure zero, we have

$$\lim_{h\to 0} f_h(x) = f(x). \tag{8}$$

For the variation of f_h we have

$$V_a^c f_h \leqslant \int_a^c |f_h'(x)|\,\mathrm{d}x = \frac{1}{h}\int_a^c |f(x+h)-f(x)|\,\mathrm{d}x \leqslant h^{-1}\omega(f;h)_1 \leqslant M.$$

Analogously

$$V_{c+0}^{b} f_h \leqslant M.$$

On the other hand

$$|f_h(c) - f_h(c+0)| = \left| h^{-1} \int_0^h f(c+t)\,\mathrm{d}t - \lim_{x \to c+0} h^{-1} \int_{-h}^0 f(x+t)\,\mathrm{d}t \right|$$

$$\leqslant \frac{1}{h} \int_0^h |f(c+t) - f(c+t-h)|\,\mathrm{d}t \leqslant \frac{1}{h}\,\omega(f;h)_1 \leqslant M.$$

Since

$$V_a^b f_h = V_a^c f_h + V_c^b f_h + |f_h(c) - f_h(c+0)|,$$

we obtain from the above estimation

$$V_a^b f_h \leqslant 3M. \tag{9}$$

On the set $[a,b]\setminus Q$ the function f has bounded variation. In fact, if $x_1 < x_2 < \cdots < x_N$ are points from $[a,b]\setminus Q$ it follows from (8) and (9) that

$$\sum_{i=1}^{N-1} |f(x_{i+1}) - f(x_i)| = \lim_{h \to 0} \sum_{i=1}^{N-1} |f_h(x_{i+1}) - f_h(x_i)| \leqslant \lim_{h \to 0} V_a^b f_h \leqslant 3M. \tag{10}$$

Let us set

$$\tilde{f}(x) = \begin{cases} f(x), & x \in \{[a,b]\setminus Q\}\setminus\{b\} = A, \\ \lim\limits_{y \to x+0,\, y \in A} f(y), & x \in Q\setminus\{b\}, \\ \lim\limits_{y \to b-0,\, y \in A} f(y), & x = b. \end{cases}$$

We note, that from the fact that Q has measure zero and f has bounded variation on A it follows that $\lim_{y \to x \pm 0}\{f(y)\colon y \in A\}$ exists. It follows from (10) that

$$V_a^b \tilde{f} \leqslant 3M.$$

Since $\tilde{f}$ is equivalent to f, the theorem is proved. □

Finally we shall consider the 2π-periodic case. For 2π-periodic functions the modifications of the definitions of the moduli of continuity, moduli of smoothness and L_p-moduli are evident.

The definition of the kth modulus of smoothness is

$$\omega_k(f;\delta) = \omega_k(f;\delta)_{C[0,2\pi]} = \sup\{|\Delta_h^k f(x)| \colon |h| \leqslant \delta, x \in [0, 2\pi]\}.$$

For the L_p-moduli we have

$$\omega_k(f;\delta)_p = \omega_k(f;\delta)_{L_p(0,2\pi)} = \sup_{|h| \leqslant \delta} \left\{ \int_0^{2\pi} |\Delta_h^k f(x)|^p \,\mathrm{d}x \right\}^{1/p}.$$

The properties of these moduli are the same as in the nonperiodic case, i.e. properties (i)–(viii) remain valid.

Finally we give without proof the following property of $\omega_1(f;\delta)_p$ (see Zygmund (1959) or Timan (1960)).

(ix) *We have* $\omega_1(f;\delta)_p \underset{\delta \to 0+}{\longrightarrow} 0$ *if and only if* $f \in L_p(a,b)$.

3.2 Direct theorems: Jackson's theorem

Let f be a function belonging to some metric space X with a distance d. Let the family $\{G_n\}_1^\infty$, be given, where every G_n is an existence set in X and an n-parameter set, $n=1,\dots,\infty$. The best approximation to f by means of elements of G_n is given by

$$E_{G_n}(f)_X = \inf\{d(f,g): g \in G_n\}.$$

As a direct theorem in approximation theory we understand an equality of the type

$$E_{G_n}(f)_X = \mathrm{O}(\phi(f;n)), \quad n \to \infty, \tag{1}$$

where $\phi(f;n)$ is a functional of f, depending only on the number n of parameters of G_n.

A classical example of such a direct theorem is the famous Jackson theorem for the best uniform approximation $E_n(f)$ to a function $f \in C[0,2\pi]$ by means of trigonometric polynomials of nth order with respect to the uniform distance:

$$E_n(f) \equiv E_n(f)_{C[0,2\pi]} = \inf\{\|f-t\|_{C[0,2\pi]}: t \in T_n\}.$$

Theorem 3.2 (Jackson, 1911). *Let* $f \in C[0,2\pi]$. *For every natural number* $n \geqslant 1$ *we have*

$$E_n(f) \leqslant c\omega(f;n^{-1}), \tag{2}$$

where c is an absolute constant.

Remark. If f is an even function then there exists an even $t \in T_n$ such that

$$\|f-t\|_{C[0,2\pi]} \leqslant c\omega(f;n^{-1}).$$

If we compare (2) with (1), we see that here the functional ϕ is the value of the modulus of continuity of f at the point $\delta = n^{-1}$ (multiplied by a constant).

The Jackson theorem has many generalizations. One of them is the following.

Let $E_n(f)_p$ be the best L_p-trigonometrical approximation to f:

$$E_n(f)_p \equiv E_n(f)_{L_p(0,2\pi)} = \inf\{\|f-t\|_{L_p(0,2\pi)}: t \in T_n\}.$$

Theorem 3.3. *Let* $f \in L_p(0,2\pi)$. *Let* $k \geqslant 1$ *be a natural number. There exists a* *constant $c(k)$ depending only on k such that for every natural number $n \geqslant 1$ and every number p, $1 \leqslant p \leqslant \infty$, we have*

$$E_n(f)_p \leqslant c(k)\omega_k(f; n^{-1})_p. \tag{3}$$

Let us mention that there exists an analog of (3) for $0 < p < 1$ (see the notes to Chapter 3).

We shall show somewhat later on that (2) as well as (3) result from the following version of Jackson's theorem, which we shall call the natural direct theorem of Jackson type.

Theorem 3.4. *Let f be a 2π-periodic function with a derivative $f' \in L_p(0, 2\pi)$. Then*

$$E_n(f)_p \leqslant c\frac{\| f' \|_p}{n}, \quad 1 \leqslant p \leqslant \infty, \tag{4}$$

where c is an absolute constant.

Remark. If f is even, then there exists an even $t \in T_n$ such that

$$\| f - t \|_p \leqslant c\frac{\| f' \|_p}{n}.$$

The proofs of (2) and (4) are very similar, but we prefer to derive (2) and (3) from (4) by means of the classical method, using intermediate functions which we shall also use later on.

Now there are many different proofs of (4), but we shall give here the original proof of Jackson (1911).

Proof of theorem 3.4. Let us consider the following operator on the function $f \in L(0, 2\pi)$:

$$\mathfrak{J}_n(f; x) = \int_{-\pi}^{\pi} f(x + t)K_n(t)\mathrm{d}t, \tag{5}$$

where

$$K_n(t) = \lambda_n^{-1}\left(\frac{\sin(nt/2)}{\sin(t/2)}\right)^4$$

is the so-called Jackson kernel.

The constant λ_n^{-1} is chosen in such a way that

$$\int_{-\pi}^{\pi} K_n(t)\mathrm{d}t = 1.$$

From this equality it is possible to obtain the exact value of λ_n, which is $\lambda_n^{-1} = 3/(2\pi n(2n^2 + 1))$, see for example Natanson (1949), but this involves some calculations. It will be sufficient for us to have an estimate for λ_n. In order to obtain this estimate let us remark that

$$t/\pi \leqslant \sin(t/2) \leqslant t/2 \quad \text{for } 0 \leqslant t \leqslant \pi. \tag{6}$$

Since the kernel K_n is even, we have

$$\lambda_n = 2\int_0^\pi \left(\frac{\sin(nt/2)}{\sin(t/2)}\right)^4 \mathrm{d}t \geqslant 2\int_0^{\pi/n} \left(\frac{\sin(nt/2)}{\sin(t/2)}\right)^4 \mathrm{d}t$$

$$\geqslant 2\int_0^{\pi/n} \left(\frac{nt/\pi}{t/2}\right)^4 \mathrm{d}t = \frac{2^5}{\pi^3} n^3. \tag{7}$$

We shall need two lemmas.

Lemma 3.1. *We have*

$$\int_0^\pi tK_n(t)\mathrm{d}t \leqslant cn^{-1}, \tag{8}$$

where c is an absolute constant.

Proof. Using the inequalities (6) and (7) we have

$$\int_0^\pi tK_n(t)\,\mathrm{d}t = \sum_{k=0}^{n-1} \int_{k\pi/n}^{(k+1)\pi/n} tK_n(t)\,\mathrm{d}t$$

$$\leqslant \frac{1}{\lambda_n}\left\{\int_0^{\pi/n} t\left(\frac{nt/2}{t/\pi}\right)^4 \mathrm{d}t + \sum_{k=1}^{n-1} \int_{k\pi/n}^{(k+1)\pi/n} \frac{(k+1)\pi}{n}\left(\frac{1}{k/n}\right)^4 \mathrm{d}t\right\}$$

$$= \frac{1}{\lambda_n}\left\{\frac{\pi^6}{2^5} n^2 + \sum_{k=1}^{n-1} \pi^2 \frac{k+1}{k^4} n^2\right\}$$

$$\leqslant \frac{n^2}{\lambda_n}\left(\frac{\pi^6}{2^5} + \sum_{k=1}^{\infty} \pi^2 \frac{k+1}{k^4}\right) = cn^{-1}$$

where c is an absolute constant. □

Lemma 3.2. *For every function $f \in L(0, 2\pi)$, $\mathfrak{J}_n(f; x)$ is a trigonometrical polynomial of order $2n-2$. If f is even, then $\mathfrak{J}_n(f; x)$ is also even.*

Proof. We have $K_n \in T_{2(n-1)}$. Indeed, from the equalities

$$1/2 + \cos x + \cdots + \cos mx = \frac{\sin((2m+1)x/2)}{2\sin(x/2)},$$

$$\sin(x/2) + \cdots + \sin((2n-1)x/2) = \frac{\sin^2(nx/2)}{\sin(x/2)},$$

it follows that

$$\left(\frac{\sin(nx/2)}{\sin(x/2)}\right)^2 \in T_{n-1}, \left(\frac{\sin(nx/2)}{\sin(x/2)}\right)^4 \in T_{2(n-1)}$$

(since the square of a trigonometrical polynomial of order m is a trigonometrical polynomial of order $2m$).

Since $K_n \in T_{2(n-1)}$, $\mathfrak{J}_n$ also belongs to $T_{2(n-1)}$ since

$$\mathfrak{J}_n(f;x) = \int_{-\pi}^{\pi} f(x+t)K_n(t)\mathrm{d}t = \int_{-\pi}^{\pi} f(t)K_n(x-t)\mathrm{d}t$$

and $K_n(x-t)$ is a trigonometric polynomial of order $2(n-1)$ with respect to x with coefficients functions (also trigonometric polynomials) of t.

The second part of the lemma is evident, since K_n is even. □

Now we continue the proof of theorem 3.4. Let us estimate $\| f - \mathfrak{J}_n f \|_p$. Using $\int_{-\pi}^{\pi} K_n(t)\mathrm{d}t = 1$ we obtain

$$\mathfrak{J}_n(f;x) - f(x) = \int_{-\pi}^{\pi} (f(x+t) - f(x))K_n(t)\,\mathrm{d}t.$$

Hence, using the generalized Minkovski inequality

$$\left\| \int g(x,t)\,\mathrm{d}t \right\|_p \leqslant \int \| g(\cdot,t) \|_p \,\mathrm{d}t$$

(see for example Nikol'skij (1969)) and the fact that $K_n(t) \geqslant 0$, we obtain

$$\| \mathfrak{J}_n f - f \|_p \leqslant \int_{-\pi}^{\pi} \| f(\cdot + t) - f \|_p K_n(t)\mathrm{d}t. \tag{9}$$

If $f' \in L_p(0, 2\pi)$, we have, again using the generalized Minkovski inequality,

$$\| f(\cdot + t) - f \|_p = \left\| \int_0^t f'(\cdot + u)\mathrm{d}u \right\|_p \leqslant |t| \, \| f' \|_p. \tag{10}$$

Using (8) (lemma 3.1), (9) and (10) and the fact that K_n is even, we obtain

$$\| \mathfrak{J}_n f - f \|_p \leqslant \int_{-\pi}^{\pi} |t| \, \| f' \|_p K_n(t)\mathrm{d}t$$

$$= 2 \| f' \|_p \int_0^{\pi} t K_n(t)\mathrm{d}t \leqslant 2c \frac{\| f' \|_p}{n}. \tag{11}$$

Since $\mathfrak{J}_n f \in T_{2(n-1)}$ (lemma 3.2), (11) gives us

$$E_{2n-2}(f)_p \leqslant 2c \frac{\| f' \|_p}{n}.$$

□

The remark after theorem 3.4 follows from (11) and the second part of lemma 3.2.

In order to obtain the classical Jackson theorem (2) from (4) we shall use intermediate approximation by means of the Steklov function.

The Steklov function f_h for $f \in L(0, 2\pi)$ is given by

$$f_h(x) = \frac{1}{h}\int_0^h f(x+t)\mathrm{d}t, \quad h > 0.$$

Using the generalized Minkovski inequality we have

$$\|f - f_h\|_p = \left\| \frac{1}{h}\int_0^h (f(x) - f(x+t))\mathrm{d}t \right\|_p$$

$$\leqslant \frac{1}{h}\int_0^h \|f - f(\cdot + t)\|_p \mathrm{d}t \leqslant \omega(f; h)_p \tag{12}$$

since $\|f - f(\cdot + t)\|_p \leqslant \omega(f; h)_p$ if $0 \leqslant t \leqslant h$.

On the other hand f'_h exists almost everywhere and

$$\|f'_h\|_p = \left\| \frac{1}{h}(f(x+h) - f(x)) \right\|_p \leqslant \frac{1}{h}\omega(f; h)_p. \tag{13}$$

From (4), (12) and (13) it follows that

$$E_n(f)_p \leqslant \|f - f_h\|_p + E_n(f_h)_p \leqslant \omega(f; h)_p + c\frac{h^{-1}\omega(f; h)_p}{n}.$$

Setting $h = n^{-1}$ we obtain theorem 3.2. □

We shall also use the generalized Steklov function for $f \in L(0, 2\pi)$ (see G. Freud, V.A. Popov (1970)),

$$f_{k,h}(x) = \frac{(-1)^k}{h^k}\int_0^h \cdots \int_0^h \Bigg\{ -f(x + t_1 + \cdots + t_k)$$

$$+ \binom{k}{1} f\left(x + \frac{k-1}{k}(t_1 + \cdots + t_k)\right) + \cdots + (-1)^k \binom{k}{k-1}$$

$$\times f\left(x + \frac{t_1 + \cdots + t_k}{k}\right)\Bigg\}\mathrm{d}t_1 \cdots \mathrm{d}t_k. \tag{14}$$

Theorem 3.5. *Let $f \in L_p(0, 2\pi)$. Then*

(i) $\|f_{k,h} - f\|_p \leqslant \omega_k(f; h)_p$,

(ii) $\|f_{k,h}^{(s)}\|_p \leqslant c(k)h^{-s}\omega_s(f; h)_p, \quad s = 1, 2, \ldots, k,$

where the constant $c(k)$ depends only on k.

Proof. From the definition (14) using the generalized Minkovski inequality we obtain

$$\|f_{k,h} - f\|_p \leqslant \left\| h^{-k}\int_0^h \cdots \int_0^h |\Delta^k_{(t_1 + \cdots + t_k)/k} f(x)| \mathrm{d}t_1 \cdots \mathrm{d}t_k \right\|_p$$

$$\leqslant \omega_k(f; h)_p, \quad \text{since } |(t_1 + \cdots + t_k)/k| \leqslant h \quad \text{if } 0 \leqslant t_i \leqslant h,\ i = 1, \ldots, k.$$

For almost all $x \in [0, 2\pi]$ we have for the sth derivative $f_{k,h}^{(s)}$ of $f_{k,h}$

$$f_{k,h}^{(s)}(x) = (-h)^{-k} \int_0^h \cdots \int_0^h \Big\{ -\Delta_h^s f(x + t_1 + \cdots + t_{k-s})$$
$$+ \binom{k}{1}\left(\frac{k}{k-1}\right)^s \Delta_{(k-1)h/k}^s f\left(x + \frac{k-1}{k}(t_1 + \cdots + t_{k-s})\right) + \cdots$$
$$+ (-1)^k \binom{k}{k-1} k^s \Delta_{h/k}^s f\left(x + \frac{t_1 + \cdots + t_{k-s}}{k}\right)\Big\} \mathrm{d}t_1 \cdots \mathrm{d}t_{k-s},$$

hence

$$\| f_{k,h}^{(s)} \|_p \leqslant h^{-k} \int_0^h \cdots \int_0^h \Big\{ \| \Delta_h^s f(\cdot + t_1 + \cdots + t_{k-s}) \|_p$$
$$+ \binom{k}{1}\left(\frac{k}{k-1}\right)^s \left\| \Delta_{(k-1)h/k}^s f\left(\cdot + \frac{k-1}{k}(t_1 + \cdots + t_{k-s})\right)\right\|_p + \cdots$$
$$+ \binom{k}{k-1} k^s \left\| \Delta_{h/k}^s f\left(\cdot + \frac{t_1 + \cdots + t_{k-s}}{k}\right)\right\|_p \Big\} \mathrm{d}t_1 \cdots \mathrm{d}t_{k-s}$$
$$\leqslant h^{-s} \Big\{ \omega_s(f; h)_p + \binom{k}{1}\left(\frac{k}{k-1}\right)^s \omega_s\left(f; \frac{k-1}{k}h\right)_p + \cdots$$
$$+ \binom{k}{k-1} k^s \omega_s\left(f; \frac{h}{k}\right)_p \Big\} \leqslant (2k)^k h^{-s} \omega_s(f; h)_p$$

(if $g \in L_p(0, 2\pi)$ then

$$\| g(\cdot + u) \|_p = \| g \|_p)$$

i.e. property (ii) with a constant $c(k) = (2k)^k$. □

Theorem 3.6. *Let $f \in C[0, 2\pi]$ have a derivative $f' \in L_p(0, 2\pi)$. Then*

$$E_n(f)_p \leqslant c \frac{E_n(f')_p}{n},$$

where c is an absolute constant.

Proof. Let $q \in T_n$ be such that $\| f' - q \|_p = E_n(f')_p$. Let

$$q(x) = a_0 + \sum_{k=1}^{n} (a_k \cos kx + b_k \sin kx) = a_0 + r(x)$$

and let $s \in T_n$ be such that $s' = r$. Then we have, using theorem 3.4,

$$E_n(f)_p = E_n(f - s)_p \leqslant c \frac{\| f' - s' \|_p}{n} = c \frac{\| f' - r \|_p}{n}$$
$$\leqslant c\left(\frac{\| f' - q \|_p + \| a_0 \|_p}{n}\right) = c\left(\frac{E_n(f')_p + (2\pi)^{1/p} |a_0|}{n}\right). \tag{15}$$

Let us now estimate $|a_0|$. Since $f' \in L_p(0, 2\pi)$ is a derivative of the 2π-periodic function f, we have

$$\int_0^{2\pi} f'(x)\mathrm{d}x = 0$$

and also obviously $\int_0^{2\pi} r(x)\mathrm{d}x = 0$. Therefore

$$\int_0^{2\pi} (f'(x) - r(x))\mathrm{d}x = 0,$$

i.e. $\int_0^{2\pi}(f'(x) - q(x) + a_0)\mathrm{d}x = 0$, or $2\pi a_0 = -\int_0^{2\pi}(f'(x) - q(x))\mathrm{d}x$.
Hence

$$2\pi|a_0| \leqslant (2\pi)^{1-1/p}\|f' - q\|_p = (2\pi)^{1-1/p}E_n(f')_p. \tag{16}$$

Theorem 3.6 follows from (15) and (16). □

Corollary 3.1. *If $f^{(k)} \in L_p(0, 2\pi)$, $k \geqslant 1$, then*

$$E_n(f)_p \leqslant c(k)\frac{\|f^{(k)}\|_p}{n^k}, \tag{17}$$

where the constant $c(k)$ depends only on k.

Proof. Using theorems 3.4 and 3.6 we obtain

$$E_n(f)_p \leqslant c\frac{E_n(f')_p}{n} \leqslant \cdots \leqslant c^{k-1}\frac{E_n(f^{(k-1)})_p}{n^{k-1}} \leqslant c^{k-1}c'\frac{\|f^{(k)}\|}{n^k},$$

i.e. (17) with a constant $c(k) = c^k$.

Finally we shall prove theorem 3.3 – the general case of the Jackson type theorem.

Let $f \in L_p(0, 2\pi)$ and $f_{k,h}$ be the function from theorem 3.5 for f. Using theorem 3.5 and (17) we have

$$\begin{aligned} E_n(f)_p &\leqslant \|f - f_{k,h}\|_p + E_n(f_{k,h})_p \leqslant \omega_k(f;h)_p + c(k)\frac{\|f_{k,h}^{(k)}\|_p}{n^k} \\ &\leqslant \omega_k(f;h)_p + c(k)c'(k)\frac{h^{-k}\omega_k(f;h)_p}{n^k}. \end{aligned} \tag{18}$$

Setting $h = n^{-1}$ in (18) we obtain theorem 3.3. □

Remark. Let us note that this method, using intermediate functions like $f_{k,h}$, cannot give a good constant c or $c(k)$. Since in the book we shall be interested mainly in the order of approximation, we shall not give here good estimates of the constants or their exact values (which are known only in a few cases).

3.3 Converse theorems

We have seen that estimation of the type

$$E_{G_n}(f) \leqslant \phi(f;n), \tag{1}$$

where n is the number of the parameters of the elements of the approximating family G_n, and ϕ is a functional depending only on f and n, is the so-called direct theorem in the theory of approximation. For example in the estimation (2) in section 3.2 we have $\phi(f;n)=c\omega(f;n^{-1})$.

The so-called converse theorems are not of so simple a type as (1). Usually it is not possible to estimate some functionals of f (of course important functionals) only by $E_{G_n}(f)$. The typical converse estimations are of the type

$$\phi(f;n) \leqslant F(E_{G_1}(f),\ldots,E_{G_{N(n)}}(f)),$$

where F is a function of $E_{G_1}(f),\ldots,E_{G_{N(n)}}(f)$.

In order to be explicit we shall present at once the classical converse theorem of Bernstein in the form given by Salem (1940) and Stechkin (1951).

Theorem 3.7. *Let $f \in L_p(0,2\pi)$. Then for every natural number n, $n \geqslant 1$, we have*

$$\omega_k(f;n^{-1})_p \leqslant \frac{c(k)}{n^k}\sum_{s=0}^{n}(s+1)^{k-1}E_s(f)_p, \tag{2}$$

where the constant $c(k)$ depends only on k, $k \geqslant 1$.

Usually we have good direct and converse theorems that give characterization of the order of the best approximation by means of some properties of the function f. In this sense theorems 3.3 and 3.7 are some of the best: it is easy to see that there results from them the following characterization of the best trigonometric approximation of the function f in L_p by means of L_p-moduli of smoothness of f.

Theorem 3.8. *Let $f \in L_p(0,2\pi)$. We have $E_n(f)_p = \mathrm{O}(n^{-\alpha})$ if and only if $\omega_k(f;\delta)_p = \mathrm{O}(\delta^\alpha)$, $k > \alpha$.*

Proof. Setting $E_s(f)_p = \mathrm{O}(s^{-\alpha})$ in (2) for $k > \alpha$ we obtain

$$\omega_k(f;n^{-1}) \leqslant \frac{c(k)}{n^k}\sum_{s=0}^{n}(s+1)^{k-1}\mathrm{O}(s^{-\alpha}) = \frac{c'(k)}{n^k}n^{k-\alpha} = \mathrm{O}(n^{-\alpha})$$

which gives $\omega_k(f;\delta)_p = \mathrm{O}(\delta^\alpha)$.

The converse follows directly from theorem 3.3. □

All converse theorems are connected with the differential properties of the elements of the approximating family G_n. Usually the basic property is an inequality of Bernstein type. We shall obtain here the classical Bernstein inequality for trigonometrical polynomials.

Let

$$t_n(x) = \frac{a_0}{2} + \sum_{k=1}^{n} (a_k \cos kx + b_k \sin kx) \tag{3}$$

be a trigonometric polynomial of nth order. If we substitute

$$\cos kx = \frac{e^{ikx} + e^{-ikx}}{2}, \quad \sin kx = \frac{e^{ikx} - e^{-ikx}}{2i}$$

we obtain

$$t_n(x) = \sum_{k=-n}^{n} c_k e^{ikx}, \quad c_k = \frac{a_k + ib_k}{2}, \quad c_{-k} = \frac{a_k - ib_k}{2}, \quad k = 1, \ldots, n, c_0 = a_0/2. \tag{4}$$

From the representation (4) it is evident that we can have at most $2n$ zeros in $[0, 2\pi)$ (or in $(0, 2\pi]$) since (4) can be written as

$$t_n(x) = e^{-inx} \sum_{k=0}^{2n} c_{k-n} e^{ikx}$$

or

$$t_n(x) = g_n(z) = z^{-n} \sum_{k=0}^{2n} c_{k-n} z^k, z = e^{ix}$$

If z_k, $k = 1, \ldots, 2n$, are the zeros of $t_n(z)$, we see that

$$t_n(z) = c_n e^{-inz} \prod_{k=1}^{2n} (e^{iz} - e^{iz_k}) = c_n e^{\frac{1}{2}i\sum_{k=1}^{2n} z_k} \prod_{k=1}^{2n} (e^{i(z-z_k)/2} - e^{i(z_k - z)/2})$$

$$= A \prod_{k=1}^{2n} \sin \frac{z - z_k}{2}, \tag{5}$$

where A is a constant.

Now we shall obtain the interpolation formula of M. Riess, following S.M. Nikol'skij (1969).

Let us set $\theta_k = (2k-1)\pi/2n$, $k = 1, \ldots, 2n$. The points θ_k, $k = 1, \ldots, 2n$, are the zeros of the polynomial $\cos n\theta$, so using (5) we have

$$\cos n\theta = A \prod_{k=1}^{2n} \sin \frac{\theta - \theta_k}{2}. \tag{6}$$

Hence

$$Q_m(\theta) = \frac{\cos n\theta}{2n} (-1)^m \cot \frac{\theta - \theta_m}{2} = (-1)^{m+1} \frac{\cos n\theta}{2n} \frac{\sin \frac{1}{2}(\theta - (\pi + \theta_m))}{\sin \frac{1}{2}(\theta - \theta_m)} \tag{7}$$

is a trigonometrical polynomial of a degree n, since we have

$$Q_m(x) = \text{const.} \left(\prod_{\substack{k=1 \\ k \neq m}}^{2n} \sin \frac{\theta - \theta_k}{2} \right) \left(\sin \frac{\theta - (\pi + \theta_m)}{2} \right).$$

We have, for the trigonometrical polynomial $Q_m(\theta)$,

$$Q_m(\theta_k) = \begin{cases} 0, & k \neq m, \\ 1, & k = m \end{cases}$$

(that $Q_m(\theta_m) = 1$ it is possible to obtain directly letting $\theta \to \theta_m$ in (7)).

Consequently, for every trigonometric polynomial $t_n \in T_n$ we have, for the trigonometric polynomial $t_n^* \in T_n$,

$$t_n^*(\theta) = \frac{\cos n\theta}{2n} \sum_{k=1}^{2n} (-1)^k \cot \frac{\theta - \theta_k}{2} t_n(\theta_k),$$

the interpolation conditions

$$t_n(\theta_k) = t_n^*(\theta_k), \quad k = 1, \ldots, 2n.$$

Since $t_n(\theta) - t_n^*(\theta)$ has zeros in the points θ_k, $k = 1, \ldots, 2n$, by (5), (6) we have $t_n(\theta) - t_n^*(\theta) = A \cos n\theta$, or

$$t_n(\theta) = A \cos n\theta + t_n^*(\theta).$$

So we obtain

Lemma 3.3. *For every $t_n \in T_n$ we have*

$$t_n(\theta) = A \cos n\theta + \frac{\cos n\theta}{2n} \sum_{k=1}^{2n} (-1)^k \cot \frac{\theta - \theta_k}{2} t_n(\theta_k). \tag{8}$$

It is possible to show that the constant A is equal to a_n in the representation (3) (see Nikol'skij (1969)). We shall not make use of this fact.

From (8) by differentiation and setting $\theta = 0$ we obtain

$$t_n'(0) = \frac{1}{4n} \sum_{k=1}^{2n} (-1)^{k+1} \frac{1}{\sin^2(\theta_k/2)} t_n(\theta_k). \tag{9}$$

Since $t_n \in T_n$ is an arbitrary trigonometric polynomial of order n, (9) applied to $t_n(\theta + x)$ gives us the M. Riess interpolation formula

$$t_n'(x) = \frac{1}{4n} \sum_{k=1}^{2n} (-1)^{k+1} \frac{1}{\sin^2(\theta_k/2)} t_n(x + \theta_k), \quad \theta_k = \frac{2k-1}{2n}\pi, \quad k = 1, \ldots, 2n. \tag{10}$$

Using this formula, it is easy to prove

Theorem 3.9 (Bernstein inequality). *If $t_n \in T_n$ then*[†]

$$\| t_n' \|_{L_p(0,2\pi)} \leqslant n \| t_n \|_{L_p(0,2\pi)}, \quad 1 \leqslant p \leqslant \infty.$$

[†] Again we set $L_\infty(0, 2\pi) = C[0, 2\pi]$.

Proof. If we set $t_n(x) = \sin nx$, from (10) we obtain, setting $x = 0$,

$$n = \frac{1}{4n} \sum_{k=1}^{2n} \frac{1}{\sin^2(\theta_k/2)} \tag{11}$$

Taking L_p norms, $1 \leqslant p \leqslant \infty$, in (10), we obtain, using (11),

$$\| t_n' \|_p \leqslant \frac{1}{4n} \sum_{k=1}^{2n} \frac{1}{\sin^2(\theta_k/2)} \| t_n \|_p = n \| t_n \|_p. \qquad \square$$

Corollary 3.2. *If* $t_n \in T_n$ *then*

$$\| t_n^{(k)} \|_p \leqslant n^k \| t_n \|_p.$$

The Bernstein inequality is exact in the following sense.

For the trigonometric polynomial $g_n(x) = A \sin(nx + \alpha)$, where A and α are constants, we have

$$\| g' \|_p = n \| g \|_p, \quad 1 \leqslant p \leqslant \infty.$$

Now we are ready to prove theorem 3.7. The method of the proof is typical for obtaining estimations using Bernstein type inequalities (compare with sections 3.4 and 3.5).

Let us estimate $\| \triangle_h^k f \|_p$. Let $Q_n \in T_n$ be such that

$$\| f - Q_n \|_p = E_n(f)_p, \quad n = 0, 1, \ldots. \tag{12}$$

For every natural number s, $s \geqslant 1$, we have

$$f = f - Q_{2^s} + (Q_{2^s} - Q_{2^{s-1}}) + \cdots + (Q_1 - Q_0) + Q_0;$$

consequently (we set $Q_{1/2} = Q_0$)

$$\begin{aligned} \| \triangle_h^k f \|_p &\leqslant \| \triangle_h^k (f - Q_{2^s}) \|_p + \sum_{m=0}^{s} \| \triangle_h^k (Q_{2^m} - Q_{2^{m-1}}) \|_p \\ &\leqslant 2^k \| f - Q_{2^s} \|_p + \sum_{m=0}^{s} \omega_k(Q_{2^m} - Q_{2^{m-1}}; h)_p. \end{aligned} \tag{13}$$

Since $Q_{2^m} - Q_{2^{m-1}} \in T_{2^m}$ it ensues from the property (vi) of ω_k and corollary 3.2 that $(E_{1/2}(f)_p = E_0(f)_p)$:

$$\begin{aligned} \omega_k(Q_{2^m} - Q_{2^{m-1}}) &\leqslant h^k \| Q_{2^m}^{(k)} - Q_{2^{m-1}}^{(k)} \|_p \leqslant h^k 2^{mk} \| Q_{2^m} - Q_{2^{m-1}} \|_p \\ &\leqslant h^k 2^{mk} \{ \| Q_{2^m} - f \|_p + \| f - Q_{2^{m-1}} \|_p \} \leqslant 2h^k 2^{mk} E_{2^{m-1}}(f)_p. \end{aligned} \tag{14}$$

From (13) and (14) it follows that

$$\| \triangle_h^k f \|_p \leqslant 2^k E_{2^s}(f)_p + 2h^k \sum_{m=0}^{s} 2^{mk} E_{2^{m-1}}(f)_p. \tag{15}$$

Now let us set $|h| \leqslant 1/n$, $n = 2^s$. Then (15) gives us

$$\| \triangle_h^k f \|_p \leqslant 2^k E_{2^s}(f)_p + \frac{2}{n^k} \sum_{m=0}^{s} 2^{mk} E_{2^{m-1}}(f)_p$$

$$\leqslant \frac{2^{2k}}{n^k} \sum_{r=0}^{2^s} (r+1)^{k-1} E_r(f)_p, \tag{16}$$

since

$$E_{2^s}(f)_p = \frac{2^{sk}}{n^k} E_{2^s}(f)_p,$$

$$2^{mk} E_{2^{m-1}}(f)_p \leqslant 2^{2k-1} \sum_{l=2^{m-2}}^{2^{m-1}} (l+1)^{k-1} E_l(f)_p.$$

The inequality (16) gives us theorem 3.7 for $n = 2^s$:

$$\omega_k(f; n^{-1})_p \leqslant \frac{c'(k)}{n^k} \sum_{r=0}^{n} (r+1)^{k-1} E_r(f)_p.$$

The transition to arbitrary n is standard. Let n be given. Choose s so that $2^s < n \leqslant 2^{s+1}$. Then

$$\omega_k(f; n^{-1})_p \leqslant \omega_k(f; 2^{-s})_p \leqslant \frac{c'(k)}{2^{sk}} \sum_{r=0}^{2^s} (r+1)^{k-1} E_r(f)_p$$

$$\leqslant \frac{2^k c'(k)}{2^{(s+1)k}} \sum_{r=0}^{n} (r+1)^{k-1} E_r(f)_p \leqslant \frac{c(k)}{n^k} \sum_{r=0}^{n} (r+1)^{k-1} E_r(f)_p. \quad \square$$

3.4 Direct and converse theorems for algebraic polynomial approximation in a finite interval

Let us consider now the problem of best approximation of functions defined on a finite interval, say $[-1, 1]$, by means of algebraic polynomials. Here the characterization of the best uniform or L_p-approximation by means of the moduli of smoothness is not so fine as in the trigonometric case (see theorem 3.8) because of the so-called 'end-effect' or 'Nikol'skij effect'.

S.M. Nikol'skij (1946) was the first to note that the algebraic approximation at the ends of the interval is better than the approximation in the middle of the interval. In other words it is possible to obtain good algebraic approximations, for functions which are 'not good', at the ends of the interval. Many authors after Nikol'skij have worked on this effect, see the books of Dzjadik (1977), Timan (1960).

It is easy to obtain an analog of theorem 3.2 for algebraic uniform approximation $E_n(f)$ on $[-1, 1]$:

$$E_n(f) = \inf\{ \| f - p \|_{C[-1,1]} : p \in P_n \}.$$

The following theorem holds.

Theorem 3.10. *Let $f \in C[-1,1]$. Then*

$$E_n(f) \leqslant c\omega(f; n^{-1})_{C[-1,1]},$$

where c is an absolute constant.

Proof. For the function $g(t) = f(\cos t)$ we have $g \in C[0, 2\pi]$, g even. From the remark after theorem 3.2 it follows that there exists an even trigonometric polynomial $t_n \in T_n$ such that

$$\| g - t_n \|_{C[0,2\pi]} \leqslant c\omega(g; n^{-1})_{C[0,2\pi]}, \tag{1}$$

c an absolute constant. Then $q_n(x) = t_n(\arccos x) \in P_n$ and

$$\| f - q_n \|_{C[-1,1]} = \| g - t_n \|_{C[0,2\pi]} \leqslant c\omega(g; n^{-1})_{C[0,2\pi]} \leqslant c\omega(f; n^{-1})_{C[-1,1]}$$

since

$$\begin{aligned} \omega(g; \delta)_{C[0,2\pi]} &= \sup\{|g(t) - g(t')| : |t - t'| \leqslant \delta\} \\ &= \sup\{|g(\arccos x) - g(\arccos x')| : |\arccos x - \arccos x'| \leqslant \delta\} \\ &\leqslant \sup\{|f(x) - f(x')| : |x - x'| \leqslant \delta\} \end{aligned}$$

$(|x - x'| = |\cos(\arccos x) - \cos(\arccos x')| \leqslant |\arccos x - \arccos x'| \leqslant \delta)$. □

But unfortunately it is not possible to obtain a direct analog of the Bernstein theorem 3.7. The reasons for this are that the analogs of the Bernstein inequality (theorem 3.9) are the following.

Theorem 3.11 (Bernstein). *Let $p \in P_n$. Then for $x \in [-1,1]$ we have*

$$|p'(x)| \leqslant \frac{n}{\sqrt{(1 - x^2)}} \| p \|_{C[-1,1]}.$$

Theorem 3.12 (Markov). *Let $p \in P_n$. Then*

$$\| p' \|_{C[-1,1]} \leqslant n^2 \| p \|_{C[-1,1]}.$$

These theorems are exact. Theorem 3.12 shows that we have an estimation with n^2 instead of n. So we cannot have an analog of the same type as theorem 3.7. Theorem 3.11 shows that we may have success if we work 'pointwise'. In fact this is the way to obtain a characterization of Lipschitz classes with the weighted uniform algebraic approximation on $[-1,1]$ (see the book by Dzjadik (1977)).

Here we shall present another characterization of $E_n(f)$ given by K. Ivanov (1983a), which seems to be better than the others. We shall consider the case $k = 1$. For $k > 1$ and $E_n(f)_p$, $1 \leqslant p < \infty$, see the notes to this chapter.

First let us give our notations. We set

$$\Delta_m(x) = \frac{1}{m}\sqrt{(1-x^2)} + \frac{1}{m^2},$$

$$\tau_1(f;\Delta_m)_{C[-1,1]} \equiv \tau_1(f;\Delta_m)$$
$$= \sup\{|f(x)-f(y)|: x, y \in [-1,1], |x-y| \leqslant \Delta_m(x)\}.$$

Theorem 3.13. *Let $f \in C[-1,1]$. Then*

$$E_m(f)_{C[-1,1]} = O(\tau_1(f;\Delta_m)_{C[-1,1]}).$$

In the proof we shall use lemmas 3.1, 3.2 and the following.

Lemma 3.4. *Let $|h| \leqslant \pi/4m$. Then*

$$|\cos y - \cos(y+h)| \leqslant \Delta_m(\cos y).$$

Proof. We have

$$|\cos y - \cos(y+h)| = 2\sin\frac{|h|}{2}\left|\sin\left(y+\frac{h}{2}\right)\right|$$

$$\leqslant \frac{2|h|}{2}\left(|\sin y| + \frac{|h|}{2}\right) \leqslant \frac{\pi}{4m}\left(\sqrt{(1-\cos^2 y)} + \frac{\pi}{8m}\right) \leqslant \Delta_m(\cos y). \qquad \square$$

Proof of theorem 3.13. We shall use the Jackson operator from section 3.2. Let us set

$$n = [m/2] + 1, \; g(t) = f(\cos t),$$

$$\mathfrak{J}_n(g;y) = \int_{-\pi}^{\pi} g(y+t)K_n(t)\,dt = \int_{-\pi}^{\pi} f(\cos(y+t))K_n(t)\,dt,$$

$$K_n(t) = \lambda_n^{-1}\left(\frac{\sin(nt/2)}{\sin(t/2)}\right)^4, \quad \int_{-\pi}^{\pi} K_n(t)\,dt = 1.$$

Since $\mathfrak{J}_n \in T_{2n-2}$ and J_n is even (g is an even function, see lemma 3.2), $\mathfrak{J}_n(\arccos x) = Q_n(x)$ will be an algebraical polynomial of degree $2n-2 = 2([m/2]+1)-2 \leqslant m$, i.e. $Q_n \in P_n$.

Let us now estimate $|f(x) - Q_n(x)|$. We have, setting $y = \arccos x$, $x \in [-1,1]$, $y \in [0,\pi]$,

$$|f(x) - Q_n(x)| = |g(y) - \mathfrak{J}_n(g;y)|$$
$$\leqslant \int_{-\pi}^{\pi} |f(\cos y) - f(\cos(y+t))|K_n(t)\,dt. \tag{2}$$

Let us set

$$h(t) = \frac{\pi}{4m}\operatorname{sign} t, \quad r(t) = \left[\frac{4m|t|}{\pi}\right].$$

Then we have from (2), using lemma 3.4 and lemma 3.1,

$$|f(x)-Q_n(x)| \leqslant \int_{-\pi}^{\pi}\left\{|f(\cos(y+r(t)h(t))-f(\cos(y+t))| + \sum_{i=1}^{r(t)}|f(\cos(y+(i-1)h))-f(\cos(y+ih))|\right\}K_n(t)\mathrm{d}t$$

$$\leqslant \int_{-\pi}^{\pi}\left(1+\frac{4m|t|}{\pi}\right)\tau_1(f;\Delta_m)K_n(t)\mathrm{d}t = \mathrm{O}(\tau_1(f;\Delta_m)),$$

since, by lemma 3.4,

$$|\cos(y+ih)-\cos(y+(i-1)h)| \leqslant \Delta_m(\cos(y+(i-1)h)),$$

and therefore

$$|f(\cos(y+ih))-f(\cos(y+(i-1)h))| \leqslant \tau_1(f;\Delta_m),$$

$$|\cos(y+r(t)h(t))-\cos(y+t)| \leqslant \Delta_m(\cos(y+t)),$$

$$|f(\cos(y+r(t)h(x)))-f(\cos(y+t))| \leqslant \tau_1(f;\Delta_m).$$

On the other hand lemma 3.1 gives us

$$\int_{-\pi}^{\pi}\left(1+\frac{4m|t|}{\pi}\right)K_n(t)\mathrm{d}t = 1+\frac{8m}{\pi}\int_0^{\pi} tK_n(t)\mathrm{d}t$$

$$= 1+\mathrm{O}(mn^{-1}) = \mathrm{O}(1).$$

Consequently

$$E_m(f) \leqslant \sup_{x\in[-1,1]} |f(x)-Q_n(x)| = \mathrm{O}(\tau_1(f;\Delta_m)). \qquad \square$$

Corollary 3.3. *We have*

$$E_n(\sqrt{(1-x^2)})_{C[-1,1]} = \mathrm{O}(n^{-1}).$$

In fact one can easily see that for $f(x)=\sqrt{(1-x^2)}$ we have $\tau_1(f;\Delta_m)=\mathrm{O}(m^{-1})$.

This corollary shows that while the function f may be bad at the ends of $[-1,1]$ (the derivative of $\sqrt{(1-x^2)}$ goes to $\pm\infty$ as $x\to\mp 1$), nevertheless the best uniform algebraic approximation may be good. The same effect exists also for rational approximation (see section 5.5).

In order to prove the converse theorem we shall need some lemmas. Let us first prove the Bernstein inequality for algebraic polynomials (theorem 3.11) and Markov inequality (theorem 3.12).

Proof of theorem 3.11. Let $p\in P_n$. Then for $q(t)=p(\cos t)$ we have $q\in T_n$.

Therefore the Bernstein inequality for trigonometrical polynomials (theorem 3.9) gives us

$$|q'(t)| \leqslant n \| q \|_{C[0,2\pi]}, \tag{3}$$

But $q'(t) = p'(\cos t) \sin t$. Setting $t = \arccos x$, we obtain from (3), since $\| q \|_{C[0,2\pi]} = \| p \|_{C[-1,1]}$, $\sin \arccos x = \sqrt{(1 - \cos^2 \arccos x)} = \sqrt{(1 - x^2)}$,

$$|p'(x)\sqrt{(1 - x^2)}| \leqslant n \| p \|_{C[-1,1]}. \qquad \square$$

Proof of theorem 3.12. Let $x_k = \cos((2k-1)\pi/2n)$, $k = 1, \ldots, n$, be the zeros of the Chebyshev polynomial $T_n(x) = \cos(n \arccos x)$. For every algebraic polynomial $p \in P_{n-1}$ we have this Lagrange interpolation formula

$$P(x) = \sum_{i=1}^{n} p(x_i) \frac{(-1)^{i-1}\sqrt{(1-x_i^2)}}{n} \frac{T_n(x)}{x - x_i}, \tag{4}$$

since we have

$$\frac{(-1)^{i-1}\sqrt{(1-x_i^2)}}{n} \frac{T_n(x_j)}{x_j - x_i} = \begin{cases} 0, & j \neq i, \\ 1, & i = j. \end{cases}$$

The first case ($i \neq j$) is evident since $T_n(x_j) = 0$. In order to prove the second case, first we remark that

$$\left(\frac{T_n(x)}{x - x_i}\right)_{x = x_i} = \lim_{x \to x_i} \frac{T_n(x) - T_n(x_i)}{x - x_i} = T_n'(x_i)$$

$$= (\cos(n \arccos x))'_{x = x_i} = \frac{n}{\sqrt{(1 - x_i^2)}} \sin(n \arccos x_i)$$

$$= \frac{n}{\sqrt{(1 - x_i^2)}} \sin n \frac{2i - 1}{2n} \pi = (-1)^{i-1} \frac{n}{\sqrt{(1 - x_i^2)}};$$

therefore

$$\frac{(-1)^{i-1}\sqrt{(1-x_i^2)}}{n} \left(\frac{T_n(x)}{x - x_i}\right)_{x = x_i} = 1.$$

Now let $p \in P_{n-1}$ and $x \in [-1, 1]$. There are two cases: (a) $|x| \leqslant x_1 = \cos(\pi/2n)$, (b) $x_1 < |x| \leqslant 1$. In the first case we have

$$\sqrt{(1 - x^2)} \geqslant \sqrt{(1 - x_1^2)} = \sqrt{\left(1 - \cos^2 \frac{\pi}{2n}\right)} = \sin \frac{\pi}{2n} \geqslant \frac{1}{n};$$

therefore

$$|p(x)| \leqslant n\sqrt{(1 - x^2)}|p(x)| \leqslant n \| \sqrt{(1 - x^2)} p(x) \|_{C[-1,1]}. \tag{5}$$

In the second case, since $-1 \leqslant x < x_n$ or $x_1 < x \leqslant 1$, all members $x - x_i$,

$i=1,\ldots,n$, have the same sign, therefore from (4) we obtain

$$\begin{aligned}|p(x)| &\leqslant \sum_{i=1}^{n} |p(x_i)| \frac{\sqrt{(1-x_i^2)}}{n} \left| \frac{T_n(x)}{x-x_i} \right| \\ &\leqslant \frac{1}{n} \| \sqrt{(1-x^2)}p(x) \|_{C[-1,1]} \sum_{i=1}^{n} \left| \frac{T_n(x)}{x-x_i} \right| \\ &= \frac{1}{n} \| \sqrt{(1-x^2)}p(x) \|_{C[-1,1]} \left| \sum_{i=1}^{n} \frac{T_n(x)}{x-x_i} \right|, \end{aligned} \tag{6}$$

since all $T_n(x)/(x-x_i)$ have the same sign.

Let us now estimate $|\sum_{i=1}^{n} T_n(x)/(x-x_i)|$. We have

$$T_n'(x) = \sum_{i=1}^{n} \frac{T_n(x)}{x-x_i} \quad \left(T_n(x) = 2^{n-1} \prod_{i=1}^{n} (x-x_i) \right),$$

therefore we must estimate $\| T_n' \|_{C[-1,1]}$. We have

$$T_n'(x) = \frac{n \sin(n \arccos x)}{\sqrt{(1-x^2)}} = \frac{n \sin n\theta}{\sin \theta}, \quad \theta = \arccos x.$$

Since $|\sin n\theta / \sin \theta| \leqslant n$ for $|\theta| \leqslant \pi$, we obtain that

$$\| T_n' \|_{C[-1,1]} \leqslant n^2, \tag{7}$$

and from (6) we obtain that

$$|p(x)| \leqslant n \| \sqrt{(1-x^2)}p(x) \|_{C[-1,1]}. \tag{8}$$

Now the inequalities (5) and (8) give us

Lemma 3.5. *Let $p \in P_{n-1}$. Then*

$$\| p \|_{C[-1,1]} \leqslant n \| \sqrt{(1-x^2)}p(x) \|_{C[-1,1]}.$$

We return to the proof of Markov's inequality.

The Bernstein inequality (theorem 3.11) gives us, for every $p \in P_n$,

$$\| \sqrt{(1-x^2)}p'(x) \|_{C[-1,1]} \leqslant n \| p(x) \|_{C[-1,1]}. \tag{9}$$

Lemma 3.5, applied to p', and (9) give us

$$\| p' \|_{C[-1,1]} \leqslant n \| \sqrt{(1-x^2)}p'(x) \|_{C[-1,1]} \leqslant n^2 \| p \|_{C[-1,1]}.$$ □

Lemma 3.6. *Let $p \in P_n$. Then*

$$\left\| \left(\frac{1}{n}\sqrt{(1-x^2)} + \frac{1}{n^2} \right) p'(x) \right\|_{C[-1,1]} \leqslant 2 \| p \|_{C[-1,1]}.$$

Proof. This inequality follows immediately from the Bernstein inequality

$$\| \sqrt{(1-x^2)}p'(x) \|_{C[-1,1]} \leqslant n \| p \|_{C[-1,1]}$$

and the Markov inequality

$$\|p'\|_{C[-1,1]} \leqslant n^2 \|p\|_{C[-1,1]}. \qquad \square$$

Lemma 3.7. *We have*

$$\tau_1(f;\Delta_m) \leqslant 2\|f\|_{C[-1,1]},$$

$$\tau_1(f+g;\Delta_m) \leqslant \tau_1(f;\Delta_m) + \tau_1(g;\Delta_m).$$

The lemma follows directly from the definition of $\tau_1(f;\Delta_m)$.

Lemma 3.8. *Let $x, x+t \in [-1,1], |t| \leqslant \Delta_m(x)$. Then*

$$\tfrac{1}{6}\Delta_m(x) \leqslant \Delta_m(x+t).$$

Proof. We set

$$\psi_m(x) = \begin{cases} \Delta_m(x), & |x| \leqslant \sqrt{(1-4m^{-2})}, \\ 3m^{-2}, & |x| > \sqrt{(1-4m^{-2})}. \end{cases}$$

Then $\|\psi'_m\|_{C[-1,1]} \leqslant \frac{1}{2}$ and

$$\Delta_m(y) \leqslant \psi_m(y) \leqslant 3\Delta_m(y) \tag{10}$$

for every $y \in [-1,1]$. We have

$$|\psi_m(x+t) - \psi_m(x)| \leqslant \tfrac{1}{2}\Delta_m(x),$$

$$\tfrac{1}{2}\Delta_m(x) \leqslant \psi_m(x) - \tfrac{1}{2}\Delta_m(x) \leqslant \psi_m(x+t).$$

Hence from (10) we obtain

$$\Delta_m(x+t) \geqslant \tfrac{1}{6}\Delta_m(x). \qquad \square$$

Lemma 3.9. *Let $f' \in L_\infty[-1,1]$. Then*

$$\tau_1(f;\Delta_m) \leqslant 6\|\Delta_m f'\|_{C[-1,1]}.$$

Proof. Let x, $x+h \in [-1,1]$ and $|h| \leqslant \Delta_m(x)$. Using lemma 3.8 we obtain

$$|f(x+h) - f(x)| = \left| \int_x^{x+h} f'(t) \frac{\Delta_m(t)}{\Delta_m(t)} \, dt \right|$$
$$\leqslant \|\Delta_m f'\|_{C[-1,1]} \int_x^{x+|h|} \frac{dt}{\Delta_m(t)} \leqslant 6\|\Delta_m f'\|_{C[-1,1]},$$

since, by lemma 3.8,

$$\int_x^{x+|h|} \frac{dt}{\Delta_m(t)} \leqslant 6 \int_x^{x+|h|} \frac{dt}{\Delta_m(x)} \leqslant 6 \frac{|h|}{\Delta_m(x)} \leqslant 6. \qquad \square$$

Now we are ready to prove the converse theorem for the best uniform algebraic approximations.

Theorem 3.14. *Let $f \in C[-1,1]$. Then for every natural number m we have*

$$\tau_1(f;\Delta_m) \leqslant \frac{100}{m} \sum_{s=0}^{m} E_s(f).$$

Proof. Let $q_\nu \in P_\nu$ be such that

$$\| q_\nu - f \|_{C[-1,1]} = E_\nu(f).$$

Let m be given. We set $n = [\ln m / \ln 2]$, then $2^n \leqslant m \leqslant 2^{n+1}$. Using lemma 3.7 we have

$$\tau_1(f;\Delta_m) \leqslant \tau_1(f - q_{2^n};\Delta_m) + \tau_1(q_{2^n};\Delta_m). \tag{11}$$

From lemma 3.7 we also obtain

$$\tau_1(f - q_{2^n};\Delta_m) \leqslant 2 \| f - q_{2^n} \|_{C[-1,1]} = 2E_{2^n}(f) \leqslant \frac{4}{m} \sum_{s=0}^{m} E_s(f). \tag{12}$$

From now on to the end of the section we shall write $\|\cdot\|$ instead of $\|\cdot\|_{C[-1,1]}$.

From lemma 3.9 we obtain

$$\tau_1(q_{2^n};\Delta_m) \leqslant 6 \| \Delta_m q'_{2^n} \| \leqslant 6 \left\{ \sum_{\nu=1}^{n} \| \Delta_m (q'_{2^\nu} - q'_{2^{\nu-1}}) \| + \| \Delta_m q'_1 \| \right\}. \tag{13}$$

Since $q_1 - q_0 \in P_1$, lemma 3.6 give us

$$\begin{aligned} \| \Delta_m q'_1 \| &= \| \Delta_m (q'_1 - q'_0) \| \leqslant \frac{1}{m} \| \Delta_1 (q'_1 - q'_0) \| \\ &\leqslant \frac{2}{m} \| q_1 - q_0 \| \leqslant 2m^{-1} (\| q_1 - f \| + \| f - q_0 \|) = 4m^{-1} E_0(f). \end{aligned} \tag{14}$$

Since for $\nu \leqslant n$ we have

$$\Delta_m(x) = \frac{1}{m}\left(\sqrt{(1-x^2)} + \frac{1}{m} \right) \leqslant \frac{1}{m} (\sqrt{(1-x^2)} + 2^{-\nu}) = 2^\nu m^{-1} \Delta_{2^\nu}(x)$$

and $q_{2^\nu} - q_{2^{\nu-1}} \in P_{2^\nu}$, lemma 3.6 gives us

$$\begin{aligned} \| \Delta_m (q'_{2^\nu} - q'_{2^{\nu-1}}) \| &\leqslant 2^\nu m^{-1} \| \Delta_{2^\nu} (q'_{2^\nu} - q'_{2^{\nu-1}}) \| \\ &\leqslant 2^{\nu+1} m^{-1} \| q_{2^\nu} - q_{2^{\nu-1}} \| \leqslant 2^{\nu+2} m^{-1} E_{2^{\nu-1}}(f). \end{aligned} \tag{15}$$

From (13)–(15) it follows that

$$\begin{aligned} \tau_1(q_{2^n};\Delta_m) &\leqslant \frac{6}{m} \left(4E_0(f) + \sum_{\nu=1}^{n} 2^{\nu+2} E_{2^{\nu-1}}(f) \right) \\ &\leqslant \frac{6}{m} \left(12E_0(f) + 16 \sum_{\nu=2}^{n} \sum_{s=2^{\nu-2}}^{2^{\nu-1}} E_s(f) \right) \leqslant \frac{96}{m} \sum_{s=0}^{m} E_s(f). \end{aligned} \tag{16}$$

From (11), (12) and (16) we obtain the statement of the theorem. □

Corollary 3.4. *We have* $\tau_1(f;\Delta_m)=O(m^{-\alpha})$, $0<\alpha<1$, *if and only if* $E_m(f)=O(m^{-\alpha})$.

This corollary follows immediately from theorems 3.13 and 3.14 for, if $E_m(f)=O(m^{-\alpha})$, $0<\alpha<1$, it follows from theorem 3.14 that

$$\tau_1(f;\Delta_m)=O\left(\frac{1}{m}\sum_{s=1}^{m}s^{-\alpha}\right)=O\left(\frac{m^{1-\alpha}}{m}\right)=O(m^{-\alpha}).$$ □

Corollary 3.4 gives us a characterization of the best uniform algebraic approximation $E_m(f)$ by means of the modulus $\tau_1(f;\Delta_m)$ for all orders $O(m^{-\alpha})$, $0<\alpha<1$.

It is possible to characterize all orders $O(m^{-\alpha})$, $\alpha>0$, using moduli $\tau_k(f;\Delta_m)$ (see K. Ivanov (1983a)) and also the best algebraic approximation in L_p (see K. Ivanov (1983b)). We shall not present here the corresponding results, since our intention is only to show the 'Nikol'skij effect', which is evident from theorems 3.13 and 3.14.

3.5 Direct and converse theorems and the K-functional of J. Peetre

A basic problem in the theory of approximation is to find direct and converse theorems for approximation by polynomials, rational functions, splines, etc. In our opinion the most natural way to obtain such theorems is to prove pairs of adjusted inequalities of Jackson and Bernstein type and then to characterize the approximation considered by the K-functional of J. Peetre generated by the appropriate spaces.

Let X_0, X_1 be two normed (or quasinormed) linear spaces and $X_1\subset X_0$. We shall denote by $\|\cdot\|_{X_i}$, $i=0,1$, the quasinorms in X_i, $i=0,1$ (we say that $\|\cdot\|_X$ is a quasinorm in X if $\|f\|_X\geqslant 0$ – we suppose that it is possible $\|f\|_X=0$ for $f\neq 0$ – and $\|f+g\|_X\leqslant c(\|f\|_X+\|g\|_X)$ for $f\in X$, $g\in X$, c a constant, $\|-f\|_X=\|f\|_X$).

For each $f\in X_0$ we set

$$K(f,t)=K(f,t;X_0,X_1)=\inf_{f=f_0+f_1}\{\|f_0\|_{X_0}+t\|f_1\|_{X_1}\}.$$

This functional was introduced by J. Peetre (1963) and is called the K-functional. The K-functional plays an essential role in many domains of the analysis, for example in the theory of interpolation spaces. This functional is also very useful in the theory of approximation. The K-functional provides an alternative way to characterize the 'smoothness' of the functions in place of the moduli of continuity and smoothness considered in the previous sections.

First we shall show that, if $X_0 = L_p[0, 2\pi]$, $X_1 = W_p^k = \{f : \| f^{(k)} \|_p < \infty\}$, the K-functional is equivalent to the kth L_p-modulus of smoothness, $1 \leq p \leq \infty$. We set $\| f \|_{X_0} = \| f \|_p$, $\| f \|_{X_1} = \| f \|_{W_p^k} = \| f^{(k)} \|_p$. $\| f \|_{W_p^k}$ is a quasi-norm in W_p^k.

Theorem 3.15. *Let $f \in L_p(0, 2\pi)$, $1 \leq p \leq \infty$. Then there exist constants $c_1(k)$ and $c_2(k)$, depending only on k, such that*

$$c_1(k)\omega_k(f; t)_p \leq K(f, t^k; L_p, W_p^k) \leq c_2(k)\omega_k(f; t)_p.$$

Proof. We shall use the generalized Steklov function $f_{k,h}$ from section 3.1. Using theorem 3.5 we obtain

$$K(f, t^k; L_p, W_p^k) \leq \| f - f_{k,h} \|_p + t^k \| f_{k,h}^{(k)} \|_p$$
$$\leq \omega_k(f; h)_p + c(k)t^k h^{-k} \omega_k(f; h)_p$$

Setting $h = t$ we obtain the right hand side of the inequality

$$K(f, t^k; L_p, W_p^k) \leq (1 + c(k))\omega_k(f; t)_p.$$

In order to prove the left hand side of the inequality, let $f_1 \in W_p^k$ be chosen arbitrarily. Using properties (ii) and (vi) of ω_k, we have

$$\omega_k(f; t)_p \leq \omega_k(f - f_1, t)_p + \omega_k(f_1; t)_p$$
$$\leq 2^k \| f - f_1 \|_p + t^k \| f_1^{(k)} \|_p.$$

Since $f_1 \in W_p^k$ is arbitrary, we get from here

$$\omega_k(f; t)_p \leq 2^k K(f, t^k; L_p, W_p^k). \qquad \square$$

Let $\{G_n\}_1^\infty$ be a family of subsets of X_0 and let each G_n be an existence set in X_0 (see section 1.1). We shall call $\{G_n\}_1^\infty$ a normal approximating family if $G_n \subset G_{n+1}$ and $G_n \pm G_{n-1} \subset G_{2n}$, i.e. for each $g_1 \in G_n$, $g_2 \in G_{n-1}$ we have $g_1 \pm g_2 \in G_{2n}$. Note that the most important approximating families, as algebraic and trigonometric polynomials, rational functions, spline functions, are normal approximating families.

The best approximation to $f \in X_0$ by means of elements of G_n we shall denote by $E_n(f)_{X_0}$:

$$E_n(f)_{X_0} = \inf\{\| f - g \|_{X_0} : g \in G_n\}.$$

We shall say that the quasinorm $\|\cdot\|_X$ satisfies the σ-condition, $0 < \sigma \leq 1$, if for each $f, g \in X$ we have

$$\| f + g \|_X^\sigma \leq \| f \|_X^\sigma + \| g \|_X^\sigma. \qquad (1)$$

The following theorem gives direct and converse theorems in terms of the K-functional.

Theorem 3.16. *Let X_0, X_1, $X_1 \subset X_0$, be two quasinormed linear spaces with quasinorms $\|\cdot\|_{X_i}$, $i=0,1$. Let the quasinorm $\|\cdot\|_{X_1}$ satisfy the σ-condition, $0<\sigma\leqslant 1$. Let $\{G_n\}_1^\infty$, $G_n\subset X_1$, $n=1,2,\ldots$, be a normal approximation family. Suppose that one of the following inequalities holds.*

(i) *Jackson type inequality: if $f\in X_1$, then*

$$E_n(f)_{X_0}\leqslant c\frac{\|f\|_{X_1}}{n^\alpha}, \tag{2}$$

where c is constant independent of n and f, $\alpha>0$ is a fixed number.

(ii) *Bernstein type inequality: if $\varphi\in G_n$ then*

$$\|\varphi\|_{X_1}\leqslant cn^\alpha\|\varphi\|_{X_0}, \tag{3}$$

where c is a constant independent of n and f.

Then, for every $f\in X_0$ and $n=1,2,\ldots$, we have, in case (i).

$$E_n(f)_{X_0}\leqslant cK(f,n^{-\alpha};X_0,X_1), \tag{4}$$

and, in case (ii),

$$K(f,n^{-\alpha};X_0,X_1)\leqslant cn^{-\alpha}\left\{\|f\|_{X_0}^\sigma+\sum_{\nu=1}^n\frac{1}{\nu}(\nu^\alpha E_\nu(f)_{X_0})^\sigma\right\}^{1/\sigma}, \tag{5}$$

where c is a constant independent of n and f.

Proof. We shall prove first that (2) implies (4). Let $f=f_0+f_1$, $f_0\in X_0, f_1\in X_1$. Then by (2) we get[†]

$$E_n(f)_{X_0}\leqslant c_0\left(\|f_0\|_{X_0}+\inf_{\varphi\in G_n}\|f_1-\varphi\|_{X_0}\right)$$
$$=c_0(\|f_0\|_{X_0}+E_n(f_1)_{X_0})\leqslant c_1(\|f_0\|_{X_0}+n^{-\alpha}\|f_1\|_{X_1}).$$

Hence

$$E_n(f)_{X_0}\leqslant cK(f,n^{-\alpha};X_0,X_1),$$

i.e. the estimate (4) holds.

Now we shall prove that (3) implies (5). Choose $\varphi_{2^\nu}\in G_{2^\nu}$ such that

$$\|f-\varphi_{2^\nu}\|_{X_0}=E_{2^\nu}(f)_{X_0}$$

(let us remember that G_n is an existence set in X_0).

Hence[†]

$$\|\varphi_{2^\nu}-\varphi_{2^{\nu-1}}\|_{X_0}\leqslant c_0(\|f-\varphi_{2^\nu}\|_{X_0}+\|f-\varphi_{2^{\nu-1}}\|_{X_0})\leqslant 2c_0E_{2^{\nu-1}}(f)_{X_0}. \tag{6}$$

Similarly as in the proof of theorem 3.7 using (1), and (3) and (6) we get

[†] $\|f+g\|_{X_0}\leqslant c_0(\|f\|_{X_0}+\|g\|_{X_0})$

for every $m \geqslant 0$

$$K(f, 2^{-m\alpha}; X_0, X_1) \leqslant \| f - \varphi_{2^m} \|_{X_0} + 2^{-m\alpha} \| \varphi_{2^m} \|_{X_1}$$

$$\leqslant E_{2^m}(f)_{X_0} + 2^{-m\alpha} \left\{ \sum_{v=1}^{m} \| \varphi_{2^v} - \varphi_{2^{v-1}} \|_{X_1}^{\sigma} + \| \varphi_1 \|_{X_1}^{\sigma} \right\}^{1/\sigma}$$

$$\leqslant E_{2^m}(f)_{X_0} + c \cdot 2^{-m\alpha} \left\{ \sum_{v=1}^{m} (2^{(v+1)\alpha} \| \varphi_{2^v} - \varphi_{2^{v-1}} \|_{X_0})^{\sigma} + \| \varphi_1 \|_{X_0}^{\sigma} \right\}^{1/\sigma \dagger}$$

$$\leqslant c_1 \left\{ E_{2^m}(f)_{X_0}^{\sigma} + 2^{-m\alpha} \left\{ \sum_{v=1}^{m} (2^{(v-1)\alpha} E_{2^{v-1}}(f)_{X_0})^{\sigma} + \| f - \varphi_1 \|_{X_0}^{\sigma} + \| f \|_{X_0}^{\sigma} \right\} \right\}^{1/\sigma}$$

$$\leqslant c_2 \cdot 2^{-m\alpha} \left\{ \sum_{v=1}^{m} \sum_{\mu=2^{v-1}}^{2v} \frac{1}{\mu} (\mu^{\alpha} E_{\mu}(f)_{X_0})^{\sigma} + (E_1(f)_{X_0})^{\sigma} + \| f \|_{X_0}^{\sigma} \right\}^{1/\sigma}$$

$$\leqslant c_3 \cdot 2^{-m\alpha} \left\{ \| f \|_{X_0}^{\sigma} + \sum_{\mu=1}^{2m} \frac{1}{\mu} (\mu^{\alpha} E_{\mu}(f)_{X_0})^{\sigma} \right\}^{1/\sigma}$$

This estimate implies (5) because of monotony of the K-functional. □

Corollary 3.5. *Let the conditions of theorem 3.16 be fulfilled and let $f \in X_0$. Then $E_n(f)_{X_0} = O(n^{-\gamma})$, $0 < \gamma < \alpha$, if and only if*

$$K(f, t; X_0, X_1) = O(t^{\gamma/\alpha}).$$

Corollary 3.5 is a special case of

Corollary 3.6. *Let the conditions of theorem 3.16 be fulfilled. Let $\omega(\delta)$ be a nonnegative and nondecreasing function on $[0, \infty)$, such that $\omega(2\delta) \leqslant 2^{\beta}\omega(\delta)$ for $\delta \geqslant 0$, $\beta > 0$. Then we have*

$$E_n(f)_{X_0} = O(n^{-\gamma}\omega(n^{-1})), \quad 0 < \beta + \gamma < \alpha, \quad \gamma \geqslant 0,$$

if and only if

$$K(f, t; X_0, X_1) = O(t^{\gamma/\alpha}\omega(t^{1/\alpha})).$$

Proof of corollary 3.6. If

$$K(f, t; X_0, X_1) = O(t^{\gamma/\alpha}\omega(t^{1/\alpha})),$$

then by (4) we obtain

$$E_n(f)_{X_0} \leqslant cK(f, n^{-\alpha}; X_0, X_1) = O(n^{-\gamma}\omega(n^{-1})).$$

Now let $E_n(f)_{X_0} = O(n^{-\gamma}\omega(n^{-1}))$. First we shall prove that

$$\omega(m\delta) \leqslant (2m)^{\beta}\omega(\delta), \quad \delta \geqslant 0, \quad m \geqslant 1. \tag{7}$$

Indeed, since $\omega(2\delta) \leqslant 2^{\beta}\omega(\delta)$, then $\omega(2^{v}\delta) \leqslant 2^{v\beta}\omega(\delta)$ for $\delta \geqslant 0$, $v \geqslant 0$. Suppose

† Since $\varphi_{2^v} - \varphi_{2^{v-1}} \in G_{2^{v-1}}$.

that $2^{\nu-1} \leqslant m < 2^{\nu}$. Then from the last inequality we get

$$\omega(m\delta) \leqslant \omega(2^{\nu}\delta) \leqslant 2^{\nu\beta}\omega(\delta) \leqslant (2m)^{\beta}\omega(\delta)$$

as required.

The inequality (7) implies immediately that

$$\omega(\lambda\delta) \leqslant (2(\lambda+1))^{\beta}\omega(\delta), \quad \delta, \lambda \geqslant 0. \tag{8}$$

Now we estimate $K(f, n^{-\alpha}; X_0, X_1)$ using (5), (8) and the fact that $\beta + \gamma < \alpha$. We get

$$\begin{aligned}
K(f, n^{-\alpha}; X_0, X_1) &\leqslant cn^{-\alpha}\left(\|f\|_{X_0}^{\sigma} + \sum_{\nu=1}^{n} \frac{1}{\nu} (\nu^{\alpha-\gamma}\omega(\nu^{-1}))^{\sigma} \right)^{1/\sigma} \\
&\leqslant c_1 n^{-\alpha}\left\{ \|f\|_{X_0}^{\sigma} + \sum_{\nu=1}^{n} \frac{1}{\nu}\left(\nu^{\alpha-\gamma}\left(\frac{n}{\nu}+1\right)^{\beta}\omega(n^{-1})\right)^{\sigma} \right\}^{1/\sigma} \\
&\leqslant c_2 n^{-\alpha}\left\{ \|f\|_{X_0}^{\sigma} + n^{\beta\sigma}(\omega(n^{-1}))^{\sigma} \sum_{\nu=1}^{n} \nu^{(\alpha-\gamma-\beta)\sigma-1} \right\}^{1/\sigma} \\
&\leqslant c_3 n^{-\alpha}\left\{ \|f\|_{X_0}^{\sigma} + n^{(\alpha-\gamma)\sigma}(\omega(n^{-1}))^{\sigma} \right\}^{1/\sigma} = \mathrm{O}(n^{-\gamma}\omega(n^{-1})).
\end{aligned}$$

□

Let us mention finally that theorem 3.16 can be considered as an abstract generalization of theorems 3.3 and 3.7 because of theorem 3.15 and the fact that we have Jackson type and Bernstein type inequalities for the trigonometrical polynomials.

Theorem 3.16 can be used successfully in more general situations than that considered, but for orders of approximation not better than $\mathrm{O}(n^{-\alpha})$.

Consider the following approximation spaces:

$$A_q^{\gamma}(X_0) = \left\{ f \in X_0 : \|f\|_{A_q^{\gamma}(X_0)} = \|f\|_{X_0} + \left(\sum_{\nu=0}^{\infty} (2^{\nu\gamma} E_{2^{\nu}}(f)_{X_0})^q \right)^{1/q} < \infty \right\},$$

when $0 < q < \infty$, and

$$A_{\infty}^{\gamma}(X_0) = \{ f \in X_0 : \|f\|_{A_{\infty}^{\gamma}(X_0)} = \|f\|_{X_0} + \sup_n n^{\gamma} E_n(f)_{X_0} < \infty \}.$$

As usually we shall denote by $(X_0, X_1)_{\theta,q}$ the real interpolation space between X_0 and X_1:

$$\begin{aligned}
&(X_0, X_1)_{\theta,q} \\
&= \left\{ f \in X_0 : \|f\|_{(X_0,X_1)_{\theta,q}} = \|f\|_{X_0} + \left(\sum_{\nu=0}^{\infty} \left(2^{\nu\theta} K\left(f, \frac{1}{2^{\nu}}; X_0, X_1\right)\right)^q \right)^{1/q} < \infty \right\},
\end{aligned}$$

when $0<q<\infty$, and

$$(X_0,X_1)_{\theta,\infty}=\left\{f\in X_0:\|f\|_{(X_0,X_1)_{\theta,\infty}}=\|f\|_{X_0}+\sup_{t>0}t^{-\theta}K(f,t;X_0,X_1)<\infty\right\}$$

(see J. Peetre (1963), J. Bergh, J. Löfström (1976)).

Corollary 3.7. *Under the assumptions* (i) *and* (ii) *of theorem 3.16 we have*

$$A_q^{\gamma}(X_0)=(X_0,X_1)_{\gamma/\alpha,q}$$

with equivalent quasi-norms provided $0<\gamma<\alpha$ *and* $0<q\leqslant\infty$.

Proof. It is readily seen that the inequality (4) from Theorem 3.16 implies the estimate

$$\|f\|_{A_q^{\gamma}(X_0)}\leqslant C\|f\|_{(X_0,X_1)_{\gamma/\alpha,q}}.$$

In order to prove an estimate in the opposite direction we shall use the inverse estimate (5) and the following discrete variant of Hardy's inequality (4) in section 7.1.

Lemma 3.10. *If* $0<p<\infty$, $\beta>0$ *and* $a_k\geqslant 0$, $k=1,2,\ldots$, *then*

$$\sum_{\nu=0}^{\infty}\left(2^{-\nu\beta}\sum_{k=0}^{\nu}a_k\right)^p\leqslant C(\beta,p)\sum_{k=0}^{\infty}(2^{-k\beta}a_k)^p.$$

Proof. Let $1\leqslant p<\infty$ and set $\alpha=\beta/2$. Then applying Hölder's inequality we get

$$\begin{aligned}\sum_{\nu=0}^{\infty}\left(2^{-\nu\beta}\sum_{k=0}^{\nu}a_k\right)^p&=\sum_{\nu=0}^{\infty}2^{-\nu\beta p}\left(\sum_{k=0}^{\nu}2^{\alpha k}2^{-\alpha k}a_k\right)^p\\&\leqslant\sum_{\nu=0}^{\infty}2^{-\nu\beta p}\left(\sum_{k=0}^{\nu}2^{\alpha kp'}\right)^{p/p'}\left(\sum_{k=0}^{\nu}2^{-\alpha kp}a_k^p\right)^1\\&\leqslant C(\beta,p)\sum_{\nu=0}^{\infty}2^{-\nu\beta p+\nu\alpha p}\sum_{k=0}^{\nu}2^{-\alpha kp}a_k^p\\&=C(\beta,p)\sum_{k=0}^{\nu}2^{-\alpha kp}a_k^p\sum_{\nu=k}^{\infty}2^{-\nu\alpha p}\\&=C_1(\beta,p)\sum_{k=0}^{\infty}2^{-k\beta p}a_k^p.\end{aligned}$$

Now consider the case $0<p<1$. Then using inequality (3) in section 7.1 and changing the order of summation we get

$$\begin{aligned}\sum_{\nu=0}^{\infty}\left(2^{-\nu\beta}\sum_{k=0}^{\nu}a_k\right)^p&\leqslant\sum_{\nu=0}^{\infty}2^{-\nu\beta p}\sum_{k=0}^{\nu}a_k^p\\&=\sum_{k=0}^{\infty}a_k^p\sum_{\nu=k}^{\infty}2^{-\nu\beta p}=C(\beta,p)\sum_{k=0}^{\infty}2^{-k\beta p}a_k^p.\end{aligned}$$ □

Now we continue the proof of corollary 3.7. Inequality (5) in theorem 3.16 as we know from the proof of the same theorem is equivalent to the inequality

$$K\left(f,\frac{1}{2^{m\alpha}};X_0,X_1\right)\leqslant\frac{c}{2^{m\alpha}}\left\{\|f\|_{X_0}+\left(\sum_{\nu=0}^{m}(2^{\nu\alpha}E_{2^\nu}(f)_{X_0})^\sigma\right)^{1/\sigma}\right\}. \tag{9}$$

On the other hand, it is easily seen that

$$\|f\|_{(X_0,X_1)_{\gamma/\alpha,q}}\leqslant C\left\{\|f\|_{X_0}+\left(\sum_{m=0}^{\infty}\left(2^{m\gamma}K\left(f,\frac{1}{2^{m\alpha}};X_0,X_1\right)\right)^q\right)^{1/q}\right\}. \tag{10}$$

Combining (9) and (10) and using lemma 3.9 we obtain

$$\begin{aligned}\|f\|_{(X_0,X_1)_{\gamma/\alpha,q}}&\leqq C\left\{\|f\|_{X_0}+\|f\|_{X_0}\left(\sum_{m=0}^{\infty}2^{-m(\alpha-\gamma)q}\right)^{1/q}\right.\\&\quad\left.+\left(\sum_{m=0}^{\infty}2^{-m(\alpha-\gamma)q}\left(\sum_{\nu=0}^{m}(2^{\nu\alpha}E_{2^\nu}(f)_{X_0})^\sigma\right)^{q/\sigma}\right)^{1/q}\right\}\\&\leqslant C_1\left\{\|f\|_{X_0}+\left(\sum_{\nu=0}^{\infty}2^{-\nu(\alpha-\gamma)q}\,2^{\nu\alpha q}E_{2^\nu}(f)^q_{X_0}\right)^{1/q}\right\}\\&=C_1\|f\|_{A_q^\gamma(X_0)}.\end{aligned}$$

□

Remark. We observe that if $\|g\|_{X_1}=0$ for each $g\in G_1$ then the term $\|f\|_{X_0}$ can be omitted in estimate (5) in theorem 3.16 and as consequence corollary 3.7 holds without $\|f\|_{X_0}$ in the definitions of spaces $A_q^\gamma(X_0)$ and $(X_0,X_1)_{\theta,q}$.

3.6 Notes

The classical works on direct and converse theorems in the theory of approximation are the works of D. Jackson (1911), Ch.de la Vallée-Poussin (1910), S.N. Bernstein (1912). The second modulus of continuity or modulus of smoothness $\omega_2(f;\delta)$ was introduced by A. Zygmund (1945). Zygmund obtained by means of ω_2 characterization of the class

$$\{f: E_n(f)_{C[-\pi,\pi]}=O(n^{-1})\}.$$

As we have mentioned, the converse theorem 3.7 of Bernstein type was given in this form by Salem (1940) and Stechkin (1951).

For the generalizations to $\omega_k(f;\delta)$, $k>2$, and L_p, $p<\infty$, see Quade (1937), A.F. Timan and M.F. Timan (1950), S.B. Stechkin (1951), M.F. Timan (1958).

For Jackson and Bernstein type theorems for best approximations in the spaces L_p, $0<p<1$, see E.A. Storoženko, V.G. Krotov, P. Osvald (1975), V.A. Ivanov (1975), E.A. Storoženko (1975, 1977, 1980), V.G. Krotov (1982).

The problem of characterization of the functions on the finite interval by means of their best polynomial approximations has a long history. We cannot give here this history and all authors who have been working in this domain. We want only to mention the works of A.F. Timan (1951), V.K. Dzjadik (1956, 1958), Yu.A. Brudnyi (1963, 1970), V.P. Motornii (1971), A.L. Fuksman (1965), M.K. Potapov (1975, 1977, 1981, 1983), A.S. Dgafarov (1977), R.L. Stens (1977), R.L. Stens, M. Wehrens (1979), P.L. Butzer, R.L. Stens, M. Wehrens (1980). Between the different characterizations we prefer the characterizations given by K. Ivanov (1983a, b) and Z. Ditzian, V. Totik (1987).

For direct and converse theorems for best trigonometric approximation using fractional derivatives and fractional moduli see P.L. Butzer, R.L. Stens (1976), P.L. Butzer, H. Dyckhoff, E. Görlich, R.L. Stens (1977).

For more details concerning the classical direct and converse theorems for trigonometric polynomial approximation see the books of I.P. Natanson (1949), G.G. Lorentz (1966), V.K. Dzjadik (1977), A.F. Timan (1960), V.M. Tihomirov (1976).

The K-functional was introduced by J. Peetre (1963, 1968). The K-functional has many applications in the theory of interpolation spaces, in approximation theory and in other domains; see the books of J. Bergh, J. Löfström (1976), H. Triebel (1978), P.L. Butzer, H. Berens (1967), J. Peetre (1976).

The role of Jackson and Bernstein type inequalities in linear approximation is well known, see P. Butzer, K. Scherer (1968). For the case of nonlinear approximation compare with J. Peetre, G. Sparr (1972) and J. Bergh, J. Peetre (1974).

4

Approximation of some important functions

The development of the theory of rational approximation has a point of discontinuity, a jump in 1964, when D.J. Newman proved that the best uniform approximation of the function $|x|$ in the interval $[-1,1]$ by means of rational function of nth order has order $O(e^{-c\sqrt{n}})$. Let us remember that the order of the best uniform polynomial approximation $E_n(|x|)_{C[-1,1]}$ is only $O(n^{-1})$ (S.N. Bernstein (1952); see also G.G. Lorentz (1966)).

In this chapter we shall consider the best uniform rational approximation of some special, but very important, functions. We begin in section 4.1 with Newman's result. After this in section 4.2 we give the exact asymptotics of the best uniform rational approximation of $|x|$ in $[-1,1]$ (Vjacheslavov, 1975). In section 4.3 we give some of the few examples where it is possible to write exactly the rational function of best uniform approximation. It is interesting that this was done more than 100 years ago (E.I. Zolotarjov, 1877). In section 4.4 we give the solution of the Meinardus conjecture for best uniform rational approximation to e^x in the interval $[-1,1]$, obtained by D. Braess (1984). We end the chapter with some remarks connected with rational approximation of e^{-x} on $[0,\infty)$ (section 4.5) and notes (section 4.6).

In this chapter we shall use the notations

$$R_n(f) \equiv R_n(f)_{C[a,b]} = \inf\{\|f-r\|_{C[a,b]}; r\in R_n \equiv R_{nn}\}.$$

4.1 Newman's theorem

Theorem 4.1. *The following estimate holds:*

$$e^{-\pi\sqrt{(n+1)}} \leqslant R_n(|x|)_{C[-1,1]} \leqslant 3e^{-\sqrt{n}}, \quad n \geqslant 5. \tag{1}$$

Remark. The rational approximation of $|x|$ on $[-1,1]$ is equivalent to the rational approximation of $\sqrt{x}$ on $[0,1]$. More precisely if the rational function

$s_n \in R_n$ is the best uniform approximation to $\sqrt{x}$ in $[0, 1]$ of order n and $r_{2n} \in R_{2n}$ the best uniform approximation to $|x|$ in $[-1, 1]$ of order $2n$, then $s_n(x^2) = r_{2n}(x)$. Consequently $R_n(\sqrt{x})_{C[0,1]} = R_{2n}(|x|)_{C[-1,1]}$.

In the proof of theorem 4.1 we shall use the following lemma.

Lemma 4.1. *Let*

$$p(x) = \prod_{k=1}^{n-1} (x + \xi^k), \quad \xi = \varepsilon^{1/n}, \quad 0 < \varepsilon < 1, \quad n \geqslant 2.$$

Then for $x \in [\varepsilon, 1]$ *we have*

$$\left| \frac{p(-x)}{p(x)} \right| \leqslant \exp\left\{ -\frac{2(\varepsilon^{1/n} - \varepsilon)n}{\ln(1/\varepsilon)} \right\}. \tag{2}$$

Proof. Let $x \in [\xi^{i+1}, \xi^i]$, $0 \leqslant i \leqslant n-1$. Since the function $(a-x)/(a+x)$, $a > 0$, is monotone decreasing, then

$$\left| \frac{p(-x)}{p(x)} \right| = \prod_{k=1}^{i} \frac{\xi^k - x}{\xi^k + x} \prod_{k=i+1}^{n-1} \frac{x - \xi^k}{x + \xi^k} \leqslant \prod_{k=1}^{i} \frac{\xi^k - \xi^{i+1}}{\xi^k + \xi^{i+1}} \prod_{k=i+1}^{n-1} \frac{1 - \xi^k}{1 + \xi^k} = \prod_{k=1}^{n-1} \frac{1 - \xi^k}{1 + \xi^k}.$$

Thus for $x \in [\varepsilon, 1]$ we have

$$\left| \frac{p(-x)}{p(x)} \right| \leqslant \prod_{k=1}^{n-1} \frac{1 - \xi^k}{1 + \xi^k}. \tag{3}$$

For $t \geqslant 0$ we have $(1-t)/(1+t) \leqslant e^{-2t}$. Using this inequality and (3), we get for $x \in [\varepsilon, 1]$

$$\left| \frac{p(-x)}{p(x)} \right| \leqslant \exp\left\{ -2 \sum_{k=1}^{n-1} \xi^k \right\} = \exp\left\{ -\frac{2\xi(1 - \xi^{n-1})}{1 - \xi} \right\}$$

$$\leqslant \exp\left\{ -\frac{2(\varepsilon^{1/n} - \varepsilon)n}{\ln(1/\varepsilon)} \right\},$$

where we have used the inequality

$$1 - \xi \leqslant \ln(1/\xi) = \frac{1}{n} \ln(1/\varepsilon). \qquad \square$$

Proof of the upper bound in theorem 4.1. Let us consider the rational function

$$r(x) = x \frac{p(x) - p(-x)}{p(x) + p(-x)},$$

where

$$p(x) = \prod_{k=1}^{n-1} (x + \xi^k), \quad \xi = \varepsilon^{1/n}, \quad \varepsilon = e^{-\sqrt{n}}, \quad n \geqslant 5.$$

Clearly $r \in R_n$. Since $2(\varepsilon^{1/n} - \varepsilon) = 2(e^{-1/\sqrt{n}} - e^{-\sqrt{n}}) > 1$ for $n \geqslant 5$, then by (2) we get

$$\left|\frac{p(-x)}{p(x)}\right| \leqslant e^{-\sqrt{n}} \quad \text{for } x \in [e^{-\sqrt{n}}, 1], \quad n \geqslant 5. \tag{4}$$

Since $|x|$ and $r(x)$ are even functions in the interval $[-1, 1]$, we can consider only the case $x \in [0, 1]$. If $x \in [0, e^{-\sqrt{n}}] = [0, \xi^n]$, then $p(-x) > 0$ and $x \geqslant r(x) > 0$. Consequently for $x \in [0, e^{-\sqrt{n}}]$ we have

$$||x| - r(x)| = x - r(x) \leqslant x \leqslant e^{-\sqrt{n}}.$$

If $x \in [e^{-\sqrt{n}}, 1]$, then by (4) we get for $n \geqslant 5$

$$||x| - r(x)| = 2x\left|\frac{p(-x)}{p(x) + p(-x)}\right| = \frac{2x|p(-x)/p(x)|}{|1 + p(-x)/p(x)|} \leqslant \frac{2|p(-x)/p(x)|}{1 - |p(-x)/p(x)|}$$

$$\leqslant \frac{2e^{-\sqrt{n}}}{1 - e^{-\sqrt{n}}} < 3e^{-\sqrt{n}},$$

which gives us the upper bound in theorem 4.1.

To prove the lower bound we shall need some lemmas.

Lemma 4.2. *If $p \in P_n$, $n \geqslant 0$, then there exists a polynomial $q \in P_n$ such that*

$$\left|\frac{p(-x)}{p(x)}\right| \geqslant \left|\frac{q(-x)}{q(x)}\right|, \quad \left|\frac{q(-x)}{q(x)}\right| \leqslant 1 \quad \textit{for } x \geqslant 0.$$

Proof. Let $\xi = u + iv$ be any complex number and $t > 0$. Then

$$\left|\frac{t + \xi}{t - \xi}\right| = \left(\frac{(t + u)^2 + v^2}{(t - u)^2 + v^2}\right)^{1/2} \geqslant \left(\frac{(t - |u|)^2 + v^2}{(t + |u|)^2 + v^2}\right)^{1/2}$$

$$\geqslant \left(\frac{(t - |u|)^2}{(t + |u|)^2}\right)^{1/2} = \left|\frac{t - |u|}{t + |u|}\right|. \tag{5}$$

If $p(x) = A\prod_{i=1}^{n}(x - \xi_i)$, then we set $q(x) = A\prod_{i=1}^{n}(x + |\mathrm{Re}\,\xi_i|)$. By (5) we get for $x \geqslant 0$

$$\left|\frac{p(-x)}{p(x)}\right| = \left|\frac{\prod_{i=1}^{n}(x + \xi_i)}{\prod_{i=1}^{n}(x - \xi_i)}\right| \geqslant \left|\prod_{i=1}^{n}\frac{(x - |\mathrm{Re}\,\xi_i|)}{(x + |\mathrm{Re}\,\xi_i|)}\right| = \left|\frac{q(-x)}{q(x)}\right|.$$

On the other hand obviously

$$|q(-x)/q(x)| \leqslant 1 \quad \text{for } x \geqslant 0. \qquad \square$$

Lemma 4.3. *Let $r \in R_n$ and let r be an even function. Then there exists $q \in P_{n+1}$ such that*

$$|x - r(x)| \geqslant x\left|\frac{q(-x)}{q(x)}\right| \quad \textit{for } x \geqslant 0.$$

Proof. Since $r \in R_n$ is an even rational function, there exist two algebraic polynomials $p_1 \in P_{[n/2]}$ and $p_2 \in P_{[n/2]}$ such that $r(x) = p_1(x^2)/p_2(x^2)$. Denote

$$F = \{x: x \geqslant 0, r(x) > 0\}, \quad E = [0, \infty) \backslash F.$$

If $x \in F$ then

$$|x - r(x)| = x\left|\frac{xp_2(x^2) - p_1(x^2)}{xp_2(x^2)}\right| > x\left|\frac{xp_2(x^2) - p_1(x^2)}{xp_2(x^2) + p_1(x^2)}\right|. \tag{6}$$

We set $p(x) = xp_2(x^2) + p_1(x^2)$. Then $p \in P_{n+1}$. By (6) we have for $x \in F$

$$|x - r(x)| > x|p(-x)/p(x)|.$$

Now by lemma 4.2 there exists $q \in P_{n+1}$ such that for $x \geqslant 0$ we have

$$\left|\frac{p(-x)}{p(x)}\right| \geqslant \left|\frac{q(-x)}{q(x)}\right| \quad \text{and} \quad \left|\frac{q(-x)}{q(x)}\right| \leqslant 1.$$

Then we have for $x \in F$

$$|x - r(x)| > x\left|\frac{q(-x)}{q(x)}\right|$$

and clearly for $x \in E$

$$|x - r(x)| \geqslant x \geqslant x|q(-x)/q(x)|. \qquad \square$$

Lemma 4.4. *Let $q \in P_n$, $n \geqslant 1$ and $q \not\equiv 0$. Then there exists a point $x \in [\mathrm{e}^{-\pi\sqrt{n}}, 1]$ where*

$$x(q(-x)/q(x)) > \mathrm{e}^{-\pi\sqrt{n}}.$$

Proof. Let $\xi = u + \mathrm{i}v$ be any complex number and $0 \leqslant a < b$. Then

$$\int_a^b \ln\left|\frac{t+\xi}{t-\xi}\right|\frac{\mathrm{d}t}{t} > -\frac{\pi^2}{2}. \tag{7}$$

Indeed we have by (5) for $t \geqslant 0$

$$\left|\frac{t+\xi}{t-\xi}\right| \geqslant \left|\frac{t-|u|}{t+|u|}\right|.$$

We may assume that $u \neq 0$. Then

$$\int_a^b \ln\left|\frac{t+\xi}{t-\xi}\right|\frac{dt}{t} \geqslant \int_a^b \ln\left|\frac{t-|u|}{t+|u|}\right|\frac{\mathrm{d}t}{t} = \int_{a/|u|}^{b/|u|} \ln\left|\frac{t-1}{t+1}\right|\frac{\mathrm{d}t}{t} > \int_0^\infty \ln\left|\frac{t-1}{t+1}\right|\frac{\mathrm{d}t}{t}$$
$$= 2\int_0^1 \ln\frac{1-t}{1+t}\frac{\mathrm{d}t}{t} = -\frac{\pi^2}{2}.$$

The last integral can be calculated by using the Taylor series for the logarithmic function and the fact that

$$\sum_{k=1}^{\infty} \frac{1}{(2k-1)^2} = \frac{\pi^2}{8}.$$

Next put $\delta = \exp\{-\pi\sqrt{n}\}$ and assume that

$$x\left|\frac{q(-x)}{q(x)}\right| \leqslant \exp\{-\pi\sqrt{n}\} \quad \text{for } \delta \leqslant x \leqslant 1.$$

Then

$$\int_\delta^1 \ln\left|t\frac{q(-t)}{q(t)}\right|\frac{dt}{t} \leqslant -\pi\sqrt{n}\int_\delta^1 \frac{dt}{t} = -\pi^2 n. \tag{8}$$

On the other hand one has

$$\ln\left|t\frac{q(-t)}{q(t)}\right| = \ln t + \sum_\xi \ln\left|\frac{t+\xi}{t-\xi}\right|,$$

where ξ runs through the zeros of q. Noting that

$$\int_\delta^1 \frac{\ln t}{t}\,dt = -\pi^2 n/2$$

and applying (7) to each term in the sum we get

$$\int_\delta^1 \ln\left|t\frac{q(-t)}{q(t)}\right|\frac{dt}{t} \geqslant \int_\delta^1 \frac{\ln t}{t}\,dt + \sum_\xi \int_\delta^1 \ln\left|\frac{t+\xi}{t-\xi}\right|\frac{dt}{t} > -\pi^2 n/2 - \pi^2 n/2 = -\pi^2 n.$$

The comparison of this inequality with (8) proves lemma 4.4. □

Proof of the lower bound in theorem 4.1. Let $r \in R_n$ be the rational function of order n of best uniform approximation to $|x|$ in $[-1, 1]$, i.e.

$$\||x| - r(x)\|_{C[-1,1]} = R_n(|x|)_{C[-1,1]}.$$

Since $|x|$ is an even function, $r(x)$ and $r(-x)$ are rational functions of order n of best uniform approximation to $|x|$ in $[-1, 1]$ and by the uniqueness of the best rational uniform approximation (theorem 2.2) it follows that $r(x) = r(-x)$, i.e. r is also an even function.

By lemma 4.3 there exists $q \in P_{n+1}$, $q \not\equiv 0$, such that

$$|x - r(x)| \geqslant x\left|\frac{q(-x)}{q(x)}\right| \quad \text{for } x \geqslant 0.$$

Now by lemma 4.4 it follows that there exists $x_0 \in [\exp(-\pi\sqrt{(n+1)}), 1]$

such that

$$x_0\left|\frac{q(-x_0)}{q(x_0)}\right| > e^{-\pi\sqrt{(n+1)}}.$$

Consequently

$$R_n(|x|)_{C[-1,1]} \geqslant |x_0 - r(x_0)| \geqslant x_0\left|\frac{q(-x_0)}{q(x_0)}\right| > e^{-\pi\sqrt{(n+1)}}. \qquad \square$$

4.2 The exact order of the best uniform approximation to $|x|$

In this paragraph we shall give the following results of N.S. Vjacheslavov (1975).

Theorem 4.2. *The following estimates hold:*

$$c_1 e^{-\pi\sqrt{n}} \leqslant R_n(|x|)_{C[-1,1]} \leqslant c_2 e^{-\pi\sqrt{n}}, \quad n \geqslant 1, \tag{1}$$

where c_1 and c_2 are absolute constants.

Instead of the inequalities of the type (1) we shall write also

$$R_n(|x|)_{C[0,1]} \asymp O(e^{-\pi\sqrt{n}}).$$

Using the remark after theorem 4.1 we obtain from (1) the following.

Corollary 4.1. *We have*

$$R_n(\sqrt{x})_{C[0,1]} \asymp O(e^{-\pi\sqrt{(2n)}}).$$

The lower bound in (1) we have by theorem 4.1.

To obtain the upper bound in (1) we shall use some lemmas.

Lemma 4.5. *Let* $\xi = e^{-\alpha}$, $0 < \alpha \leqslant 1/2$, *and* $\mu \geqslant (1/\alpha)\ln(1/\beta)$, $0 < \beta \leqslant 1/2$, μ *an integer. Then*

$$\prod_{k=0}^{\mu-1}\frac{1-\xi^{k+1/2}}{1+\xi^{k+1/2}} < 2\exp\left\{-\frac{\pi^2}{4}\frac{1}{\alpha} + 2.1\frac{\beta}{\alpha}\right\}.$$

Proof. For $|x| < 1$ we have

$$\ln\frac{1-x}{1+x} = \ln(1-x) - \ln(1+x) = -2\sum_{s=1}^{\infty}\frac{x^{2s-1}}{2s-1}$$

and therefore

$$\prod_{k=0}^{\mu-1}\frac{1-\xi^{k+1/2}}{1+\xi^{k+1/2}} = \exp\left\{\sum_{k=0}^{\mu-1}\ln\frac{1-\xi^{k+1/2}}{1+\xi^{k+1/2}}\right\} = \exp\left\{-2\sum_{s=1}^{\infty}\frac{1}{2s-1}\sum_{k=0}^{\mu-1}\xi^{(k+1/2)(2s-1)}\right\}$$

$$< \exp\left\{-2\sum_{s=1}^{2[1/\alpha]}\frac{1-\xi^{\mu(2s-1)}}{(2s-1)(\xi^{-(2s-1)/2}-\xi^{(2s-1)/2})}\right\} = \exp\{A_1 + A_2\},$$

where

$$A_1 = -\frac{2}{\alpha}\sum_{s=1}^{2[1/\alpha]} \frac{\dfrac{1}{(2s-1)^2}(2s-1)\alpha}{\exp\left\{\dfrac{(2s-1)\alpha}{2}\right\} - \exp\left\{-\dfrac{(2s-1)\alpha}{2}\right\}},$$

$$A_2 = \frac{2}{\alpha}\sum_{s=1}^{2[1/\alpha]} \frac{\dfrac{1}{(2s-1)^2}(2s-1)\alpha\xi^{\mu(2s-1)}}{\exp\left\{\dfrac{(2s-1)\alpha}{2}\right\} - \exp\left\{-\dfrac{(2s-1)\alpha}{2}\right\}}.$$

We shall estimate A_1 and A_2 using the following inequalities:

$$\frac{2}{\alpha}\sum_{s=2[1/\alpha]+1}^{\infty} \frac{1}{(2s-1)^2} \leqslant 0.43, \quad 0<\alpha\leqslant 1/2, \tag{2}$$

$$1-0.06x^2 < \varphi(x) < 1, \quad 0<x<4, \tag{3}$$

where

$$\varphi(x) = \frac{x}{e^{x/2} - e^{-x/2}}.$$

Indeed we have for $0<\alpha\leqslant 1/2$

$$\frac{2}{\alpha}\sum_{s=2[1/\alpha]+1}^{\infty} \frac{1}{(2s-1)^2} < \frac{2}{\alpha}\int_{2[1/\alpha]}^{\infty} \frac{dt}{(2t-1)^2} = \frac{1}{\alpha(4[1/\alpha]-1)} < \frac{[1/\alpha]+1}{4[1/\alpha]-1} \leqslant \frac{3}{7} < 0.43.$$

On the other hand

$$\varphi(x) = \frac{x}{2\sum_{k=0}^{\infty} \dfrac{1}{(2k+1)!}\left(\dfrac{x}{2}\right)^{2k+1}} = \frac{1}{1+y},$$

where we set

$$y = x^2 \sum_{k=1}^{\infty} \frac{x^{2k-2}}{(2k+1)!\cdot 2^{2k}}.$$

Then for $0<x<4$ we have

$$y < x^2 \sum_{k=1}^{\infty} \frac{4^{2k-2}}{(2k+1)!\cdot 2^{2k}} = \frac{x^2}{32}\left(\sum_{k=0}^{\infty} \frac{2^{2k+1}}{(2k+1)!} - 2\right) = \frac{x^2}{32}(\sinh 2 - 2) < 0.06x^2.$$

Since $(1+y)^{-1} > (1-y)$ for $y>0$,

$$\varphi(x) > 1-y > 1-0.06x^2.$$

Thus (2) and (3) are proved.

Now we shall estimate A_1 using (3) with $x=(2s-1)\alpha$, the equality $\sum_{s=1}^{\infty}(2s-1)^{-2}=\pi^2/8$ and (2). We get

$$A_1=-\frac{2}{\alpha}\sum_{s=1}^{2[1/\alpha]}\frac{1}{(2s-1)^2}\varphi((2s-1)\alpha)<-\frac{2}{\alpha}\sum_{s=1}^{2[1/\alpha]}\frac{1}{(2s-1)^2}(1-0.06(2s-1)^2\alpha^2)$$

$$=-\frac{2}{\alpha}\left(\frac{\pi^2}{8}-\sum_{s=2[1/\alpha]+1}^{\infty}\frac{1}{(2s-1)^2}-0.06\cdot 2[1/\alpha]\alpha^2\right)$$

$$<-\frac{\pi^2}{4}\frac{1}{\alpha}+0.43+0.24=-\frac{\pi^2}{4}\frac{1}{\alpha}+0.67.$$

It remains to estimate A_2. Since $\mu\geqslant(1/\alpha)\ln(1/\beta)$,

$$\xi^{\mu}=\exp\{-\mu\alpha\}\leqslant\exp\{-\ln(1/\beta)\}=\beta\leqslant 1/2.$$

Moreover $\varphi((2s-1)\alpha)<1$. Then we obtain

$$A_2=\frac{2}{\alpha}\sum_{s=1}^{2[1/\alpha]}\frac{1}{(2s-1)^2}\xi^{\mu(2s-1)}\varphi((2s-1)\alpha)<\frac{2}{\alpha}\sum_{s=1}^{2[1/\alpha]}\frac{1}{(2s-1)^2}\xi^{\mu(2s-1)}$$

$$<\frac{2}{\alpha}\left(\xi^{\mu}+\frac{1}{9}\xi^{3\mu}+\sum_{s=3}^{\infty}\frac{\xi^{\mu(2s-1)}}{(2s-1)^2}\right)$$

$$<\frac{2}{\alpha}\xi^{\mu}\left(1+\frac{1}{9}\xi^{2\mu}+\xi^{4\mu}\sum_{s=3}^{\infty}\frac{1}{(2s-1)^2}\right)$$

$$<\frac{\beta}{\alpha}\left(2+\frac{2}{9}\cdot\frac{1}{4}+\frac{2}{16}\left(\frac{\pi^2}{8}-1\right)\right)<2.1\beta/\alpha.$$

Using the estimates given above for A_1 and A_2 we get

$$\prod_{k=0}^{\mu-1}\frac{1-\xi^{k+1/2}}{1+\xi^{k+1/2}}<\exp\{A_1+A_2\}<\exp\left\{-\frac{\pi^2}{4}\frac{1}{\alpha}+2.1\frac{\beta}{\alpha}+0.67\right\}$$

$$<2\exp\left\{-\frac{\pi^2}{4}\frac{1}{\alpha}+2.1\frac{\beta}{\alpha}\right\}.\qquad\square$$

Next we shall assume that $n>2n_0$ where n_0 is a sufficiently large number. Let us denote

$$m=\left[\frac{1}{\pi}\ln\sqrt{n}+1\right],\quad r=[2\sqrt{n}+1],$$

$$A_\nu=\begin{cases}\left(1-\dfrac{1}{e}\right)\dfrac{\pi}{2\sqrt{n}}, & 0\leqslant\nu\leqslant 2r-1,\\[2ex] \left(1-\dfrac{1}{e^i}\right)\dfrac{\pi}{2\sqrt{n}}, & (i+1)r\leqslant\nu<(i+2)r,\quad 1\leqslant i\leqslant m-2,\\[2ex] \dfrac{\pi}{2\sqrt{(n-\nu)}}, & mr\leqslant\nu\leqslant n-1,\end{cases}$$

$$l = l(v) = \left[\frac{1}{A_v}\ln\frac{1}{A_v}\right], \quad mr \leqslant v \leqslant n - n_0.$$

Note that $A_0 \leqslant A_1 \leqslant A_2 \leqslant \cdots$.

We shall denote by c absolute positive constants.

Lemma 4.6. *The following inequalities hold true:*

$$A_{2r} + A_{2r+1} + \cdots + A_v + \frac{\pi^2}{2A_{v+1}} > \pi\sqrt{n} - c, \tag{4}$$

where $mr \leqslant v < n - 1$;

$$\left(\frac{\pi^2}{4} - \frac{2.1}{e^{i+1}}\right)\frac{1}{A_{(i+2)r}} + \frac{\pi^2}{4A_{2mr}} > \pi\sqrt{n} + \ln n, \tag{5}$$

where $1 \leqslant i \leqslant m - 2$;

$$A_{v-k} + A_{v-k+1} + \cdots + A_{v+k} \leqslant (2k+1)A_v(1 + cA_v), \tag{6}$$

where $0 \leqslant k \leqslant l(v)$, $mr \leqslant v < n - n_0$.

Proof. Evidently

$$\sqrt{(1-x)} \geqslant 1 - \tfrac{1}{2}x - \tfrac{1}{2}x^2 \quad \text{for } |x| \leqslant 1. \tag{7}$$

We have for $1 \leqslant k \leqslant s$

$$2\sqrt{(s+1)} - 2\sqrt{k} < \frac{1}{\sqrt{s}} + \frac{1}{\sqrt{(s-1)}} + \cdots + \frac{1}{\sqrt{k}} < 2\sqrt{(s+1/2)} - 2\sqrt{(k-1/2)} < 2\sqrt{s} - 2\sqrt{(k-1)}. \tag{8}$$

Indeed

$$\frac{1}{\sqrt{s}} + \frac{1}{\sqrt{(s-1)}} + \cdots + \frac{1}{\sqrt{k}} < \frac{2}{\sqrt{(s+1/2)} + \sqrt{(s-1/2)}} + \frac{2}{\sqrt{(s-1/2)} + \sqrt{(s-3/2)}}$$
$$+ \cdots + \frac{2}{\sqrt{(k+1/2)} + \sqrt{(k-1/2)}} = 2(\sqrt{(s+1/2)} - \sqrt{(s-1/2)})$$
$$+ 2(\sqrt{(s-1/2)} - \sqrt{(s-3/2)}) + \cdots + 2(\sqrt{(k+1/2)} - \sqrt{(k-1/2)})$$
$$= 2\sqrt{(s+1/2)} - 2\sqrt{(k-1/2)} < 2\sqrt{s} - 2\sqrt{(k-1)}.$$

The lower bound in (8) can be proved similarly.

By (8) we get

$$\frac{\pi}{2}\left(\frac{1}{\sqrt{n}} + \frac{1}{\sqrt{(n-1)}} + \cdots + \frac{1}{\sqrt{(n-v)}} + 2\sqrt{(n-v-1)}\right) > \pi\sqrt{n} - c, \tag{9}$$

where $0 \leqslant v \leqslant n - 1$.

Now we prove (4). We have

$$A_{2r}+A_{2r+1}+\cdots+A_{mr-1}=(m-2)r\frac{\pi}{2\sqrt{n}}-\frac{\pi}{2\sqrt{n}}\sum_{i=1}^{m-2}\frac{r}{e^i}$$

$$=\frac{\pi r}{2\sqrt{n}}\left(m-2-\frac{1-e^{-m+2}}{e-1}\right)$$

$$>\frac{\pi 2\sqrt{n}}{2\sqrt{n}}(m-c)>\pi m-c.$$

On the other hand by (7) and (8) we get

$$\frac{\pi}{2}\left(\frac{1}{\sqrt{n}}+\frac{1}{\sqrt{(n-1)}}+\cdots+\frac{1}{\sqrt{(n-mr-1)}}\right)<\pi(\sqrt{n}-\sqrt{(n-mr)})$$

$$=\pi\sqrt{n}\left(1-\sqrt{\left(1-\frac{mr}{n}\right)}\right)<\pi\sqrt{n}\left(1-1+\frac{1}{2}\frac{mr}{n}+\frac{1}{2}\left(\frac{mr}{n}\right)^2\right)$$

$$<\frac{\pi mr}{2\sqrt{n}}+c<\pi m+c.$$

Consequently

$$A_{2r}+A_{2r+1}+\cdots+A_{mr-1}>\frac{\pi}{2}\left(\frac{1}{\sqrt{n}}+\frac{1}{\sqrt{(n-1)}}+\cdots+\frac{1}{\sqrt{(n-mr-1)}}\right)-c. \tag{10}$$

Then using (9) and (10) we obtain for $mr\leqslant v\leqslant n-1$

$$A_{2r}+A_{2r+1}+\cdots+A_v+\frac{\pi^2}{2A_{v+1}}=A_{2r}+\cdots+A_{mr-1}+A_{mr}+\cdots+A_v+\frac{\pi^2}{2A_{v+1}}$$

$$>\frac{\pi}{2}\left(\frac{1}{\sqrt{n}}+\cdots+\frac{1}{\sqrt{(n-mr-1)}}+\frac{1}{\sqrt{(n-mr)}}\right.$$

$$\left.+\cdots+\frac{1}{\sqrt{(n-v)}}+2\sqrt{(n-v-1)}\right)$$

$$-c>\pi\sqrt{n}-c.$$

The inequality (4) is proved.

We shall prove (5) using (7). We have for $1\leqslant i\leqslant m-2$

$$\left(\frac{\pi^2}{4}-\frac{2.1}{e^{i+1}}\right)\frac{1}{A_{(2+i)r}}+\frac{\pi^2}{4A_{2mr}}\geqslant\left(\frac{\pi^2}{4}-\frac{2.1}{e^{i+1}}\right)\frac{2\sqrt{n}}{\pi\left(1-\frac{1}{e^{i+1}}\right)}+\frac{\pi}{2}\sqrt{(n-2mr)}$$

$$=\frac{\pi}{2}\sqrt{n}+\left(\frac{\pi^2}{4}-2.1\right)\frac{2\sqrt{n}}{\pi(e^{i+1}-1)}+\frac{\pi}{2}\sqrt{n}\sqrt{\left(1-\frac{2mr}{n}\right)}$$

$$> \frac{\pi}{2}\sqrt{n} + c\frac{\sqrt{n}}{e^m} + \frac{\pi\sqrt{n}}{2}\left(1 - \frac{mr}{n} - 2\left(\frac{mr}{n}\right)^2\right)$$
$$> \pi\sqrt{n} + cn^{1/2 - 1/2\pi} - c\ln n - c > \pi\sqrt{n} + \ln n,$$

where $n > 2n_0$, n_0 sufficiently large. Thus (5) is proved.

It remains to prove (6). Using (7), (8) we obtain for $0 \leqslant k \leqslant l(v)$, $mr \leqslant v \leqslant n - n_0$,

$$A_{v-k} + A_{v-k+1} + \cdots + A_{v+k}$$
$$\leqslant \frac{\pi}{2}\left(\frac{1}{\sqrt{(n-v+k)}} + \frac{1}{\sqrt{(n-v+k-1)}} + \cdots + \frac{1}{\sqrt{(n-v-k)}}\right)$$
$$\leqslant \pi(\sqrt{(n-v+k+1/2)} - \sqrt{(n-v-k-1/2)})$$
$$= \frac{\pi(2k+1)}{\sqrt{(n-v+k+1/2)} + \sqrt{(n-v-k-1/2)}}$$
$$= \frac{2(2k+1)A_v}{\sqrt{\left(1 + \frac{k+1/2}{n-v}\right)} + \sqrt{\left(1 - \frac{k+1/2}{n-v}\right)}} \leqslant \frac{(2k+1)A_v}{1 - \frac{1}{2}\frac{(k+1/2)^2}{(n-v)^2}}$$
$$\leqslant \frac{(2k+1)A_v}{1 - \frac{1}{2}\frac{(l(v)+1/2)^2}{(n-v)^2}} \leqslant \frac{(2k+1)A_v}{1 - \frac{1}{2}\frac{(\sqrt{(n-v)}\ln\sqrt{(n-v)}+1/2)^2}{(n-v)^2}} < \frac{(2k+1)A_v}{1 - \frac{\pi}{4\sqrt{(n-v)}}}$$
$$= \frac{(2k+1)A_v}{1 - A_v/2} < (2k+1)A_v(1 + A_v),$$

where $n - v > n_0$. The inequality (6) is proved. □

Next we put

$$p(x) = \prod_{j=0}^{n-1}(x + \xi_j), \quad \xi_j = \exp\{-(A_0 + A_1 + \cdots + A_j) + 2rA_0\}. \qquad (11)$$

Note that $1 = \xi_{2r-1} > \xi_{2r} > \cdots > \xi_{n-1} > 0$.

Lemma 4.7. *We have*

$$x\left|\frac{p(-x)}{p(x)}\right| < ce^{-\pi\sqrt{n}}, \quad 0 \leqslant x \leqslant 1, \qquad (12)$$
$$\left|\frac{p(-x)}{p(x)}\right| < \pi/4, \qquad \xi_{n-1} \leqslant x \leqslant 1. \qquad (13)$$

Proof. Obviously if $x = \xi_j$ then the inequalities (12) and (13) are true. In what follows we shall use the fact that $(a - x)/(a + x)$, $a > 0$, is a decreasing function and for $0 < a < x < b$ we have

$$\frac{(b-x)(x-a)}{(b+x)(x+a)} \leqslant \frac{(1-\sqrt{(a/b)})^2}{(1+\sqrt{(a/b)})^2}. \tag{14}$$

If $x \in (\xi_{v+1}, \xi_v)$, $0 \leqslant v \leqslant n-2$, then (14) gives us

$$\left|\frac{p(-x)}{p(x)}\right| < \frac{(\xi_v - x)(x - \xi_{v+1})}{(\xi_v + x)(x + \xi_{v+1})} \leqslant \left(\frac{1-\sqrt{(\xi_v/\xi_{v+1})}}{1+\sqrt{(\xi_v/\xi_{v+1})}}\right)^2$$
$$\leqslant \frac{1-\exp(-\frac{1}{2}A_{v+1})}{1+\exp(-\frac{1}{2}A_{v+1})} < A_{v+1}/2 \leqslant \pi/4.$$

The inequality (13) is proved.

Now we prove (12). If $x \in (\xi_{v+1}, \xi_v)$ then

$$x\left|\frac{p(-x)}{p(x)}\right| = x\prod_{j=0}^{v}\frac{\xi_j - x}{\xi_j + x}\prod_{j=v+1}^{n-1}\frac{x-\xi_j}{x+\xi_j}. \tag{15}$$

Let us consider the case when $(i+1)r \leqslant v < (i+2)r$, $1 \leqslant i \leqslant m-2$. By (15) it follows that

$$x\left|\frac{p(-x)}{p(x)}\right| < \prod_{j=0}^{v}\frac{\xi_j - \xi_{v+1}}{\xi_j + \xi_{v+1}}\prod_{j=v+1}^{2mr}\frac{\xi_v - \xi_j}{\xi_v + \xi_j}$$
$$= \prod_{j=0}^{v}\frac{1-\xi_{v+1}/\xi_j}{1+\xi_{v+1}/\xi_j}\prod_{j=v+1}^{2mr}\frac{1-\xi_j/\xi_v}{1+\xi_j/\xi_v}.$$

Since $A_0 \leqslant A_1 \leqslant A_2 \leqslant \cdots$, for $1 \leqslant j \leqslant v$ we have

$$\xi_{v+1}/\xi_j = \exp\{-(A_{j+1} + \cdots + A_{v+1})\} \geqslant \exp\{-(v+1-j)A_{v+1}\}$$
$$= \exp(-kA_{v+1}) > \exp\{-(k+1/2)A_{(j+2)r}\},$$

where $k = v + 1 - j$, $1 \leqslant k \leqslant v$.

Similarly if $v + 1 \leqslant j \leqslant 2mr$ then

$$\xi_j/\xi_v = \exp\{-(A_{v+1} + \cdots + A_j)\} \geqslant \exp\{-(j-v)A_j\}$$
$$> \exp\{-kA_{2mr}\} > \exp\{-(k+1/2)A_{2mr}\},$$

where $k = j - v$, $1 \leqslant v \leqslant 2mr - v$.

Consequently

$$x\left|\frac{p(-x)}{p(x)}\right| < cn\prod_{k=0}^{v}\frac{1-\exp\{-(k+1/2)A_{(j+2)r}\}}{1+\exp\{-(k+1/2)A_{(j+2)r}\}}\prod_{k=v+1}^{2mr-v}\frac{1-\exp\{-(k+\frac{1}{2})A_{2mr}\}}{1+\exp\{-(k+\frac{1}{2})A_{2mr}\}}, \tag{16}$$

where we have used the fact that

$$\frac{1+\exp\{-A_{(j+2)r}/2\}}{1-\exp\{-A_{(j+2)r}/2\}}\cdot\frac{1+\exp\{-A_{2mr}/2\}}{1-\exp\{-A_{2mr}/2\}}<\left(\frac{1+\exp\{-\frac{1}{2}A_{(j+2)r}\}}{1-\exp\{-\frac{1}{2}A_{(j+2)r}\}}\right)^2$$

$$=\frac{c}{\left(\exp\left\{\frac{1}{2}\left(1-\frac{1}{c}\right)\frac{\pi}{2\sqrt{n}}\right\}-1\right)^2}<cn.$$

In our case

$$v\geqslant(j+1)r>2\sqrt{n}\ln e^{j+1}>\frac{2\sqrt{n}}{\pi(1-e^{-j-1})}\ln e^{j+1}=\frac{\ln e^{j+1}}{A_{(j+2)r}}$$

and

$$2mr-v\geqslant mr>\frac{2}{\pi}\sqrt{n}\ln\sqrt{n}>\frac{2}{\pi}\sqrt{(n-2mr)}\ln\frac{2}{\pi}\sqrt{(n-2mr)}=\frac{1}{A_{2mr}}\ln\frac{1}{A_{2mr}}.$$

Then using lemma 4.5 on the product in (15) we obtain

$$x\left|\frac{p(-x)}{p(x)}\right|<4cn\exp\left\{-\frac{\pi^2}{4}\frac{1}{A_{(j+2)r}}+2.1\frac{1}{e^{j+1}A_{(j+2)r}}-\frac{\pi^2}{4A_{2mr}}+2.1\right\}$$

and by (5) we get

$$x\left|\frac{p(-x)}{p(x)}\right|\leqslant cn\exp\{-\pi\sqrt{n}-\ln n\}=c\exp\{-\pi\sqrt{n}\}.$$

Thus (12) is proved in the considered case.

Now let $mr\leqslant v\leqslant n-n_0$. If we put $a=\xi_{v+1+k}$ and $b=\xi_{v-k}$, $0\leqslant k\leqslant l$, in (14) we get by (15)

$$x\left|\frac{p(-x)}{p(x)}\right|<x\prod_{j=v-l}^{v}\frac{\xi_j-x}{\xi_j+x}\prod_{j=v+1}^{v+l+1}\frac{x-\xi_j}{x+\xi_j}$$

$$<x\prod_{k=0}^{l}(1-\sqrt{(\xi_{v+1+k}/\xi_{v-k})})^2(1+\sqrt{(\xi_{v+1+k}/\xi_{v-k})})^{-2}. \tag{17}$$

By (6) we have

$$\begin{aligned}\xi_{v+1+k}/\xi_{v-k}&=\exp\{-(A_{v-k+1}+\cdots+A_{v+k+1})\}\\&>\exp\{-(2k+1)A_{v+1}(1+cA_{v+1})\}\end{aligned}$$

and

$$x<\xi_v<c\exp\{-(A_{2r}+A_{2r+1}+\cdots+A_v)\}.$$

Then by (17) we get

$$\begin{aligned}x\left|\frac{p(-x)}{p(x)}\right|&<c\exp\{-(A_{2r}+A_{2r+1}+\cdots+A_v)\}\\&\quad\cdot\prod_{k=0}^{l}\frac{1-\exp\{-(k+1/2)A_{v+1}(1+cA_{v+1})\}}{1+\exp\{-(k+1/2)A_{v+1}(1+cA_{v+1})\}}.\end{aligned} \tag{18}$$

We have

$$l(v)+1>\frac{1}{A_v}\ln\frac{1}{A_v}\geqslant\frac{1}{A_{v+1}}\ln\frac{1}{A_{v+1}}>\frac{1}{A_{v+1}(1+cA_{v+1})}\ln\frac{1}{A_{v+1}(1+cA_{v+1})}.$$

Then by (18), (4), using lemma 4.5 we get

$$x\left|\frac{p(-x)}{p(x)}\right|\leqslant c\exp\{-(A_{2r}+\cdots+A_v)\}\cdot 2\exp\left\{-\frac{2\pi^2}{4}\frac{1}{A_{v+1}(1+cA_{v+1})}+2\cdot 2.1\right\}$$

$$\leqslant c\exp\left\{-\left(A_{2r}+\cdots+A_v+\frac{\pi^2}{2A_{v+1}}\right)\right\}\leqslant c\exp(-\pi\sqrt{n}).$$

Now let $0\leqslant x\leqslant\xi_{n-n_0}$. Then by (4) we obtain

$$x\left|\frac{p(-x)}{p(x)}\right|\leqslant x\leqslant\xi_{n-n_0}=\exp\{-(A_0+\cdots+A_{n-n_0})-2rA_0\}$$

$$\leqslant c\exp\left\{-\left(A_{2r}+A_{2r+1}+\cdots+A_{n-n_0}+\frac{\pi^2}{2A_{n-n_0+1}}\right)\right\}$$

$$<c\exp(-\pi\sqrt{n}).$$

□

We can now complete the proof of the upper bound in theorem 4.2. Let us consider Newman's rational function

$$r(x)=x\frac{p(x)-p(-x)}{p(x)+p(-x)},$$

where p is the polynomial from (11). Obviously r is an even function and $r\in R_{n+1}$.

Since $p(-x)\geqslant 0$ for $x\in[0,\xi_{n-1}]$, for $x\in[0,\xi_{n-1}]$ we have

$$|x-r(x)|=x-r(x)\leqslant x\leqslant\xi_{n-1}<c\exp(-\pi\sqrt{n}).$$

If $x\in[\xi_{n-1},1]$ then by (12) and (13) we get

$$|x-r(x)|=\frac{2x|p(-x)/p(x)|}{|1+p(-x)/p(x)|}<ce^{-\pi\sqrt{n}}.$$

Consequently

$$R_{n+1}(|x|)_{C[-1,1]}\leqslant ce^{-\pi\sqrt{n}}$$

which implies the upper bound in theorem 4.2. □

4.3 Zolotarjov's results

There are few examples where we can solve the problem of finding exactly the best uniform rational approximation. Some, but very important, examples were given by E.I. Zolotarjov more than a hundred years ago (see Zolotarjov (1877)).

Following N.I. Ahiezer (1965, 1970), we shall consider here four of the problems which can be solved by the technique developed by Zolotarjov. This technique involves elliptic functions.

Problem 1. *Let* $0 < k < 1$. *Find* $\tilde{r}_n \in R_n$ *such that*

$$\| \operatorname{sign} x - \tilde{r}_n(x) \|_{C(G(k))} = \inf \{ \| \operatorname{sign} x - r(x) \|_{C(G(k))} : r \in R_n \} = R_n(k), \qquad (1)$$

where $G(k) = [-1, -k] \cup [k, 1]$.

This problem leads to the rational approximation of the step function and rational uniform approximation of $|x|$, problems which are basic in the theory of rational approximation of functions.

Problem 1 can be formulated by linear transformation in the following equivalent way.

Problem 1′. *Let* $0 < k < 1$. *Find* $r_n \in R_n$ *such that*

$$\| \operatorname{sign} x - r_n(x) \|_{C(\Delta(k))} = \inf \{ \| \operatorname{sign} x - r(x) \|_{C(\Delta(k))} : r \in R_n \} = R_n(k),$$

where $\Delta(k) = [-1/k, -1] \cup [1, 1/k]$.

Problem 2. *Let* $0 < \kappa < 1$. *Let us set*

$$\tilde{R}_n = \{ r : r \in R_n, |r(x)| \geq 1 \text{ for } |x| \geq 1/\kappa \}.$$

Find $q_n \in \tilde{R}_n$ *such that*

$$\| q_n \|_{C[-1,1]} = \inf \{ \| r \|_{C[-1,1]} : r \in \tilde{R}_n \}.$$

Problem 3. *Let* $0 < \theta < 1$. *Find* $s_n \in R_n$ *such that*

$$\| 1 - \sqrt{x} s_n(x) \|_{C[\theta,1]} = \inf \{ \| 1 - \sqrt{x} r(x) \|_{C[\theta,1]} : r \in R_n \}.$$

Problem 4. *Let* $0 < k < 1$. *Find* $\bar{r}_n \in R_n$ *such that*

$$\| 1 - \sqrt{(1 - k^2 x)} \bar{r}_n(x) \|_{C[0,1]} = \inf \{ \| 1 - \sqrt{(1 - k^2 x)} r(x) \|_{C[0,1]} : r \in R_n \}.$$

Obviously problems 3 and 4 are equivalent – setting $u = 1 - k^2 x$, $\theta = 1 - k^2$ we obtain from problem 4 problem 3 and setting $1 - k^2 u = x$, $k^2 = 1 - \theta$, we obtain from problem 3 problem 4.

We shall prove now that problems 1 and 2 are also equivalent.

Theorem 4.3. *Problems 1 and 2 are equivalent.*

Proof. It will be sufficient to prove that problems 1′ and 2 are equivalent. Let us show that problem 1′ is equivalent to problem 2. Let $q_n = p/q$ be a solution of problem 2.[†] Then p has degree exactly n. If not, then for the rational function $q_n^*(x) = \kappa x p(x)/q(x)$ we have

(a) $q_n^* \in \tilde{R}_n$, since $|q_n^*(x)| \geqslant \kappa |x| |q_n(x)| \geqslant |q_n(x)| \geqslant 1$ for $|x| > 1/\kappa$,
(b) $\max_{x\in[-1,1]} |q_n^*(x)| = \kappa \max_{x\in[-1,1]} |q_n(x)| < \| q_n \|_{C[-1,1]}$,

which contradicts the definition of q_n.

It is also evident that we must have

$$\min\{|q_n(x)|: |x| \geqslant 1/\kappa\} = 1.$$

Now let $m = \max_{x\in[-1,1]}|q_n(x)|$. Obviously $0 < m < 1$.

Let us consider the rational function $r_n \in R_n$ given parametrically by

$$\left.\begin{aligned} r_n(x) &= \frac{(1-m)(q_n(t)-\sqrt{m})}{(1+m)(q_n(t)+\sqrt{m})}, \\ x &= \frac{(1+\sqrt{\kappa})(t\sqrt{\kappa}-1)}{(1-\sqrt{\kappa})(t\sqrt{\kappa}+1)}, \end{aligned}\right\} \tag{2}$$

and let us set $k = (1-\sqrt{\kappa})^2/(1+\sqrt{\kappa})^2$.

We have, when $-1 \leqslant t \leqslant 1$, then $-1/k \leqslant x \leqslant -1$, and when $|t| \geqslant 1/\kappa$, then $1 \leqslant x \leqslant 1/k$. On the other hand

$$r_n(x) + 1 = \frac{2(q_n(t) + m\sqrt{m})}{(1+m)(q_n(t)+\sqrt{m})}$$

and, since $\| q_n \|_{C[-1,1]} = m$,

$$\max_{x\in[-1/k,-1]} |r_n(x) + 1| = \max_{t\in[-1,1]} \left| \frac{2(q_n(t) + m\sqrt{m})}{(1+m)(q_n(t)+\sqrt{m})} \right| = \frac{2\sqrt{m}}{1+m}. \tag{3}$$

Since $\min_{|t|\geqslant 1/\kappa} |q_n(t)| = 1$, we have by simple calculations

$$\max_{x\in[1,1/k]} |r_n(x) - 1| = \max_{|t|\geqslant 1/\kappa} \left| \frac{-2(mq_n(t) + \sqrt{m})}{(1+m)(q_n(t)+\sqrt{m})} \right| = \frac{2\sqrt{m}}{1+m}. \tag{4}$$

From (3) and (4) it follows that

$$\max_{x\in\Delta(k)} |\operatorname{sign} x - r_n(x)| = \frac{2\sqrt{m}}{1+m} = \mu(m). \tag{5}$$

Let us remark now that $\mu(m)$ is a monotone increasing function for $m \in [0, 1]$. From here and (5) it follows that r_n is a solution of problem 1′. Indeed, if we assume that there exists another $\bar{r}_n \in R_n$ such that

$$\max_{x\in\Delta(k)} |\operatorname{sign} x - \bar{r}_n(x)| = \mu(m_1) < \mu(m), \quad m_1 < m. \tag{6}$$

[†] Problems 1–4 have solutions – the proof is the same as in theorem 2.1.

Then, determinating $\bar{q}_n$ by means of $\bar{r}_n$ and m_1 instead of r_n and m by the converse formulas of (2), we obtain $\bar{q}_n \in \tilde{R}_n$ with $\|\bar{q}_n\|_{C[-1,1]} = m_1 < m$, which contradicts the definition of q_n (this can be seen using the first equality in (4) for $\bar{r}_n, \bar{q}_n, m_1$). So r_n is a solution of problem 1'.

We have shown that if q_n is a solution of problem 2 then r_n given by (2) is a solution of problem 1'. In the same way it is evident that if r_n is a solution of problem 1' then q_n given by (2) is a solution of problem 2. The theorem is proved. □

We shall give also a connection between problems 1 and 3, i.e. between all four problems. But before doing this let us consider some of the properties of the solutions of these problems.

Theorem 4.4. *If $\tilde{r}_n \in R_n$ is a solution of problem 1 then $\tilde{r}_n$ realizes an alternation in $[-1,-k] \cup [k,1]$, more exactly there exist $2n+2$ points $x_1 < x_2 < \cdots < x_{2n+2}$, $x_i \in [-1,-k] \cup [k,1]$, $i = 1, \ldots, 2n+2$, such that*

$$\operatorname{sign} x_i - \tilde{r}_n(x_i) = \varepsilon(-1)^i \|\operatorname{sign} x - \tilde{r}_n(x)\|_{C(G(k))} = \varepsilon(-1)^i \mu_n, \tag{7}$$

where $\varepsilon = \pm 1$, $G(k) = [-1,-k] \cup [k,1]$, $\|\operatorname{sign} x - \tilde{r}_n\|_{C(G(k))} = \mu_n$.

The proof of this lemma is exactly the same as the proof of the second part of the characterization theorem for rational uniform approximation (theorem 2.2).

From this characterization theorem we derive the following.

Corollary 4.1. *If $\tilde{r}_n = \varphi/\psi \in R_n$ is a solution of problem 1, then the rational function $(1-\mu_n)\psi/\varphi \in R_n$ is also a solution of problem 1.*

Corollary 4.2. *Let us consider the equation*

$$(\tilde{r}_n(x) - \operatorname{sign} x)^2 = \mu_n^2. \tag{8}$$

The points $-1, -k, k, 1$ are simple zeros of (8). In $(-1,-k) \cup (k,1)$ (8) has $2n-2$ double zeros.

Corollary 4.3. *Let $\tilde{r}_n = \varphi/\psi \in R_n$ be a solution of problem 1. Then in the interval $(-k,k)$ only one of the polynomials φ and ψ can have zeros.*

In fact, if it is not so, the function $|\tilde{r}_n|$ must take the values $1-\mu_n$, $1+\mu_n$ in the interval $(-k,k)$ so the number of zeros of (8) in $G(k)$ can not be $2n+2$.

Corollary 4.4. *There exists one unique solution of problem 1 which is bounded in $[-k,k]$.*

Indeed, from corollaries 4.1 and 4.3 it follows that there exists a solution bounded in $[-k,k]$. The uniqueness follows from the alternation theorem 4.4 similarly to the uniqueness in theorem 2.2.

Corollary 4.5. *The bounded solution in $[-k,k]$ of problem 1 is odd.*

In fact, if $\tilde{r}_n = \varphi/\psi$ is a bounded solution of problem 1, then $-\tilde{r}_n(-x) =$

$-\varphi(-x)/\psi(-x)$ is also a bounded solution of problem 1, and by corollary 4.4 we must have $\tilde{r}_n(x) = -\tilde{r}_n(-x)$, i.e. $\tilde{r}_n$ is odd.

For characterization of the solutions of problems 3 and 4 we shall use the following modification of theorem 2.2 (see Ahiezer (1965)).

Theorem 4.5. *Let $[a, b]$ be a finite closed interval, $f \in C[a, b]$, $s \in C[a, b]$, $s(x) > 0$ for $x \in [a, b]$. Let us consider*

$$R_n(f; s) = \inf\{\| f - sr \|_{C[a,b]}: r \in R_n\}.$$

There exists a unique $r^* \in R_n$ such that

$$\| f - sr^* \|_{C[a,b]} = R_n(f; s).$$

Moreover r^* realizes an alternation for f, i.e. there exist $N = 2n + 2 - d(r^*)$ points x_i, $i = 1, \ldots, N$, $a \leqslant x_1 < \cdots < x_N \leqslant b$, $d(r^*)$ the defect of r^* (see chapter 2), *such that*

$$f(x_i) - s(x_i)r^*(x_i) = \varepsilon(-1)^i \| f - sr^* \|_{C[a,b]}, \quad i = 1, \ldots, N,$$

where $\varepsilon = \pm 1$.

The proof of theorem 4.5 is similar to the proof of theorem 2.2.

Now let us see the connection between problems 1 and 3. Let n be odd, $n = 2m + 1$, and let $\tilde{r}_n \in R_n$ be the bounded solution in $[-k, k]$ of problem 1. Since $\tilde{r}_n$ is odd, $\tilde{r}_n$ must have the representation $\tilde{r}_n(x) = x\varphi(x^2)/\psi(x^2)$, where $\varphi \in R_m$ and $\psi \in R_m$. By theorem 4.4 the difference $\operatorname{sign} x - \tilde{r}_n(x)$ must alternate $4m + 2 + 2$ times at least in $[-1, -k] \cup [k, 1]$, and since $\tilde{r}_n$ is odd this difference alternates $2m + 2$ times in $[k, 1]$. Setting $x = \sqrt{u}$ we see that the difference $1 - \sqrt{u}\,\varphi(u)/\psi(u)$ alternates $2m + 2$ times at least in $[k^2, 1]$ and therefore by theorem 4.5 the rational function $\varphi(u)/\psi(u) \in R_m$ is the solution of problem 3. We have obtained the following.

Theorem 4.6. *Problems 1 and 2 for rational functions of degree $2m + 1$ are equivalent to problems 3 and 4 for rational functions of degree m.*

Now we shall give the solution of problems 1 and 4. This solution uses elliptic functions, so we shall give some facts from the theory of elliptic functions.

The Jacobian elliptic function $\xi = \operatorname{sn}(u; k)$ is defined by the elliptic integral

$$u = \int_0^{\xi} \frac{dt}{\sqrt{((1 - t^2)(1 - k^2 t^2))}}, \quad 0 < k < 1,$$

where k is called a modulus of the elliptic function.

The Jacobian elliptic function $\operatorname{dn}(u; k)$ is given by

$$\operatorname{dn}(u; k) = \sqrt{(1 - k^2 \operatorname{sn}^2(u; k))}; \quad d(0; k) = 1.$$

As usual we set

$$K = \int_0^1 \frac{\mathrm{d}t}{\sqrt{((1-t^2)(1-k^2t^2))}}.$$

K is called the complete elliptic integral of the first kind (for the modulus k).

Let us set $x = \operatorname{sn}^2(u;k)$. When u changes from 0 to K, x changes from 0 to 1.

The function $x = \operatorname{sn}^2(u;k)$ is now defined for $u \in [0, k]$. If we set $\operatorname{sn}(u;k) = -\operatorname{sn}(-u;k)$ for $u \in [-K, 0]$, we can consider $x = \operatorname{sn}^2(u;k)$ as a $2K$-periodic function if we set $\operatorname{sn}^2(u + 2K; k) = \operatorname{sn}^2(u;k)$.

In order to give the solution of problem 4, we need some notation.

Let us set

$$y = \frac{2}{1 + \sqrt{(1-\lambda^2)}} \operatorname{dn}\left(\frac{u}{M}; \lambda\right), \tag{9}$$

where

$$\left.\begin{aligned} \lambda &= k^{2m+1} \prod_{r=1}^{m} c_{2r-1}^2, \\ M &= \prod_{r=1}^{m} c_{2r-1}/c_{2r}, \\ c_r &= \operatorname{sn}^2\left(\frac{rK}{2m+1}; k\right), \quad r = 0, \ldots, 2m+1. \end{aligned}\right\} \tag{10}$$

From the theory of elliptic functions it follows that the complete elliptic integral of the first kind for the modulus λ is

$$L = \frac{K}{(2m+1)M}. \tag{11}$$

Using one more fact from the theory of elliptic functions – the equality

$$\operatorname{dn}\left(\frac{u}{M}; \lambda\right) = \operatorname{dn}(u;k) \prod_{r=1}^{m} \frac{1 - k^2 c_{2r-1} \operatorname{sn}^2(u;k)}{1 - k^2 c_{2r} \operatorname{sn}^2(u;k)}$$

where λ, M, c_r are given by (10) (see Ahiezer (1970)) – we can write (9) as

$$y(x) = \frac{2}{1 + \sqrt{(1-\lambda^2)}} \sqrt{(1 - k^2 x)} \prod_{r=1}^{m} \frac{1 - k^2 c_{2r-1} x}{1 - k^2 c_{2r} x}. \tag{12}$$

If u changes from 0 to $K = (2m+1)ML$, x will change from 0 to 1, and $\operatorname{dn}(u/M; \lambda)$ will be between $\operatorname{dn}(0;\lambda) = 1$ and $\operatorname{dn}((2m+1)L; \lambda)$, which equals $\sqrt{(1-\lambda^2)}$ since $\operatorname{dn}(L;\lambda) = \sqrt{(1-\lambda^2)}$ (see the definition of $\operatorname{dn}(u;k)$) and $\operatorname{dn}(u;\lambda)$ has period $2L$. From here and (9) it follows that when $x \in [0, 1]$ for $y(x)$ we have

$$1 - \mu = \frac{2\sqrt{(1-\lambda^2)}}{1 + \sqrt{(1-\lambda^2)}} \leqslant y(x) \leqslant \frac{2}{1 + \sqrt{(1-\lambda^2)}} = 1 + \frac{1 - \sqrt{(1-\lambda^2)}}{1 + \sqrt{(1-\lambda^2)}} = 1 + \mu. \tag{13}$$

On the other hand in the points $u = 0, 2ML, 4ML, \ldots, 2mML$, $y(x) = y(x(u))$ has the value $1 + \mu$, since

$$y(x(2sML)) = (1+\mu)\mathrm{dn}\left(\frac{2sML}{M}; \lambda\right) = (1+\mu)\,\mathrm{dn}\,(2sL; \lambda) = 1 + \mu,$$
$$i = 0, \ldots, m,$$

and at the points $u = ML, 3ML, \ldots, (2m+1)ML$, $y(x) = y(x(u))$ has the value $1 - \mu$, since y in these points is equal to $(1+\mu)\,\mathrm{dn}\,((2s+1)L; \lambda) = (1+\mu)\sqrt{(1-\lambda^2)} = 1 - \mu$.

We have from (10) and (11)

$$c_r = \mathrm{sn}^2\left(\frac{rK}{2m+1}; k\right) = \mathrm{sn}^2(rML; k).$$

This shows that at the points $x_r = c_r$, $r = 0, \ldots, 2m+1$, $0 \leqslant x_0 < x_1 < \cdots < x_{2m+1} \leqslant 1$, the function $y(x)$ alternately has values $1 + \mu$ and $1 - \mu$. Together with (13) and (12) this gives

$$1 - \sqrt{(1 - k^2 x_r)}\bar{r}_m(x_r) = (-1)^{r+1} \| 1 - \sqrt{(1 - k^2 x)}\bar{r}_m(x) \|_{C[0,1]},$$
$$r = 0, \ldots, 2m+1, \qquad (14)$$

$$\bar{r}_m(x) = \frac{2}{1 + \sqrt{(1-\lambda^2)}} \prod_{r=1}^{m} \frac{1 - k^2 c_{2r-1} x}{1 - k^2 c_{2r} x}. \qquad (15)$$

From (15) we obtain the result that $\bar{r}_m \in R_m$. Therefore (14) and theorem 4.5 give us that $\bar{r}_m$ is the solution of problem 4 for $n = m$.

Theorem 4.7. *The solution $\bar{r}_m$ of problem 4 ($n = m$) is given by* (15), *where λ, c_r are given by* (10).

Now we shall give the solution of problem 1.

Theorem 4.8 (Zolotarjov, 1877). *The solution of problem 1 is given by*

$$\left.\begin{aligned} \tilde{r}_n(x) &= (1 - l)\,\mathrm{sn}\left(\frac{u}{M}; \lambda\right), \\ x &= k\,\mathrm{sn}\,(u; k), \\ \lambda &= \frac{1-l}{1+l}, \quad L' = \frac{K'}{nM}, \quad L = \frac{K}{M}, \end{aligned}\right\} \qquad (16)$$

where K, L are the complete elliptic integrals of the first kind for the modulus k and λ respectively, and K', L' are the complete elliptic integrals for the complementary moduli k', λ', and $l = R_n(k)$ (see (1)).

Remark. Let us remember that the complementary moduli for k, λ are given by

$$k^2 + k'^2 = 1, \quad \lambda^2 + \lambda'^2 = 1.$$

Proof of theorem 4.8. We shall show that $\tilde{r}_n \in R_n$ and the difference $\operatorname{sign} x - \tilde{r}_n(x)$ alternates $2n+2$ times on $G(k) = [-1, -k] \cup [k, 1]$.

We shall use the following formula from the theory of elliptic functions (see Ahiezer (1970)):

$$\operatorname{sn}\left(\frac{u}{M}; \lambda\right) = \frac{\operatorname{sn}(u;k)}{M} \prod_{r=1}^{[n/2]} \frac{1 + \dfrac{\operatorname{sn}^2(u;k)}{c_{2r}}}{1 + \dfrac{\operatorname{sn}^2(u;k)}{c_{2r-1}}}, \tag{17}$$

where

$$c_r = \frac{\operatorname{sn}^2(rK'/n; k')}{1 - \operatorname{sn}^2(rK'/n; k')}. \tag{18}$$

Since $1 - l = 2\lambda/(1+\lambda)$, we obtain from (16) and (17)

$$\tilde{r}_n(x) = \frac{p(x)}{q(x)} = \frac{2\lambda}{1+\lambda} \frac{x}{kM} \prod_{r=0}^{[n/2]} \frac{1 + x^2/k^2 c_{2r}}{1 + x^2/k^2 c_{2r-1}}.$$

From the properties of $\operatorname{sn}(u;k')$ (see below) we have $\operatorname{sn}(K';k') = 1$, therefore $1/c_n = 0$. From here and (17), (18) we obtain that, if n is odd, then $p \in P_n$, $q \in P_{n-1}$, and, if n is even, then $p \in P_{n-1}$, $q \in P_n$. Therefore for all n we have $\tilde{r}_n \in R_n$.

Now we shall prove that the difference $\operatorname{sign} x - \tilde{r}_n(x)$ alternates at least $2n+2$ times in $G(k)$ (it alternates exactly $2n+2$ times). To show this we need some more properties of $\operatorname{sn}(u;k)$. From the theory of Jacobian elliptic functions (see Ahiezer (1970)) we have that $\operatorname{sn}(u;k)$ has two periods: $4K$ and $2iK'$. We shall need the following values of $\operatorname{sn}(u;k)$:

$$\operatorname{sn}(0;k) = \operatorname{sn}(2K;k) = 0, \quad \operatorname{sn}(K;k) = 1,$$
$$\operatorname{sn}(K + iK';k) = 1/k, \quad \operatorname{sn}(K';k') = 1.$$

And evidently, since L is the complete elliptic integral of the first kind for modulus λ, $4L$ and $2iL'$ are the periods of $\operatorname{sn}(u;\lambda)$ and $\operatorname{sn}(0;\lambda) = \operatorname{sn}(2L;\lambda) = 0$;

$$\operatorname{sn}(L;\lambda) = 1, \quad \operatorname{sn}(L + iL';\lambda) = 1/\lambda.$$

Since $x = k\operatorname{sn}(u;k)$ we obtain from here that when v changes in $[0, K']$, $u = K + iv$, x changes in $[k, 1]$. We have, if we set $w = v/M$,

$$\tilde{r}_n(x) = (1-l)\operatorname{sn}\left(\frac{u}{M};\lambda\right) = (1-l)\operatorname{sn}\left(\frac{K}{M} + i\frac{v}{M};\lambda\right) = (1-l)\operatorname{sn}(L + iw;\lambda).$$

Since λ is the modulus for L, $\tilde{r}_n$ changes in the interval $[1-l, 1+l]$ and takes alternately the values $1-l$ and $1+l$ at $n+1$ points $w = 0$, L', $2L', \ldots, nL'$, or, what is the same, in the points

$$v = 0, \quad \frac{K'}{n}, \quad \frac{2K'}{n}, \ldots, \frac{nK'}{n}.$$

The analogous situation is in the interval $[-1, -k]$, so $\tilde{r}_n$ alternates $2n+2$ times in $G(k)$. So $\tilde{r}_n$ given by (16) is the solution of problem 1. □

Let us mention that from (16), using the asymptotics developments of the elliptic functions, it is possible by means of long calculations to obtain the Newman result for $|x|$, or the Gonchar result for sign x:

$$R_n(\operatorname{sign} x)_{C(G(k))} = O(e^{-cn/\ln(1/k)})$$

which gives $R_n(|x|)_{C[-1,1]} = O(e^{-c\sqrt{n}})$.

But nobody before 1964, when Newman's result appeared, had made these calculations. So in our opinion the exact solution of problem 1 has not the same value for the development of the quantitative theory of rational approximations as Newman's result.

4.4 Uniform approximation of e^x on $[-1, 1]$: Meinardus conjecture

In this section we shall consider the problem of uniform rational approximation of the function e^x on the interval $[-1, 1]$. We shall consider best rational approximation of order (n, m) (see chapter 2). The Meinardus conjecture for this approximation (see Meinardus (1967)) was proved by D. Braess (1984). We shall give here his elegant proof.

We shall use the notation

$$R_{nm}(e^x) \equiv R_{nm}(e^x)_{C[-1,1]} = \inf\{\|f - r\|_{C[-1,1]} : r \in R_{nm}\}.$$

Theorem 4.9. *We have*

$$R_{nm}(e^x) = \frac{2^{-m-n} n!\, m!}{(n+m)!\,(n+m+1)!}(1 + o(1))^{\dagger} \tag{1}$$

as $n + m \to \infty$.

Proof. The crucial point in the proof is Newman's trick (see D. Newman (1979b)). It gives a connection between the rational approximations on the interval and on the circle. Let the rational function $p/q \in R_{nm}$. Given $x \in [-1, 1]$, put $z = (x + iy)/2$, where $x^2 + y^2 = 1$. Then

$$r(x) = \frac{p(z)p(\bar{z})}{q(z)q(\bar{z})} \tag{2}$$

is again a rational function and $r \in R_{nm}$, i.e. the degree is not doubled when products of this special form are taken. In order to understand this let us consider the product of two linear expressions. We have

$$(az + b)(a\bar{z} + b) = ab(z + \bar{z}) + a^2 z\bar{z} + b^2 = abx + a^2/4 + b^2$$

† $o(1) \xrightarrow[n+m\to\infty]{} 0$.

if $|z| = 1/2$.

If α and β are complex numbers, then

$$\alpha\bar{\alpha} - \beta\bar{\beta} = 2\,\mathrm{Re}\,\{\bar{\alpha}(\alpha - \beta)\} - |\alpha - \beta|^2.$$

By applying these equalities to the product $e^x = e^z e^{\bar{z}}$ we obtain

$$e^x - r(x) = 2\,\mathrm{Re}\left\{e^{\bar{z}}\left(e^z - \frac{p(z)}{q(z)}\right)\right\} - \left|e^z - \frac{p(z)}{q(z)}\right|^2. \tag{3}$$

Let us assume that $qe^z - p$ has $m + n + 1$ zeros in the disk $|z| < \frac{1}{2}$, counted with their multiplicities, but q has none. Then

$$\begin{aligned} 2\min_{|z|=1/2}\left|e^{\bar{z}}\left(e^z - \frac{p}{q}\right)\right| - \max_{|z|=1/2}\left|e^z - \frac{p}{q}\right|^2 &\leqslant R_{nm}(e^x) \\ \leqslant 2\max_{|z|=1/2}\left|e^{\bar{z}}\left(e^z - \frac{p}{q}\right)\right| + \max_{|z|=1/2}\left|e^z - \frac{p}{q}\right|^2&. \end{aligned} \tag{4}$$

Indeed, the upper estimate is obvious from (3). The lower bound will be derived by using de la Vallée-Poussin's theorem 2.3. As usual we denote by $\arg w$ the argument of the complex number w. Note that

$$\mathrm{Re}\left\{e^{\bar{z}}\left(e^z - \frac{p}{q}\right)\right\} = \begin{cases} \left|e^{\bar{z}}\left(e^z - \frac{p}{q}\right)\right|, \arg\left\{e^{\bar{z}}\left(e^z - \frac{p}{q}\right)\right\} \equiv 0 \pmod{2\pi}, \\ -\left|e^{\bar{z}}\left(e^z - \frac{p}{q}\right)\right|, \arg\left\{e^{\bar{z}}\left(e^z - \frac{p}{q}\right)\right\} \equiv \pi \pmod{2\pi}. \end{cases} \tag{5}$$

Let us denote

$$h(z) = \frac{e^{-z}}{q(z)}(q(z)e^z - p(z)).$$

By assumption h has $n + m + 1$ zeros in $|z| < 1/2$ (every zero counted with its multiplicity) and h has no poles. Consequently h has winding number $n + m + 1$ for the circle $|z| = 1/2$. Hence, when an entire circuit has been completed,

$$\arg\{h(z)\} = \arg\left\{e^{\bar{z}}\left(e^z - \frac{p(z)}{q(z)}\right)\right\}$$

is increased by $(n + m + 1)\cdot 2\pi$. The argument is increased by $(n + m + 1)\pi$ as z traverses the upper half of the circle, since $h(x)$ is real for x on the real line.

Thus h has real values on $n + m + 2$ points $z_k = (x_k + iy_k)/2$ with $1 = x_0 > x_1 > \cdots > x_{n+m+1} = -1$, $y_k \geqslant 0$, $k = 0, \ldots, n + m + 1$, and the sign changes between each pair of consecutive points. The same is true for $e^{\bar{z}}(e^z - p/q)$.

Then by (3) and (5) it follows that

$$e^{x_k} - r(x_k) = (-1)^k 2\left|e^{\bar{z}_k}\left(e^{z_k} - \frac{p(z_k)}{q(z_k)}\right)\right| - \left|e^{z_k} - \frac{p(z_k)}{q(z_k)}\right|^2,$$

$$k = 0, \ldots, n + m + 1.$$

From de la Vallée-Poussin's theorem 2.3 we get

$$R_{nm}(e^x) \geqslant \min_{|z|=1/2} 2\left|e^{\bar{z}}\left(e^z - \frac{p(z)}{q(z)}\right)\right| - \min_{|z|=1/2}\left|e^z - \frac{p(z)}{q(z)}\right|^2,$$

The lower bound in (4) is proved.

In order to apply the estimates (4) we need a rational approximation to e^z on the circle $|z| = 1/2$. We shall use the (n, m)-Padé-approximation (compare with chapter 12). We shall use the polynomials (see O. Perron (1957))

$$\left.\begin{aligned} p(z) &= \int_0^\infty t^m(t+z)^n e^{-t}\,dt, \\ q(z) &= \int_0^\infty t^n(t-z)^m e^{-t}\,dt. \end{aligned}\right\} \tag{6}$$

We shall show that the polynomials from (6) give the corresponding Padé-approximation. We have $p/q \in R_{nm}$, $q(0) = (m+n)! \neq 0$. Let us consider the remainder term:

$$\begin{aligned} e^z q(z) - p(z) &= \int_0^\infty t^n(t-z)^m e^{z-t}\,dt - \int_0^\infty t^m(t+z)^n e^{-t}\,dt \\ &= \int_0^\infty (t-z)^m t^n e^{z-t}\,dt - \int_z^\infty (t-z)^m t^n e^{z-t}\,dt \\ &= \int_0^z (t-z)^m t^n e^{z-t}\,dt \\ &= z^{n+m+1}\int_0^1 (u-1)^m u^n e^{(1-u)z}\,du \\ &= (-1)^m z^{n+m+1}\int_0^1 u^m(1-u)^n e^{uz}\,du, \end{aligned}$$

i.e.

$$e^z q(z) - p(z) = (-1)^m z^{n+m+1}\int_0^1 u^m(1-u)^n e^{uz}\,du. \tag{7}$$

Since the integral in (7) is bounded for $|z| \leqslant 1$, we have

$$e^z q(z) - p(z) = O(z^{n+m+1})$$

and (6) provides the (n,m)-Padé-approximation for e^z (compare with the definition in chapter 12).

In order to estimate the integral in (7) we note that

$$|e^{uz-u_0z}-1-z(u-u_0)|\leqslant \tfrac{1}{2}(u-u_0)^2|z|^2e^{|z|} \tag{8}$$

whenever $|u-u_0|\leqslant 1$. This estimate follows by the Taylor's series for the function e^z. By choosing $u_0=(m+1)/(n+m+2)$ we get

$$\int_0^1 u^m(1-u)^n\,du=\frac{m!n!}{(n+m+1)!},$$

$$\int_0^1 u^m(1-u)^n(u-u_0)\,du=0,$$

and

$$\int_0^1 u^m(1-u)^n(u-u_0)^2\,du=\frac{(n+1)!}{n+m+2}\,\frac{(m+1)!}{(n+m+3)!}.$$

Then by (8) we get

$$e^zq(z)-p(z)=\frac{(-1)^m n!m!}{(n+m+1)!}z^{n+m+1}e^{(m+1)z/(n+m+2)}(1+o(1)) \tag{9}$$

as $n+m\to\infty$, $|z|\leqslant 1$.

Now we shall estimate $q(z)$. By (6) we have

$$q(-z)=\int_0^\infty (t+z)^m t^n e^{-t}\,dt=\sum_{k=0}^m\int_0^\infty\binom{m}{k}z^k t^{n+m-k}e^{-t}\,dt$$

$$=\sum_{k=0}^m\binom{m}{k}(n+m-k)!\,z^k=(n+m)!\sum_{k=0}^m\frac{(m-k+1)_k}{(n+m-k+1)_k}\frac{z^k}{k!}$$

where the Pochhammer symbol

$$(a)_k\equiv a(a+1)\cdots(a+k-1),\quad (a)_0=1,$$

appears.

Evidently for $k=2,3,\ldots,m$ we have

$$\frac{m^k}{(n+m)^k}\geqslant\frac{(m-k+1)_k}{(n+m-k+1)_k}=\frac{(m-k+1)(m-k+2)\cdots m}{(n+m-k+1)\cdots(n+m)}$$

$$>\frac{m^k}{(n+m)^k}\left(1-\frac{1}{m}\right)\left(1-\frac{2}{m}\right)\cdots\left(1-\frac{k-1}{m}\right)$$

$$>\frac{m^k}{(n+m)^k}\left(1-\sum_{i=1}^{k-1}\frac{i}{m}\right)=\frac{m^k}{(n+m)^k}\left(1-\frac{k(k-1)}{2m}\right).$$

We have used the inequality $\prod_{i=1}^s(1-\varepsilon_i)>1-\sum_{i=1}^s\varepsilon_i$, $0<\varepsilon_i<1$.

Therefore for each $k = 2, \ldots, m$ there is a θ_k, $0 \leqslant \theta_k \leqslant 1$, such that

$$\frac{(m-k+1)_k}{(n+m-k+1)_k} = \frac{m^k}{(n+m)^k}\left(1 - \theta_k \frac{k(k-1)}{2m}\right).$$

Using these equalities we have

$$\begin{aligned} q(-z) &= (n+m)!\left\{\sum_{k=0}^{m} \frac{1}{k!}\left(\frac{mz}{n+m}\right)^k - \sum_{k=2}^{m} \theta_k \frac{k(k-1)}{k!2m}\left(\frac{mz}{n+m}\right)^k\right\} \\ &= (n+m)!\left\{\sum_{k=0}^{m} \frac{1}{k!}\left(\frac{mz}{n+m}\right)^k - \frac{mz^2}{2(n+m)^2}\sum_{k=2}^{m} \frac{\theta_k}{(k-2)!}\left(\frac{mz}{n+m}\right)^{k-2}\right\}, \end{aligned} \tag{10}$$

where the second sum disappears when $m = 1$.

By (10) we get

$$\begin{aligned} &q(-z) - (n+m)!\mathrm{e}^{-mz/(n+m)} \\ &= (n+m)!\left\{-\frac{mz^2}{2(n+m)^2}\sum_{k=2}^{m} \frac{\theta_k}{(k-2)!}\left(\frac{mz}{n+m}\right)^{k-2} - \sum_{k=m+1}^{\infty} \frac{1}{k!}\left(\frac{mz}{n+m}\right)^k\right\}. \end{aligned}$$

Obviously for $k \geqslant m+1$ we have

$$\begin{aligned} \frac{1}{k!}\left(\frac{mz}{n+m}\right)^k &= \frac{mz^2}{2(n+m)^2}\,\frac{2m}{k(k-1)}\,\frac{1}{(k-2)!}\left(\frac{mz}{n+m}\right)^{k-2} \\ &= \frac{mz^2}{2(n+m)^2}\,\frac{\theta_k}{(k-2)!}\left(\frac{mz}{n+m}\right)^{k-2}, \quad 0 \leqslant \theta_k \leqslant 1. \end{aligned}$$

Hence

$$q(-z) - (n+m)!\mathrm{e}^{mz/(n+m)} = -(n+m)!\sum_{k=2}^{\infty} \frac{\theta_k}{(k-2)!}\left(\frac{mz}{n+m}\right)^{k-2}\frac{mz^2}{2(n+m)^2}$$

and therefore

$$|q(-z) - (n+m)!\,\mathrm{e}^{mz/(n+m)}| \leqslant (n+m)!\frac{|z|^2}{2(n+m)}\mathrm{e}^{|z|}.$$

This estimate implies

$$q(z) = (n+m)!\,\mathrm{e}^{-mz/(n+m)}(1 + \mathrm{o}(1)) \tag{11}$$

as $n+m \to \infty, |z| \leqslant 1$.

From (9) and (11) it follows that

$$\mathrm{e}^z - \frac{p(z)}{q(z)} = \frac{(-1)^m n!m!}{(m+n)!(m+n+1)!}z^{n+m+1}\mathrm{e}^{\alpha z}(1 + \mathrm{o}(1)) \tag{12}$$

as $n+m \to \infty$; $|z| \leqslant 1$, where $\alpha = 2m/(n+m)$.

From (12) we obtain

$$e^z\left(e^z - \frac{p(z)}{q(z)}\right) = \frac{(-1)^m n!m!}{(n+m)!(n+m+1)!} z^{n+m+1} e^{\beta z}(1 + o(1)), \tag{13}$$

with $\beta = 1 + 2m/(n+m)$, $1 \leqslant \beta \leqslant 3$.

The modulus of this expression is not constant on the circle $|z| = 1/2$ mainly because $e^{\beta z} z^{n+m+1}$ is not constant. We shall see that by choosing z_0 appropriately one can make $|e^{\beta z}(z - z_0)^{n+m+1}|$ deviate very little from a constant on this circle.

From the Taylor series for the logarithmic function we get

$$e^{-3/2N} \leqslant \left| e^{\beta z}\left(1 - \frac{\beta z}{N}\right)^N \right| \leqslant e^{3/2N}, \quad |\beta z| \leqslant \tfrac{3}{2}, \tag{14}$$

$N \geqslant 6$. Indeed, if $|w| \leqslant 3/2$ and $N \geqslant 6$ we have

$$e^w\left(1 - \frac{w}{N}\right)^N = e^{w + N\ln(1 - w/N)} = \exp\left\{w - N\sum_{k=1}^{\infty} \frac{1}{k}\left(\frac{w}{N}\right)^k\right\}$$

$$= \exp\left\{-\frac{w^2}{2N}\left(1 + \frac{2}{3}\frac{w}{N} + \frac{2}{4}\left(\frac{w}{N}\right)^2 + \cdots\right)\right\},$$

where $\ln z = -\int_z^1 (du/u)$ and the path of integration does not pass through the origin and does not cross the negative real axis.

Then the inequalities in (14) follow by

$$e^{-|z|} \leqslant |e^z| \leqslant e^{|z|}$$

and

$$\left|\frac{w^2}{2N}\left(1 + \frac{2}{3}\frac{w}{N} + \frac{2}{4}\left(\frac{w}{N}\right)^2 + \cdots\right)\right| \leqslant \frac{|w|^2}{2N}\left(1 + \frac{|w|}{N} + \frac{|w|^2}{N^2} + \cdots\right)$$

$$= |w|^2\left(2N\left(1 - \frac{|w|}{N}\right)\right)^{-1}$$

$$\leqslant \frac{(3/2)^2}{2N\left(1 - \frac{3/2}{6}\right)} \leqslant \frac{3}{2N}.$$

From (14) it follows for $|z|^2 = z\bar{z} = 1/4$ that

$$e^{-3/2N} \leqslant 2^N \left| e^{\beta z}\left(\bar{z} - \frac{\beta}{4N}\right)^N \right| \leqslant e^{3/2N}, \quad |z| = 1/2, \quad |\beta| \leqslant 3, \quad N \geqslant 6.$$

Therefore

$$e^{-3/2N} \leqslant 2^N \left| e^{\beta z}\left(z - \frac{\beta}{4N}\right)^N \right| \leqslant e^{3/2N}, \quad |z| = 1/2, \quad |\beta| \leqslant 3, \quad N \geqslant 6.$$

Consequently if we put $N = n + m + 1$ and $z_0 = \beta/4(n + m + 1)$ we get:

$$|e^{\beta z}(z - z_0)^{n+m+1}| = \frac{1}{2^{n+m+1}}(1 + o(1)), \quad |z| = 1/2. \tag{15}$$

Let z_0 be as above. Then

$$\frac{\tilde{p}(z)}{\tilde{q}(z)} = e^{z_0}\frac{p(z - z_0)}{q(z - z_0)}$$

is Padé-approximant to e^z at the point z_0.

From (13), replacing z by $z - z_0$ we get

$$\begin{aligned} e^z\left(e^z - \frac{\tilde{p}(z)}{\tilde{q}(z)}\right) &= \frac{(-1)^m m!n!}{(n+m)!(n+m+1)!}(z - z_0)^{n+m+1}e^{\beta z + z_0 - \beta z_0}(1 + o(1)) \\ &= \frac{(-1)^m n!m!}{(n+m)!(n+m+1)!}(z - z_0)^{n+m+1}e^{\beta z}(1 + o(1)). \end{aligned} \tag{16}$$

Now using (4), (12), (15) and (16) we conclude that the estimate (1) is true. □

4.5 Uniform approximation of e^{-x} on $[0, \infty)$

The problem of rational approximation of e^{-x} on $[0, \infty)$ is one of the most interesting problems in the theory of rational approximations. The problem has many applications, for example in numerical analysis (see Cody, Meinardus and Varga (1969)), but it is also interesting as a mathematical problem itself.

Let us first adopt our notations. We set

$$\lambda_{nm} = R_{nm}(e^{-x})_{C[0,\infty)} = \inf\{\|e^{-x} - r(x)\|_{C[0,\infty)}: r \in R_{nm}\}.$$

There are two cases of special interest: when $n = 0$ and when $n = m$. It is possible to show that there exist constants q_i, $i = 1, \ldots, 4$, $0 < q_i < 1$, such that

(a) $q_2^m \leqslant \lambda_{0m} \leqslant q_1^m$,
(b) $q_4^n \leqslant \lambda_{nn} \leqslant q_3^n$.

The most essential problem here is what we can say about the constants q_i, $i = 1, \ldots, 4$.

The situation is quite different in these two cases. In the first case there exists a solution of the problem – the nice theorem of Schönhage (1973), see theorem 4.10 below. The exact solution of the second case has been obtained analytically only in 1986 (see the end of the chapter).

Let us consider estimates of the type (a). The case $n = 0$ is really approximation of e^{-x} by inverse polynomials of degree m:

$$\lambda_{0m} = \inf\left\{\left\|e^{-x} - \frac{1}{p(x)}\right\|_{C[0,\infty)} : p \in P_m\right\}.$$

Using $p_m(x) = \sum_{k=0}^{m} (x^k/k!)$ (the mth partial Taylor sum for e^x) it is easy to see that $\limsup (\lambda_{0m})^{1/m} \leqslant 1/2$ (Cody, Meinardus, Varga (1969)).

We shall prove the following exact estimation.

Theorem 4.10 (Schönhage, 1973). *We have*

$$\lim_{m\to\infty} (\lambda_{0m})^{1/m} = 1/3.$$

We shall obtain theorem 4.10 from the following more exact result of Schönhage.

Theorem 4.11. *We have*

$$\frac{1}{6((4m+4)\ln 3 + \ln 4)^{1/2}} \leqslant 3^m \lambda_{0m} \leqslant \sqrt{2}.$$

To prove theorem 4.11 let us first remark that we can consider the best uniform approximation to the function $e^{-x/4}$ instead of e^{-x}; evidently

$$\lambda_{0m} = \inf\left\{\left\|e^{-x/4} - \frac{1}{p(x)}\right\|_{C[0,\infty)} : p \in P_m\right\}.$$

Before proving theorem 4.11 we shall prove one lemma concerning the best uniform approximation to $e^{x/4}$ on the interval $[0, a]$, $a > 0$, with a weight $e^{-x/2}$ by means of algebraic polynomials of mth degree. Let $q_{m,a} \in P_m$ be the polynomial of such best uniform approximation:

$$\max_{x\in[0,a]} e^{-x/2}|e^{x/4} - q_{m,a}(x)| = \inf\left\{\max_{x\in[0,a]} e^{-x/2}|e^{x/4} - p(x)| : p \in P_m\right\} = \lambda(m; a). \tag{1}$$

Lemma 4.8. *We have*

(a) $e^{x/4} \geqslant q_{m,a}(x)$, $\quad x \geqslant a$,
(b) $q'_{m,a}(x) \geqslant 0$, $\quad x \geqslant a$.

Proof. It is easy to see, as in theorem 1.6, that the difference $\varepsilon(x) = e^{x/4} - q_{m,a}(x)$ must alternate at least $m+2$ times in the interval $[0, a]$, i.e. $\varepsilon(x)$ must have at least $m+1$ zeros in the interval $[0, a]$. Since $1, x, \ldots, x^m, e^{x/4}$ is a Chebyshev system $\varepsilon(x)$ must have exactly $m+1$ zeros in $[0, a]$. Let these zeros be $0 \leqslant x_1 < x_2 < \cdots < x_{m+1} \leqslant a$.

Let us prove (a). Evidently $e^{x/4} - q_{m,a}(x) > 0$ for sufficiently large x, and if we assume that there is $\tilde{x} > a$ such that $e^{\tilde{x}/4} - q_{m,n}(\tilde{x}) = \varepsilon(\tilde{x}) < 0$, then there must exist $\bar{x} > a$ such that $\varepsilon(\bar{x}) = 0$, but this is impossible since all zeros of $\varepsilon(x)$ are in the interval $[0, a]$.

To prove (b), let us assume the converse, that there exists $\alpha_1 > a$ such that $q'_{m,a}(\alpha_1) < 0$. Since $\varepsilon(x_j) = 0, j = 1, \ldots, m+1$, we must have m zeros of ε' in the interval $[0, a]$; let them be $x_j^{(1)}, j = 1, \ldots, m$. We have $\varepsilon'(x_m^{(1)}) = 0$, i.e. $q'_{m,a}(x_m^{(1)}) - \frac{1}{4}e^{x_m^{(1)}/4} = 0$, or $q'_{m,a}(x_m^{(1)}) > 0$. This, together with $q'_{m,a}(\alpha_1) < 0$, gives us that there exists $\alpha_2 > x_m^{(1)}$ such that $q''_{m,a}(\alpha_2) < 0$, since $q'_{m,a}(\alpha_1) - q'_{m,a}(x_m^{(1)}) = q''_{m,a}(\alpha_2)(\alpha_1 - x_m^{(1)}) < 0, \alpha_2 \in (x_m^{(1)}, \alpha_1)$. Continuing thus we obtain that there exist $x_1^{(m)}$ and α_m such that $\varepsilon^{(m)}(x_1^{(m)}) = 0$ and $q_{m,a}^{(m)}(\alpha_m) < 0$, $\alpha_m > x_1^{(m)}$. But this is a contradiction, since $q_{m,a}^{(m)}$ is always a positive constant. □

Proof of theorem 4.11. We shall use the Laguerre polynomials

$$L_n(x) = \frac{e^x}{n!}\frac{d^n}{dx^n}(e^{-x}x^n), \quad n = 0, 1, 2, \ldots. \tag{2}$$

It is well-known (it is easy to verify by means of integration by parts) that the Laguerre polynomials are orthogonal on $[0, \infty)$ with a weight e^{-x}:

$$\int_0^\infty L_n(x)L_m(x)e^{-x}dx = \begin{cases} 1, & n = m, \\ 0, & n \neq m. \end{cases}$$

Let us consider the expansion of the function $e^{x/4}$ with respect to the orthogonal system $\{L_n\}_{n=0}^\infty$. We have

$$e^{x/4} \sim \sum_{n=0}^\infty c_n L_n(x),$$

where the coefficients c_n are given by

$$c_n = \int_0^\infty e^{-x}e^{x/4}L_n(x)dx = \frac{1}{n!}\int_0^\infty (e^{-x}x^n)^{(n)}e^{x/4}dx = \frac{4(-1)^n}{3^{n+1}}$$

(again using n integrations by parts).

From the theory of orthogonal polynomials it is well-known that then

$$g_m(x) = 4\sum_{n=0}^m \frac{(-1)^n}{3^{n+1}}L_n(x) \tag{3}$$

is the algebraic polynomial of mth degree of best approximation to $e^{x/4}$ in $L_2[0, \infty)$ with weight e^{-x} and

$$\int_0^\infty |e^{x/4} - g_m(x)|^2 e^{-x}dx = \sum_{n=m+1}^\infty c_n^2 = \frac{2}{3^{2m+2}}. \tag{4}$$

We have

$$\tfrac{3}{4}e^x\int_x^\infty e^{t/4}e^{-t}dt = e^{x/4}. \tag{5}$$

Using (2) and (3) it is not difficult to see that

$$p_m(x) = \tfrac{3}{4}e^x \int_x^\infty g_m(t)e^{-t}\,dt \tag{6}$$

is an algebraic polynomial of mth degree.

We have from (5) and (6)

$$|e^{x/4} - p_m(x)| \leqslant \tfrac{3}{4}e^x \int_x^\infty |e^{t/4} - g_m(t)|e^{-t}\,dt.$$

Using the Cauchy–Schwarz inequality we get from here and (4)

$$|e^{x/4} - p_m(x)| \leqslant \tfrac{3}{4}e^x \left(\int_x^\infty e^{-t}\,dt\right)^{1/2} \left(\int_x^\infty |e^{t/4} - g_m(t)|^2 e^{-t}\,dt\right)^{1/2}$$

$$\leqslant \tfrac{3}{4}e^x e^{-x/2} \frac{\sqrt{2}}{3^{m+1}} = \frac{\sqrt{2}}{4\cdot 3^m} e^{x/2}, \quad x \geqslant 0. \tag{7}$$

From here we obtain that $\lambda(m; a) \leqslant \sqrt{2}/(4\cdot 3^m)$ for every $a > 0$.

On the other hand (1), (7) give us that for every $x \in [0, a]$ we have

$$|q_{m,a}(x) - e^{x/4}|e^{-x/2} \leqslant \frac{\sqrt{2}}{4\cdot 3^m}$$

and for $x \in [0, a]$ we have

$$q_{m,a}(x) \geqslant e^{x/4}\left(1 - \frac{\sqrt{2}}{4\cdot 3^m}e^{x/4}\right) \geqslant e^{x/4}\left(1 - \frac{\sqrt{2}}{4\cdot 3^m}e^{a/4}\right).$$

If we set $a = 4m\ln 3 + \ln 4$ we get that for this a and $x \in [0, a]$ we have

$$q_{m,a}(x) \geqslant \tfrac{1}{2}e^{x/4}. \tag{8}$$

Now we can obtain an upper bound for λ_{0m}. Let $a = 4m\ln 3 + \ln 4$. Then for $x \in [0, a]$ we have

$$\left|e^{-x/4} - \frac{1}{q_{m,a}(x)}\right| = \frac{|e^{x/4} - q_{m,a}(x)|}{e^{x/4}|q_{m,a}(x)|} \leqslant 2e^{-x/2}|e^{x/4} - q_{m,a}(x)| \leqslant \frac{\sqrt{2}}{2\cdot 3^m}.$$

By lemma 4.8 $q'_{m,a}(x) \geqslant 0$ for $x \geqslant a$; this, together with (8) and part (a) of lemma 4.8 gives us

$$\tfrac{1}{2}e^{a/4} \leqslant q_{m,a}(x) \leqslant e^{x/4}$$

for $x \geqslant a$. Therefore

$$e^{-x/4} \leqslant \frac{1}{q_{m,a}(x)} \leqslant 2e^{-a/4}, \quad x \geqslant a,$$

and

$$0 \leqslant \frac{1}{q_{m,a}(x)} - e^{-x/4} \leqslant 2e^{-a/4} = \frac{\sqrt{2}}{3^m}.$$

We have obtained that $\lambda_{0m} \leqslant \sqrt{2}/3^m$.

To obtain a lower bound estimate we write

$$\left| e^{-x/4} - \frac{1}{p_m(x)} \right| \leqslant \lambda_{0m}, \quad x \geqslant 0, \tag{9}$$

where $p_m \in P_m$ and $1/p_m$ is the rational function of order $(0, m)$ of best uniform approximation to $e^{-x/4}$. We have from (9)

$$|p_m(x) - e^{x/4}| \leqslant \lambda_{0m} e^{x/4} p_m(x). \tag{10}$$

Let us assume that

$$\lambda_{0m} < \frac{1}{2((4m+4)\ln 3 + \ln 4)^{1/2} 3^{m+1}}. \tag{11}$$

Then for $x \in [0, a]$, $a = (4m+4)\ln 3 + \ln 4$ we obtain from (10)

$$p_m(x) \leqslant \frac{e^{x/4}}{1 - \lambda_{0m} e^{x/4}} \leqslant 2e^{x/4} \tag{12}$$

and therefore

$$|p_m(x) - e^{x/4}| \leqslant 2\lambda_{0m} e^{x/2}, \quad x \in [0, a]. \tag{13}$$

Evidently the same inequality holds for the polynomial $q_{m,a}$ (see (1)):

$$|q_{m,a}(x) - e^{x/4}| \leqslant 2\lambda_{0m} e^{x/2}, \quad x \in [0, a]. \tag{14}$$

Exactly in the same way as in part (b) of lemma 4.8 it is possible to show that $q_{m,a}(x) \geqslant 0$ for $x \geqslant a$ and this together with (a) of lemma 4.8 gives us

$$0 \leqslant q_{m,a}(x) \leqslant e^{x/4}, \quad x \geqslant a. \tag{15}$$

From (14), (15) and (4) we get $(a = (4m+4)\ln 3 + \ln 4)$

$$\frac{2}{3^{2m+2}} \leqslant \int_0^\infty e^{-t}(q_{m,a}(t) - e^{t/4})^2 \, dt \leqslant \int_0^a e^{-t}(2\lambda_{0m} e^{t/2})^2 \, dt + \int_a^\infty e^{-t/2} \, dt$$
$$= 4a\lambda_{0m}^2 + 2e^{-a/2} = 4a\lambda_{0m}^2 + \frac{1}{3^{2m+2}}.$$

This inequality gives us

$$\frac{1}{2((4m+4)\ln 3 + \ln 4)^{1/2} 3^{m+1}} \leqslant \lambda_{0m}$$

which contradicts (11). Therefore

$$3^m \lambda_{0m} \geqslant \frac{1}{6((4m+4)\ln 3 + \ln 4)^{1/2}}. \qquad \square$$

We shall consider briefly also the results connected with λ_{nn}. The result of Schönhage (theorem 4.10) gives some reasons for the so-called $\frac{1}{9}$ conjecture,

$$\lim (\lambda_{nn})^{1/n} = \tfrac{1}{9}. \tag{16}$$

But this conjecture is not true: first Schönhage (1982) and Trefethen and Gutknecht (1983b) observed that the conjecture (16) is numerically false; after this Opitz and Scherer (1984b) rigorously proved that

$$\limsup (\lambda_{nn})^{1/n} \leqslant \frac{1}{9.037}$$

The best estimate for lim inf is given by Schönhage (1982):

$$\liminf (\lambda_{nn})^{1/n} > \frac{1}{13.928}$$

Very strong calculations, made by Carpenter, Rutan and Varga (1984) give

$$\lim_{n\to\infty} \lambda_{nn}^{1/n} \approx \frac{1}{9.280\,025\,491\,92081}.$$

4.6 Notes

The basic result of this chapter – the famous Newman theorem 4.1 (D. Newman, 1964a) – is the starting point of the modern theory of rational approximation.

Before the final result of Vjacheslavov (1975) – theorem 4.2 – we want to mention the results of Gonchar (1967b),

$$\exp\{-\pi(\sqrt{2}+\varepsilon)\sqrt{n}\} < R_n[|x|]_{C[-1,1]} < \exp\{-\tfrac{1}{2}\pi(\sqrt{2}-\varepsilon)\sqrt{n}\}$$

for every $\varepsilon > 0$ and $n > n(\varepsilon)$, and the result of A.P. Bulanov (1975a),

$$\exp\{-\pi\sqrt{(n+1)}\} \leqslant R_n(|x|)_{C[-1,1]} < \exp\{-\pi\sqrt{n}(1 - O(n^{-1/4}))\}.$$

The problem of best uniform rational approximation to the function x^α on the interval $[0,1]$ was posed by D. Newman (1964b). Gonchar (1974) proved the following estimate:

$$\pi\sqrt{\alpha} \leqslant \liminf n^{-1/2} \ln R_n^{-1}(x^\alpha)_{C[0,1]} \leqslant \limsup n^{-1/2} \ln R_n^{-1}(x^\alpha)_{C[0,1]} \leqslant 4\pi\sqrt{\alpha}$$

for any positive non-integer α.

Gonchar conjectured that the limit exists and is $2\pi\sqrt{\alpha}$.

Gonchar's conjecture was proved by T. Ganelius (1979) (see also T. Ganelius (1982)):

$$\lim_{n\to\infty} (R_n(x^\alpha)_{C[0,1]})^{1/\sqrt{n}} = \mathrm{e}^{-2\pi\sqrt{\alpha}}.$$

T. Ganelius (1979) also proved the following result

Let $\alpha = p/q$ be a positive rational number. We have

$$\exp(-2\pi(\alpha+2))|\sin \pi\alpha| \leqslant \exp(2\pi\sqrt{(\alpha n)})R_n(x^2)_{C[0,1]} \leqslant B(p,q)$$

where the constant $B(p,q)$ depends on p and q.

Theorem 4.2 is in our opinion a very interesting fact – it connects two basic constants in mathematics – e and π

$$\lim_{n\to\infty} (R_n(|x|)_{C[-1,1]})^{1/\sqrt{n}} = \mathrm{e}^{-\pi}.$$

This connection is very important, we believe, but it is not very clear at present.

There exist many unsolved problems connected with the Zolotarjov's problems – for example how to obtain easily the Newman result from the exact solution, given by means of elliptic functions.

Meinardus' conjecture,

$$R_{nm}(\mathrm{e}^x)_{C[-1,1]} = \frac{n!m!}{2^{n+m}(n+m)!(n+m+1)!}(1+\mathrm{o}(1)),$$

is given in his book (Meinardus, 1967).

The exact value of $\lim_{n\to\infty} \lambda_{nn}^{1/n}$ is $\exp(-\pi K'/K)$, where K, K' are the complete elliptic integrals of the first kind for the moduli $k, k' = \sqrt{(1-k^2)}$ (see section 4.3), where k is the solution of the equation $K(k) = 2E(k)$, E the complete elliptic integral of the second kind. This is proved by A.A. Gonchar and E.A. Rahmanov; the number $\exp(-\pi K'/K)$ is given by A. Mangnus, who proposed a method for the proof.

5

Uniform approximation of some function classes

There exist function classes which can be approximated by rational functions in uniform or L_p-metric better than by algebraic polynomials. In this chapter we shall investigate the uniform rational approximation of some classes of this kind. We apply certain methods of approximation which allow us to obtain exact estimates (with respect to the order).

We begin in section 5.1 with some preliminaries. In section 5.2 rational uniform approximation is considered of the basic class V_r of all functions with rth derivative of bounded variation. In section 5.3 we deal with certain classes whose order of approximation is not better than $O(n^{-1})$, such as Sobolev classes W_p^1, $p > 1$, the class of absolutely continuous functions with derivative in $L \log L$, the class of all functions of bounded variation with a prescribed modulus of continuity. Section 5.4 is devoted to the study of the method of R. DeVore which is illustrated on the classes W_p^1, $p > 1$, and V_2. Section 5.5 investigates the uniform rational approximation of convex functions and convex $\operatorname{Lip} \alpha$ functions. Finally, in section 5.6 we give two theorems for approximation of functions with singularities.

5.1 Preliminaries

As we noted, in this chapter we shall consider the uniform rational approximation of certain function classes. We shall be interested in obtaining exact estimates with respect to the order of approximation. A characteristic particularity of the rational approximation is the appearance of the so-called o-effect on the order of approximation of individual functions of a given class. So, for instance, for the class Lip 1 on $[a, b]$ from the Jackson theorem (see theorem 3.10) we have

$$\sup_{f \in \mathrm{Lip}\, 1} R_n(f)_C = O(n^{-1})$$

but, as we shall see in corollary 10.2, for every individual function $f \in \mathrm{Lip}\, 1$,

$$R_n(f)_C = o(n^{-1}) \quad \text{(a conjecture of D. Newman).}$$

Both estimates are of exact order.

The presence of the o-effect for some function classes will be established and investigated in chapter 10. In chapter 11 the exactness of these estimates will be shown and commented on.

To avoid some possible terminological ambiguities we shall give precise definitions for exactness of a given estimate. Let X be a given function class and $R_n(f)$ the best approximation to $f \in X$ by means of rational functions of order n in a certain metric.

Definition 5.1. *The estimate*

$$\sup_{f \in X} R_n(f) = O(\varphi(n)), \; \varphi(n) > 0, \quad n \geqslant 1,$$

will be called exact with respect to the order, or shortly exact for the class X if

$$\limsup_{n \to \infty} \left\{ \sup_{f \in X} R_n(f)/\varphi(n) \right\} > 0.$$

Sometimes we shall simply say that the estimate $R_n(f) = O(\varphi(n))$ is exact for a given class X.

Definition 5.2. *The estimate*

$$R_n(f) = O(\varphi(n))$$

will be called exact with respect to the order in the class or shortly exact in the class X, if there exists a function $f \in X$ such that.

$$\limsup_{n \to \infty} \{R_n(f)/\varphi(n)\} > 0.$$

Definition 5.3. *We shall say that the estimate*

$$R_n(f) = o(\varphi(n))$$

is exact (with respect to the order) in the class X if for each sequence $\{\eta_n\}_{n=1}^{\infty}, \eta_n \searrow 0, \eta_n > 0$, there exists a function $f \in X$ such that

$$\limsup_{n \to \infty} \{R_n(f)/\eta_n \varphi(n)\} \geqslant 1.$$

It is worthwhile to observe that the set of all rational functions of degree n is not linear. This fact and some other particularities of this class as a tool for approximation require special and sometimes difficult methods of approximation.

The fundamental statement in most of the real methods for rational approximation which we shall use is the following lemma.

Lemma 5.1. *Let $\alpha, \beta, \gamma > 0$ and $\alpha \leqslant \beta$. Then there exists a rational function σ such that*

$$|\sigma(x)| \leqslant \gamma, \quad x \in [-\beta, -\alpha],$$
$$|1 - \sigma(x)| \leqslant \gamma, \quad x \in [\alpha, \beta],$$
$$0 \leqslant \sigma(x) \leqslant 1, \quad x \in (-\infty, \infty)$$

and

$$\deg \sigma \leqslant B \ln\left(\mathrm{e} + \frac{\beta}{\alpha}\right) \ln\left(\mathrm{e} + \frac{1}{\gamma}\right),$$

where $B > 1$ is an absolute constant, $\deg \sigma$ denotes the degree of σ.

This lemma provides a good rational approximation of the jump-function. In fact, it is equivalent to the upper estimate in theorem 4.1 (D. Newman's theorem) for rational approximation of $|x|$.

To prove lemma 5.1 we need the following lemma.

Lemma 5.2. *Let $0 < \varepsilon \leqslant \frac{1}{2}$ and $n \geqslant 1$. Then for the rational function*

$$s(x) = \frac{p(-x)}{p(x)}, \quad p(x) = \prod_{i=1}^{n} (x + \varepsilon^{i/n}),$$

we have

$$|s(x)| \leqslant c_1 \exp\left\{-\frac{c_2 n}{\ln(1/\varepsilon)}\right\}, \quad x \in [\varepsilon, 1], \tag{1}$$

$$|s(x)| \geqslant \frac{1}{c_1} \exp\left\{\frac{c_2 n}{\ln(1/\varepsilon)}\right\}, \quad x \in [-1, -\varepsilon], \tag{2}$$

where $c_1 = \mathrm{e}^{1/\mathrm{e}}$, $c_2 = \mathrm{e}^{-1}$. In (1), (2) we can put $c_1 = 1$ when $n \geqslant \ln(1/\varepsilon)$.

Proof. Exactly as in the proof of lemma 4.1 we obtain for $x \in [\varepsilon^{1/n}, 1]$

$$\left|\frac{p(-x)}{p(x)}\right| \leqslant \prod_{i=1}^{n} \frac{1 - \varepsilon^{i/n}}{1 + \varepsilon^{i/n}} \leqslant \exp\left\{-2 \sum_{i=1}^{n} \varepsilon^{i/n}\right\}$$

$$= \exp\left\{-\frac{2\varepsilon^{1/n}(1-\varepsilon)}{1 - \varepsilon^{1/n}}\right\} \leqslant \exp\left\{-\frac{2\varepsilon^{1/n}(1-\varepsilon)n}{\ln(1/\varepsilon)}\right\}. \tag{3}$$

Let $n \geqslant \ln(1/\varepsilon)$. Since $0 < \varepsilon \leqslant 1/2$, $2\varepsilon^{1/n}(1-\varepsilon) \geqslant \mathrm{e}^{-1} = c_2$ and by (3) we get

$$|s(x)| \leqslant \exp\left\{-\frac{c_2 n}{\ln(1/\varepsilon)}\right\}, \quad x \in [\varepsilon, 1]. \tag{4}$$

If $1 \leqslant n < \ln(1/\varepsilon)$, then clearly we have for $x \in [\varepsilon^{1/n}, 1]$

$$|s(x)| \leqslant 1 \leqslant \exp\left\{ -\frac{c_2 n}{\ln(1/\varepsilon)} + c_2 \right\} = c_1 \exp\left\{ -\frac{c_2 n}{\ln(1/\varepsilon)} \right\}. \tag{5}$$

The estimates (4) and (5) imply (1). Since $s(-x) = 1/s(x)$, (2) follows by (1). □

Proof of lemma 5.1. Let s be the rational function of lemma 5.2 with

$$\varepsilon = \frac{1}{e + \beta/\alpha}, \quad n = \left[\frac{1 + c_1^2}{2c_2} \ln\frac{1}{\varepsilon} \ln\left(e + \frac{1}{\gamma} \right) + 1 \right], \tag{6}$$

where $c_1 = e^{1/e}$, $c_2 = 1/e$ are the constants in lemma 5.2. Consider the rational function

$$\sigma_1(x) = 1/(s^2(x) + 1).$$

Clearly, by (6) we have

$$\deg \sigma_1 = 2n \leqslant B \ln\left(e + \frac{\beta}{\alpha} \right) \ln\left(e + \frac{1}{\gamma} \right), \tag{7}$$

where $B = \text{constant} > 1$. In view of (2) and (6) it can be verified that

$$|\sigma_1(x)| < \frac{1}{s^2(x)} \leqslant c_1^2 \exp\left\{ -\frac{2c_2 n}{\ln(1/\varepsilon)} \right\} \leqslant \gamma \tag{8}$$

for $x \in [-1, -\varepsilon] \supset [-1, -(\alpha/\beta)]$. Similarly, by (1) and (6)

$$|1 - \sigma_1(x)| < s^2(x) \leqslant \gamma, \quad x \in [\varepsilon, 1] \supset \left[\frac{\alpha}{\beta}, 1 \right]. \tag{9}$$

Obviously

$$0 \leqslant \sigma_1(x) \leqslant 1, \quad x \in (-\infty, \infty). \tag{10}$$

Finally, set $\sigma(x) = \sigma_1(\beta x)$. The assertion of lemma 5.1 follows by (7)–(10). □

The basic idea in most real methods for rational approximation is the following. First a good approximation by piece-wise rational functions is constructed. Then, 'joining' the pieces by a good rational approximation of the jump-function a single rational function is obtained. This function realizes the required rational approximation of the given function. The difficulty consists in the optimization of the process of 'joining' pieces. To overcome this difficulty we shall often use the following lemma.

Lemma 5.3. *Let $f \in C_\Delta$, $\Delta = [a, b]$. Let there exist compact subintervals Δ_1 and Δ_2 such that $\Delta = \Delta_1 \cup \Delta_2$, $|\Delta_1 \cap \Delta_2| > 0$, and rational functions r_1 and r_2 such* *that*

$$\| f - r_i \|_{C(\Delta_i)} \leqslant \varepsilon_1, \tag{11}$$

$$\| r_i \|_{C(-\infty,\infty)} \leqslant A \tag{12}$$

and

$$\deg r_i \leqslant k_i, \quad i = 1, 2, \tag{13}$$

where $\varepsilon_1 > 0$, $A > 0$, $k_i \geqslant 0$, $i = 1, 2$, *are given numbers. Then for each* $\varepsilon_2 > 0$ *there is a rational function* r *such that*

$$\| f - r \|_{C(\Delta)} \leqslant \varepsilon_1 + \varepsilon_2, \tag{14}$$

$$\| r \|_{C(-\infty,\infty)} \leqslant A \tag{15}$$

and

$$\deg r \leqslant k_1 + k_2 + B_1 \ln\left(\mathrm{e} + \frac{|\Delta|}{|\Delta_1 \cap \Delta_2|} \right) \ln\left(\mathrm{e} + \frac{A}{\varepsilon_2} \right), \tag{16}$$

where $B_1 > 1$ *is an absolute constant,* $|\Delta|$ *is the length of* Δ.

Proof. Let $\Delta_1 \cap \Delta_2 = [u, v]$ and $\Delta_1 = [a, v]$, $\Delta_2 = [u, b]$. Consider the rational function

$$r(x) = \left(1 - \sigma\left(x - \frac{u+v}{2} \right) \right) r_1(x) + \sigma\left(x - \frac{u+v}{2} \right) r_2(x),$$

where σ is the rational function of lemma 5.1 with

$$\alpha = |\Delta_1 \cap \Delta_2|/2, \quad \beta = |\Delta|, \quad \gamma = \varepsilon_2/2A.$$

We shall prove that the rational function r satisfies (14)–(16). At first we estimate $\deg r$. By lemma 5.1 we have

$$\begin{aligned} \deg \sigma &\leqslant B \ln\left(\mathrm{e} + \frac{\beta}{\alpha} \right) \ln\left(\mathrm{e} + \frac{1}{\gamma} \right) \\ &\leqslant B \ln\left(\mathrm{e} + \frac{2|\Delta|}{|\Delta_1 \cap \Delta_2|} \right) \ln\left(\mathrm{e} + \frac{2A}{\varepsilon_2} \right) \leqslant B_1 \ln\left(\mathrm{e} + \frac{|\Delta|}{|\Delta_1 \cap \Delta_2|} \right) \ln\left(\mathrm{e} + \frac{A}{\varepsilon_2} \right), \end{aligned}$$

where $B_1 = 4B =$ constant. Hence

$$\begin{aligned} \deg r &\leqslant \deg r_1 + \deg r_2 + \deg \sigma \\ &\leqslant k_1 + k_2 + B_1 \ln\left(\mathrm{e} + \frac{|\Delta|}{|\Delta_1 \cap \Delta_2|} \right) \ln\left(\mathrm{e} + \frac{A}{\varepsilon_2} \right), \end{aligned}$$

i.e. r satisfies (16).

Now we estimate $\| r \|_{C(-\infty,\infty)}$. Since $0 \leqslant \sigma(x) \leqslant 1$, $x \in (-\infty, \infty)$, by (12) we get for $x\in(-\infty,\infty)$,

$$|r(x)|\leqslant\left(1-\sigma\left(x-\frac{u+v}{2}\right)\right)\|r_1\|_{C(-\infty,\infty)}+\sigma\left(x-\frac{u+v}{2}\right)\|r_2\|_{C(-\infty,\infty)}\leqslant A$$

and therefore (15) holds.

It remains to estimate $\|f-r\|_{C(\Delta)}$. To this end we shall use (11), (12) and the properties of σ from lemma 5.1. There are the following cases.

(i) If $x\in\Delta_1\setminus\Delta_2$ then

$$|f(x)-r(x)|\leqslant|f(x)-r_1(x)|+\sigma\left(x-\frac{u+v}{2}\right)\left(\|r_1\|_{C(-\infty,\infty)}+\|r_2\|_{C(-\infty,\infty)}\right)$$

$$\leqslant\varepsilon_1+\frac{\varepsilon_2}{2A}\cdot 2A=\varepsilon_1+\varepsilon_2.$$

(ii) If $x\in\Delta_1\cap\Delta_2$ then

$$|f(x)-r(x)|\leqslant\left(1-\sigma\left(x-\frac{u+v}{2}\right)\right)|f(x)-r_1(x)|$$

$$+\sigma\left(x-\frac{u+v}{2}\right)|f(x)-r_2(x)|\leqslant\varepsilon_1.$$

(iii) If $x\in\Delta_2\setminus\Delta_1$ then

$$|f(x)-r(x)|\leqslant|f(x)-r_2(x)|+\left(1-\sigma\left(x-\frac{u+v}{2}\right)\right)(\|r_1\|_{C(-\infty,\infty)}$$

$$+\|r_2\|_{C(-\infty,\infty)})\leqslant\varepsilon_1+\varepsilon_2.$$

Consequently

$$\|f-r\|_{C(\Delta)}\leqslant\varepsilon_1+\varepsilon_2,$$

i.e. estimate (14) holds. □

Lemma 5.3 implies the following more specific lemma for 'joining' of rational functions.

Lemma 5.4. *Let $f\in C[a,b]$ and let there exist intervals $\Delta_1=[a,c]$ and $\Delta_2=[c,b]$, $a<c<b$, and rational functions r_1 and r_2 such that*

$$\|f-r_i\|_{C(\Delta_i)}\leqslant\varepsilon_1,\quad i=1,2,\tag{17}$$

where $\varepsilon_1>0$ is a given number.

Then for each $\varepsilon_2>0$ and $\delta>0$ there exists a rational function r such that

$$\|f-r\|_{C[a,b]}\leqslant\varepsilon_1+\varepsilon_2+\omega(f;\delta)_C\tag{18}$$

$$\deg r\leqslant 2\deg r_1+2\deg r_2+B_2\ln\left(e+\frac{b-a}{\delta}\right)\ln\left(e+\frac{\|f\|_C}{\varepsilon_2}\right),\tag{19}$$

where $\omega(f;\delta)_C$ is the modulus of continuity of f, $B_2 > 1$ is an absolute constant.

Proof. Let $\varepsilon_2 > 0$ and $\delta > 0$ be arbitrary numbers. Denote $d = [a, b]$, $d_1 = [a, c+\delta]$ and $d_2 = [c-\delta, b]$. Let λ_i, $i = 1, 2$, be the increasing linear function which maps d_i onto Δ_i. Clearly

$$\| x - \lambda_i(x) \|_{C(d_i)} = \delta, \quad i = 1, 2. \tag{20}$$

Set $\tilde{r}_i = r_i(\lambda_i)$, $i = 1, 2$. By (17) and (18) it follows that

$$\| f - \tilde{r}_i \|_{C[d_i \cap d]} \leqslant \| f - f(\lambda_i) \|_{C[d_i \cap d]} + \| f - r_i \|_{C(\Delta_i)} \leqslant \varepsilon_1 + \omega(f;\delta)_C. \tag{21}$$

If $d \subset d_1$ or $d \subset d_2$, then lemma 5.4 follows from (21) immediately.

Consider the opposite case. Then $d = d_1 \cup d_2$ and $|d_1 \cap d_2| = 2\delta$, where $|\Delta|$ is the length of the interval Δ. If $\| \tilde{r}_i \|_{C(d_i)} \leqslant 2 \| f \|_{C(d_i)}$, then we set

$$q_i = \frac{\tilde{r}_i}{1 + \eta_i \tilde{r}_i^2}, \quad \eta_i = \frac{\varepsilon_2}{16 \| f \|_{C(d)}^3},$$

when $\| f \|_{C(d)} > 0$ (the case $\| f \|_C = 0$ is trivial).

By (21) we get

$$\begin{aligned} \| f - q_i \|_{C(d_i)} &\leqslant \| f - \tilde{r}_i \|_{C(d_i)} + \eta_i \| f \|_C \| \tilde{r}_i \|_{C(d_i)}^2 \\ &\leqslant \varepsilon_1 + \omega(f;\delta)_C + 4\eta_i \| f \|_C^3 < \varepsilon_1 + \varepsilon_2/2 + \omega(f;\delta)_C, \end{aligned}$$

i.e.

$$\| f - q_i \|_{C(d_i)} \leqslant \varepsilon_1 + \varepsilon_2/2 + \omega(f;\delta)_C, \quad i = 1, 2. \tag{22}$$

Obviously we have

$$\| q_i \|_{C(-\infty,\infty)} \leqslant \frac{1}{2\sqrt{\eta_i}} = 2 \| f \|_{C(d)}^{3/2} \varepsilon_2^{-1/2} \tag{23}$$

and

$$\deg q_i \leqslant 2 \deg r_i. \tag{24}$$

In the case when $\| \tilde{r}_i \|_{C(d_i)} > 2 \| f \|_{C(d_i)}$ we have in view of (21)

$$\| f \|_{C(d_i)} \leqslant \| \tilde{r}_i \|_{C(d_i)} - \| f \|_{C(d_i)} \leqslant \| f - \tilde{r}_i \|_{C(d_i)} \leqslant \varepsilon_1 + \omega(f;\delta)_C$$

and therefore the rational function $q_i = 0$ satisfies (22)–(24). Thus for $i = 1, 2$ there exists a rational function q_i which satisfies (22)–(24).

Now we apply lemma 5.3 with ε_2 replaced by $\varepsilon_2/2$. In view of (22)–(24) we conclude that there exists a rational function r such that

$$\| f - r \|_{C(d)} \leqslant \varepsilon_1 + \varepsilon_2 + \omega(f;\delta)_C$$

and

$$\deg r \leqslant 2 \deg r_1 + 2 \deg r_2 + B_1 \ln\left(e + \frac{|d|}{|d_1 \cap d_2|} \right) \ln\left(e + \frac{2 \| f \|_C^{3/2} \varepsilon_2^{-1/2}}{\varepsilon_2/2} \right)$$

$$\leqslant 2\deg r_1 + 2\deg r_2 + B_1 \ln\left(e + \frac{b-a}{2\delta}\right)\ln\left(e + 4\left(\frac{\|f\|_C}{\varepsilon_2}\right)^{3/2}\right)$$

$$\leqslant 2\deg r_1 + 2\deg r_2 + 3B_1 \ln\left(e + \frac{b-a}{\delta}\right)\ln\left(e + \frac{\|f\|_C}{\varepsilon_2}\right),$$

which imply (18) and (19). □

5.2 Functions with rth derivative of bounded variation

We shall investigate the rational uniform approximation of the class $V_r = V_r(M, [a,b])$, $r \geqslant 1$, of all functions defined on $[a,b]$ for which $f^{(r-1)}$ is absolutely continuous and is an integral of a function $f^{(r)}$ with variation bounded by $M(V_a^b f^{(r)} \leqslant M)$. The class V_r is basic for the rational and spline approximations of functions. Historically V_r was one of the first function classes which was approximated by rational functions better than by polynomials. It was the first class for which the exact order of rational uniform approximation was found.

The foundations were laid by P. Szűsz and P. Turan (1966) who showed that the convex Lip 1 functions can be approximated uniformly by rational functions better than algebraic polynomials. Later on G. Freud (1966) improved and generalized their result for V_r, $r \geqslant 1$. The final estimate was obtained by V. Popov (1976a, 1977).

Theorem 5.1. *Let $r \geqslant 1$, $M \geqslant 0$ and $[a,b]$ be an arbitrary compact interval. Then for $n \geqslant r$*

$$\sup_{f \in V_r(M,[a,b])} R_n(f)_C \leqslant C(r)\frac{M(b-a)^r}{n^{r+1}}, \tag{1}$$

where $C(r) = D^r$, $D > 1$ is an absolute constant.

Remarks. The estimate (1) is equivalent to the following.

If $f^{(r)}$, $r \geqslant 1$, is absolutely continuous, then for $n \geqslant r$

$$R_n(f)_C \leqslant C(r)\frac{(b-a)^r \|f^{(r+1)}\|_{L_1}}{n^{r+1}}. \tag{2}$$

Indeed, the set of all functions f with absolutely continuous rth derivative $f^{(r)}$ on $[a,b]$ such that $\|f^{(r+1)}\|_{L_1[a,b]} \leqslant M$ is dense in $V_r(M,[a,b])$. From this follows the equivalence of estimates (1) and (2).

The estimate (1) is exact (see definition 5.1). This fact will be established in 11.1.3, theorem 11.4. The existence of o-effect for the class V_r will be shown in 10.1. Also this effect will be characterized there.

Note that the exact estimate for uniform polynomial approximation of f in V_r is the following: $E_n(f)_C = O(n^{-r})$.

The fact that the constant $C(r)$ in the estimate (1) is of the form $C(r) = D^r$ will allow us to prove in 9.3 a fundamental theorem in the theory of rational approximation in Hausdorff metric.

The proof of theorem 5.1 is based on the following theorem for 'joining' of rational functions.

Theorem 5.2. *Let $f \in V_r(M,[a,b])$, $r \geqslant 1$, $M \geqslant 0$. Let there exist a subdivision Ω of $[a,b]$ into m $(m \geqslant 1)$ compact subintervals with disjoint interiors and rational functions $r_\Delta, \Delta \in \Omega$, such that for each $\Delta \in \Omega$*

$$\| f - r_\Delta \|_{C(\Delta)} \leqslant \varepsilon \tag{3}$$

and

$$\deg r_\Delta \leqslant k_\Delta, \tag{4}$$

where $\varepsilon > 0$, $k_\Delta \geqslant 0$ are given numbers.

Then there is a rational function R such that

$$\| f - R \|_{C[a,b]} \leqslant 3\varepsilon \tag{5}$$

and

$$\deg R \leqslant 2 \sum_{\Delta \in \Omega} k_\Delta + Dm \ln^2 \left(\mathrm{e} + \frac{M(b-a)^r}{\varepsilon m^{r+1}} \right), \tag{6}$$

where $D = D(r) = D_1 r^2$, $D_1 > 1$ is an absolute constant.

Proof. Without loss of generality we shall assume that $f^{(r)}$ is continuous. Clearly, if theorem 5.2 is true in the special case when $m = 2^s$, s an integer, then it is true in the general case also with another absolute constant D_1. Thus we shall suppose that $m = 2^s$, s an integer.

Next we shall use the following notations:

$$f_\Delta(x) = f(x) - \sum_{\nu=0}^{r} f^{(\nu)}(u)(x-u)^\nu/\nu!, \tag{7}$$

where $\Delta = [u, v] \subset [a,b]$,

$$N(\mu, M, \Delta)$$

$$= \sum_{\nu=0}^{\mu} 2^\nu \{ 36 B_1 r^2 \ln^2 (\mathrm{e} + 2^\mu 4^{-\nu} (M|\Delta|^r)^{1/(r+1)} \varepsilon^{-1/(r+1)}) + 4r \}, \tag{8}$$

where $B_1 > 1$ is the constant from lemma 5.3, $r \geqslant 1$ and $\varepsilon > 0$ are from the assumptions of theorem 5.2, μ, M and the interval Δ are parameters.

For brevity we shall write $\| \cdot \|_\Delta = \| \cdot \|_{C(\Delta)}$.

We need the following two lemmas where we shall use the assumptions and notations introduced above.

Lemma 5.5. *If the conditions of theorem 5.2 are satisfied then for each interval $\Delta \in \Omega$ there is a rational function q_Δ such that*

$$\| f_\Delta - q_\Delta \|_\Delta \leqslant 2\varepsilon,$$

$$\| q_\Delta \|_{(-\infty,\infty)} \leqslant (V_\Delta f^{(r)} |\Delta|^r)^{3/2} \varepsilon^{-1/2}$$

and

$$\deg q_\Delta \leqslant 2k_\Delta + 2r.$$

Proof. Let $\Delta \in \Omega$ and $\Delta = [u, v]$. If $V_\Delta f^{(r)} = 0$, then by (7) it follows that $f_\Delta(x) = 0$ for $x \in \Delta$. Hence, the rational function $q_\Delta \equiv 0$ satisfies the assertion of lemma 5.5.

Denote

$$\tilde{r}_\Delta(x) = r_\Delta(x) - \sum_{\nu=0}^{r} f^{(\nu)}(u)(x-u)^\nu/\nu!.$$

By (4) we have $\deg \tilde{r}_\Delta \leqslant k_\Delta + r$. By (3) we have

$$\| f_\Delta - \tilde{r}_\Delta \|_\Delta \leqslant \varepsilon. \tag{9}$$

If $\| \tilde{r}_\Delta \|_\Delta > 2 \| f_\Delta \|_\Delta$, then by (9) we get

$$\| f_\Delta \|_\Delta \leqslant \| \tilde{r}_\Delta \|_\Delta - \| f_\Delta \|_\Delta \leqslant \| f_\Delta - \tilde{r}_\Delta \|_\Delta \leqslant \varepsilon$$

and therefore the rational function $q_\Delta \equiv 0$ satisfies the conditions of lemma 5.5.

Let $V_\Delta f^{(r)} > 0$ and $\| \tilde{r}_\Delta \|_\Delta \leqslant 2 \| f_\Delta \|_\Delta$. Then we set $q_\Delta = \tilde{r}_\Delta/(1 + \eta_\Delta \tilde{r}_\Delta^2)$ where $\eta_\Delta = \frac{1}{4}(V_\Delta f^{(r)} |\Delta|^r)^{-3} \varepsilon$. Obviously $\deg q_\Delta \leqslant 2 \deg \tilde{r}_\Delta \leqslant 2k_\Delta + 2r$. One easily verifies that

$$\| q_\Delta \|_{(-\infty,\infty)} \leqslant \frac{1}{2\sqrt{\eta_\Delta}} = (V_\Delta f^{(r)} |\Delta|^r)^{3/2} \varepsilon^{-1/2}.$$

It remains to estimate $\| f_\Delta - q_\Delta \|_\Delta$. To this end we shall use (9), our assumptions and the fact that $\| f_\Delta \|_\Delta \leqslant V_\Delta f^{(r)} |\Delta|^r$, since $f_\Delta^{(\nu)}(u) = 0$, $\nu = 0, 1, \ldots, r$, and $V_\Delta f_\Delta^{(r)} = V_\Delta f^{(r)}$. We get

$$\begin{aligned} \| f_\Delta - q_\Delta \|_\Delta &\leqslant \| f_\Delta - \tilde{r}_\Delta \|_\Delta + \eta_\Delta \| f_\Delta \|_\Delta \| \tilde{r}_\Delta \|_\Delta^2 \\ &\leqslant \varepsilon + 4\eta_\Delta \| f_\Delta \|_\Delta^3 \leqslant \varepsilon + 4\eta_\Delta (V_\Delta f^{(r)} |\Delta|^r)^3 = 2\varepsilon. \end{aligned}$$ □

By the assumption of theorem 5.2 there exist points $x_i, i = 0, 1, \ldots, m$, $a = x_0 < x_1 < \cdots < x_m = b$, such that $\Omega = \{[x_i, x_{i+1}]: i = 0, 1, \ldots, m-1\}$. Denote for $0 \leqslant \mu \leqslant s$

$$\Omega_\mu = \{[x_i, x_{i+2^\mu}]: i = 0, 1, \ldots, m - 2^\mu\} \tag{10}$$

and for $\Delta \subset [a, b]$

$$\Omega_\Delta = \{\Delta^*: \Delta^* \in \Omega, \Delta^* \subset \Delta\}. \tag{11}$$

Lemma 5.6. *Let $0 \leqslant \mu \leqslant s-1$ and for each $\Delta \in \Omega_\mu$ (see (10)) suppose there is a rational function q_Δ such that*

$$\| f_\Delta - q_\Delta \|_\Delta \leqslant \varphi(\mu)\varepsilon, \tag{12}$$

where $\varphi(\mu) \geqslant 1$ depends only on μ,

$$\| q_\Delta \|_{(-\infty,\infty)} \leqslant 2^\mu (V_\Delta f^{(r)} |\Delta|^r)^{3/2} \varepsilon^{-1/2} \tag{13}$$

and

$$\deg q_\Delta \leqslant 2 \sum_{\Delta^* \in \Omega_\Delta} k_{\Delta^*} + N(\mu, V_\Delta f^{(r)}, \Delta), \tag{14}$$

where $N(\mu, M, \Delta)$ is given by (8).

Then for every $\Delta \in \Omega_{\mu+1}$ there is a rational function r_Δ such that

$$\| f_\Delta - r_\Delta \|_\Delta \leqslant (\varphi(\mu) + 2^{-\mu-1})\varepsilon, \tag{15}$$

$$\| r_\Delta \|_{(-\infty,\infty)} \leqslant 2^{\mu+1} (V_\Delta f^{(r)} |\Delta|^r)^{3/2} \varepsilon^{-1/2} \tag{16}$$

and

$$\deg r_\Delta \leqslant 2 \sum_{\Delta^* \in \Omega_\Delta} k_{\Delta^*} + N(\mu+1, V_\Delta f^{(r)}, \Delta), \tag{17}$$

where the last sum is taken over all intervals Δ^* which belong to the set Ω_Δ defined in (11).

Proof. Let $\Delta \in \Omega_{\mu+1}$ and $\Delta = [z_1, z_3]$. If $V_\Delta f^{(r)} |\Delta|^r \leqslant \varepsilon$, then $\| f_\Delta \|_\Delta \leqslant V_\Delta f^{(r)} |\Delta|^r \leqslant \varepsilon$, since $f_\Delta^{(\nu)}(z_1) = 0$, $\nu = 0, 1, \ldots, r$, and $V_\Delta f_\Delta^{(r)} = V_\Delta f^{(r)}$. Then the rational function $r_\Delta \equiv 0$ satisfies (15)–(17).

Now let $V_\Delta f^{(r)} |\Delta|^r > \varepsilon$. Obviously, there is $z_2 \in \Delta$ such that the intervals $\Delta_1 = [z_1, z_2]$ and $\Delta_2 = [z_2, z_3]$ are in Ω_μ. Next we shall denote for short $M_1 = V_{\Delta_1} f^{(r)}$, $M_2 = V_{\Delta_2} f^{(r)}$, $M = V_\Delta f^{(r)}$, $\eta = 2^{-\mu-3}(M|\Delta|^{r-1})^{-1}\varepsilon$, $d_1 = [z_1, z_2 + \eta]$, $d_2 = [z_2 - \eta, z_3]$. Clearly $M_1 + M_2 = M$.

By (12)–(14) there are rational functions q_{Δ_1} and q_{Δ_2} such that

$$\| f_{\Delta_i} - q_{\Delta_i} \|_{\Delta_i} \leqslant \varphi(\mu)\varepsilon, \tag{18}$$

$$\| q_{\Delta_i} \|_{(-\infty,\infty)} \leqslant 2^\mu (M_i |\Delta_i|^r)^{3/2} \varepsilon^{-1/2}, \tag{19}$$

$$\deg q_{\Delta_i} \leqslant 2 \sum_{\Delta^* \in \Omega_{\Delta_i}} k_{\Delta^*} + N(\mu, M_i, \Delta_i), \quad i = 1, 2. \tag{20}$$

We need an approximation to $f_\Delta = f_{\Delta_1}$ on the interval Δ_2. Denote $p_{\Delta_i}(x) = \sum_{\nu=0}^r f^{(\nu)}(z_i)(x - z_i)^\nu/\nu!$, $i = 1, 2$. Since $p_{\Delta_1} \in P_r$, then $p_{\Delta_1}(x) = \sum_{\nu=0}^r p_{\Delta_1}^{(\nu)}(z_2) \times (x - z_2)^\nu/\nu!$. We have

$$\begin{aligned} f_{\Delta_2}(x) &= f(x) - p_{\Delta_2}(x) = f_\Delta(x) + p_{\Delta_1}(x) - p_{\Delta_2}(x) \\ &= f_\Delta(x) - \sum_{\nu=0}^r (f^{(\nu)}(z_2) - p_{\Delta_1}^{(\nu)}(z_2))(x - z_2)^\nu/\nu! \\ &= f_\Delta(x) - \sum_{\nu=0}^r f_\Delta^{(\nu)}(z_2)(x - z_2)^\nu/\nu!. \end{aligned}$$

Then putting

$$p(x) = \sum_{\nu=0}^{r} f_{\Delta}^{(\nu)}(z_2)(x - z_2)^{\nu}/\nu!$$

we have by (18) with $i = 2$

$$\| f_{\Delta} - p - q_{\Delta_2} \|_{\Delta_2} \leqslant \varphi(\mu)\varepsilon. \tag{21}$$

Consider the rational function

$$q = \frac{p}{1 + \tau p^2}, \quad \tau = 2^{-\mu-2}(M|\Delta|^r)^{-3}\varepsilon.$$

Since $f_{\Delta}^{(\nu)}(z_1) = 0$, $\nu = 0, 1, \ldots, r$, and $V_{\Delta_1} f_{\Delta}^{(r)} = V_{\Delta_1} f^{(r)} = M_1$, then $|f_{\Delta}^{(\nu)}(z_2)| \leqslant M_1 |\Delta_1|^{r-\nu}$, $\nu = 0, 1, \ldots, r$. Hence

$$\begin{aligned}
\| p \|_{\Delta_2} &\leqslant \sum_{\nu=0}^{r} |f_{\Delta}^{(\nu)}(z_2)|(z_3 - z_2)^{\nu}/\nu! \\
&\leqslant M_1 \sum_{\nu=0}^{r} |\Delta_1|^{r-\nu}|\Delta_2|^{\nu}/\nu! \leqslant M_1 \sum_{\nu=0}^{r} \binom{r}{\nu} |\Delta_1|^{r-\nu}|\Delta_2|^{\nu} \\
&= M_1(|\Delta_1| + |\Delta_2|)^r = M_1|\Delta|^r.
\end{aligned}$$

From this it follows that

$$\| p - q \|_{\Delta_2} \leqslant \tau \| p \|_{\Delta_2}^3 \leqslant \tau(M_1|\Delta|^r)^3 \leqslant 2^{-\mu-2}\varepsilon, \tag{22}$$

$$\| q \|_{(-\infty,\infty)} \leqslant \frac{1}{2\sqrt{\tau}} \leqslant 2^{\mu}(M|\Delta|^r)^{3/2}\varepsilon^{-1/2} \tag{23}$$

and

$$\deg q = 2r. \tag{24}$$

Set $\tilde{q}_2 = q_{\Delta_2} + q$. Combining (21) and (22) we obtain

$$\| f_{\Delta} - \tilde{q}_2 \|_{\Delta_2} \leqslant \| f_{\Delta} - p - q_{\Delta_2} \|_{\Delta_2} + \| p - q \|_{\Delta_2} \leqslant (\varphi(\mu) + 2^{-\mu-2})\varepsilon. \tag{25}$$

By (19) and (23) we get

$$\begin{aligned}
\| \tilde{q}_2 \|_{(-\infty,\infty)} &\leqslant \| q_{\Delta_2} \|_{(-\infty,\infty)} + \| q \|_{(-\infty,\infty)} \\
&\leqslant 2^{\mu}(M_1|\Delta|^r)^{3/2}\varepsilon^{-1/2} + 2^{\mu}(M|\Delta|^r)^{3/2}\varepsilon^{-1/2} \leqslant 2^{\mu+1}(M|\Delta|^r)^{3/2}\varepsilon^{-1/2}.
\end{aligned} \tag{26}$$

By (20) and (24) we get

$$\deg \tilde{q}_2 \leqslant 2 \sum_{\Delta^* \in \Omega_{\Delta_2}} k_{\Delta^*} + N(\mu, M_2, \Delta_2) + 2r. \tag{27}$$

We need an estimate of the modulus of continuity of f_{Δ} on Δ:

$$\omega(f_{\Delta}, \Delta; \delta) = \sup\{|f_{\Delta}(x') - f_{\Delta}(x'')| : x', x'' \in \Delta, |x' - x''| \leqslant \delta\}.$$

Since $f_\Delta^{(\nu)}(z_1)=0$, $\nu=0,1,\ldots,r$, and $V_\Delta f_\Delta^{(r)}=V_\Delta f^{(r)}=M$, $\|f'_\Delta\|_\Delta\leqslant M|\Delta|^{r-1}$ and therefore

$$\omega(f_\Delta,\Delta;\delta)\leqslant M|\Delta|^{r-1}\delta,\quad \delta\geqslant 0. \tag{28}$$

Let λ_i $(i=1,2)$ be the linear increasing function mapping the interval d_i onto Δ_i. It is easy to see that

$$\|\lambda_i(x)-x\|_{d_i}=\eta,\quad i=1,2. \tag{29}$$

Set

$$q_1(x)=q_{\Delta_1}(\lambda_1(x))\quad\text{and}\quad q_2(x)=\tilde q_2(\lambda_2(x)).$$

Using (18) with q_{Δ_1}, (28) and (29) we obtain

$$\begin{aligned}\|f_\Delta-q_1\|_{d_1\cap\Delta}&\leqslant\|f_\Delta-f_\Delta(\lambda_1)\|_{d_1\cap\Delta}+\|f_\Delta(\lambda_1)-q_{\Delta_1}(\lambda_1)\|_{d_1\cap\Delta}\\&\leqslant\omega(f_\Delta,\Delta;\eta)+\|f_{\Delta_1}-q_{\Delta_1}\|_{\Delta_1}\leqslant(\varphi(\mu)+2^{-\mu-3})\varepsilon.\end{aligned}$$

Similarly, by (25), (28) and (29) we get

$$\|f_\Delta-q_2\|_{d_2\cap\Delta}\leqslant(\varphi(\mu)+2^{-\mu-2}+2^{-\mu-3})\varepsilon.$$

Consequently, we have

$$\|f_\Delta-q_i\|_{d_i\cap\Delta}\leqslant(\varphi(\mu)+3.2^{-\mu-3})\varepsilon,\quad i=1,2. \tag{30}$$

It follows by (19) and (26) that

$$\|q_i\|_{(-\infty,\infty)}\leqslant 2^{\mu+1}(M|\Delta|^r)^{3/2}\varepsilon^{-1/2},\quad i=1,2, \tag{31}$$

and by (20) and (27) that

$$\deg q_i\leqslant 2\sum_{\Delta^*\in\Omega_{\Delta_i}}k_{\Delta^*}+N(\mu,M_i,\Delta_i)+2r,\quad i=1,2. \tag{32}$$

If $\Delta\subset d_1$ or $\Delta\subset d_2$, then it follows by (30)–(32) that the rational function $r_\Delta=q_1$ or $r_\Delta=q_2$ satisfies (15)–(17).

In the opposite case we have $|d_1\cap d_2\cap\Delta|>\eta$. Now, in view of (30)–(32), we are ready to apply lemma 5.3 with parameters from (30)–(32). Setting $\varepsilon_2=2^{-\mu-3}\varepsilon$ we conclude that there exists a rational function r_Δ such that

$$\|f_\Delta-r_\Delta\|_\Delta\leqslant(\varphi(\mu)+2^{-\mu-1})\varepsilon,$$

$$\|r_\Delta\|_{(-\infty,\infty)}\leqslant 2^{\mu+1}(M|\Delta|^r)^{3/2}\varepsilon^{-1/2}$$

and

$$\begin{aligned}\deg r_\Delta&\leqslant\deg q_1+\deg q_2\\&\quad+B_1\ln\left(e+\frac{|\Delta|}{\eta}\right)\ln(e+2^{2\mu+4}(M|\Delta|^r\varepsilon^{-1})^{3/2}).\end{aligned} \tag{33}$$

The rational function r_Δ satisfies (15) and (16). It remains to prove that r_Δ satisfies (17). First we note that

$$\begin{aligned}B_1 \ln\left(e+\frac{|\Delta|}{\eta}\right)&\ln(e+2^{2\mu+4}(M|\Delta|^r\varepsilon^{-1})^{3/2})\\ &=B_1\ln(e+2^{\mu+3}M|\Delta|^r\varepsilon^{-1})\ln(e+2^{2\mu+4}(M|\Delta|^r\varepsilon^{-1})^{3/2})\\ &\leqslant 9B_1\ln^2(e+2^{\mu+1}M|\Delta|^r\varepsilon^{-1})\\ &\leqslant 36\,B_1r^2\ln^2(e+2^{\mu+1}(M|\Delta|^r)^{1/(r+1)}\varepsilon^{-1/(r+1)}).\end{aligned}$$

From this, (32) and (33) we get

$$\begin{aligned}\deg r_\Delta \leqslant 2\sum_{\Delta^*\in\Omega_\Delta} k\Delta^* &+ N(\mu, M_1, \Delta_1) + N(\mu, M_2, \Delta_2)\\ &+36\,B_1r^2\ln^2(e+2^{\mu+1}(M|\Delta|^r)^{1/(r+1)}\varepsilon^{-1/(r+1)})+4r. \qquad (34)\end{aligned}$$

Next, we shall estimate $N(\mu, M_1, \Delta_1) + N(\mu, M_2, \Delta_2)$. To this end, we shall use the following inequality

$$\begin{aligned}&\ln^2(e+(x_1y_1^r)^{1/(r+1)})+\ln^2(e+(x_2y_2^r)^{1/(r+1)})\\ &\leqslant 2\ln^2\left(e+\left(\frac{x_1+x_2}{2}\left(\frac{y_1+y_2}{2}\right)^r\right)^{1/(r+1)}\right),\quad x_1,x_2,y_1,y_2\geqslant 0. \qquad (35)\end{aligned}$$

This inequality is equivalent to the fact that the function $F(x, y) = -\ln^2(e+(xy^r)^{1/(r+1)})$ is convex on the set $D = \{(x, y): x, y \geqslant 0\}$. The function F is convex on D since $\partial^2F/\partial x^2$, $\partial^2F/\partial y^2$ and $\partial^2F/\partial x^2\cdot\partial^2F/\partial y^2 - (\partial^2F/\partial x\partial y)^2$ are nonnegative in D. The same fact follows also from the convexity of the function $F_1(x) = -\ln^2(e+x)$ on $[0, \infty)$ and convexity of the function $F_2(x, y) = -(xy^r)^{1/(r+1)}$ on D. Thus by the definition of $N(\mu, M, \Delta)$ in (8) and the inequality (35) we get

$$\begin{aligned}N(\mu, M_1, \Delta_1)+N(\mu, M_2, \Delta_2)&\leqslant\sum_{v=0}^{\mu}2^{v+1}\left\{36B_1r^2\ln^2\left(e\right.\right.\\ &\left.\left.+2^\mu 4^{-v}\left(\frac{M_1+M_2}{2}\left(\frac{|\Delta_1|+|\Delta_2|}{2}\right)^r\right)^{1/(r+1)}\varepsilon^{-1/(r+1)}\right)+4r\right\}\\ &=\sum_{j=1}^{\mu+1}2^j\{36\,B_1r^2\ln^2(e+2^{\mu+1}4^{-j}(M|\Delta|^r)^{1/(r+1)}\varepsilon^{-1/(r+1)})+4r\}.\end{aligned}$$

Combining this with (34) we see that r_Δ satisfies (17), which completes the proof of lemma 5.6. □

Completion of the proof of theorem 5.2. Starting from lemma 5.5 and applying lemma 5.6 s times, we obtain that there is a rational function r_Δ, $\Delta = [a, b]$, such that

$$\|f_\Delta - r_\Delta\|_\Delta \leqslant \left(2+\sum_{\mu=1}^{s}2^{-\mu}\right)\varepsilon < 3\varepsilon$$

and

$$\deg r_\Delta \leqslant 2 \sum_{\Delta^* \in \Omega} k_{\Delta^*} + N(s, V_\Delta f^{(r)}, \Delta).$$

Putting $R(x) = r_\Delta(x) - \sum_{\nu=0}^{r} f^{(\nu)}(a)(x-a)^\nu/\nu!$ we have

$$\| f - R \|_{[a,b]} < 3\varepsilon$$

and

$$\deg R \leqslant 2 \sum_{\Delta^* \in \Omega} k_{\Delta^*} + N(s, M, [a, b]) + r,$$

where $M = V_a^b f^{(r)}$. It remains to prove that

$$N(s, M, \Delta) \leqslant D_1 r^2 m \ln^2 \left(\mathrm{e} + \frac{M|\Delta|^r}{\varepsilon m^{r+1}} \right) \tag{36}$$

where $D_1 =$ constant. By (8) we have

$$\begin{aligned}
N(s, M, \Delta) &= \sum_{\nu=0}^{s} 2^\nu \{36 B_1 r^2 \ln^2 (\mathrm{e} + 2^s 4^{-\nu} (M|\Delta|^r)^{1/(r+1)} \varepsilon^{-1/(r+1)}) + 4r\} \\
&\leqslant 8r \cdot 2^s + Cr^2 \sum_{\nu=0}^{s} 2^\nu \ln^2 ((\mathrm{e} + 2^{-s(r+1)} M|\Delta|^r \varepsilon^{-1}) \cdot 4^{s-\nu}) \\
&= 8r \cdot 2^s + Cr^2 \left\{ \sum_{\nu=0}^{s} 2^\nu \ln^2 (\mathrm{e} + 2^{-s(r+1)} M|\Delta|^r \varepsilon^{-1}) \right. \\
&\quad + \sum_{\nu=0}^{s} 2^{\nu+1} \ln (\mathrm{e} + 2^{-s(r+1)} M|\Delta|^r \varepsilon^{-1}) (\ln 4)(s-\nu) \\
&\quad \left. + \sum_{\nu=0}^{s} 2^\nu (\ln 4)^2 (s-\nu)^2 \right\}.
\end{aligned}$$

Since the function $2^x(s + 1 + 2/\ln 2 - x)^2$ is increasing on $[0, s+1]$,

$$\begin{aligned}
\sum_{\nu=0}^{s} 2^\nu (s-\nu) \leqslant \sum_{\nu=0}^{s} 2^\nu (s-\nu)^2 &< \sum_{\nu=0}^{s} 2^\nu \left(s + 1 + \frac{2}{\ln 2} - \nu \right)^2 \\
&< \int_0^{s+1} 2^x \left(s + 1 + \frac{2}{\ln 2} - x \right)^2 \mathrm{d}x < \frac{16}{(\ln 2)^3} \cdot 2^s.
\end{aligned}$$

Therefore

$$\begin{aligned}
N(s, M, \Delta) &\leqslant D_1 r^2 \cdot 2^s \ln^2 \left(\mathrm{e} + \frac{M|\Delta|^r}{\varepsilon \cdot 2^{s(r+1)}} \right) \\
&= D_1 r^2 m \ln^2 \left(\mathrm{e} + \frac{M|\Delta|^r}{\varepsilon m^{r+1}} \right),
\end{aligned}$$

i.e. estimate (36) holds. □

Proof of theorem 5.1. If we are not interested in the form of the constant $C(r)$ in the estimate (1), then (1) follows as in theorem 5.2. But we need to prove the estimate (1) with $C(r)$ of the kind $C(r)=D^r$, $D=$ constant. To this end, we shall use the following lemma.

Lemma 5.7. *If the function f is defined on $\Delta=[a,b]$ and $f^{(r)}\in C_\Delta$, $r\geqslant 1$, then for $n\geqslant r$*

$$E_n(f)_C \leqslant C^r \frac{|\Delta|^r \|f^{(r)}\|_{C(\Delta)}}{n^r}, \tag{37}$$

where $C>0$ is an absolute constant, $E_n(f)_C$ is the best uniform approximation to f by means of all algebraic polynomials of degree n.

Proof. It follows by theorem 3.10 that for $n\geqslant 1$

$$E_n(f)_C \leqslant C\frac{|\Delta|\,\|f'\|_{C(\Delta)}}{n}. \tag{38}$$

Choose $p\in P_{n-1}$ such that $\|f'-p\|_{C(\Delta)}=E_{n-1}(f')_C$ and $q\in P_n$ such that $q'=p$. Applying (38) we get

$$E_n(f)_C = E_n(f-q)_C \leqslant C\frac{|\Delta|\,\|f'-q'\|_{C(\Delta)}}{n} \leqslant C\frac{|\Delta|\,E_{n-1}(f')_C}{n}.$$

From this and (38) it follows for $n\geqslant r$ that

$$\begin{aligned} E_n(f)_C &\leqslant C\frac{|\Delta|\,E_{n-1}(f')_C}{n} \leqslant C^2\frac{|\Delta|^2 E_{n-2}(f'')_C}{n(n-1)} \\ &\leqslant \cdots \leqslant C^r \frac{|\Delta|^r\|f^{(r)}\|_{C(\Delta)}}{n(n-1)\cdots(n-r+1)}. \end{aligned} \tag{39}$$

On the other hand from Stirling's formula

$$\left(k! = \sqrt{2\pi k}\cdot\left(\frac{k}{\mathrm{e}}\right)^k \mathrm{e}^{\theta_k/12}, \quad 0<\theta_k<1\right)$$

it follows that for $n\geqslant r\geqslant 1$

$$\frac{n^r}{n(n-1)\cdots(n-r+1)} = \frac{n^r(n-r)!}{n!} < \mathrm{e}^{2r}.$$

This inequality together with (39) implies (37). □

Now we are able to prove theorem 5.1. Let $f\in V_r(M,[a,b])$, $r\geqslant 1$, $M\geqslant 0$. Without loss of generality we shall assume that $f^{(r)}\in C_{[a,b]}$ and $M>0$.

Let $r\leqslant n\leqslant Ar^8$, where $A=D_1^2\mathrm{e}^{10}$, $D_1>1$, is the absolute constant from theorem 5.2. Putting $p(x)=f^{(r)}(a)x^r/r!$ we get by lemma 5.7

$$R_n(f)_C \leqslant E_n(f)_C = E_n(f-p)_C \leqslant C^r \frac{(b-a)^r \| f^{(r)} - p^{(r)} \|_C}{n^r}$$

$$= C^r \frac{(b-a)^r \| f^{(r)} - f^{(r)}(a) \|_C}{n^r} \leqslant C^r A r^8 \frac{M(b-a)^r}{n^{r+1}}.$$

Hence

$$R_n(f)_C \leqslant C_1^r \frac{M(b-a)^r}{n^{r+1}}, \quad r \leqslant n \leqslant Ar^8, \tag{40}$$

where $C_1 = \text{constant}$.

Let $n > Ar^8$. Choose m integer so that

$$\frac{n}{Ar^8} \leqslant m \leqslant \frac{2n}{Ar^8}. \tag{41}$$

Clearly, there is a division Ω of $[a,b]$ into $2m$ compact subintervals Δ, disjoint except for the end points, such that

$$|\Delta| \leqslant \frac{b-a}{m} \quad \text{and} \quad V_\Delta f^{(r)} \leqslant \frac{M}{m}. \tag{42}$$

We put for $\Delta \in \Omega$, $\Delta = [u,v]$, $p_\Delta(x) = f^{(r)}(u)x^r/r!$. It follows by (37), (41) and (42) that for $\Delta \in \Omega$

$$E_{r^8}(f)_{C(\Delta)} = E_{r^8}(f-p_\Delta)_{C(\Delta)}$$

$$\leqslant C^r \frac{|\Delta|^r \| f^{(r)} - f^{(r)}(u) \|_{C(\Delta)}}{r^{8r}} \leqslant C^r \frac{M|\Delta|^r}{r^{8r} m^{r+1}}$$

$$\leqslant C^r A^{r+1} r^8 \frac{M(b-a)^r}{n^{r+1}}.$$

Consequently, for each $\Delta \in \Omega$ there is a polynomial q_Δ such that $\| f - q_\Delta \|_{C(\Delta)} \leqslant C_2^r(M(b-a)^r/n^{r+1})$, $C_2 > 1$ and $\deg q_\Delta \leqslant r^8$.

Now we are able to apply theorem 5.2 with $\varepsilon = C_2^r(M(b-a)^r/n^{r+1})$ and $k_\Delta = r^8$, $\Delta \in \Omega$. We obtain that there is a rational function R such that

$$\| f - R \|_{C[a,b]} \leqslant 3C_2^r \frac{M(b-a)^r}{n^{r+1}}$$

and

$$\deg R \leqslant 2 \sum_{\Delta \in \Omega} r^8 + D_1 r^2 \cdot 2m \ln^2\left(\mathrm{e} + \frac{M(b-a)^r}{C_2^r M(b-a)^r n^{-r-1}(2m)^{r+1}} \right)$$

$$\leqslant 4r^8 m + 2D_1 r^2 m \ln^2\left(\mathrm{e} + \left(\frac{n}{2m}\right)^{r+1} \right).$$

Using (41), the fact that $n > Ar^8 = D_1^2 e^{10} r^8$ and the inequality $\ln^2 x < \sqrt{x}$ for $x \geqslant e^{10}$ we obtain

$$\ln^2\left(e + \frac{n}{2m}\right) \leqslant \ln^2\left(e + \frac{Ar^8}{2}\right) \leqslant \ln^2(Ar^8) \leqslant \sqrt{Ar^8}$$

and therefore

$$\deg R \leqslant \frac{n}{2} + \frac{8D_1 n}{Ar^4}\sqrt{Ar^8} \leqslant n.$$

Consequently

$$R_n(f)_C \leqslant 3C_2^r \frac{M(b-a)^r}{n^{r+1}}, \quad n > Ar^8.$$

This estimate together with (40) implies estimate (1). □

5.3 Some classes of absolutely continuous functions and functions of bounded variation

It is not difficult to show that (see 11.1.2, theorem 11.3) the estimate $R_n(f)_C = 0(1)$ is exact in the class of all functions absolutely continuous on $[a, b]$. It turns out that if we consider classes of absolutely continuous functions and functions of bounded variation which satisfy additional conditions, then it is possible to obtain a uniform rate of approximation which is better than the approximation by polynomials. In this section we shall investigate some classes of this kind.

5.3.1 One technical result

The following technical theorem is basic for the uniform rational approximation of functions with order of approximation not greater than $O(n^{-1})$.

Theorem 5.3. *Let the function f be bounded on $[a, b]$ and let there exist $m+1$ points x_i, $i = 0, 1, \ldots, m$, $a = x_0 < x_1 < \cdots < x_m = b$ $(m \geqslant 1)$ such that*

$$\| f - f(x_i) \|_{C(\Delta_i)} \leqslant \varepsilon, \quad i = 0, 1, \ldots, m-1, \tag{1}$$

where $\Delta_i = [x_i, x_{i+1}]$, $\varepsilon > 0$ is a given number. Then there exists a rational function r such that

$$\| f - r \|_{C[a,b]} \leqslant D\varepsilon \tag{2}$$

and

$$\deg r \leqslant D \sum_{i=0}^{m-1} \ln\left(e + \frac{b-a}{m|\Delta_i|}\right), \tag{3}$$

where $|\Delta_i| = x_{i+1} - x_i$, $D > 1$ is an absolute constant.

Proof. There is no loss of generality in assuming that $m=2^s$ with s a positive integer and

$$|\Delta_{\max}| = \max\{|\Delta_i|: i=0,1,\ldots,m-1\} \leqslant \frac{2(b-a)}{m}.$$

Indeed, suppose that theorem 5.3 holds true in this situation. Consider the general case. Set $f(x)=f(b)$ for $x>b$. Add at most m new points $u_i \in (a,b)$ with the property that each interval Δ_i, $|\Delta_i|>(b-a)/m$, is divided by means of u_i, $i=1,2,\ldots$, into subintervals with length contained in $[(b-a)/2m, (b-a)/m]$. Also, take if necessary, some points $v_i=b+i(b-a)/m$, $i=1,2,\ldots$, so that the set $N=\{x_i\}\cup\{u_i\}\cup\{v_i\}$ contains exactly 2^s+1 points, where $2m\leqslant 2^s<4m$. Finally, we renumerate the points of the set N in increasing order and denote them again by x_i, $i=0,1,\ldots,2^s$. Clearly,

$$\|f-f(x_i)\|_{C(\Delta_i)} \leqslant 2\varepsilon, \quad i=0,1,\ldots,2^s-1, \quad \Delta_i=[x_i,x_{i+1}].$$

Then by our assumption it follows that theorem 5.3 holds true in the general case, eventually with another absolute constant D.

Thus we shall suppose that $m=2^s$, s integer and $|\Delta_{\max}|\leqslant 2(b-a)/m$. Set

$$f(x)=\begin{cases} f(a), & x<a, \\ f(b), & x>b, \end{cases}$$

and

$$u_i=\begin{cases} x_i-(b-a), & i=0,1,\ldots,m, \\ x_{i-m}, & i=m+1,m+2,\ldots,2m, \\ x_{i-2m}+(b-a), & i=2m+1,2m+2,\ldots,3m. \end{cases}$$

Denote $d_i=[u_i,u_{i+1}]$, $i=0,1,\ldots,3m-1$, $d_{\max}$ an interval d_i with maximum length,

$$N(k,i)=\sum_{\nu=0}^{k-1}\sum_{\mu=1}^{2^{k-\nu}-1} 12B_1(\nu+1)\ln\left(e+\frac{2^\nu|d_{\max}|}{|d_{i+\mu\cdot 2^\nu}|}\right), \tag{4}$$

where $B_1>1$ is the absolute constant from lemma 5.3, k and i are parameters. Also, denote

$$f_i(x)=f(x)-f(x_i), \quad i=0,1,\ldots,3m-1.$$

Next, we shall denote for short $\|\cdot\|_\Delta=\|\cdot\|_{C(\Delta)}$.

Theorem 5.3 we shall prove applying s times the following lemma, where we use the assumptions and notations introduced above.

Lemma 5.8. *Let $0\leqslant k\leqslant s$, k integer. Let there be, for each i, $0\leqslant i\leqslant 3m-2^k-1$, a rational function q_i such that*

$$\|f_i-q_i\|_{[u_i,u_{i+2^k+1}]}\leqslant\varphi(k)\varepsilon, \tag{5}$$

where $\varphi(k)\geqslant 1$, $\varphi(k)$ *depends only on* k,

$$\|q_i\|_{(-\infty,\infty)}\leqslant(2^k+1)\varepsilon \tag{6}$$

and

$$\deg q_i\leqslant N(k,i), \tag{7}$$

where $N(k,i)$ *is defined in* (4).

Then for each i, $0\leqslant i\leqslant 3m-2^{k+1}-1$, *there exists a rational function* r_i *such that*

$$\|f_i-r_i\|_{[u_i,u_{i+2^{k+1}+1}]}\leqslant\left(\varphi(k)+\frac{1}{2^{k+1}}\right)\varepsilon, \tag{8}$$

$$\|r_i\|_{(-\infty,\infty)}\leqslant(2^{k+1}+1)\varepsilon \tag{9}$$

and

$$\deg r_i\leqslant N(k+1,i). \tag{10}$$

Proof. Let $0\leqslant i\leqslant 3m-2^{k+1}-1$. Denote for short

$$\begin{aligned}
&z_1=u_i,\quad z_2=(u_{i+2^k}+u_{i+2^k+1})/2,\quad z_3=u_{i+2^{k+1}+1},\\
&\eta=|d_{i+2^k}|/2,\quad \Delta_1=[z_1,z_2+\eta],\quad \Delta_2=[z_2-\eta,z_3],\\
&\Delta=[z_1,z_3],\quad \tilde q_1=q_i,\quad \tilde q_2=q_{i+2^k}+f(x_i)-f(x_{i+2^k}).
\end{aligned}$$

By (1) and (5)–(7) we get

$$\begin{aligned}
&\|f_i-\tilde q_1\|_{\Delta_1}\leqslant\varphi(k)\varepsilon,\quad \|\tilde q_1\|_{(-\infty,\infty)}\leqslant(2^k+1)\varepsilon,\quad \deg\tilde q_1\leqslant N(k,i),\\
&\|f_i-\tilde q_2\|_{\Delta_2}=\|f_{i+2^k}-q_{i+2^k}\|_{[u_{i+2^k},u_{i+2^{k+1}+1}]}\leqslant\varphi(k)\varepsilon,\\
&\|\tilde q_2\|_{(-\infty,\infty)}\leqslant\|q_{i+2^k}\|_{(-\infty,\infty)}+|f(x_i)-f(x_{i+2^k})|\\
&\qquad\leqslant(2^k+1)\varepsilon+\sum_{\nu=1}^{2^k}|f(x_{i+\nu-1})-f(x_{i+\nu})|\leqslant(2^{k+1}+1)\varepsilon
\end{aligned}$$

and

$$\deg\tilde q_2\leqslant N(k,i+2^k).$$

Now we are in a position to apply lemma 5.3 to the function f_i with the intervals Δ_1,Δ_2, rational functions $\tilde q_1,\tilde q_2$ and $\varepsilon_1=\varphi(k)\varepsilon$, $A=(2^{k+1}+1)\varepsilon$, $k_{\Delta_1}=N(k,i)$, $k_{\Delta_2}=N(k,i+2^k)$. Set $\varepsilon_2=\varepsilon/2^{k+1}$. It follows that there exists a rational function r_i such that

$$\|f_i-r_i\|_{\Delta}\leqslant\left(\varphi(k)+\frac{1}{2^{k+1}}\right)\varepsilon,$$

$$\|r_i\|_{(-\infty,\infty)}\leqslant(2^{k+1}+1)\varepsilon$$

and

$$\deg r_i \leqslant N(k,i) + N(k,i+2^k) + B_1 \ln\left(e + \frac{|\Delta|}{2\eta}\right) \ln(e + 2^k(2^{k+1}+1)). \tag{11}$$

Consequently, the rational function r_i satisfies (8) and (9). It remains to prove that $\deg r_i \leqslant N(k+1,i)$. Clearly, we have

$$\begin{aligned} B_1 \ln\left(e + \frac{|\Delta|}{2\eta}\right) \ln(e + 2^k(2^{k+1}+1)) &\leqslant B_1 \ln\left(e + \frac{(2^{k+1}+1)|d_{\max}|}{|d_{i+2^k}|}\right) \ln(e + 2^{2k+2}) \\ &\leqslant 12B_1(k+1)\ln\left(e + \frac{2^k|d_{\max}|}{|d_{i+2^k}|}\right). \end{aligned}$$

From this, (11) and the definition of $N(k,i)$ in (4) it follows that

$$\begin{aligned} \deg r_i \leqslant & \sum_{\nu=0}^{k-1} \sum_{\mu=1}^{2^{k-\nu}-1} 12\,B_1(\nu+1)\ln\left(e + \frac{2^\nu|d_{\max}|}{|d_{i+\mu\cdot 2^\nu}|}\right) \\ & + \sum_{\nu=0}^{k-1} \sum_{\mu=1}^{2^{k-\nu}-1} 12B_1(\nu+1)\ln\left(e + \frac{2^\nu|d_{\max}|}{|d_{i+2^k+\mu\cdot 2^\nu}|}\right) \\ & + 12B_1(k+1)\ln\left(e + \frac{2^k|d_{\max}|}{|d_{i+2^k}|}\right) \\ \leqslant & \sum_{\nu=0}^{k} \sum_{\mu=1}^{2^{k+1-\nu}-1} 12B_1(\nu+1)\ln\left(e + \frac{2^\nu|d_{\max}|}{|d_{i+\mu\cdot 2^\nu}|}\right) = N(k+1,i). \end{aligned}$$

Thus r_i satisfies (10). □

Now we continue the proof of theorem 5.3. By (1) it follows that $\|f - f(x_i)\|_{[x_i,x_{i+2}]} \leqslant 2\varepsilon$ for $i = 0, 1, \ldots, m-2$ and therefore

$$\|f_i - 0\|_{[u_i,u_{i+2}]} \leqslant 2\varepsilon, \quad i = 0, 1, \ldots, 3m-2.$$

Consequently, the assumptions of lemma 5.8 hold with $k = 0$, $q_i \equiv 0$ and $\varphi(0) = 2$.

Starting from there and applying lemma 5.8 $s+1$ times we obtain that for every i, $0 \leqslant i \leqslant m-1$, there exists a rational function r_i such that

$$\|f_i - r_i\|_{[u_i,u_{i+2m+1}]} \leqslant \left(2 + \sum_{k=1}^{s+1} \frac{1}{2^{k+1}}\right)\varepsilon < 3\varepsilon$$

and

$$\deg r_i \leqslant N(s+1,i).$$

Choose i_0 such that $\deg r_{i_0} = \min_i \deg r_i$ and set $r = r_{i_0}$. Since $[a,b] \subset [u_i, u_{i+2m+1}]$ for every $i = 0, 1, \ldots, m-1$, $\| f - r \|_{[a,b]} < 3\varepsilon$, i.e. r satisfies (2) with $D = 3$. From the choice of r it follows that

$$\deg r \leqslant \frac{1}{2^s} \sum_{i=0}^{2^s - 1} N(s+1, i).$$

Using the definition of $N(k,i)$ in (4) and the fact $|d_{\max}| = |\Delta_{\max}| \leqslant 2(b-a)/m$ we get

$$\begin{aligned}
\deg r &\leqslant \frac{1}{2^s} \sum_{i=0}^{2^s-1} \sum_{v=0}^{s} \sum_{\mu=1}^{2^{s+1-v}-1} 12B_1(v+1) \ln\left(e + \frac{2^v |d_{\max}|}{|d_{i+\mu \cdot 2^v}|} \right) \\
&\leqslant \frac{1}{2^s} \sum_{v=0}^{s} \sum_{\mu=1}^{2^{s+1-v}-1} 24B_1(v+1) \sum_{j=0}^{m-1} \ln\left(e + \frac{2^v |\Delta_{\max}|}{|\Delta_j|} \right) \\
&\leqslant \frac{24B_1}{2^s} \sum_{v=0}^{s} \sum_{\mu=1}^{2^{s+1-v}-1} (v+1)^2 \sum_{j=0}^{m-1} \ln\left(e + \frac{|\Delta_{\max}|}{|\Delta_j|} \right) \\
&\leqslant 24B_1 \sum_{j=0}^{m-1} \ln\left(e + \frac{|\Delta_{\max}|}{|\Delta_j|} \right) \frac{1}{2^s} \sum_{v=0}^{s} \sum_{\mu=1}^{2^{s+1-v}-1} (v+1)^2 \\
&\leqslant 24B_1 \sum_{j=0}^{m-1} \ln\left(e + \frac{2(b-a)}{m|\Delta_j|} \right) \frac{1}{2^s} \sum_{v=0}^{s} (v+1)^2 \cdot 2^{s+1-v}.
\end{aligned}$$

Since the function $F(x) = 2^x(s + 2 + 2/\ln 2 - x)^2$ is increasing in $[1, s+3]$,

$$\begin{aligned}
\frac{1}{2^s} \sum_{v=0}^{s} 2^{s+1-v}(v+1)^2 &= \frac{1}{2^s} \sum_{j=1}^{s+1} 2^j (s+2-j)^2 \\
&\leqslant \frac{1}{2^s} \int_1^{s+2} 2^x \left(s + 2 + \frac{2}{\ln 2} - x \right)^2 dx \leqslant C,
\end{aligned}$$

where C is an absolute constant. (This is the most essential point in the proof!)

Consequently, there exists an absolute constant D such that

$$\deg r \leqslant D \sum_{i=0}^{m-1} \ln\left(e + \frac{b-a}{m|\Delta_i|} \right),$$

i.e. estimate (3) holds. □

5.3.2 Sobolev classes $W_p^1, p > 1$

Consider the Sobolev class $W_p^1[a,b]$, $p > 1$, of all functions f absolutely continuous on $[a,b]$ such that $f' \in L_p[a,b]$. First Yu.A. Brudnyi (1979) announced that for the functions $f \in W_p^1, p > 1$, the estimate $R_n(f)_C = O(n^{-1})$ holds. The first proof of this result was given by V.A. Popov (1980). Note that for $f \in W_p^1, 1 < p < \infty$, we have only $E_n(f)_C = o(1)$. We shall next prove this result as a consequence of theorem 5.3.

Theorem 5.4. *If $f \in W_p^1[a,b]$, $p > 1$, then the following estimate holds for $n \geqslant 1$:*

$$R_n(f)_C \leqslant C\frac{\|f'\|_{L_p}}{n}, \tag{12}$$

where $C = C_1 p(p-1)^{-1}(b-a)^{1-1/p}$, C_1 = constant.

Remark. The estimate (12) is exact for the class $W_p^1, p > 1$. For every individual function $f \in W_p^1$, $p > 1$, the o-effect appears (see 10.2).

Proof. To prove theorem 5.4 it is enough to consider the case $[a,b] = [0,1]$ and $\|f'\|_p \Rightarrow 1$. Indeed, let theorem 5.4 be true in that case. Suppose $f' \in L_p[a,b]$. The case when $\|f'\|_p = 0$ is trivial. Let $\|f'\|_p > 0$. Consider the function $g(x) = f(a+(b-a)x)/\|f'\|_p(b-a)^{1-1/p}$. Obviously, we have $\|g'\|_{L_p[0,1]} = 1$. Hence by our assumptions $R_n(g)_C \leqslant C(p)/n$, $n \geqslant 1$. Then we get

$$R_n(f)_C = \|f'\|_p(b-a)^{1-1/p}R_n(g)_C \leqslant C(p)(b-a)^{1-1/p}\frac{\|f'\|_p}{n}, \quad n \geqslant 1.$$

Thus we shall suppose that $[a,b] = [0,1]$ and $\|f'\|_p = 1$. Put $f(x) = f(1)$ for $x > 1$. Clearly, there are points x_i, $i = 1, 2, \ldots, m$, where $n \leqslant m \leqslant 2n$, such that

$$0 = x_0 < x_1 < \cdots < x_{m-1} < 1 \leqslant x_m < 1 + \frac{1}{n}$$

and for each $i, 0 \leqslant i \leqslant m-1$, one of the following two conditions is fulfilled:

(i) $\int_{\Delta_i} |f'(x)|\,dx = 1/n$ and $|\Delta_i| \leqslant 1/n$,
(ii) $\int_{\Delta_i} |f'(x)|\,dx \leqslant 1/n$ and $|\Delta_i| = 1/n$,

where $\Delta_i = [x_i, x_{i+1}]$.

Indeed, set $x_0 = 0$ and define by induction $x_{i+1} = \max\{x: x_i < x \leqslant x_i + 1/n$ and $\int_{x_i}^x |f'(t)|\,dt \leqslant 1/n\}$, $i \geqslant 1$. Finally, denote by x_m the first point $x_i \geqslant 1$. Because $\|f'\|_{L_1} \leqslant \|f'\|_{L_p} = 1$ these points x_i satisfy the required conditions.

Obviously, we have

$$\|f - f(x_i)\|_{C(\Delta_i)} \leqslant \int_{\Delta_i} |f'(x)|\,dx \leqslant \frac{1}{n}, \quad i = 0, 1, \ldots, m-1. \tag{13}$$

Denote by N_1 and N_2 the sets of all indices i, $0 \leqslant i \leqslant m-1$, which satisfy (i) and (ii) respectively.

If $i \in N_1$, then by Hölder's inequality we get

$$\frac{1}{n} = \int_{\Delta_i} |f'(x)|\,dx \leqslant \left(\int_{\Delta_i} |f'(x)|^p\,dx\right)^{1/p} |\Delta_i|^{1-1/p}.$$

Hence for $i\in N_1$

$$\frac{1}{n^p|\Delta_i|^{p-1}}\leqslant\int_{\Delta_i}|f'(x)|^p\mathrm{d}x$$

and since $\|f'\|_p=1$,

$$\sum_{i\in N_1}\frac{1}{(n|\Delta_i|)^{p-1}}\leqslant n. \tag{14}$$

One easily verifies the following inequality:

$$\ln\left(\mathrm{e}+\frac{2}{x}\right)\leqslant\frac{2p}{p-1}\frac{1}{x^{p-1}},\quad 0<x\leqslant 1, p>1. \tag{15}$$

By (14) and (15) we get

$$\sum_{i=0}^{m-1}\ln\left(\mathrm{e}+\frac{2}{m|\Delta_i|}\right)\leqslant\sum_{i=0}^{m-1}\ln\left(\mathrm{e}+\frac{2}{n|\Delta_i|}\right)\leqslant\sum_{i\in N_1}\ln\left(\mathrm{e}+\frac{2}{n|\Delta_i|}\right)+\sum_{i\in N_2}\ln(\mathrm{e}+2)$$
$$\leqslant\frac{2p}{p-1}\sum_{i\in N_1}\frac{1}{(n|\Delta|_i)^{p-1}}+2m\leqslant\left(\frac{2p}{p-1}+4\right)n\leqslant\frac{6p}{p-1}n.$$

From this and (13), applying theorem 5.3 we conclude that there is a rational function r such that

$$\|f-r\|_{C[0,1]}\leqslant\frac{2D}{n}$$

and

$$\deg r\leqslant D\sum_{v=0}^{m-1}\ln\left(\mathrm{e}+\frac{2}{m|\Delta_v|}\right)\leqslant\frac{6Dp}{p-1}n,\quad n\geqslant 1.$$

These estimates imply (12). □

As a consequence of theorem 5.1 and theorem 5.4, we shall obtain relations between rational uniform approximation of functions and polynomial L_p-approximation of their derivatives. First P. Turan noted that there should exist relations of this kind. Here we shall present the results of V.A. Popov (1980).

Theorem 5.5. *Let $f\in W_p^r[0,1]$, $r\geqslant 1$, $p\geqslant 1$, i.e. $f^{(r)}\in L_p[0,1]$. Then the estimate*

$$R_{2n+r}(f)_C\leqslant C(p,r)\frac{E_n(f^{(r)})_p}{n^r},\quad n\geqslant 1,$$

holds true in the following situations:

(i) $r=1$, $p>1$,
(ii) $r\geqslant 2$, $p=1$,

where $E_n(f^{(r)})_p$ *denotes the best* L_p*-approximation to* $f^{(r)}$ *by means of algebraic polynomials of degree n.*

Proof. Consider the case when $r=1$ and $p>1$. Choose $p_1 \in P_n$ such that $\| f' - p_1 \|_{L_p} = E_n(f')_p$ and $q \in P_{n+1}$ such that $q' = p_1$. In view of theorem 5.4 we get

$$R_{2n+1}(f)_C \leqslant R_n(f-q)_C \leqslant C(p)\frac{\| f' - q' \|_{L_p}}{n} = C(p)\frac{E_n(f')_p}{n}.$$

In the other case we proceed similarly, applying theorem 5.1. □

5.3.3 Absolutely continuous functions with derivative in Orlicz space *L* log *L*

The function ϕ is called an Orlicz function if it is continuous, strictly increasing and convex in $[0, \infty)$, $\phi(0) = 0$ and $\lim_{x \to \infty}(\phi(x)/x) = \infty$.

The function

$$\psi(y) = \max_{x \geqslant 0} \{xy - \phi(x)\}, \quad y \geqslant 0,$$

is called the complementary function to ϕ.

The Orlicz space $L_\phi[a, b]$, generated by the Orlicz function ϕ, is defined as the set of all functions f measurable in $[a, b]$ such that there exists a constant $K > 0$ with the property

$$\int_a^b \phi(|f(x)|/K)\mathrm{d}x < \infty.$$

The Orlicz space $L_\phi[a, b]$ is a Banach space under the norm

$$\| f \|_{L_\phi} = \inf_{K > 0} \left\{ K + K \int_a^b \phi(|f(x)|/K)\mathrm{d}x \right\}.$$

We need some simple facts concerning Orlicz spaces which will be given without proofs (for a detailed study of the theory of Orlicz spaces see Krasnoselskii, Rutitskii (1958)).

If χ_E is the characteristic function of a measurable set $E \subset [a, b]$, i.e.

$$\chi_E(x) = \begin{cases} 1, & x \in E, \\ 0, & x \in [a, b] \backslash E, \end{cases}$$

then

$$\| \chi \|_{L_\phi} = \text{mes}\, E \cdot \psi^{-1}\left(\frac{1}{\text{mes}\, E}\right), \tag{16}$$

where ψ^{-1} is the inverse function to ψ.

Hölder's inequality is valid: if $f \in L_\phi[a, b]$ and $g \in L_\psi[a, b]$, then $f \cdot g \in L_1[a, b]$

and

$$\int_a^b |f(x)g(x)|\,dx \leqslant \|f\|_{L_\phi}\|g\|_{L_\psi}. \tag{17}$$

If $f \in L_\phi[a,b]$, then

$$\int_a^b \phi\left(\frac{|f(x)|}{\|f\|_{L_\phi}}\right)dx \leqslant 1. \tag{18}$$

Next, we shall consider the Orlicz space $L_{\phi_0}[0,1]$, where

$$\phi_0(x) = x\ln_+ x, \quad \ln_+ x = \begin{cases} \ln x & x \geqslant 1, \\ 0, & 0 < x < 1. \end{cases}$$

This space is denoted usually by $L\log L$. A light computation establishes that the function

$$\psi_0(y) = \begin{cases} y, & 0 \leqslant y \leqslant 1, \\ e^{y-1}, & y > 1, \end{cases}$$

is complementary to ϕ_0.

Next we shall prove the following result of A.A Pekarskii (1982).

Theorem 5.6. *If the function f is absolutely continuous on $[0,1]$ and $f' \in L\log L$, then for $n \geqslant 1$*

$$R_n(f)_C \leqslant c\frac{\|f'\|_{L\log L}}{n}, \tag{19}$$

where c is an absolute constant.

Remark. This theorem is an improvement of theorem 5.4.

Note that $L\log L$ consists exactly of those functions f in L_1 for which $\|Mf\|_{L_1} < \infty$, where Mf is the Hardy–Littlewood maximal function for f (see section 5.4). Moreover, the Orlicz norm $\|\cdot\|_{L\log L}$ is equivalent to the norm $\|f\|$ defined with $\|f\| = \|Mf\|_{L_1}$. Now the estimate (19) is equivalent to the following: if f is absolutely continuous on $[0,1]$ and $Mf' \in L_1[0,1]$, then for $n \geqslant 1$

$$R_n(f)_C \leqslant c\frac{\|Mf'\|_{L_1}}{n}.$$

Proof of theorem 5.6. Without loss of generality we shall suppose that

$$\|f'\|_{L\log L} = \|f'\|_{L_{\phi_0}[0,1]} = 1, \quad \phi_0(x) = x\ln_+ x.$$

Then by (16) and (17) it follows that

$$\int_0^1 |f'(x)|\,dx \leqslant \|f'\|_{L_{\phi_0}}\|\chi_{[0,1]}\|_{L_{\psi_0}} \leqslant 1. \tag{20}$$

Let $n \geqslant 1$. Set $f(x) = f(1)$ for $x > 1$. Exactly as in the proof of theorem 5.4, in view of (20) there exist points x_i, $i = 0, 1, \dots, m$, where $n \leqslant m \leqslant 2n$, such that

$$0 = x_0 < x_1 < \cdots < x_{m-1} < 1 \leqslant x_m < 1 + \frac{1}{n}$$

and for each i, $0 \leqslant i \leqslant m-1$, one of the following two conditions is fulfilled:

(i) $\int_{\Delta_i} |f'(x)| \mathrm{d}x = \frac{1}{n}$ and $|\Delta_i| \leqslant 1/n$

(ii) $\int_{\Delta_i} |f'(x)| \mathrm{d}x \leqslant 1/n$ and $|\Delta_i| = 1/n$,

where $\Delta_i = [x_i, x_{i+1}]$.

Denote by N_1 and N_2 the sets of indices i, $0 \leqslant i \leqslant m-1$, which satisfy (i) and (ii) respectively. We need the following inequality:

$$\sum_{i \in N_1} |\Delta_i| \phi_0\left(\frac{1}{|\Delta_i|} \int_{\Delta_i} |f'(x)| \mathrm{d}x \right) \leqslant \int_0^1 \phi_0(|f'(x)|) \mathrm{d}x, \tag{21}$$

where $\Delta_i = [x_i, x_{i+1}]$.

Since ϕ_0 is convex, then by Jensen's inequality we have for $i \in N_1$

$$\phi_0\left(\frac{1}{|\Delta_i|} \int_{\Delta_i} |f'(x)| \mathrm{d}x \right) \leqslant \frac{1}{|\Delta_i|} \int_{\Delta_i} \phi_0(|f'(x)|) \mathrm{d}x.$$

Multiplying by $|\Delta_i|$ and summing over $i \in N_1$ we obtain (21).

By (18) we have

$$\int_0^1 \phi_0(|f'(x)|) \mathrm{d}x = \int_0^1 \phi_0\left(\frac{|f'(x)|}{\|f'\|_{L_{\phi_0}}} \right) \mathrm{d}x \leqslant 1.$$

From this, (21) and the fact that condition (i) holds for $i \in N_1$ it follows that

$$\sum_{i \in N_1} \ln_+ \left(\frac{1}{n|\Delta_i|} \right) \leqslant n.$$

Consequently, we have

$$\begin{aligned} \sum_{i=0}^{m-1} \ln\left(\mathrm{e} + \frac{2}{m|\Delta_i|} \right) &\leqslant \sum_{i=0}^{m-1} \ln\left(\mathrm{e} + \frac{2}{n|\Delta_i|} \right) \\ &\leqslant \sum_{i \in N_1} \ln\left(\mathrm{e} + \frac{2}{n|\Delta_i|} \right) + \sum_{i \in N_2} \ln(\mathrm{e} + 2) \leqslant cn. \end{aligned} \tag{22}$$

On the other hand

$$\|f - f(x_i)\|_{C(\Delta_i)} \leqslant \int_{\Delta_i} |f'(x)| \mathrm{d}x \leqslant \frac{1}{n}, \quad i = 0, 1, \dots, m-1. \tag{23}$$

Now we are ready to apply theorem 5.3 for the function f on $[0, x_m]$, $1 \leqslant x_m \leqslant 2$. In view of (22) and (23) we conclude that there exists a rational function r such that

$$\| f - r \|_{C[0,1]} \leqslant \frac{D}{n}$$

and

$$\deg r \leqslant D \sum_{i=0}^{m-1} \ln\left(\mathrm{e} + \frac{2}{m|\Delta_i|} \right) \leqslant Dcn,$$

$D = \text{constant}$.

Theorem 5.6 follows from this immediately. □

5.3.4 Functions with bounded variation and given modulus of continuity

Denote by $V(\omega) = V(M, [a,b], \omega)$ the set of all functions f continuous on $[a,b]$ for which $V_a^b f \leqslant M$ and $\omega(f;\delta) \leqslant \omega(\delta)$, $\delta \geqslant 0$, where $\omega(f;\delta)$ is the modulus of continuity of f and ω is a given modulus of continuity. A function ω is called a modulus of continuity if ω is a nondecreasing function on $[0, \infty)$, $\lim_{\delta \to 0} \omega(\delta) = \omega(0) = 0$ and $\omega(\delta_1 + \delta_2) \leqslant \omega(\delta_1) + \omega(\delta_2)$ for $\delta_1, \delta_2 \geqslant 0$.

First G. Freud (1966) and E. P. Dolženko and A.A. Abdulgaparov have shown that the class $V(c\delta^\alpha)$, $0 < \alpha < 1$, can be approximated uniformly by rational functions better than by polynomials. Later on A.P. Bulanov (1975b) considered the rational approximation of the class $V(\omega)$ for arbitrary ω. The final estimate was obtained by A.A. Pekarskii (1978b) and P.P. Petrushev (1977).

Theorem 5.7. *Let $f \in V(M, [a,b], \omega)$, where $M \geqslant 0$, ω is an arbitrary modulus of continuity. Then we have for $n \geqslant 1$*

$$R_n(f)_C \leqslant c \inf_{1 \leqslant t \leqslant n} \left\{ \frac{M}{t} + \omega\left(\frac{b-a}{t\mathrm{e}^{n/t}} \right) \right\}, \tag{24}$$

where c is an absolute constant.

Corollary 5.1. (i) *If $f \in V(c\delta^\alpha)$, $c > 0$, $0 < \alpha < 1$, then $R_n(f)_C = \mathrm{O}(\ln n/n)$.*

(ii) *If $f \in V(c(\ln(1/\delta))^{-\gamma})$, $c, \gamma > 0$, then $R_n(f)_C = \mathrm{O}(n^{-\gamma/(1+\gamma)})$.*

(iii) *If $f \in V(c(\underbrace{\ln \ln \cdots \ln}_{k}(1/\delta))^{-\gamma})$, $c, \gamma > 0$, $k \geqslant 2$, then $R_n(f)_C = \mathrm{O}((\underbrace{\ln \ln \cdots \ln}_{k-1} n)^{-\gamma})$.*

Remark. The estimate (24) is exact with respect to the order (see definition 5.2 in section 5.1) in the class $V(\omega)$, where the modulus of continuity ω satisfies $\lim_{\delta \to 0}(\omega(\delta)/\delta) = \infty$. This nontrivial fact, to be established in section 11.2, shows that there is no o-effect in this case. However, we shall prove in section

10.2 that the o-effect appears for the uniform rational approximation of all absolutely continuous functions f in $V(\omega)$ for some ω. Note that for polynomial approximation of such functions we have only $E_n(f)_C = O(\omega(n^{-1}))$.

Proof of theorem 5.7. We shall use theorem 5.3. First, consider the trivial case $1 \leqslant n \leqslant 2D$, where $D > 1$ is the constant from theorem 5.3. Obviously, we have

$$R_n(f)_C \leqslant V_a^b f \leqslant M \leqslant \frac{2DM}{n}.$$

This estimate implies (24) in the case $1 \leqslant n \leqslant 2D$.

Now, let $n > 2D$. We shall denote for short $\omega(\delta) = \omega(f;\delta)$, $\delta \geqslant 0$. We shall consider three situations.

(i) If $M < \omega((b-a)/e^{n/2D})$, then

$$R_n(f)_C \leqslant V_a^b f \leqslant M < \omega\left(\frac{(b-a)}{e^{n/2D}}\right)$$

which implies (24).

(ii) Let $2DM/n > \omega(2D(b-a)/ne)$. Then by Jackson's theorem (see theorem 3.10)

$$R_n(f)_C \leqslant E_n(f)_C \leqslant c\omega\left(\frac{b-a}{n}\right) \leqslant c\omega\left(\frac{2D(b-a)}{ne}\right) < \frac{2DcM}{n}$$

which implies (24).

(iii) Let $M \geqslant \omega((b-a)/e^{n/2D})$ and $2DM/n \leqslant \omega(2D(b-a)/ne)$. The function $F_1(t) = M/t$ is strictly decreasing and the function $F_2(t) = \omega((b-a)/te^{n/2Dt})$ is nondecreasing on $[1, n/2D]$. Consequently, there exists exactly one point $t_0 \in [1, n/2D]$ such that

$$\frac{M}{t_0} = \omega\left(\frac{b-a}{t_0 e^{n/2Dt_0}}\right). \tag{25}$$

Now we define a partition of $[a,b]$ in the following way. Pick $x_0 = a$ and suppose that $x_0, x_1, \ldots, x_k$ are defined. To define x_{k+1} we consider the set

$$A_{k+1} = \left\{x \in [a,b]: x > x_k, |f(x) - f(x_k)| = \frac{M}{t_0}\right\}.$$

If $A_{k+1} \neq \varnothing$ we put $x_{k+1} = \inf\{x: x \in A_{k+1}\}$.

Clearly, after a finite number $m+1$ of steps we get $A_{m+1} = \varnothing$. Set $x_m = b$. So we have a division of $[a,b]$,

$$a = x_0 < x_1 < \cdots < x_m = b,$$

with the properties

$$\frac{M}{t_0} \leqslant |f(x_i) - f(x_{i+1})| \leqslant \frac{2M}{t_0}, \tag{26}$$

$$\| f(x) - f(x_i) \|_{C[x_i, x_{i+1}]} \leqslant \frac{2M}{t_0}, \quad i = 0, 1, \ldots, m-1. \tag{27}$$

Now we are able to apply theorem 5.3. Using (27) we conclude that there exists a rational function r such that

$$\| f - r \|_{C[a,b]} \leqslant \frac{2DM}{t_0} \tag{28}$$

and

$$\deg r \leqslant D \sum_{i=0}^{m-1} \ln\left(e + \frac{b-a}{m|\Delta_i|} \right) \tag{29}$$

where $\Delta_i = [x_i, x_{i+1}]$, $D > 1$.

It follows from (25) and our assumptions that

$$\frac{M}{t_0} = \omega\left(\frac{b-a}{t_0 e^{n/2Dt_0}} \right) \leqslant \inf_{1 \leqslant t \leqslant n/2D} \left\{ \frac{M}{t} + \omega\left(\frac{b-a}{t e^{n/2Dt}} \right) \right\}.$$

From this and (28) it follows that

$$\| f - r \|_{C[a,b]} \leqslant c \inf_{1 \leqslant t \leqslant n} \left\{ \frac{M}{t} + \omega\left(\frac{b-a}{t e^{n/t}} \right) \right\}.$$

It remains to estimate $\deg r$. Since $V_a^b f \leqslant M$, then by (26)

$$\frac{mM}{t_0} \leqslant \sum_{i=0}^{m-1} |f(x_i) - f(x_{i+1})| \leqslant M$$

and therefore $m \leqslant t_0$. Also, by (25) and (26) we have for $i = 0, 1, \ldots, m-1$

$$\omega\left(\frac{b-a}{t_0 e^{n/2Dt_0}} \right) = \frac{M}{t_0} \leqslant |f(x_i) - f(x_{i+1})|$$

and consequently

$$|\Delta_i| \geqslant \frac{b-a}{t_0 e^{n/2Dt_0}}.$$

Then by (29) we get

$$\deg r \leqslant Dm \ln\left(e + \frac{t_0}{m} e^{n/2Dt_0} \right)$$

$$\leqslant Dt_0 \ln(e + e^{n/2Dt_0}) \leqslant n,$$

where we have used the fact that the function $F(x) = x \ln(e + 1/x)$ is increasing on $(0, \infty)$. This implies (24). □

Proof of corollary 5.1. The assertions (i), (ii) and (iii) in corollary 5.1 follow immediately setting consecutively $t = n/(\alpha \ln n)$, $t = n^{\gamma/(1+\gamma)}$ and $t = (\underbrace{\ln \ln \cdots \ln}_{k-1} n)^{\gamma}$ in (24).

5.4 DeVore's method

5.4.1 Hardy–Littlewood maximal function

It is well-known that the Hardy–Littlewood maximal function and its modifications have many important applications in analysis – for example in the differentiation theory of multivariate functions (see E. Stein (1970), E. Stein and G. Weiss (1971)). Ronald DeVore (1983) was the first who showed that the maximal functions may be very useful in the theory of rational approximations. Before describing his method, we shall give the properties of the Hardy–Littlewood maximal function we need.

Let f be an integrable function on the interval $[a, b]$ or $(-\infty, \infty)$. For every $x \in [a, b]$ (or $(-\infty, \infty)$) we define the function

$$(Mf)(x) = \sup_{\Delta : x \in \Delta} \frac{1}{|\Delta|} \int_{\Delta} |f(t)| \, dt,$$

where the sup is taken over all subintervals $\Delta = [c, d] \subset [a, b]$ $((-\infty, \infty))$, which contain the point x, and $|\Delta| = d - c$ is the length of the interval Δ.

The most essential property of the Hardy–Littlewood maximal function is perhaps the following one.

Theorem 5.8. *Let f be an integrable function on $(-\infty, \infty)$ (i.e. $f \in L_1(-\infty, \infty)$). Then for every $\alpha > 0$ we have*

$$\text{mes}\{x : (Mf)(x) > \alpha\} \leqslant \frac{2}{\alpha} \| f \|_{L_1(-\infty,\infty)},$$

where $\text{mes}\, A = |A|$ *denotes the Lebesgue measure of the set A.*

Remark. Usually this property of the maximal function Mf is expressed as follows. *The operator Mf is of the weak type* $(1, 1)$.

To prove theorem 5.8 we shall need a covering lemma of Vitali type.

Lemma 5.9. *Let $\Delta_1, \ldots, \Delta_n$ be a finite family of open intervals in $(-\infty, \infty)$. Then there exists a subfamily of intervals $\delta_1, \delta_2, \ldots, \delta_m$ such that δ_i are pairwise disjoint and*

$$\sum_{i=1}^{m} |\delta_i| \geqslant \frac{1}{2} \left| \bigcup_{i=1}^{n} \Delta_i \right|.$$

Proof. Obviously we can consider only the case when the intervals Δ_i, $i = 1, \ldots, n$, are such that no interval Δ_i is contained in the union of the others.

Let $\Delta_i = [a_i, b_i]$ and let us assume that we have ordered the intervals in such a way that

$$a_1 < a_2 < \cdots < a_n.$$

Then $b_{i+1} > b_i$ holds since otherwise $\Delta_{i+1} \subset \Delta_i$. Also $a_{i+1} > b_{i-1}$ since otherwise $\Delta_i \subset \Delta_{i-1} \cup \Delta_{i+1}$. Therefore the even-numbered intervals and odd-numbered intervals are disjoint. Obviously

$$\sum_{i \text{ even}} |\Delta_i| + \sum_{i \text{ odd}} |\Delta_i| \geqslant \left| \bigcup_{i=1}^{n} \Delta_i \right|. \tag{1}$$

For $\{\delta_i\}$ we take $\{\Delta_i : i \text{ even}\}$ or $\{\Delta_i : i \text{ odd}\}$ in dependence on which one of the sums $\sum_{i \text{ even}}$ or $\sum_{i \text{ odd}}$ is bigger. Then from (1) it follows that

$$\sum_{i=1}^{m} |\delta_i| \geqslant \frac{1}{2} \left| \bigcup_{i=1}^{n} \Delta_i \right|. \qquad \square$$

Proof of theorem 5.8. Let $f \in L(-\infty, \infty)$ and $\alpha > 0$ be given. Using the definition of Mf it is not difficult to see that the set $G_\alpha = \{x : (Mf)(x) > \alpha\}$ is open and therefore measurable. Again using the definition of Mf we get that for every $x \in G_\alpha$ we have an open interval Δ_x containing x such that

$$\frac{1}{|\Delta_x|} \int_{\Delta_x} |f(t)| \, \mathrm{d}t > \alpha. \tag{2}$$

For every compact subset K of G_α we can choose a finite subset of intervals Δ_{x_i}, $i = 1, \ldots, n$, $x_i \in K$, such that $K \subset \bigcup_{i=1}^{n} \Delta_{x_i}$. Applying lemma 5.8 to the intervals Δ_{x_i}, $i = 1, \ldots, n$, we obtain a subset of pairwise disjoint intervals Δ_{y_i}, $i = 1, \ldots, m$, such that

$$\sum_{i=1}^{m} |\Delta_{y_i}| \geqslant \frac{1}{2} \left| \bigcup_{i=1}^{m} \Delta_{x_i} \right| \geqslant \tfrac{1}{2} |K|. \tag{3}$$

Since the intervals Δ_{y_i} are pairwise disjoint, from (2) it follows that

$$\begin{aligned} \sum_{i=1}^{m} \left| \Delta_{y_i} \right| &< \frac{1}{\alpha} \sum_{i=1}^{m} \int_{\Delta_{y_i}} |f(t)| \, \mathrm{d}t \\ &= \frac{1}{\alpha} \int_{\bigcup_{i=1}^{m} \Delta_{y_i}} |f(t)| \, \mathrm{d}t \leqslant \frac{1}{\alpha} \int_{-\infty}^{\infty} |f(t)| \, \mathrm{d}t = \frac{1}{\alpha} \| f \|_{L(-\infty,\infty)}. \end{aligned} \tag{4}$$

The inequalities (3) and (4) give us

$$|K| \leqslant \frac{2}{\alpha} \| f \|_{L(-\infty,\infty)}.$$

Since the compact subset $K \subset G_\alpha$ was an arbitrary one, we have

$$|G_\alpha| = \operatorname{mes} \{x : (Mf)(x) > \alpha\} \leqslant \frac{2}{\alpha} \| f \|_{L(-\infty,\infty)}. \qquad \square$$

Remark. If $f \in L_1[a,b]$, then setting $f(x)=0$ if $x \notin [a,b]$ we obtain the estimate

$$\text{mes}\{x:(Mf)(x)>\alpha\} \leqslant \frac{2}{\alpha}\int_a^b |f(t)|\,dt.$$

The function

$$m(g;\alpha)=\text{mes}\{x:|g(x)|>\alpha\}$$

is called a distribution function of g. Obviously if g is an integrable function then $m(g;\alpha)$ is defined for $\alpha>0$ and is a decreasing function of α.

Lemma 5.10. *Let $f \in L_p(-\infty,\infty)$, $0<p<\infty$. Then*

$$\int_{-\infty}^{\infty} |f(x)|^p\,dx = \int_0^{\infty} p\alpha^{p-1} m(f;\alpha)\,d\alpha. \tag{5}$$

Proof. Let

$$G=\{(x,\alpha): x \in (-\infty,\infty), \quad 0<\alpha<|f(x)|^p\}.$$

Using Fubini's theorem we obtain

$$\begin{aligned}\int_{-\infty}^{\infty} |f(x)|^p\,dx &= \int_G\int dx\,d\alpha = \int_{-\infty}^{\infty}\int_0^{|f(x)|} p\alpha^{p-1}\,d\alpha\,dx \\ &= \int_0^{\infty} p\alpha^{p-1}\,\text{mes}\{x:|f(x)|>\alpha\}\,d\alpha \\ &= \int_0^{\infty} p\alpha^{p-1} m(f;\alpha)\,d\alpha.\end{aligned}$$

□

Now we shall prove the results for the Hardy–Littlewood maximal function we need. The first one is the following

Theorem 5.9. *Let $f \in L_1[a,b]$. For $0<p<1$ we have*

$$\left(\int_a^b |(Mf)(x)|^p\,dx\right)^{1/p} \leqslant c(p,b-a)\|f\|_{L_1[a,b]},$$

where the constant $c(p,b-a)$ depends only on p and the length $b-a$ of the interval $[a,b]$.

Remark. The constant $c(p,b-a)$ tends to ∞ as p tends to $1-0$.

Assume that $\|f\|_{L_1[a,b]}=1$. We have

$$\int_a^b ((Mf)(x))^p\,dx = \int_{\{x:(Mf)(x)>1\}} + \int_{\{x:(Mf)(x)\leqslant 1\}} = I_1+I_2.$$

For the first integral I_1 we have from lemma 5.10 and theorem 5.8 (see the remark after the theorem)

$$I_1 = \int_{\{x:(Mf)(x)>1\}} ((Mf)(x))^p \mathrm{d}x = p\int_1^\infty \alpha^{p-1} m(Mf;\alpha)\mathrm{d}\alpha$$
$$\leqslant p\int_1^\infty \alpha^{p-1}\frac{2}{\alpha}\int_a^b |f(t)|\mathrm{d}t\,\mathrm{d}\alpha \leqslant \frac{2p}{1-p}$$

since $\| f \|_{L_1[a,b]} = 1$.

For I_2 we have trivially $I_2 \leqslant b-a$, therefore

$$\int_a^b ((Mf)(x))^p \mathrm{d}x \leqslant \frac{2p}{1-p} + b - a = c'(p, b-a).$$

If $\| f \|_{L_1[a,b]} \neq 1$, we set $g = f/\| f \|_{L_1[a,b]}$. Then $\| g \|_{L_1[a,b]} = 1$ and using the fact that $M(\lambda g) = |\lambda| Mg$ we obtain

$$\int_a^b ((Mg)(x))^p \mathrm{d}x = \int_a^b \left(\left(M\frac{f}{\| f \|_{L_1[a,b]}} \right)(x) \right)^p \mathrm{d}x \leqslant c'(p, b-a),$$

i.e.

$$\int_a^b ((Mf)(x))^p \mathrm{d}x \leqslant c'(p, b-a) \| f \|_{L_1[a,b]}^p,$$

therefore

$$\left(\int_a^b ((Mf)(x))^p \mathrm{d}x \right)^{1/p} \leqslant (c'(p, b-a))^{1/p} \| f \|_{L_1[a,b]}. \qquad \square$$

The second result that we need is the famous M. Rieze theorem.

Theorem 5.10. *Let $f \in L_p(-\infty, \infty)$, $1 < p \leqslant \infty$. Then*

$$\| Mf \|_p \leqslant c(p) \| f \|_p,$$

where the constant $c(p)$ depends only on p.

Proof. If $p = \infty$, then obviously $\| Mf \|_\infty \leqslant \| f \|_\infty$. Now let $1 < p < \infty$. For every fixed $\alpha \geqslant 0$ let us set

$$f_\alpha(x) = \begin{cases} f(x), & |f(x)| > \alpha/2, \\ 0, & |f(x)| \leqslant \alpha/2. \end{cases}$$

Then $|f(x)| \leqslant |f_\alpha(x)| + \alpha/2$ for every x and therefore $(Mf)(x) \leqslant (Mf_\alpha)(x) + \alpha/2$. We obtain from this inequality

$$A_\alpha = \{x: (Mf)(x) > \alpha\} \subset \{x: (Mf_\alpha)(x) > \alpha/2\} = B_\alpha$$

and therefore $|A_\alpha| \leqslant |B_\alpha|$.

Using theorem 5.8 we obtain:

$$|A_\alpha| = m(Mf;\alpha) \leqslant |B_\alpha| = m(Mf_\alpha;\alpha/2) \leqslant \frac{4}{\alpha}\|f_\alpha\|_1. \tag{6}$$

But from the definition of f_α we get

$$\|f_\alpha\|_1 = \int_{|f(x)|\geqslant\alpha/2} |f(x)|\,\mathrm{d}x. \tag{7}$$

The inequalities (6) and (7) give us

$$m(Mf;\alpha) \leqslant \frac{4}{\alpha}\int_{|f(x)|\geqslant\alpha/2} |f(x)|\,\mathrm{d}x. \tag{8}$$

Now we shall use lemma 5.10. We have

$$\int_{-\infty}^{\infty} |(Mf)(x)|^p\,\mathrm{d}x = \int_0^\infty p\alpha^{p-1}m(Mf;\alpha)\,\mathrm{d}\alpha \leqslant \int_0^\infty \int_{|f(x)|\geqslant\alpha/2} p\alpha^{p-1}\frac{4}{\alpha}|f(x)|\,\mathrm{d}x\,\mathrm{d}\alpha$$

$$= 4p\int_{-\infty}^{\infty}|f(x)|\int_0^{2|f(x)|}\alpha^{p-2}\,\mathrm{d}\alpha\,\mathrm{d}x$$

$$= 4p\int_{-\infty}^{\infty}|f(x)|\frac{(2|f(x)|)^{p-1}}{p-1}\,\mathrm{d}x = \frac{2^{p+1}p}{p-1}\int_{-\infty}^{\infty}|f(x)|^p\,\mathrm{d}x.$$

Hence the theorem follows with a constant $c(p) = (2^{p+1}p/(p-1))^{1/p}$. □

Remark 1. We see that $c(p)\to\infty$ when $p\to 1$, more exactly $c(p) = \mathrm{O}(1/(p-1))$, $p\to 1$.

Remark 2. We give theorem 5.10 for the interval $(-\infty,\infty)$. If $f\in L_p[a,b]$, $1<p\leqslant\infty$, we also have

$$\|Mf\|_{L_p[a,b]} \leqslant c(p)\|f\|_{L_p[a,b]}. \tag{9}$$

This follows immediately, since we can set $f(x)=0$ for $x\notin[a,b]$, then $\|f\|_{L_p(-\infty,\infty)} = \|f\|_{L_p[a,b]}$, $\|Mf\|_{L_p[a,b]} \leqslant \|Mf\|_{L_p(-\infty,\infty)}$.

We shall use M. Rieze's theorem in the form (9).

Lemma 5.11. *Let $f\in L(-\infty,\infty)$ and $[a,b]$ be a given finite interval. Then Mf attains its* inf *in $[a,b]$.*

Proof. Indeed, using the definition of Mf it is not difficult to see that the set

$$\{x: x\in[a,b], (Mf)(x)\geqslant\alpha\}$$

is closed. □

5.4.2 More on the class $W_p^1, p>1$

First we shall demonstrate the method of R. DeVore on the functions belonging to the class W_p^1, $p>1$. We proved the following theorem in 5.3.2:

Theorem 5.4. *Let $f \in W_p^1[0,1]$, $p > 1$. Then*

$$R_n(f)_{C[0,1]} \leqslant c(p) n^{-1} \| f' \|_p,$$

where the constant $c(p)$ depends only on p.

Here we present another proof of this theorem.

Proof. We can assume that $\| f' \|_p = 1$. Let us divide the interval $[0,1]$ into $2n$ subintervals $\Delta_j = [x_{j-1}, x_j]$, $j = 1, \dots, 2n$, $0 = x_0 < x_1 < \cdots < x_{2n} = 1$, such that

$$\left.\begin{array}{l} \text{(i) } |\Delta_j| \leqslant n^{-1}, \quad |\Delta_j| = x_j - x_{j-1}, \quad j = 1, \dots, 2n. \\ \text{(ii) } \int_{\Delta_j} |f'(t)|^p \mathrm{d}t \leqslant n^{-1}, \quad j = 1, \dots, 2n. \end{array}\right\} \tag{1}$$

Since $\| f' \|_p = 1$, such intervals can be got by first finding n intervals which satisfy (i) and then subdividing them so as to guarantee (ii).

Now let us choose $\xi_j \in \Delta_j$ so that

$$(Mf')(\xi_j) = \inf \{ (Mf')(x) : x \in \Delta_j \},$$

where Mf' is the Hardy–Littlewood maximal function for f'. The inf is attained by lemma 5.10.

We set

$$\phi_j(x) = (1 + |\Delta_j|^{-2}(x - \xi_j)^2)^{-2}, \quad j = 1, \dots, 2n,$$

$$\phi = \sum_{j=1}^{2n} \phi_j, \quad R_j = \phi_j / \phi,$$

$$R(x) = \sum_{j=1}^{2n} f(\xi_j) R_j(x).$$

We have the following properties of the functions ϕ_j, ϕ and R:

$\phi_j(x) \geqslant 2^{-2}$, for $x \in \Delta_j$, $j = 1, \dots, 2n$, therefore

$$\phi(x) = \sum_{j=1}^{2n} \phi_j(x) \geqslant 2^{-2}, \quad x \in [0,1]; \tag{2}$$

ϕ_j is a rational function of degree 4, therefore

$$\phi \in R_{8n}, \quad R \in R_{16n}.$$

Let us mention also that

$$\left.\begin{array}{l} \sum_{j=1}^{2n} R_j(x) = 1, \quad R_j(x) \geqslant 0, \quad x \in [0,1], \\ R_j(x) \leqslant 4(1 + |\Delta_j|^{-2}(x - \xi_j)^2)^{-2}, \quad x \in [0,1]. \end{array}\right\} \tag{3}$$

Let us estimate now $|f(x) - R(x)|$, $x \in [0,1]$. First we have, using the choice of ξ_j, theorem 5.10 (more exactly remark 2 after the theorem), and (1) (ii),

$$
\begin{aligned}
|f(x)-f(\xi_j)| &\leqslant |x-\xi_j|\frac{1}{|x-\xi_j|}\left|\int_{\xi_j}^{x} f'(t)\mathrm{d}t\right| \leqslant |x-\xi_j|(Mf')(\xi_j)\\
&\leqslant |x-\xi_j|\left(\frac{1}{|\Delta_j|}\int_{\Delta_j}((Mf')(t))^p\mathrm{d}t\right)^{1/p}\\
&\leqslant |x-\xi_j|\,|\Delta_j|^{-1/p}c(p)\left(\int_{\Delta_j}|f'(t)|^p\mathrm{d}t\right)^{1/p}\\
&\leqslant c(p)|x-\xi_j|(n|\Delta_j|)^{-1/p},
\end{aligned}
\tag{4}
$$

where $c(p)$ is a constant depending only on p ($c(p)$ is the constant from theorem 5.10: $c(p)=2^{1+1/p}(p/(p-1))^{1/p}$).

Since $\sum_{j=1}^{2n} R_j(x)=1$ for $x\in[0,1]$, we have, using (3) and (4),

$$
\begin{aligned}
|f(x)-R(x)| &\leqslant \sum_{j=1}^{2n}|f(x)-f(\xi_j)||R_j(x)|\\
&\leqslant 4c(p)\sum_{j=1}^{2n}|x-\xi_j|(n\Delta_j)^{-1/p}(1+|\Delta_j|^{-2}(x-\xi_j)^2)^{-2}\\
&= 4c(p)\sum_{\nu=1}^{\infty}s_\nu(x),
\end{aligned}
\tag{5}
$$

where $s_\nu(x)$ is the sum of those terms for intervals Δ_j which satisfy $2^{-\nu}n^{-1}<|\Delta_j|\leqslant 2^{-\nu+1}n^{-1}$ (recall that $|\Delta_j|\leqslant n^{-1}$, because of (1) (i)).

It is easy to see that

$$
|x-\xi_j|\leqslant|\Delta_j|(1+|\Delta_j|^{-2}(x-\xi_j)^2)^{1/2}. \tag{6}
$$

Indeed, $|\Delta_j|\;(1+|\Delta_j|^{-2}(x-\xi_j)^2)^{1/2}=(|\Delta_j|^2+(x-\xi_j)^2)^{1/2}\geqslant|x-\xi_j|$.

Using (6) and the definition of s_ν we obtain

$$
\begin{aligned}
s_\nu(x) &\leqslant (n\cdot 2^{-\nu}n^{-1})^{-1/p}2^{-\nu+1}n^{-1}\sum_{\Delta_j\in A_\nu}(1+|\Delta_j|^{-2}(x-\xi_j)^2)^{-3/2}\\
&= 2\cdot 2^{\nu(1/p-1)}n^{-1}\sum_{\Delta_j\in A_\nu}(1+|\Delta_j|^{-2}(x-\xi_j)^2)^{-3/2},
\end{aligned}
\tag{7}
$$

where A_ν is the set of those Δ_j which appear in s_ν. Each $\Delta_j\in A_\nu$ has a length $\geqslant 2^{-\nu}n^{-1}$ and the intervals are disjoint. Therefore for any integer $k\geqslant 0$ there are four ξ_j at most which satisfy

$$
k\cdot 2^{-\nu}n^{-1}\leqslant|x-\xi_j|\leqslant(k+1)2^{-\nu}n^{-1}.
$$

Using this in (7) we obtain

$$
s_\nu(x)\leqslant 8n^{-1}\cdot 2^{-\nu(1-1/p)}\sum_{k=0}^{\infty}(1+(k/2)^2)^{-3/2}\leqslant c\cdot 16n^{-1}2^{-\nu(1-1/p)}, \tag{8}
$$

where c is an absolute constant.

From (5) and (8) we obtain, using that $p > 1$,

$$|f(x) - R(x)| \leqslant 4c(p)c \cdot 16n^{-1} \sum_{\nu=1}^{\infty} 2^{-\nu(1-1/p)} = c'(p)n^{-1},$$

where the constant $c'(p)$ depends only on p. Since $R \in R_{16n}$ (see (2)), the theorem follows. □

5.4.3 Functions with derivative of bounded variation

Here we shall apply the method of R. DeVore for rational uniform approximation of functions with derivative with bounded variation. In section 5.2 was considered the general case of rational uniform approximation of the class V_r (theorem 5.1). For $r = 1$ we have the following.

Theorem 5.1′. *Let $f \in V_1[a,b]$. Then*

$$R_n(f)_{C[a,b]} \leqslant c \frac{(b-a)V_a^b f'}{n^2},$$

where c is an absolute constant.

Proof. (DeVore, 1983). If we consider the function $g(x) = f(a + (b-a)x)$ we see that it is sufficient to take only the case when $[a,b] = [0,1]$. We see also that we can consider only the case when $V_0^1 f' = 1$. Since any function f with $V_0^1 f' < \infty$ can be approximated uniformly by functions g with $\|g''\|_{L(0,1)} = \|g''\|_1 \leqslant V_0^1 f'$, it will be sufficient to prove the following estimate only.

If $f'' \in L(0,1)$ and $\|f''\|_1 = 1$ then

$$R_n(f)_{C[0,1]} \leqslant cn^{-2}, \tag{1}$$

where c is an absolute constant.

Let f be such that $\|f''\|_1 = 1$. Using theorem 5.9 and putting $p = \frac{3}{4}$ (every p, $\frac{1}{2} < p < 1$, will work infact), we have the estimate

$$\left(\int_0^1 ((Mf'')(x))^{3/4} \, dx \right)^{4/3} \leqslant c \|f''\|_1 = c, \tag{2}$$

where c is an absolute constant and Mf'' is the Hardy–Littlewood maximal function for f''.

Now let us choose the intervals $\Delta_j = [x_{j-1}, x_j]$, $j = 1, \ldots, 2n$, $0 = x_0 < x_1 < \cdots < x_{2n} = 1$, such that

$$\left.\begin{array}{l} \text{(i)}\ |\Delta_j| \leqslant n^{-1}, \quad |\Delta_j| = x_j - x_{j-1}, \quad j = 1, \ldots, 2n, \\ \text{(ii)}\ \int_{\Delta_j} ((Mf'')(x))^{3/4} \, dx \leqslant c^{3/4} n^{-1}, \end{array}\right\} \tag{3}$$

where c is the constant from (2).

Since by (2) $\int_0^1 ((Mf'')(x))^{3/4} \, dx \leqslant c^{3/4}$, such intervals can be obtained by first finding n intervals which satisfy (i) and further subdividing them so as to guarantee (ii).

Later on we shall work as in subsection 5.4.2. Choose $\xi_j \in \Delta_j$ so that

$$(Mf'')(\xi_j) = \inf\{(Mf'')(x): x \in \Delta_j\}.$$

The inf is attained by lemma 5.11. Now let us set

$$P_{\xi_j}(x) = f(\xi_j) + f'(\xi_j)(x - \xi_j)$$
$$\phi_j(x) = (1 + |\Delta_j|^{-2}(x - \xi_j)^2)^{-2}, \quad j = 1, \ldots, 2n$$
$$\phi = \sum_{j=1}^{2n} \phi_j, \quad R_j = \phi_j/\phi$$
$$R = \sum_{j=1}^{2n} P_{\xi_j} R_j.$$

As in subsection 5.4.2 we have the following properties:

$$\phi_j(x) \geqslant 2^{-2} \text{ for } x \in \Delta_j,\ j = 1, \ldots, 2n, \text{ therefore}$$

$$\phi(x) = \sum_{j=1}^{2n} \phi_j(x) \geqslant 2^{-2}, \quad x \in [0, 1]; \tag{4}$$

$$\phi_j \in R_4, \quad \phi \in R_{8n}, \quad R \in R_{16n}$$

(since P_{ξ_j} are linear).

Let us remark also that

$$\left.\begin{aligned} &\sum_{j=1}^{2n} R_j(x) = 1, \quad R_j(x) \geqslant 0, \quad x \in [0, 1], \\ &R_j(x) \leqslant 4(1 + |\Delta_j|^{-2}(x - \xi_j)^2)^{-2}, \quad x \in [0, 1]. \end{aligned}\right\} \tag{5}$$

Let us estimate now $|f(x) - R(x)|$. Using the Taylor formula with integral remainder, the choice of ξ_j and (3), (ii) we get

$$\begin{aligned} |f(x) - P_{\xi_j}(x)| &= \left| f(x) - f(\xi_j) - \frac{x - \xi_j}{1!} f'(\xi_j) \right| = \left| \int_{\xi_j}^{x} (x - t) f''(t) \mathrm{d}t \right| \\ &\leqslant |x - \xi_j|^2 \frac{1}{|x - \xi_j|} \left| \int_{\xi_j}^{x} |f''(t)| \mathrm{d}t \right| \leqslant |x - \xi_j|^2 (Mf'')(\xi_j) \\ &= (x - \xi_j)^2 \min_{t \in \Delta_j} (Mf'')(t) \\ &\leqslant (x - \xi_j)^2 \left(\frac{1}{|\Delta_j|} \int_{\Delta_j} ((Mf'')(t))^{3/4} \mathrm{d}t \right)^{4/3} \\ &\leqslant c(x - \xi_j)^2 (n|\Delta_j|)^{-4/3}. \end{aligned}$$

Using this together with (5), we get

$$\begin{aligned} |f(x) - R(x)| &\leqslant \sum_{j=1}^{2n} |f(x) - P_{\xi_j}(x)| |R_j(x)| \\ &\leqslant 4c \sum_{j=1}^{2n} (x - \xi_j)^2 (n|\Delta_j|)^{-4/3} (1 + |\Delta_j|^{-2}(x - \xi_j)^2)^{-2} = 4c \sum_{\nu=0}^{\infty} s_\nu(x), \end{aligned} \tag{6}$$

where $s_\nu(x)$ is the sum of those terms for intervals Δ_j which satisfy $2^{-\nu}n^{-1} < |\Delta_j| \leqslant 2^{-\nu+1}n^{-1}$.

We have

$$(x-\xi_j)^2 \leqslant |\Delta_j|^2(1+|\Delta_j|^{-2}(x-\xi_j)^2). \tag{7}$$

Using (7) and the definition of s_ν we get

$$s_\nu(x) \leqslant 4(2^\nu)^{4/3}2^{-2\nu}n^{-2} \sum_{\Delta_j\in A_\nu} (1+|\Delta_j|^{-2}(x-\xi_j)^2)^{-1}, \tag{8}$$

where A_ν is the set of those Δ_j which appear in s_ν. Each $\Delta_j \in A_\nu$ has a length $\geqslant 2^{-\nu}n^{-1}$ and the Δ_j are disjoint. Therefore for any integer $k \geqslant 0$ there are four ξ_j at most such that

$$k\cdot 2^{-\nu}n^{-1} < |x-\xi_j| \leqslant (k+1)2^{-\nu}n^{-1}.$$

Using this in (8) we obtain

$$s_\nu(x) \leqslant 16n^{-2}2^{-2\nu/3} \sum_{k=0}^{\infty} (1+(k/2)^2)^{-1} \leqslant c'n^{-2}\cdot 2^{-2\nu/3}, \tag{9}$$

where c' is an absolute constant.

From (6) and (9) we obtain

$$|f(x)-R(x)| \leqslant 4cc'n^{-2} \sum_{\nu=0}^{\infty} 2^{-2\nu/3} = c''n^{-2}. \qquad \square$$

5.5 Convex functions

One of the interesting classes of functions for rational approximations is the class $\mathrm{Conv}_M[a,b]$ of all convex and continuous functions f on the interval $[a,b]$ such that $\|f\|_{C[a,b]} \leqslant M$. A.P. Bulanov (1969) showed that there exists a universal order $\mathrm{O}(\ln^2 n/n)$ for the uniform rational approximation of the class $\mathrm{Conv}_M[a,b]$, while we have only $E_n(f)_{C[a,b]} = \mathrm{o}(1)$ for $f \in \mathrm{Conv}_M[a,b]$. We shall prove first the exact order $\mathrm{O}(n^{-1})$ obtained by V.A. Popov and P.P. Petrushev (1977). There are many results concerning the uniform rational approximation of convex functions with a given modulus of continuity (see the notes at the end of the chapter). We shall consider here only the class of convex functions with modulus of continuity of the type $\omega(f;\delta)_C = \mathrm{O}(\delta^\alpha)$. The rational approximation in uniform and L_p-metric of some classes of piecewise convex functions and functions with piecewise convex derivatives will be investigated in chapter 7.

Theorem 5.11. *Let $M>0$ and let $[a,b]$ be an arbitrary compact interval. Then*

$$\sup_{f\in \mathrm{Conv}_M[a,b]} R_n(f)_{C[a,b]} \leqslant cMn^{-1},$$

where c is an absolute constant.

Proof. Theorem 5.11 we shall prove applying another method for rational approximation. Note that this method was used for obtaining the final results for uniform rational approximation of the class V_r (see also the notes at the end of the chapter).

Let us denote by $K[a,b]$ the set of all convex nondecreasing functions on the interval $[a,b]$, continuous at the point b, and by $K_M[a,b]$ the subset of $K[a,b]$ consisting of those functions $f \in K[a,b]$ for which $\|f\|_{C[a,b]} \leqslant M$, $f(x) \geqslant 0$ for $x \in [a,b]$.

Let us set

$$\phi_n = \sup\{R_n(f)_{C[0,1]} : f \in K_1[0,1]\},$$

$$\phi_{n,A} = \sup_{f \in K_1[0,1]} \inf\{\|f - r\|_{C[0,1]} : r \in R_n, 0 \leqslant r(x) \leqslant A \text{ for every } x\}.$$

Lemma 5.12. *We have the equality*

$$\sup\{R_n(f)_{C[a,b]} : f \in K_M[a,b]\} = M\phi_n.$$

Proof. Let $f \in K_M[a,b]$. Then $g(x) = M^{-1} f(a + (b-a)x)$ belongs to $K_1[a,b]$ and consequently there exists $r \in R_n$ such that $\|g - r\|_{C[0,1]} \leqslant \phi_n$, i.e.

$$\phi_n \geqslant \max_{x \in [0,1]} |M^{-1} f(a + (b-a)x) - r(x)| = M^{-1} \max_{x \in [a,b]} \left| f(x) - Mr\left(\frac{x-a}{b-a}\right)\right|.$$

This shows that

$$\sup\{R_n(f)_{C[a,b]} : f \in K_M[a,b]\} \leqslant M\phi_n.$$

The converse is established similarly. □

Lemma 5.13. *We have the inequality*

$$\sup_{\substack{f \in K_1[a,b] \\ V_a^b f \leqslant M}} \inf_{\substack{r \in R_n \\ 0 \leqslant r(x) \leqslant AM+1}} \|f - r\|_{C[a,b]} \leqslant M\phi_{n,A}.$$

Proof. Let $f \in K_1[a,b]$ and $V_a^b f \leqslant M$, $M > 0$. Then the function $g(x) = M^{-1}(f(a + (b-a)x) - f(a))$ belongs to $K_1[0,1]$. Let $r \in R_n$ be such that $0 \leqslant r(x) \leqslant A$ for every x and $\|g - r\|_{C[0,1]} \leqslant \phi_{n,A}$, i.e.

$$\phi_{n,A} \geqslant \max_{x \in [0,1]} |g(x) - r(x)| = M^{-1} \max_{x \in [a,b]} \left| f(x) - \left(f(a) + Mr\left(\frac{x-a}{b-a}\right)\right)\right|.$$

Since $0 \leqslant f(a) + Mr((x-a)/(b-a)) \leqslant 1 + MA$ and $f(a) + r((x-a)/(b-a)) \in R_n$, the lemma follows. □

Lemma 5.14. *Let $x_0 \in (0,1)$. Then for arbitrary δ, $0 \leqslant \delta \leqslant 1/3$, the fractional-linear function $\alpha(x_0; x) = (x - x_0)/(1 - xx_0)$ satisfies*

$$\alpha(x_0; -1) = -1, \quad \alpha(x_0; 1) = 1, \quad \alpha(x_0; x_0 - \varepsilon) \leqslant -\delta,$$
$$\alpha(x_0; x_0 + \varepsilon) \geqslant \delta, \quad \varepsilon = 3\delta(1 - x_0).$$

Proof. In fact

$$\alpha(x_0; x_0 - \varepsilon) = -3\delta/(1 + x_0 + 3x_0\delta) \leqslant -\delta,$$
$$\alpha(x_0; x_0 + \varepsilon) = 3\delta/(1 + x_0 - 3x_0\delta) \geqslant \delta. \qquad \square$$

Lemma 5.15. *There exists an absolute constant $d > 0$ such that for every $x_0 \in (0, 1)$ and every positive integer $n > 1$ there is a rational function $\sigma_n(x_0; x) \in R_N$, $N \leqslant d \ln^2 n$, such that*

$$|\sigma_n(x_0; x)| \leqslant n^{-2}, \quad -1 \leqslant x \leqslant x_0 - n^{-2}(1 - x_0),$$
$$|1 - \sigma_n(x_0; x)| \leqslant n^{-2}, \quad x_0 + n^{-2}(1 - x_0) \leqslant x \leqslant 1,$$

$0 \leqslant \sigma_n(x_0; x) \leqslant 1$ *for every* x.

Proof. In view of lemma 5.1 (setting $\alpha = n^{-2}/3$, $\beta = 1$, $\gamma = n^{-2}/2$) there exists a constant $d > 0$ such that for every integer $n > 1$ there is a rational function $\sigma_n \in R_N$, $N \leqslant d \ln^2 n$, such that

$$|\sigma_n(x)| \leqslant \tfrac{1}{2} n^{-2}, \quad -1 \leqslant x \leqslant -n^{-2}/3,$$
$$|1 - \sigma_n(x)| \leqslant \tfrac{1}{2} n^{-2}, \quad n^{-2}/3 \leqslant x \leqslant 1,$$

$0 \leqslant \sigma_n(x) \leqslant 1$ for every x.

Then the rational function $\sigma_n(x_0; x) = \sigma_n(\alpha(x_0; x))$, $\alpha(x_0; x)$, the function from lemma 5.14, satisfies the requirements of lemma 5.15. $\square$

Lemma 5.16. *Let $f \in K_1[0, 1]$. Then*

$$\omega(f, x; \delta) = \max\{|f(x) - f(y)| : |x - y| \leqslant \delta\} \leqslant \delta(1 - x)^{-1}.$$

Proof. The lemma follows from the convexity and monotony of f and the inequalities $0 \leqslant f(x) \leqslant f(1) \leqslant 1$.

Lemma 5.17 (Fundamental lemma). *There is an integer n_0 such that if $n > n_0$ and if for $k = [\frac{1}{2}(n - d \ln^2 n)]$* ($d$ the constant from lemma 5.15) *we have $\phi_{k,3} \leqslant \varphi(k)/k$, $\varphi(k) \geqslant 1$, then*

$$\phi_{n,3} \leqslant \left(1 + \frac{4d \ln^2 n}{n}\right) \frac{\varphi(k)}{n}.$$

Proof. Let $f \in K_1[0, 1]$. Take a point $x_0 \in (0, 1)$ such that $V_0^{x_0} f = V_{x_0}^1 f = \frac{1}{2} V_0^1 f \leqslant \frac{1}{2}$. The function f is convex and nondecreasing, so $x_0 \geqslant \frac{1}{2}$. Put $\varepsilon = n^{-2}(1 - x_0)$, $\Delta_1 = [0, x_0 + \varepsilon]$ and $\Delta_2 = [x_0 - \varepsilon, 1]$. By lemma 5.16 and the assumption $f \in K_1[0, 1]$ we have

$$\left.\begin{aligned} V_0^{x_0 + \varepsilon} f &= V_0^{x_0} f + V_{x_0}^{x_0 + \varepsilon} f \leqslant \tfrac{1}{2} + \frac{\varepsilon}{1 - x_0} = \tfrac{1}{2} + n^{-2}, \\ V_{x_0 - \varepsilon}^1 f &= V_{x_0 - \varepsilon}^{x_0} f + V_{x_0}^1 f \leqslant \frac{\varepsilon}{1 - x_0} + \tfrac{1}{2} = \tfrac{1}{2} + n^{-2}. \end{aligned}\right\} \quad (1)$$

By (1) and lemma 5.13 there are rational functions $r_i \in R_k$, $i = 1, 2$, such that

$$\left.\begin{aligned} &\| f - r_i \|_{C[\Delta_i]} \leqslant (1 + 2n^{-2})\phi_{k,3}/2, \\ &0 \leqslant r_i(x) \leqslant 3\left(\frac{1}{2} + \frac{1}{n^2}\right) + 1 \text{ for every } x,\ i = 1, 2. \end{aligned}\right\} \tag{2}$$

Let $\sigma_n(x_0; x)$ be the rational function of degree $\leqslant d \ln^2 n$ from lemma 5.15, corresponding to the point x_0.

Consider the rational function

$$Q(x) = r_1(x) + \sigma_n(x_0; x)(r_2(x) - r_1(x)).$$

We have $Q \in R_N$, $N \leqslant 2k + d \ln^2 n$.

Let $x \in [0, 1]$. Using lemma 5.15 and (2) we obtain the following.

(a) If $0 \leqslant x \leqslant x_0 - \varepsilon$, then

$$\begin{aligned} |f(x) - Q(x)| &\leqslant |f(x) - r_1(x)| + |\sigma_n(x_0; x)||r_2(x) - r_1(x)| \\ &\leqslant (1 + 2n^{-2})\phi_{k,3}/2 + n^{-2}\left(3\left(\frac{1}{2} + \frac{1}{n^2}\right) + 1\right). \end{aligned} \tag{3}$$

(b) If $x_0 - \varepsilon \leqslant x \leqslant x_0 + \varepsilon$, then

$$\begin{aligned} |f(x) - Q(x)| &\leqslant (1 - \sigma_n(x_0; x))|f(x) - r_1(x)| + \sigma_n(x_0; x)|f(x) - r_2(x)| \\ &\leqslant (1 + 2n^{-2})\phi_{k,3}/2. \end{aligned} \tag{4}$$

(c) If $x_0 + \varepsilon \leqslant x \leqslant 1$, then

$$\begin{aligned} |f(x) - Q(x)| &\leqslant |f(x) - r_2(x)| + |1 - \sigma_n(x_0; x)||r_1(x) - r_2(x)| \\ &\leqslant (1 + 2n^{-2})\phi_{k,3}/2 + n^{-2}\left(3\left(\frac{1}{2} + \frac{1}{n^2}\right) + 1\right). \end{aligned} \tag{5}$$

In addition from lemma 5.15 and (2) it follows that for every x, $-\infty < x < \infty$, we have

$$|Q(x)| \leqslant (1 - \sigma_n(x_0; x))|r_1(x)| + \sigma_n(x_0; x)|r_2(x)| \leqslant 3\left(\frac{1}{2} + \frac{1}{n^2}\right) + 1,$$

i.e.

$$\| Q(x) \|_{C(-\infty, \infty)} \leqslant 3\left(\frac{1}{2} + \frac{1}{n^2}\right) + 1. \tag{6}$$

Now let n_0 be so large that for $n > n_0$ we have

(i) $n > 2$,
(ii) $\frac{1}{2}(n - d \ln^2 n) \geqslant 1$, $\quad d \ln^2 n > 4$,
(iii) $\left(1 + \frac{2}{n^2}\right)\frac{1}{2[\frac{1}{2}(n - d \ln^2 n)]} + \frac{6}{n^2} \leqslant \left(1 + \frac{4d \ln^2 n}{n}\right)n^{-1}$.

Then if $n > n_0$, $k = [\frac{1}{2}(n - d \ln^2 n)]$ and $\phi_{k,3} \leqslant \varphi(k)/k$, $\varphi(k) \geqslant 1$, we find from

(3)–(6) that

$$\| f-Q \|_{C[0,1]} \leqslant \left(1+\frac{4d\ln^2 n}{n}\right)\frac{\varphi(k)}{n}, \quad \| Q \|_{C(-\infty,\infty)} \leqslant 3, \quad Q \in R_N,$$

$$N = 2k + d\ln^2 n \leqslant n.$$

This establishes the lemma. □

Lemma 5.18. *We have the equality*

$$\sup_{f \in \mathrm{Conv}_M[a,b]} R_n(f)_{C[a,b]} = M \sup_{f \in \mathrm{Conv}_1[0,1]} R_n(f)_{C[0,1]}.$$

The proof is like the proof of lemma 5.12.

Now we shall prove a result a little bit stronger than theorem 5.11.

Theorem 5.12. *There exists an absolute constant $c_1 > 0$ such that*

$$\phi_{n,3} \leqslant c_1 n^{-1}.$$

Proof. Let N_0 be such that

(a) $N_0 > n_0$, where n_0 is the constant from lemma 5.17,
(b) the function $d\ln^2 x/x$ decreases in (N_0, ∞)

(here d is the constant from lemma 5.15).

Then there is obviously a constant $c_2 \geqslant 1$ such that for $n \leqslant N_0$ we have $\phi_{n,3} \leqslant c_2 n^{-1}$. Put

$$y(x) = [\tfrac{1}{2}(x - d\ln^2 x)], \quad y^s(x) = y(y^{s-1}(x)), \quad y^0(x) = x.$$

Clearly $y(x) \leqslant x/2$ for $x \geqslant N_0$. Then for each $n > N_0$ there exists an s_0 such that $y^{s_0}(n) \leqslant N_0$ and $y^{s_0-1}(n) > N_0$. Using lemma 5.17 we obtain successively

$$\phi_{y^{s_0-1}(n),3} \leqslant \left(1+\frac{4d\ln^2(y^{s_0-1}(n))}{y^{s_0-1}(n)}\right)\frac{c_2}{y^{s_0-1}(n)},$$

$$\phi_{y^{s_0-2}(n),3} \leqslant \left(1+\frac{4d\ln^2(y^{s_0-1}(n))}{y^{s_0-1}(n)}\right)\left(1+\frac{4d\ln^2(y^{s_0-2}(n))}{y^{s_0-2}(n)}\right)\frac{c_2}{y^{s_0-2}(n)},$$

$$\phi_{n,3} = \phi_{y^0(n),3} \leqslant \frac{c_2}{n}\prod_{i=0}^{s_0-1}\left(1+\frac{4d\ln^2(y^i(n))}{y^i(n)}\right).$$

Using the inequality $y^{s_0-1} > N_0$, the choice of N_0 and the inequalities $y^k(n) \leqslant y^{k-1}(n)/2$, $k \leqslant s_0 - 1$, we obtain

$$\prod_{i=0}^{s_0-1}\left(1+\frac{4d\ln^2(y^i(n))}{y^i(n)}\right) \leqslant \prod_{i=0}^{s_0-1}\left(1+\frac{4d\ln^2(2^{s_0-1-i}y^{s_0-1}(n))}{2^{s_0-1-i}y^{s_0-1}(n)}\right)$$

$$\leqslant \prod_{j=0}^{\infty}\left(1+\frac{4d\ln^2(2^j N_0)}{2^j N_0}\right) = c_3 < \infty,$$

where c_3 is a constant, depending on d and N_0, but not on n. Consequently $\phi_{n,3} \leqslant c_2 c_3 n^{-1}$. □

Proof of theorem 5.11. By lemma 5.18 it is enough to establish the existence of an absolute constant c such that

$$\sup_{f \in \mathrm{Conv}_1[0,1]} R_n(f)_{C[0,1]} \leqslant \frac{c}{n}.$$

If we take into account that f can be represented in the form $f(x) = d + f_1(x) + f_2(1-x)$, where $f_i \in K_2[0,1]$, $i = 1, 2$, then by lemma 5.12 theorem 5.11 will be established if we show that there is an absolute constant c_1 such that

$$\phi_n \leqslant c_1 n^{-1}.$$

But $\phi_n \leqslant \phi_{n,3}$, so theorem 5.12 gives us the needed inequality. □

Now we shall consider rational uniform approximation of convex functions with modulus of continuity $\omega(f;\delta)_C = O(\delta^\alpha)$, $0 < \alpha \leqslant 1$.

We shall denote by $\mathrm{Conv}_M(\alpha, [a,b])$ the set of all convex functions on the interval $[a,b]$ for which

$$\omega(f;\delta)_{C[a,b]} \leqslant M\delta^\alpha, \quad \delta \geqslant 0.$$

Theorem 5.13. *Let $0 < \alpha \leqslant 1$, $M > 0$ and let $[a,b]$ be an arbitrary compact interval. Then for $n \geqslant 1$ we have*

$$\sup_{f \in \mathrm{Conv}_M(\alpha,[a,b])} R_n(f)_{C[a,b]} \leqslant c(\alpha) \frac{M(b-a)^\alpha}{n^2},$$

where $c(\alpha)$ depends only on α.

Proof. It is easy to see (compare with lemma 5.12) that for $M > 0$ we have

$$\sup_{f \in \mathrm{Conv}_M(\alpha,[a,b])} R_n(f)_{C[a,b]} = M(b-a)^\alpha \sup_{f \in \mathrm{Conv}_1(\alpha,[0,1])} R_n(f)_{C[0,1]}.$$

Moreover every function $f \in \mathrm{Conv}_1(\alpha,[0,1])$ can be represented in the form $f(x) = d - f_1(x) - f_2(1-x)$, where f_i, $i = 1, 2$, is a nondecreasing and concave function on $[0,1]$, $f_i(0) = 0$ and $\omega(f_i;\delta)_C \leqslant \delta^\alpha$, $\delta \geqslant 0$, $i = 1, 2$. Thus we shall suppose that f is nondecreasing and concave on $[0,1]$, $f(0) = 0$ and $\omega(f;\delta)_C \leqslant \delta^\alpha$, $\delta \geqslant 0$, $0 < \alpha \leqslant 1$. Without loss of generality we shall suppose also that $f' \in C[0,1]$.

To prove theorem 5.13 it is sufficient to prove that

$$R_n(f)_{C[0,1]} \leqslant \frac{c(\alpha)}{n^2}, \quad n \geqslant 1. \tag{7}$$

Next we shall prove by induction the following lemma, where f is the function introduced above.

Lemma 5.19. *Let $s \geqslant 1$. For each ν, $0 \leqslant \nu \leqslant s$, there is a rational function r_ν such that*

$$\| f - r_\nu \|_{C[0,2^{-\nu}]} \leqslant \frac{4c}{2^{s\alpha}} + \frac{s - \nu + 1}{2^{2s}}, \tag{8}$$

$$\| r_\nu \|_{C(-\infty,\infty)} \leqslant 2^s, \tag{9}$$

$$\deg r_\nu \leqslant 4 \sum_{i=\nu}^{s} 2^{(s-i)\alpha/2} + 8B_1(s - \nu)s, \tag{10}$$

where $c = c(1)$ is the constant from theorem 5.1 and D_1 is the constant from lemma 5.3.

Proof. Putting $r_s \equiv 0$ we have

$$\| f - r_s \|_{C[0,2^{-s}]} \leqslant \omega(f; 2^{-s})_C \leqslant 2^{-s\alpha}$$

and therefore r_s satisfies (8)–(10) for $\nu = s$.

Suppose that $1 \leqslant \nu \leqslant s$ and there is a rational function r_ν which satisfies (8)–(10). Now we shall find a rational function $r_{\nu-1}$ which satisfies (8)–(10) with ν replaced by $\nu - 1$.

Denote $\Delta = [0, 2^{-\nu+1}]$, $\Delta_1 = [0, 2^{-\nu}]$, $\Delta_2 = [2^{-\nu-1}, 2^{-\nu+1}]$. We need to approximate f on Δ_2. Since f is nondecreasing and concave on $[0, 1]$, $f(0) = 0$ and $\omega(f; \delta)_C \leqslant \delta^\alpha$, $\delta \geqslant 0$, then

$$V_{\Delta_2} f' \leqslant f(2^{-\nu-1})/2^{-\nu-1} \leqslant 2^{\nu+1}\omega(f; 2^{-\nu-1}) \leqslant 2^{(\nu+1)(1-\alpha)}. \tag{11}$$

Choose the integer m so that

$$2^{(s-\nu+1)\alpha/2} \leqslant m \leqslant 2 \cdot 2^{(s-\nu+1)\alpha/2}. \tag{12}$$

It follows by theorem 5.1, (11) and (12) that there exists a rational function $q \in R_m$ such that

$$\| f - q \|_{C(\Delta_2)} \leqslant c \frac{V_{\Delta_2} f' |\Delta_2|}{m^2} \leqslant c \frac{2^{(\nu+1)(1-\alpha)} \cdot 2^{-\nu+1}}{2^{(s-\nu+1)\alpha}} = \frac{4c}{2^{s\alpha}}. \tag{13}$$

Without loss of generality we shall suppose that

$$\| q \|_{C(\Delta_2)} \leqslant 2 \| f \|_{C[0,1]} \leqslant 2\omega(f; 1) \leqslant 2.$$

Set $q_2 = q/(1 + \eta q^2)$, $\eta = 2^{-2s-2}$. Clearly, we have

$$\| f - q_2 \|_{C(\Delta_2)} \leqslant \| f - q \|_{C(\Delta_2)} + \eta \| f \|_{C(\Delta_2)} \| q \|^2_{C(\Delta_2)} \leqslant \frac{4c}{2^{s\alpha}} + \frac{4}{2^{2s+2}} = \frac{4c}{2^{s\alpha}} + \frac{1}{2^{2s}}, \tag{14}$$

$$\| q_2 \|_{C(-\infty,\infty)} \leqslant \frac{1}{2\sqrt{\eta}} = 2^s, \tag{15}$$

$$\deg q_2 \leqslant 2m \leqslant 4 \cdot 2^{(s-\nu-1)\alpha/2}. \tag{16}$$

Now we apply lemma 5.3 to f on Δ with $\Delta_1 = [0, 2^{-\nu}]$, $\Delta_2 = [2^{-\nu-1}, 2^{-\nu+1}]$, the rational functions r_ν and q_2, $\varepsilon_1 = 4c \cdot 2^{-s\alpha} + (s - \nu + 1) \cdot 2^{-2s}$ and $A = 2^s$. Note that by our assumption r_ν satisfies (8)–(10), q_2 satisfies (14)–(16). Putting $\varepsilon_2 = 2^{-2s}$ we conclude that there is a rational function $r_{\nu-1}$ such that

$$\| f - r_{\nu-1} \|_{C[0,2^{1-\nu}]} \leqslant \frac{4c}{2^{s\alpha}} + \frac{s - \nu + 1}{2^{2s}} + \frac{1}{2^{2s}}$$

$$= \frac{4c}{2^{s\alpha}} + \frac{s - (\nu - 1) + 1}{2^{2s}}, \qquad \| r_{\nu-1} \|_{C(-\infty,\infty)} \leqslant 2^s,$$

and

$$\deg r_{\nu-1} \leqslant \deg r_\nu + \deg q_2 + B_1 \ln\left(\mathrm{e} + \frac{|\Delta|}{|\Delta_1 \cap \Delta_2|} \right) \ln\left(\mathrm{e} + \frac{A}{\varepsilon_2} \right)$$

$$\leqslant 4 \sum_{i=\nu}^{s} 2^{(s-i)\alpha/2} + 4 \cdot 2^{(s-\nu-1)\alpha/2} + B_1 \ln\left(\mathrm{e} + \frac{2^{-\nu+1}}{2^{-\nu-1}} \right) \ln(\mathrm{e} + 2^{3s})$$

$$\leqslant 4 \sum_{i=\nu-1}^{s} 2^{(s-i)\alpha/2} + 8B_1(s - \nu + 1)s.$$

Consequently the rational function $r_{\nu-1}$ satisfies (8)–(10) with ν replaced by $\nu - 1$. □

Now we are in position to complete the proof of theorem 5.13. It follows by lemma 5.19 with $\nu = 0$ that for each $s \geqslant 1$ there exists a rational function r_0 such that

$$\| f - r_0 \|_{C[0,1]} \leqslant \frac{4c}{2^{s\alpha}} + \frac{s+1}{2^{2s}} \leqslant \frac{4c+1}{2^{s\alpha}}$$

and

$$\deg r_0 \leqslant 4 \sum_{i=0}^{s} 2^{(s-i)\alpha/2} + 8B_1 s^2 \leqslant B_1(\alpha) 2^{s\alpha/2}.$$

Clearly from these estimates (7) follows. □

5.6 Functions with singularities

In this section we shall prove a theorem for rational approximation of functions continuous in the interval $[0, 1]$ with bounded analytic continuation in the disk $\{z: |z - 1| < 1\}$. This result was obtained by A.A. Gonchar (1967a).

Theorem 5.14. *Let f be a continuous real function on the interval $[0, 1]$ and let there exist a bounded analytic function in the disk $D = \{z: |z - 1| < 1\}$ which coincides with f on $[0, 1]$. Denote this function also by f. Then for $n \geqslant 1$ we have*

$$R_n(f)_{C[0,1]} \leqslant c_1 \inf_{1 \leqslant t < \infty} \{ Mt\mathrm{e}^{-c_2 n/t} + \omega(f; \mathrm{e}^{-t})_C \},$$

where $M = \sup_{z\in D} |f(z)|$, *and* $\omega(f;\delta)_C$ *is the modulus of continuity of* f *on* $[0, 1]$, c_1, c_2 *are absolute constants.*

As a consequence of this theorem we shall prove the following theorem for rational approximation of functions with singularities in the middle point of $[-1, 1]$.

If we consider continuous functions of the type

$$f_\varphi(x) = \begin{Bmatrix} 0, & x\in[-1,0], \\ \varphi(x), & x\in(0,1]. \end{Bmatrix} \tag{1}$$

then we have the following.

Theorem 5.15. *Let* f_φ *be a continuous function of type* (1) *and let us assume that there exists a bounded analytic function in the disk* $D = \{z: |z-1| < 1\}$, *which coincides with* φ *on* $(0, 1]$. *Denote this function also by* φ. *Then for* $n \geqslant 1$ *we have*

$$R_n(f_\varphi)_{C[-1,1]} \leqslant c_1 \inf_{1\leqslant t<\infty} \{Mt\mathrm{e}^{-c_2 n/t} + \omega(f_\varphi; \mathrm{e}^{-t})_C\}, \tag{2}$$

where $M = \sup_{z\in D} |\varphi(z)|$, $\omega(f_\varphi;\delta)_C$ *is the modulus of continuity of* f_φ *on the interval* $[-1, 1]$ *and* c_1, c_2 *are absolute constants.*

Corollary 5.2. (i) *If* $\varphi(x) = x^\alpha$, $\alpha > 0$, *then* $R_n(f_\varphi)_{C[-1,1]} \leqslant \mathrm{e}^{-c(\alpha)\sqrt{n}}$.

(ii) *If* $\varphi(x) = \exp\{-\ln^\beta(1/x)\}$, $0 < \beta < 1$, *then* $R_n(f_\varphi)_{C[-1,1]} \leqslant \exp\{-c(\beta)n^{\beta/(\beta+1)}\}$.

(iii) *If* $\varphi(x) = (\ln(1/x))^{-\gamma}$, $\gamma > 0$, *then* $R_n(f_\varphi)_{C[-1,1]} \leqslant c(\gamma)(\ln n/n)^\gamma$.

Remark. Theorem 5.14 and theorem 5.15, as corollary 5.2 shows, allow us to obtain order of approximation of the form $\exp\{-cn^\lambda\}$. However, they do not give always the exact order of approximation.

Proof of theorem 5.14. Let $0 < h < \mathrm{e}^{-1}$. Define the functional-linear function $w(z) = (az+b)/(z+d)$ by the conditions

$$w(h/2) = 0, \quad w(1) = 1, \quad w(3/2) = \infty.$$

We get $w(z) = (z - h/2)(z - 3/2)^{-1}(h-2)^{-1}$. Set

$$\varepsilon(w(h)) = \frac{h}{(2-h)(3-2h)}.$$

Obviously $h/6 < \varepsilon < h < \mathrm{e}^{-1}$.

Denote by Γ the preimage of the imaginary axis $\mathrm{Re}\, w = 0$ under the mapping

$w = w(z)$. It is readily seen that Γ is a circle symmetric with respect to the real axis which intersects the real axis at the points $h/2$ and $\frac{3}{2}$. Consequently $\Gamma \subset D$ and f is holomorphic on the closed disk determined by Γ and $\sup_{z\in\Gamma}|f(z)| \leqslant M$.

Let S be the rational function from lemma 5.2 in section 5.1 with a given $n \geqslant 1$ and $\varepsilon = w(h)$. We have

$$|S(w)| \leqslant c_1 \exp\left\{-\frac{c_2 n}{\ln(1/\varepsilon)}\right\}, \quad w \in [\varepsilon, 1]$$

and

$$|S(w)| = 1, \quad \operatorname{Re} w = 0.$$

Set $Q(z) = S(w(z))$. From the above and the choice of $w(z)$ and ε we obtain that $\deg Q = n$,

$$|Q(z)| \leqslant c_1 \exp\left\{-\frac{c_2 n}{\ln(\varepsilon/h)}\right\} \leqslant c_1 \exp\left\{-\frac{c_3 n}{\ln(1/h)}\right\}, \quad z \in [h, 1] \tag{3}$$

and

$$|Q(z)| = 1, \quad z \in \Gamma. \tag{4}$$

It follows by the definition of S in lemma 5.2 in section 5.1 that Q has only single zeros in the interval $[h, 1]$. Denote them by $\alpha_1, \alpha_2, \ldots, \alpha_n$. The poles of Q are at the points $\beta_1, \beta_2, \ldots, \beta_n$, which are symmetrical to the zeros $\alpha_1, \alpha_2, \ldots, \alpha_n$ with respect to Γ.

Let r be the rational function of degree n which interpolates f at the points $\alpha_1, \ldots, \alpha_n$, $\alpha_{n+1} = 1$ and has single poles at $\beta_1, \ldots, \beta_n$. This means that

$$r(z) = \frac{a_0 + a_1 z + \cdots + a_n z^n}{(z - \beta_1)\cdots(z - \beta_n)} \tag{5}$$

and

$$r(\alpha_i) = f(\alpha_i), \quad i = 1, 2, \ldots, n+1. \tag{6}$$

It is readily seen that there exists exactly one rational function r of kind (5) which satisfies (6). Indeed, consider the conditions (6) as a system for obtaining $a_0, a_1, \ldots, a_n$. This system has exactly one solution since each rational function of the type (5) cannot vanish at $n+1$ different points $\alpha_1, \alpha_2, \ldots, \alpha_{n+1}$.

One easily verifies that for z in the disk determined by Γ we have

$$r(z) = \frac{1}{2\pi i}\int_\Gamma \left(1 - \frac{(z-\alpha_1)\cdots(z-\alpha_{n+1})(\xi-\beta_1)\cdots(\xi-\beta_n)}{(\xi-\alpha_1)\cdots(\xi-\alpha_{n+1})(z-\beta_1)\cdots(z-\beta_n)}\right)\frac{f(\xi)d\xi}{\xi - z}$$

and

$$f(z)-r(z)=\frac{1}{2\pi i}\int_{\Gamma}\frac{(z-\alpha_1)\cdots(z-\alpha_n)(z-1)(\xi-\beta_1)\cdots(\xi-\beta_n)f(\xi)\mathrm{d}\xi}{(\xi-\alpha_1)\cdots(\xi-\alpha_n)(\xi-1)(z-\beta_1)\cdots(z-\beta_n)(\xi-z)}$$
$$=\frac{1}{2\pi i}\int_{\Gamma}\frac{Q(z)(z-1)f(\xi)\mathrm{d}\xi}{Q(\xi)(\xi-1)(\xi-z)}$$

(for details see J.L. Walsh (1960)).

From the last identity, (3) and (4) we get for $z=[h,1]$

$$|f(z)-r(z)|\leqslant cMI\exp\left\{-\frac{c_3 n}{\ln(1/h)}\right\},$$

where $I=\int_{\Gamma}|\xi-h|^{-1}|\mathrm{d}\xi|$. Quick computation shows that $I\leqslant c\ln(1/h)$, c a constant. Consequently

$$|f(z)-r(z)|\leqslant c_1 M\ln(1/h)\cdot\exp\left\{-\frac{c_3 n}{\ln(1/h)}\right\} \tag{7}$$

for $z\in[h,1]$.

Let $z(x)=(1-h)x+h$. Clearly, $z(x)$ linearly maps $[0,1]$ onto the interval $[h,1]$ and $|z(x)-x|\leqslant h$ for $x\in[0,1]$. Then by (7) it follows that for $x\in[0,1]$ we have

$$|f(x)-r(z(x))|\leqslant|f(x)-f(z(x))|+|f(z(x))-r(z(x))|\leqslant\omega(f;h)_C$$
$$+c_1 M\ln(1/h)\cdot\exp\left\{-\frac{c_3 n}{\ln(1/h)}\right\}.$$

Set $t=t(h)=\ln(1/h)$. Obviously $t(h)$ is a one-to-one mapping of $(0,\mathrm{e}^{-1})$ onto $(1,\infty)$ and therefore for $n\geqslant 1$ we have

$$R_n(f)_{C[0,1]}\leqslant c_1\inf_{1<t<\infty}\left\{Mt\exp\left\{-\frac{c_3 n}{t}\right\}+\omega(f;\mathrm{e}^{-t})_C\right\},$$

which establishes theorem 5.14. □

Remark. The proof of theorem 5.14 shows that if f is a real valued function then the approximating rational function r has real coefficients.

Proof of theorem 5.15. Let $n\geqslant 1$ and $\mathrm{e}^{-n-1}<h<\mathrm{e}^{-1}$. From the proof of theorem 5.14 it follows (see the estimate (7)) that there exists a rational function $r\in R_n$ such that

$$\|f_\varphi-r\|_{C[h,1]}\leqslant c_1 M\ln(1/h)\cdot\exp\left\{-\frac{c_3 n}{\ln(1/h)}\right\}, \tag{8}$$

where $M=\sup_{z\in D}|\varphi(z)|$, c_1, c_3 are absolute constants.

Without loss of generality we shall suppose that $0 < c_3 < 1$, $M > 0$ and

$$\|r\|_{C[h,1]} \leqslant 2M. \tag{9}$$

Let us set

$$q = \frac{r}{1+\eta^2 r^2}, \qquad \eta = M^{-1}\exp\left\{-\frac{n}{\ln(1/h)}\right\}.$$

Then by (8) and (9) we get

$$\begin{aligned}\|f_\varphi - q\|_{C[h,1]} &\leqslant \|f_\varphi - r\|_{C[h,1]} + \eta^2\|f_\varphi\|_{C[h,1]}\|r\|^2_{C[h,1]}\\ &\leqslant \|f_\varphi - r\|_{C[h,1]} + 4\eta^2 M^3\\ &\leqslant (c_1+4)M\ln(1/h)\cdot\exp\left\{-\frac{c_3 n}{\ln(1/h)}\right\}.\end{aligned}$$

Thus we have

$$\|f_\varphi - q\|_{C[h,1]} \leqslant cM\ln(1/h)\cdot\exp\left\{-\frac{c_3 n}{\ln(1/h)}\right\}. \tag{10}$$

It is easy to see that

$$\|q\|_{C(-\infty,\infty)} \leqslant \frac{1}{2\eta} < M\exp\left\{\frac{n}{\ln(1/h)}\right\} \tag{11}$$

and

$$\deg q \leqslant 2n. \tag{12}$$

Set $\lambda(x) = (1-h)(x+h)/(1+h)+h$. Obviously the function $\lambda(x)$ linearly maps $[-h,1]$ onto $[h,1]$ and $|\lambda(x)-x| \leqslant 2h$.

Consider the rational functions $q_1 \equiv 0$ and $q_2(x) = q(\lambda(x))$. By (10) and the last arguments we get

$$\begin{aligned}\|f_\varphi - q_2\|_{C[-h,1]} &\leqslant \|f_\varphi - f_\varphi(\lambda)\|_{C[-h,1]} + \|f_\varphi(\lambda) - q(\lambda)\|_{C[-h,1]}\\ &\leqslant \omega(f_\varphi;2h)_C + \|f_\varphi - q\|_{C[h,1]}\end{aligned}$$

and therefore

$$\|f_\varphi - q_2\|_{C[-h,1]} \leqslant 2\omega(f_\varphi;h)_C + cM\ln(1/h)\exp\left\{-\frac{c_3 n}{\ln(1/h)}\right\}. \tag{13}$$

Clearly, by (11) and (12) we obtain

$$\|q_2\|_{C(-\infty,\infty)} \leqslant M\exp\left\{\frac{n}{\ln(1/h)}\right\}, \tag{14}$$

$$\deg q_2 \leqslant 2n. \tag{15}$$

Now we are in a position to apply lemma 5.3 from section 5.1 for the function f_φ using the rational functions q_1, q_2 and the parameters ε_1, A, k_i, $i=1, 2$, determined by (13)–(15). Setting $\varepsilon_2 = M\exp\{-n/\ln(1/h)\}$ we conclude that there exists a rational function r such that

$$\|f_\varphi - r\|_{C[-1,1]} \leqslant c_1 \left\{ M\ln(1/h)\exp\left\{-\frac{c_2 n}{\ln(1/h)}\right\} + \omega(f_\varphi; h)_C \right\}$$

and

$$\deg r \leqslant 2n + B_1 \ln\left(\mathrm{e} + \frac{2}{h}\right)\ln\left(\mathrm{e} + \exp\left\{\frac{2n}{\ln(1/h)}\right\}\right) \leqslant cn.$$

Since h is an arbitrary number in the interval $(0, \mathrm{e}^{-1})$, from here follows the assertion of theorem 5.15. □

Proof of corollary 5.2. The assertions (i), (ii) and (iii) of corollary 5.2 follow immediately setting successively: $\varphi(x) = x^\alpha$, $t = \sqrt{n}$; $\varphi(x) = \exp\{-\ln^\beta(1/x)\}$, $t = n^{1/(1+\beta)}$; and $\varphi(x) = (\ln(1/x))^{-\gamma}$, $t = c_2 n/((\gamma+1)\ln n)$ in the estimate (2). □

5.7 Notes

The idea of the basic lemma 5.1 is due to D. Newman (1964a). In a form similar to that given here lemma 5.1 was obtained by A.A. Gonchar (1967a, 1967b).

The result of P. Szüsz and P. Turan (1966) for best uniform approximation of the functions of the class V_1 (more exactly the convex functions of the class Lip1) was the following:

$$\sup_{V_a^b f' \leqslant 1} R_n(f)_{C[a,b]} = \mathrm{O}\left(\frac{\ln^4 n}{n^2}\right).$$

G. Freud (1966) obtained that

$$\sup_{V_0^1 f^{(r)} \leqslant 1} R_n(f)_{C[0,1]} = \mathrm{O}\left(\frac{\ln^2 n}{n^{r+1}}\right), \quad r \geqslant 1.$$

Before the final result – theorem 5.1 (Popov, 1976a), Popov (1974a) obtained that

$$R_n(f)_{C[0,1]} \leqslant c(k,r)\frac{\overbrace{\ln\cdots\ln}^{k} n}{n^{r+1}} V_0^1 f^{(r)}.$$

The result of G. Freud (1966) and E.P. Dolženko, A.A. Abdulgaparov (the lecture of Dolženko at IMC, Moscow, 1966) for rational uniform approximation of the functions of the class $V(c\delta^\alpha)$, $0<\alpha<1$, is the following.

If $f \in V(c\delta^\alpha)$ then

$$R_n(f)_C = \mathrm{O}\left(\frac{\ln^2 n}{n}\right),$$

and also if $f \in V((\ln(1/\delta))^{-1})$ then

$$R_n(f)_C = \mathrm{O}(n^{-1/3}).$$

The result of Bulanov (1975b) is the following.

If $f \in V(M, [a,b], \omega)$ and $\omega(\delta)$ is strictly increasing in $[0, b-a]$, then

$$\frac{R_n(f)_{C[a,b]}}{|\ln R_n(f)_{C[a,b]}|\,|\ln \omega^{-1}(R_n(f)_{C[a,b]})|} \leqslant \frac{c(M, b-a, \omega)}{n},$$

where the constant $c(M, b-a, \omega)$ depends only on M, $b-a$ and ω.

For application of the Hardy–Littlewood maximal function to multivariate rational uniform approximation see the work of R. DeVore and Xiang-ming Yu (1986).

For the best uniform rational approximation of convex functions we want to add also the following historical remarks.

For the class $\mathrm{Conv}_M(\alpha, [a,b])$, $0 < \alpha < 1$, A.P. Bulanov (1969) obtained the following estimate

$$\sup_{f \in \mathrm{Conv}_M(\alpha, [a,b])} R_n(f)_{C[a,b]} = \mathrm{O}\left(\frac{\ln^6 n}{n^2}\right).$$

After this A.A. Hatamov (1975a) obtained that

$$\sup_{f \in \mathrm{Conv}_M(\alpha, [a,b])} R_n(f)_{C[a,b]} = \mathrm{O}\left(\frac{\overbrace{\ln \ln \cdots \ln}^{k} u}{n^2}\right).$$

The final estimate (theorem 5.13) belongs to P. Petrushev (1976b).

For rational uniform approximation of a convex function with a given modulus of continuity ω ($\omega(f;\delta) \leqslant \omega(\delta), \delta \geqslant 0$) there are the following results.

A.P. Bulanov and A.A. Hatamov (1978):

$$R_n(f)_{C[a,b]} \leqslant c\frac{\ln^2 n}{n^2} \max_{\theta \in [\mathrm{e}-n, 1]} \{\omega(\theta) \ln(1/\theta)\}.$$

A.A. Pekarskii (1977):

$$R_n(f)_C \leqslant c \inf_{\substack{1 \leqslant \lambda \leqslant n/\ln^2 n \\ 0 < \delta \leqslant 1}} \left\{\frac{\omega(\delta)}{n} + \frac{\lambda}{n^2}\int_\delta^1 \frac{\omega(y)}{y^{1+1/\lambda}}\,\mathrm{d}y\right\}.$$

A.A. Pekarskii (1980a):

$$R_n(f)_C \leqslant \frac{c}{n^2}\left(\int_{\mathrm{e}^{-n}}^1 \frac{1}{x}\sqrt{\left(\frac{\omega(x)}{|\ln x|}\right)}\,\mathrm{d}x\right)^2.$$

Some other works which have contributed to rational approximation on the interval and on the whole real axis are G. Freud (1967, 1968, 1970), G. Freud, J. Szabados (1967a, b, 1978), J. Szabados (1967a, b).

V.N. Russak (1974, 1977, 1979, 1984) introduced rational operators which are analogs of the operators of Fejer and Jackson. He obtained the exact order of rational approximation of classes of functions similar to V_r by means of such Jackson–Russak operators.

There are many works on the approximation of analytic functions by means of rational functions, but we shall not consider such problems in this book. We only want to mention some papers which are near to the questions considered in this chapter: A.A. Gonchar (1972, 1974), G. Fichera (1970, 1974), J. Karlsson (1982), V.K. Dzjadik (1966), G. Somorjai (1976), J.E. Anderson (1980), T. Ganelius (1982).

6

Converse theorems for rational approximation

In this chapter we shall consider some converse theorems for rational approximation. In Chapter 3 in the polynomial case we have shown how the converse theorems are connected with the direct theorems – they give full characterization of the best polynomial approximation in uniform or L_p-norm by means of the smoothness properties of the function, more precisely by means of the moduli of smoothness in C or in L_p. Unfortunately till now we have not such a nice characterization of the best rational uniform approximation. Characterization of the best rational approximation in L_p will be given on the basis of the connection between best rational and best spline approximation in L_p, $1 < p < \infty$, in Chapter 8. This characterization use the converse results given in section 6.3.

First we consider some classical converse results. In section 6.1 we give the classical results of Gonchar and Dolženko. The Bernstein type inequality of Dolženko (theorem 6.1) gives an estimate of $\omega(f; \delta)_L$ by means of the best uniform rational approximations. In section 6.2 we give Russak's inequality, which is a Bernstein type inequality for the function f and the conjugate function $\tilde{f}$ of f. In section 6.3 we give Pekarskii's inequality for the norm of $r^{(s)}$ in $L_\sigma(-1, 1)$, by means of the norm of r in $L_p(-1, 1)$, $1 < p \leqslant \infty$, $1/\sigma = s + 1/p$. On the basis of this inequality an estimate for the modulus of variation in L_p of the function f by means of the best approximation to f in L_p by rational functions is given. Using this estimate the connection between best spline approximation with free knots and best rational approximations in L_p, $1 < p \leqslant \infty$, will be given in section 8.2. The chapter ends with notes.

6.1 Gonchar's and Dolženko's results

It is well known that it is impossible to estimate the value of the derivative of the rational function $r \in R_n$ in the interval $[a, b]$ by means of the Chebyshev norm of r in $[a,b]$. This is the great difference from the polynomial case, where we have the inequalities of Markov and Bernstein (see theorems 3.12 and 3.9).

But, as A.A. Gonchar was the first to note, if we omit a suitable subset with a small measure, this type of estimates becomes possible. The first results of Gonchar (1955, 1959) were of the following type.

If $R_n(f)_{C[a,b]} \leqslant cn^{-1-\delta}, \delta > 0$, then the function f is differentiable almost everywhere on the interval $[a,b]$.

Let $r \in R_n$. For every $\delta > 0$ there exists a set $E \subset [a,b]$, $\text{mes}\, E = |E| < \delta$, such that for every $x \in [a,b] \backslash E$ we have

$$|r'(x)| \leqslant 2\frac{n \ln n}{\delta} \| r \|_{C[a,b]}$$

(see also the notes at the end of the chapter).

We shall prove here some later results of Dolženko (1962, 1963), which improve this theorem.

Theorem 6.1. *Let $r \in R_n$. Then*

(a) $$\| r' \|_{L_1(a,b)} \leqslant 2n \| r \|_{C[a,b]},$$

(b) *for every interval $[a,b]$ and every $\delta > 0$ there exists a subset $E = E(\delta, r)$ such that $|E| \leqslant \delta$ and for every $x \in [a,b] \backslash E$ we have*

$$| r'(x)| \leqslant 2n\delta^{-1} \| r \|_{C[a,b]}.$$

Proof. Since the function $r - \alpha$ has at most n zeros for every $\alpha \in [-\| r \|_C, \| r \|_C]$ and has no zeros for $\alpha \notin [-\| r \|_C, \| r \|_C]$, we have

$$V_a^b r \leqslant 2n \| r \|_{C[a,b]}.$$

But we have

$$V_a^b r = \int_a^b |r'(x)| \mathrm{d}x = \| r' \|_{L_1(a,b)}$$

which proves (a).

Let E be the set of the points in the interval $[a,b]$ such that $|r'(x)| > 2n\delta^{-1} \| r \|_C$ for $x \in E$. Since

$$\int_a^b |r'(x)| \mathrm{d}x \leqslant 2n \| r \|_{C[a,b]}$$

we have

$$2n \| r \|_{C[a,b]} \geqslant \int_a^b |r'(x)| \mathrm{d}x \geqslant 2n\delta^{-1} \| r \|_{C[a,b]} |E|,$$

i.e. $|E| \leqslant \delta$. □

It is evident from the proof that in this theorem only the property of piecewise monotony of the rational function is used. Nevertheless it is impossible to improve this theorem: there exist rational functions $r \in R_n$ such that $\| r' \|_{L_1} \geqslant c' n \| r \|_C$, c' an absolute constant.

Theorem 6.2 (Dolženko (1962, 1966a)).

(a) *Let us have for the function* f

$$\sum_{n=0}^{\infty} R_n(f)_{C[a,b]} < \infty .$$

Then f is absolutely continuous on $[a,b]$ and almost everywhere on $[a,b]$ we have $f'(x) = \lim_{k\to\infty} r'_{2^k}(x)$,

$$\| f - r_{2^k} \|_{C[a,b]} = R_{2^k}(f)_{C[a,b]} .$$

(b) *We have*

$$\omega(f; n^{-1})_{L_1[a,b]} \leqslant \frac{c}{n} \sum_{m=0}^{n} R_m(f)_{C[a,b]},$$

where c is an absolute constant.

Proof. We have

$$f(x) = \sum_{k=0}^{\infty} (r_{2^k}(x) - r_{2^{k-1}}(x)), \quad r_{-1/2} \equiv 0,$$

where the series converges uniformly in $C[a,b]$, since

$$\| r_{2^k} - r_{2^{k-1}} \|_{C[a,b]} \leqslant \| r_{2^k} - f \|_{C[a,b]} + \| r_{2^{k-1}} - f \|_{C[a,b]} \leqslant 2R_{2^{k-1}}(f)_{C[a,b]}$$

and $\sum_{k=0}^{\infty} R_{2^k}(f)_C < \infty$.

Since $r_{2^k} - r_{2^{k-1}} \in R_{2^{k+1}}$, using theorem 6.1(a) we obtain

$$\sum_{k=0}^{\infty} \| r'_{2^k} - r'_{2^{k-1}} \|_{L_1(a,b)} \leqslant \sum_{k=0}^{\infty} 2 \cdot 2^{k+1} \| r_{2^k} - r_{2^{k-1}} \|_C \leqslant 32 \sum_{n=0}^{\infty} R_n(f)_{C[a,b]} < \infty,$$

i.e. $\sum_{k=0}^{\infty} (r'_{2^k} - r'_{2^{k-1}})$ is convergent in $L_1(a,b)$, therefore almost everywhere

$$f'(x) = \sum_{k=0}^{\infty} (r'_{2^k}(x) - r'_{2^{k-1}}(x))$$

and f is absolutely continuous, which proves (a).

To prove (b) we consider the K-functionals between $L_1(a,b)$ and $W_1^1(a,b)$ and between $C(a,b)$ and $W_1^1(a,b)$† (see Chapter 3). By theorem 3.15 we have:

$$\omega\left(f; \frac{1}{n}\right)_{L(a,b)} \leqslant cK(f, n^{-1}; L_1, W_1^1) \leqslant c'K(f, n^{-1}; C, W_1^1), \tag{1}$$

† In the space W_1^1 we use the same quasi-norm as in section 3.5: $\| f' \|_{L_1}$.

where c is an absolute constant; evidently $K(f,t;L_1,W_1^1)\leqslant \max\{1,(b-a)\}K(f,t;C,W_1^1)$.

By theorem 3.16 and the Bernstein type inequality from theorem 6.1(a) we have

$$K(f,n^{-1};C,W_1^1)\leqslant\frac{c'}{n}\sum_{m=0}^{n}R_m(f)_{C[a,b]}. \tag{2}$$

Then (b) follows from (1) and (2).

Remark. The inequality (2) is stronger than statement (b) of the theorem.

6.2 Estimates for L_1-norms for the derivatives of rational functions and their Hilbert transforms

Here we shall give the method and results of V.N. Russak (1979) for obtaining estimates for $L_1(-\infty,\infty)$ norm of the derivative of a rational function and the derivative of its Hilbert transform.

Definition 6.1. *We define the Hilbert transform Hf of the function $f\in L_1(-\infty,\infty)$ as follows:*

$$(Hf)(x)=-\frac{1}{\pi}\int_0^\infty\frac{f(x+t)-f(x-t)}{t}\,dt$$

$$=\lim_{\varepsilon\to+0}-\frac{1}{\pi}\int_{t>\varepsilon}\frac{f(x+t)-f(x-t)}{t}\,dt \tag{1}$$

if the limit exists.

Very often Hf is denoted $\tilde{f}$ and is called the conjugate function to f.

The following facts are well known (see for example Zygmund (1959)).

If $f\in L_1(-\infty,\infty)$, then $(Hf)(x)$ exists almost everywhere.

If $f\in L_p(-\infty,\infty)$, $p>1$, then

$$f(x)=\frac{1}{\pi}\int_0^\infty\frac{\tilde{f}(x+t)-\tilde{f}(x-t)}{t}\,dt,$$

i.e. $H^{-1}=-H$.

We give these results without proofs because we shall not use them in this book.

The most essential corollary of Russak's result is the following theorem.

Theorem 6.3 (Russak (1973)). *Let $r_{2n}\in R_{2n}$ and all poles of r_{2n} be a complex and conjugate.*[†] *Then*

$$\|r'_{2n}\|_{L_1(-\infty,\infty)}\leqslant 2\pi n\|r_{2n}\|_{C(-\infty,\infty)}, \tag{2}$$

[†] In this section the rational functions can have complex coefficients.

$$\| Hr'_{2n} \|_{L_1(-\infty,\infty)} \leqslant 2\pi n \| r_{2n} \|_{C(-\infty,\infty)}. \tag{3}$$

Remark. The inequality (2) follows from theorem 6.1 (a) of Dolženko, but with another constant on the right-hand side. We give it here for completeness.

We want also to formulate this theorem in another equivalent way.

Definition 6.2. *We shall say that the function f belongs to the Hardy space $H_1(-\infty,\infty)$ if $f \in L_1(-\infty,\infty)$ and $\tilde{f} \in L_1(-\infty,\infty)$ $(Hf \in L_1(-\infty,\infty))$.*

One of the facts of analysis is that the Hardy space $H_1(-\infty,\infty)$ is a Banach space with a norm

$$\| f \|_{H_1} = \| f \|_{L_1(-\infty,\infty)} + \| \tilde{f} \|_{L_1(-\infty,\infty)}.$$

Now we can reformulate theorem 6.3 as follows.

Theorem 6.3'. *Let $r_{2n} \in R_{2n}$ and all poles of r_{2n} be complex and conjugate. Then*

$$\| r'_{2n} \|_{H_1} \leqslant 4\pi n \| r_{2n} \|_{C(-\infty,\infty)}.$$

We shall need some notations and lemmas for the proof of theorem 6.3.

Let us denote $\mathscr{H} = \{\operatorname{Im} z > 0\}$ and let z_k, $k = 1, \ldots, n$, $z_k = \alpha_k + i\beta_k \in \mathscr{H}$, $k = 1, \ldots, n$, i.e. $\beta_k > 0$ for $k = 1, \ldots, n$.

Let us consider the Blaschke product

$$B(z) = \prod_{k=1}^{\pi} \frac{z - z_k}{z - \bar{z}_k}.$$

Obviously $|B(x)| = 1$ for $x \in \mathbb{R} = (-\infty,\infty)$, $B(z)$ is analytic in $\mathscr{H}$ and continuous in $\bar{\mathscr{H}} = \mathscr{H} \cup \mathbb{R}$.

On the other hand we have for the logarithmic derivative of B (multiplied by $1/i$):

$$\frac{B'(x)}{iB(x)} = \frac{1}{i} \sum_{k=1}^{n} \left(\frac{1}{x - z_k} - \frac{1}{x - \bar{z}_k} \right) = \frac{1}{i} \sum_{k=1}^{n} \frac{z_k - \bar{z}_k}{|x - z_k|^2}$$

$$= 2 \sum_{k=1}^{n} \frac{\beta_k}{(x - \alpha_k)^2 + \beta_k^2} = |B'(x)| > 0.$$

Let us define

$$\phi(x) = \sum_{k=1}^{n} (\arg(z_k - x) - \arg(\bar{z}_k - x)).$$

It is evident that for $x \in (-\infty,\infty)$

$$\left. \begin{aligned} & B(x) = e^{i\phi(x)}, \quad \phi'(x) = \frac{B'(x)}{iB(x)} = |B'(x)| > 0, \\ & \phi(x) \underset{x \to -\infty}{\longrightarrow} 0, \quad \phi(x) \underset{x \to +\infty}{\longrightarrow} 2\pi n. \end{aligned} \right\} \tag{4}$$

Let us define also

$$F_\varphi(x) = \operatorname{Re}(e^{i\varphi}B(x)).$$

The following properties of F_φ follow immediately from the definition and (4):

$$\text{(i)} \qquad |F_\varphi(x)| \leqslant 1, \tag{5}$$

$$\left.\begin{aligned} &\text{(ii)} \qquad F_{\varphi+\pi/2}(x) = -\operatorname{Im}(e^{i\varphi}B(x)),\\ &\text{(iii)} \qquad F_\varphi(x) - iF_{\varphi+\pi/2}(x) = e^{i\varphi}B(x).\end{aligned}\right\} \tag{6}$$

For every α we have

$$\text{(iv)} \qquad F_\varphi(x)\cos\alpha + F_{\varphi+\pi/2}(x)\sin\alpha = F_{\varphi+\alpha}(x), \tag{7}$$

$$\text{(v)} \qquad F'_\varphi(x) = F_{\varphi+\pi/2}(x)|B'(x)|. \tag{8}$$

(vi) The function

$$F_\varphi(x) = \operatorname{Re}(e^{i\varphi}B(x)) = \operatorname{Re}(e^{i(\varphi+\phi(x))}). \tag{9}$$

for $\varphi \neq \pi/2 + k\pi$ has exactly $2n$ real zeros $x_1 < x_2 < \cdots < x_{2n}$.

The last fact follows from $F_\varphi(x) = \cos(\varphi + \phi(x))$ and (4)

$$\phi(x) \underset{x\to-\infty}{\longrightarrow} 0, \quad \phi(x) \underset{x\to+\infty}{\longrightarrow} 2\pi n.$$

From property (vi) we obtain that for $\varphi \neq \pi/2 + k\pi$ we have

$$F_\varphi(x) = \cos\varphi \cdot \frac{(x-x_1)\cdots(x-x_{2n})}{\prod_{k=1}^n |x-z_k|^2}. \tag{10}$$

Let now r be an arbitrary (complex-valued) rational function without poles on $\mathbb{R}$. Let there exist z_k, $k = 1, \ldots, n$, $z_k \in \mathscr{H}$, $k = 1, \ldots, n$, such that

$$r(x) = \frac{p_{2n}(x)}{\prod_{i=1}^n((x-\alpha_i)^2 + \beta_i^2)} = \frac{p_{2n}(x)}{\prod_{k=1}^n |x-z_k|^2},$$

where the algebraic polynomial p_{2n} has degree $2n$ at most.

Lemma 6.1. *Let $x_k, k = 1, \ldots, 2n, x_1 < x_2 \cdots < x_{2n}$, be the zeros of $F_\varphi(x)$. The following equality holds.*

$$r(x) = F_\varphi(x)\left\{\sum_{k=1}^{2n} r(x_k)\frac{1}{(x-x_k)F'_\varphi(x_k)} + c_1\right\}, \tag{11}$$

where c_1 is a constant.

Proof. Let us set

$$r^*(x) = F_\varphi(x)\sum_{k=1}^{2n} r(x_k)\frac{1}{(x-x_k)F'_\varphi(x_k)}.$$

It follows from (10) that $r^*(x_k) = r(x_k)$, $k = 1, \ldots, 2n$. Hence $r(x) - r^*(x) = c_1 F_\varphi(x)$, since the denominators are the same and the numerators are algebraic polynomials of degree $\leqslant 2n$. □

Lemma 6.2. *Let r be a rational function with poles only in $\mathbb{C}\setminus\mathscr{H} = \{\operatorname{Im} z < 0\}$. Then for the Hilbert transform Hr we have*

$$\left.\begin{aligned}(H(\operatorname{Re} r))(x) &= \operatorname{Im} r(x) + c_2,\\ (H(\operatorname{Im} r))(x) &= -\operatorname{Re} r(x) + c_3,\end{aligned}\right\} \tag{12}$$

where c_2 and c_3 are constants.

Proof. The direct calculation gives us

$$Hc = 0, H\left(\frac{1}{x - (\alpha - \mathrm{i}\beta)}\right) = H\left(\frac{1}{x - \bar{z}}\right) = -\frac{\mathrm{i}}{x - \bar{z}}, \quad \beta > 0. \tag{13}$$

From (13), since H is a linear operator, the lemma follows in view of

$$\operatorname{Re}\frac{1}{x - (\alpha - \mathrm{i}\beta)} = \frac{x - \alpha}{|x - \bar{z}|^2}, \quad \operatorname{Im}\frac{1}{x - \bar{z}} = \frac{-\beta}{|x - \bar{z}|^2},$$

$$\operatorname{Re}\left(-\frac{\mathrm{i}}{x - \bar{z}}\right) = \frac{-\beta}{|x - \bar{z}|^2}, \quad \operatorname{Im}\left(-\frac{\mathrm{i}}{x - \bar{z}}\right) = -\frac{x - \alpha}{|x - \bar{z}|^2}, \quad \bar{z} = \alpha - \mathrm{i}\beta. \ \square$$

Lemma 6.3. *With the same notation as in lemma 6.2 we have*

$$(Hr)(x) \equiv \tilde{r}(x) = -c_1 F_{\varphi + \pi/2}(x) - \sum_{k=1}^{2n} r(x_k)\frac{F_{\varphi + \pi/2}(x) - F_{\varphi + \pi/2}(x_k)}{(x - x_k)F'_\varphi(x_k)} + c_4, \tag{14}$$

c_4 is a constant.

Proof. If we apply lemma 6.2 to $\mathrm{e}^{\mathrm{i}\varphi}B(x)$ we obtain, using (6),

$$(HF_\varphi)(x) = \operatorname{Im}(\mathrm{e}^{\mathrm{i}\varphi}B(x)) + c_5 = -F_{\varphi + \pi/2}(x) + c_5, \tag{15}$$

c_5 a constant.

Again from lemma 6.2, applied to $\mathrm{e}^{\mathrm{i}\varphi}(B(x) - B(x_k))/(x - x_k)$, it follows that

$$\begin{aligned} H\left(\frac{F_\varphi(x)}{x - x_k}\right) &= \operatorname{Im}\left(\mathrm{e}^{\mathrm{i}\varphi}\frac{B(x) - B(x_k)}{x - x_k}\right) + c_6 \\ &= -\frac{F_{\varphi + \pi/2}(x) - F_{\varphi + \pi/2}(x_k)}{x - x_k} + c_6 \end{aligned} \tag{16}$$

since $\operatorname{Re}(\mathrm{e}^{\mathrm{i}\varphi}B(x_k)) = 0$ in view of (10) and $\operatorname{Im}(\mathrm{e}^{\mathrm{i}\varphi}B(x)) = -F_{\varphi + \pi/2}(x)$ in view of (6).

Now (14) follows from lemma 6.1, (11), if we apply the Hilbert transform and use the linearity of H, (15) and (16). □

Let $\varphi \neq \pi/2 + k\pi$ be given and let us choose α, $\alpha \neq \pi/2 + k\pi$,

$\alpha \neq \pi/2 + k\pi - \varphi$. Let us set

$$r^*(x) = r(x)\cos\alpha - \tilde{r}(x)\sin\alpha.$$

From (7), lemma 6.1 and lemma 6.2 we obtain

$$r^*(x) = c_1 F_{\varphi+\alpha}(x) + \sum_{k=1}^{2n} \frac{r(x_k)}{F'_\varphi(x_k)} \frac{F_{\varphi+\alpha}(x) - F_{\varphi+\pi/2}(x_k)\sin\alpha}{x - x_k} - c_4 \sin\alpha. \quad (17)$$

Let y_i, $i = 1, \ldots, 2n$, be the zeros of $F_{\varphi+\alpha+\pi/2}$, i.e. $F'_{\alpha+\varphi}(y_i) = 0$ (see (8), (9)). We have from (17) and (8),

$$r^{*\prime}(y_i) = -\sum_{k=1}^{2n} \frac{r(x_k)}{F'_\varphi(x_k)} \frac{F_{\varphi+\alpha}(y_i) - F_{\varphi+\pi/2}(x_k)\sin\alpha}{(y_i - x_k)^2} = \sum_{k=1}^{2n} r(x_k)\gamma_{ki}. \quad (18)$$

If we choose the function $F_{\varphi+\pi/2}$ for r (we can do this in view of (10)), we obtain from (18), (8) and the definition of r^*

$$\tilde{r}^*(x) = F_{\varphi+\pi/2}(x)\cos\alpha - \tilde{F}_{\varphi+\pi/2}(x)\sin\alpha,$$

$$\tilde{r}^*(x) = F_{\varphi+\alpha+\pi/2}(x) - c_4\sin\alpha,$$

$$\begin{aligned} \tilde{r}^{*\prime}(y_i) &= -F_{\varphi+\alpha}(y_i)|B'(y_i)| \\ &= -\sum_{k=1}^{2n} \frac{F_{\varphi+\pi/2}(x_k)}{F'_\varphi(x_k)} \frac{F_{\varphi+\alpha}(y_i) - F_{\varphi+\pi/2}(x_k)\sin\alpha}{(y_i - x_k)^2} \\ &= -\sum_{k=1}^{2n} \frac{F_{\varphi+\alpha}(y_i)}{|B'(x_k)|} \frac{1 - F_{\varphi+\pi/2}(x_k)F_{\varphi+\alpha}(y_i)\sin\alpha}{(y_i - x_k)^2} \end{aligned} \quad (19)$$

since $F^2_{\varphi+\alpha}(y_i) = 1$ (from (6) it follows that $|F_{\varphi+\alpha}(y_i)| = |B(y_i)| = 1$).

From (19) we obtain that

$$|B'(y_i)| = \sum_{k=1}^{2n} \frac{1}{|B'(x_k)|} \frac{1 - F_{\varphi+\pi/2}(x_k)F_{\varphi+\alpha}(y_i)\sin\alpha}{(y_i - x_k)^2}. \quad (20)$$

Since $|F_{\varphi+\pi/2}(x_k)| \leqslant 1$, $|F_{\varphi+\alpha}(y_i)| \leqslant 1$ (see (5)), we have

$$1 - F_{\varphi+\pi/2}(x_k)F_{\varphi+\alpha}(y_i)\sin\alpha \geqslant 0;$$

therefore (20) and (18) give us

$$|B'(y_i)| = \sum_{k=1}^{2n} |\gamma_{ki}|,$$

$$\begin{aligned} |r^{*\prime}(y_i)| &= |r'(y_i)\cos\alpha - \tilde{r}'(y_i)\sin\alpha| \\ &\leqslant \sum_{k=1}^{2n} |\gamma_{ki}| \max_{i=1,\ldots,2n} |r(x_k)| \leqslant |B'(y_i)| \, \|r\|_{C(-\infty,\infty)}. \end{aligned} \quad (21)$$

The inequality (21) is proved for $\varphi \neq \pi/2 + k\pi$, $\alpha \neq \pi/2 + k\pi$, $\alpha \neq \pi/2 + k\pi - \varphi$, but since the two sides of (21) are continuous functions of α (obviously

y_i depends continuously on α), (21) remains valid for all α. Let x be such that $\phi(x) \neq -\alpha$. If we choose $\varphi = -\alpha - \phi(x) - \pi/2$ then $x = y_i$ for some i, $1 \leqslant i \leqslant 2n$. From (21) we now obtain the following.

Theorem 6.4. *Let*

$$r(x) = p(x) \Big/ \prod_{k=1}^{n} |x - z_k|^2, \quad p \in P_{2n}, \quad z_k \in \mathscr{H}.$$

Then for every x and α we have

$$|r'(x) \cos \alpha - \tilde{r}'(x) \sin \alpha| \leqslant |B'(x)| \, \| r \|_{C(-\infty,\infty)}, \tag{22}$$

where B is the Blaschke product

$$B(z) = \prod_{k=1}^{n} \frac{z - z_k}{z - \bar{z}_k}. \tag{23}$$

Proof. For $\phi(x) \neq \alpha$ this was proved. Since (22) is continuous with respect to α, (22) follows for all α. □

From theorem 6.4 follows immediately, since α is arbitrary, corollary 6.1.

Corollary 6.1. *Let $r \in R_{2n}$ be such that all poles of r are complex and conjugate. Then*

$$\{(r'(x))^2 + (\tilde{r}'(x))^2\}^{1/2} \leqslant |B'(x)| \, \| r \|_{C(-\infty,\infty)},$$

where B is the Blaschke product (23) corresponding to the poles of r.

From (4) it follows that

$$\int_{-\infty}^{\infty} |B'(x)| \, dx = 2\pi n. \tag{24}$$

Therefore from corollary 6.1 and (24) follows theorem 6.3. □

Let us remark, that if $r \in R_n$ and has real coefficients, then the condition $\| r \|_{C(-\infty,\infty)} < \infty$ gives us that r has only complex and conjugate poles. Theorefore we obtain from theorem 6.4 and (24) the following.

Corollary 6.2. *Let $r \in R_n$ have real coefficients. Then*

$$\| r' \|_{L_1(-\infty,\infty)} \leqslant \pi n \, \| r \|_{C(-\infty,\infty)},$$

$$\| \tilde{r}' \|_{L_1(-\infty,\infty)} \leqslant \pi n \, \| r \|_{C(-\infty,\infty)},$$

$$\| r' \|_{H_1(-\infty,\infty)} \leqslant 2\pi n \, \| r \|_{C(-\infty,\infty)}.$$

6.3 Estimation for higher derivatives of rational functions and its applications

In this section we shall obtain an estimate for $r^{(s)}$, $s \geqslant 1$, s a natural number, $r \in R_n$, in the metric $L_\sigma(-1, 1)$, $\sigma = (s + 1/p)^{-1}$, by means of $\| r \|_p = \| r \|_{L_p(-1,1)}$,

$1 < p \leqslant \infty$. From this estimation will follow a connection between the best spline approximations with free knots and the best rational approximations. This connection for $s = 1$, $p = \infty$, was given by Popov (1974b), for $s > 1$, $p = \infty$, by Pekarskii (1980b), and for $s \geqslant 1$, $p > 1$, by Pekarskii (1986).

Theorem 6.5 (A.A. Pekarskii, 1986). *Let $r \in R_n$, $n \geqslant 1$, and r have no poles on the interval $[-1, 1]$. For $1 < p \leqslant \infty$, $1/\sigma = s + 1/p$, we have*

$$\| r^{(s)} \|_\sigma \leqslant c(s, p) n^s \| r \|_p,$$

where the constant $c(s, p)$ depends only on s and p.

Following Pekarskii (1984, 1986) for the proof of theorem 6.5 we shall need some lemmas. We shall use the following notations: $D_+ = \{z: z \in \mathbb{C}, |z| < 1\}$, $D_- = \{z: z \in \mathbb{C}, |z| > 1\}$, $T = \{z: z \in \mathbb{C}, |z| = 1\}$, $T_\pm = \{z: z \in T, \pm \operatorname{Im} z > 0\}$, where $\mathbb{C}$ is the complex plane; if $a_1, \ldots, a_n$ belong to D_+, we set

$$B(z) = \prod_{k=0}^{n} \frac{z - a_k}{1 - \bar{a}_k z}, \quad a_0 = 0,$$

$$Q_\alpha(z, \xi) = \left| \frac{B(z) - B(\xi)}{z - \xi} \right|^\alpha, \quad Q(z, \xi) = \frac{B(z) - B(\xi)}{z - \xi},$$

$$\lambda(z, \beta) = \sum_{k=0}^{n} \left(\frac{1 - |a_k|}{|z - a_k|} \right)^\beta \frac{1}{|z - a_k|}, \quad \beta > 0.$$

If S is a rectifiable curve in the complex plane we set

$$\| f \|_{p,S} = \left\{ \int_S |f(z)|^p |\mathrm{d}z| \right\}^{1/p}.$$

Lemma 6.4. *Let $z \in T$ and let l be a natural number > 0. Then*

$$\int_T Q_{2l}(z, \xi) |\mathrm{d}\xi| = \frac{2\pi}{(2l - 1)!} z^l \sum_{j=1}^{l} \binom{2l}{l - j} (-1)^{l-j} B^{-j}(z) (B^j(z) z^{j-1})^{(2l-1)}.$$

Proof. If $z \in T$ and $\xi \in T$, we have $|\mathrm{d}\xi| = \mathrm{d}\xi / i\xi$ and

$$Q_2(z, \xi) = Q(z, \xi) \overline{Q(z, \xi)} = \frac{\xi z}{B(\xi) B(z)} Q^2(z, \xi)$$

since for $a, b \in T$ we have $(a - b)\overline{(a - b)} = -(a - b)^2 / ab$. Consequently

$$\int_T Q_{2l}(z, \xi) |\mathrm{d}\xi| = z^l B^{-l}(z) I_l(z), \tag{1}$$

where

$$I_l(z) = \frac{1}{\mathrm{i}} \int_T Q^{2l}(z, \xi) B^{-l}(\xi) \xi^{l-1} \mathrm{d}\xi.$$

Since $I_l(z)$ is continuous in $D_+ \cup T$, it is sufficient to calculate $I_l(z)$ only for

$z \in D_+$. We obtain for $z \in D_+$

$$I_l(z) = \sum_{j=-l}^{l} \binom{2l}{l-j} (-B(z))^{l-j} I_{l,j}(z), \tag{2}$$

where

$$I_{l,j}(z) = \frac{1}{\mathrm{i}} \int_T \frac{B^j(\xi)\xi^{l-1}}{(\xi - z)^{2l}} \mathrm{d}\xi.$$

Using the Cauchy formula we obtain

$$I_{l,j}(z) = \frac{2\pi}{(2l-1)!} (B^j(z) z^{l-j})^{(2l-1)} \tag{3}$$

for $j = 1, \ldots, l$. For $-l \leqslant j \leqslant 0$ the point $\xi = \infty$ is a zero of order at least 2 for the function $B^j(\xi)\xi^{l-1}(\xi - z)^{-2l}$; therefore we have $I_{l,j} = 0$ for $j = -l, \ldots, 0$. From equalities (1)–(3) we obtain the statement of the lemma. □

Lemma 6.5. *Let $z \in T$ and s be a natural number > 0. Then*

$$|B^{(s)}(z)| \leqslant 2^s \cdot s! \, \lambda^s(z, 1/s).$$

Proof. Setting $b_k(z) = (z - a_k)(1 - \bar{a}_k z)^{-1}$ we obtain

$$B^{(s)}(z) = \sum \frac{s!}{j_0! j_1! \cdots j_n!} b_0^{(j_0)}(z) b_1^{(j_1)}(z) \cdots b_n^{(j_n)}(z), \tag{4}$$

where the summation is over all collections $j_0, j_1, \ldots, j_n$ of nonnegative numbers satisfying $j_0 + j_1 + \cdots + j_n = s$. For $z \in T, 0 \leqslant k \leqslant n$, we have

$$b_k^{(j)}(z) = \frac{j!(1 - |a_k|^2)\bar{a}_k^{j-1}}{(1 - \bar{a}_k z)^{j+1}}, \qquad j > 0.$$

For $a_k \in D_+, z \in T$, we have $|1 - |a_k|| \leqslant |z - a_k|, |1 - \bar{a}_k z| = |z - a_k|$. Therefore for $1 \leqslant j \leqslant s$ we obtain that

$$|b_k^{(j)}(z)| \leqslant 2j! \left(\left(\frac{1 - |a_k|}{|z - a_k|} \right)^{1/s} \frac{1}{|z - a_k|} \right)^j. \tag{5}$$

The lemma follows from (4) and (5). □

Lemma 6.6. *If $z \in T$ and $\alpha > 0$ then*

$$\| Q(\cdot, z) \|_{1+\alpha, T} \leqslant c(\alpha) \lambda^{\alpha/(\alpha+1)}(z, 1/(\alpha + 2)).$$

Proof. Since $\lambda^{s_1}(z, 1/s_1) \leqslant \lambda^{s_2}(z, 1/s_2)$ for $s_2 \geqslant s_1$, from lemmas 6.4 and 6.5 it follows that for $z \in T$, $l > 0$, l a natural number, we have

$$\int_T |Q(\xi, z)|^{2l} |\mathrm{d}\xi| \leqslant c(l) \lambda^{2l-1}(z, 1/(2l-1)). \tag{6}$$

Let m be the smallest odd number such that $m > \alpha$. Let us set

$$p = (m+1)(\alpha+1)^{-1}, \quad q = (m+1)(m-\alpha)^{-1},$$
$$S(z) = \{\xi : \xi \in T, |\arg \xi - \arg z| \leqslant \lambda^{-1}(z, m^{-1})\}.$$

Using Hölder's inequality we obtain from (6), since $1/p + 1/q = 1$,

$$\int_{S(z)} |Q(\xi, z)|^{\alpha+1} |\mathrm{d}\xi| \leqslant \| 1 \|_{q,S(z)} \| Q_{\alpha+1}(\cdot, z) \|_{p,S(z)}$$
$$\leqslant c_1(m)\lambda^{-1/q}\left(z, \frac{1}{m}\right) \| Q_{m+1}(\cdot, z) \|_{1,T}^{1/p} \leqslant c_2(m)\lambda^{\alpha}\left(z, \frac{1}{m}\right). \quad (7)$$

On the other hand we have, using the choice of $S(z)$,

$$\int_{T\setminus S(z)} |Q(\xi, z)|^{\alpha+1} |\mathrm{d}\xi| \leqslant 2^{1+\alpha} \int_{T\setminus S(z)} |\xi - z|^{-1-\alpha} |\mathrm{d}\xi|$$
$$\leqslant c_3(\alpha)\lambda^{\alpha}\left(z, \frac{1}{m}\right). \quad (8)$$

Since $\lambda(z, \beta)$ does not increase in β for fixed $z \in T$, we have $\lambda(z, 1/m) \leqslant \lambda(z, 1/(\alpha+2))$ and the lemma follows from (7) and (8). □

Lemma 6.7. *For $z \in T\setminus\{\pm 1\}$ and $\alpha > 0$ we have*

$$\int_T Q_{\alpha+1}(\xi, z) Q_{\alpha+1}(\xi, \bar{z}) |1 - \xi^2| |\mathrm{d}\xi| \leqslant c(\alpha) |1 - z^2|^{-\alpha} \left(\lambda^{\alpha}\left(z, \frac{1}{\alpha+2}\right) + \lambda^{\alpha}\left(\bar{z}, \frac{1}{\alpha+2}\right)\right).$$

Proof. We shall show first that for every $z, \xi \in T_+$ we have

$$\frac{|1-\xi^2|}{|\xi - \bar{z}|^{\alpha+1}} \leqslant 2^{\alpha+1} |1 - z^2|^{-\alpha}. \quad (9)$$

In fact, if we consider the quadrilateral with vertices at the points $z, \bar{z}, \xi$ and $\bar{\xi}$, we obtain that $|1 - z^2| = |z - \bar{z}|$, $|1 - \xi^2| = |\xi - \bar{\xi}|$ are its bases, and $|\xi - \bar{z}|$ is its diagonal, therefore

$$|\xi - \bar{z}| \geqslant |1 - z^2|/2, \quad |\xi - \bar{z}| \geqslant |1 - \xi^2|/2$$

from which (9) follows.

We can suppose that $z \in T_+$. Using lemma 6.6 and (9) we obtain

$$\int_{T_+} Q_{\alpha+1}(\xi, z) Q_{\alpha+1}(\xi, \bar{z}) |1 - \xi^2| |\mathrm{d}\xi| \leqslant 2^{2\alpha+2} |1 - z^2|^{-\alpha} \int_{T_+} Q_{\alpha+1}(\xi, z) |\mathrm{d}\xi|$$
$$\leqslant c(\alpha) |1 - z|^{-\alpha} \lambda^{\alpha}\left(z, \frac{1}{\alpha+2}\right) \quad (10)$$

since $Q_{\alpha+1}(\xi,\bar{z}) \leqslant 2^{\alpha+1}|\xi-\bar{z}|^{-\alpha-1}$ for $z,\xi\in T$.

Analogously

$$\int_{T_-} Q_{\alpha+1}(\xi,z)Q_{\alpha+1}(\xi,\bar{z})|1-\xi^2||\mathrm{d}\xi| \leqslant c(\alpha)|1-z^2|^{-\alpha}\lambda^{\alpha}\left(\bar{z},\frac{1}{\alpha+1}\right). \quad (11)$$

From (10) and (11) the lemma follows. □

Let G be the exterior of the interval $[-1,1]$ ($G=\mathbb{C}\backslash[-1,1]$) and let Γ be the boundary of G. Γ consists of two intervals $[-1,1]$, one upper, the other lower. Let

$$\varphi(\eta)=\eta+\sqrt{(\eta^2-1)}$$

be the function which maps G into D_- conformally. If $r\in R_n$ is a rational function with poles η_k, $k=1,\ldots,n$, $\eta_k\in G$, $k=1,\ldots,n$, we set $a_k=1/\overline{\varphi(\eta_k)}$. We shall consider again the Blaschke product

$$B(\varphi(\eta))=\prod_{k=0}^{n}\frac{\varphi(\eta)-a_k}{1-\bar{a}_k\varphi(\eta)}, \quad \eta\in\Gamma.$$

Let us set

$$K_{\alpha}(\eta,x)=\frac{|B(\varphi(\eta))-B(\varphi(x))|^{\alpha+1}|B(\varphi(\eta))-B(\overline{\varphi(x)})|^{\alpha+1}}{|\eta-x|^{\alpha+1}}, \quad \eta,x\in\Gamma,\ \alpha>0.$$

Lemma 6.8. *We have*

$$\int_{\Gamma}\left(\int_{\Gamma}K_{\alpha}(\eta,x)|\mathrm{d}\eta|\right)^{1/\alpha}|\mathrm{d}x| \leqslant c(\alpha)n. \quad (12)$$

Proof. If we make in the integral in (12) the transform

$$x=\frac{1}{2}\left(z+\frac{1}{z}\right),$$

$$\eta=\frac{1}{2}\left(\xi+\frac{1}{\xi}\right),$$

where $z,\ \xi\in T$, we obtain that we must estimate the integral

$$A=\int_{T}\left(\int_{T}Q_{\alpha+1}(\xi,z)Q_{\alpha+1}(\xi,\bar{z})|1-\xi^2||\mathrm{d}\xi|\right)^{1/\alpha}|1-z^2||\mathrm{d}z|.$$

Using lemma 6.7 we obtain that

$$A\leqslant c_1(\alpha)\left(\int_{T}\lambda\left(z,\frac{1}{\alpha+2}\right)|\mathrm{d}z|+\int_{T}\lambda\left(\bar{z},\frac{1}{\alpha+2}\right)|\mathrm{d}z|\right).$$

But it is easy to see that

$$\int_{T}\left(\frac{1-|a_k|}{|z-a_k|}\right)^{\beta}\frac{|\mathrm{d}z|}{|z-a_k|}=(1-|a_k|)^{\beta}\int_{T}\frac{|\mathrm{d}z|}{|z-a_k|^{1+\beta}}\leqslant c(\beta),$$

where the constant $c(\beta)$ depends only on β. Therefore

$$\int_T \lambda\left(z, \frac{1}{\alpha+2}\right)|dz| = \int_T \lambda\left(\bar{z}, \frac{1}{\alpha+2}\right)|dz| \leqslant c'(\alpha)n$$

and we obtain the statement of the lemma. □

Lemma 6.9. *Let $\alpha > 0$, $1 < p \leqslant \infty$, $\sigma = (\alpha + 1/p)^{-1}$, $f \in L_p$. If*

$$g(x) = \int_\Gamma f(\eta) K_\alpha(\eta, x) |d\eta|$$

then

$$\| g \|_{\sigma,\Gamma} \leqslant c(\alpha, p) n^\alpha \| f \|_{p,\Gamma}.$$

Proof. If $p = \infty$, using lemma 6.8 we obtain

$$\| g \|_{1/\alpha,\Gamma} \leqslant \left(\int_\Gamma \left(\int_\Gamma |f(\eta)| K_\alpha(\eta, x) |d\eta| \right)^{1/\alpha} |dx| \right)^\alpha \leqslant c(\alpha) \eta^\alpha \| f \|_{\infty,\Gamma}. \quad (13)$$

Let $p \in (1, \infty)$ and $\alpha = 1 - p^{-1}$. Then $\sigma = 1$ and using again lemma 6.8, Hölder's inequality and (13) we obtain

$$\| g \|_{1,\Gamma} \leqslant \int_\Gamma \| K_\alpha(\eta, \cdot) \|_{1,\Gamma} |f(\eta)| d\eta$$

$$\leqslant \left(\int_\Gamma \| K_\alpha(\eta, \cdot) \|_1^{1/\alpha} |d\eta| \right)^\alpha \left(\int_\Gamma |f(\eta)|^p |d\eta| \right)^{1/p} \leqslant c(\alpha) n^\alpha \| f \|_{p,\Gamma}$$

since $1/p + \alpha = 1$, $1/\alpha > 1$.

Now let α be arbitrary. Choose positive numbers γ, τ, l, s satisfying the conditions $l \in (1, p)$, $l^{-1} + s^{-1} = 1$, $\gamma + \tau = \alpha$, $l\tau = 1 - l/p$. Then, using Hölder's inequality we obtain for every $x \in \Gamma$

$$|g(x)| \leqslant \| K_{s\gamma}(\cdot, x) \|_{1,\Gamma}^{1/s} \| K_{l\tau}(\cdot, x) |f(\cdot)|^l \|_{1,\Gamma}^{1/l} = \varphi(x)\psi(x). \quad (14)$$

From lemma 6.8 and (13) we obtain

$$\| \varphi \|_{1/\gamma,\Gamma} \leqslant c(s, \gamma) n^\gamma. \quad (15)$$

Using the fact that the lemma has already been proved for $\alpha = 1 - p^{-1}$ $(l\tau = 1 - (p/l)^{-1})$, we obtain

$$\| \psi \|_{l,\Gamma} \leqslant c(l, p) n^{1/l - 1/p} \| f \|_{p,\Gamma}. \quad (16)$$

Therefore we obtain the statement of the lemma in the case $p \in (1, \infty)$, $\alpha > 0$, from the inequalities (14)–(16) and Hölder's inequality. □

Proof of theorem 6.5. We shall use the notations of the previous lemmas. Let $r \in R_n$ have no poles in the interval $[-1, 1]$. Let $\rho > 0$ be such that all poles of r are in the exterior of the ellipse $\Gamma_\rho = \{\eta : |\varphi(\eta)| = \rho\}$, where

$\varphi(\eta)=\eta+\sqrt{(\eta^2-1)}$. From Cauchy's formula we obtain

$$r^{(s)}(x)=\frac{s!}{2\pi i}\int_{\Gamma_\rho}\frac{r(\eta)d\eta}{(\eta-x)^{s+1}},\quad x\in\Gamma. \tag{17}$$

Let $\eta_1,\ldots,\eta_n$ be the poles of r (every pole written so many times as its multiplicity). Then the function $g(\eta)=r(\eta)/B(\varphi(\eta))$ is analytic in G and ∞ is a zero for g at least of order 2 ($a_0=0$). Therefore

$$\frac{s!}{2\pi i}\int_{\Gamma_\rho}\frac{r(\eta)}{(\eta-x)^{s+1}}\left(1-\left(1-\frac{B(\varphi(x))}{B(\varphi(\eta))}\right)^{s+1}\left(1-\frac{\overline{B(\varphi(x))}}{B(\varphi(\eta))}\right)^{s+1}\right)d\eta=0. \tag{18}$$

Letting ρ tend to $+0$, from (17) and (18) we obtain

$$|r^{(s)}(x)|\leqslant\frac{s!}{2\pi}\int_\Gamma|r(\eta)|K_s(\eta,x)|d\eta|,\quad x\in\Gamma.$$

From here and lemma 6.9 the theorem follows. □

Now we shall use the Bernstein type inequality from theorem 6.5 to obtain some inverse results for rational approximation and to obtain a connection between best rational approximation and best spline approximation with free knots in L_p.

Let I be a finite interval and $f\in L_p(I)$. Let $E_{s-1}(f)_{p,I}$ be the best approximation in $L_p(I)$ of f by means of algebraic polynomials of degree $s-1$.

Definition 6.3. *Let $f\in L_p(-1,1)$ and let $s>0$ be a natural number and $1/\sigma=s+1/p$. Modulus of variation of f of order s in L_p is the following function of n:*

$$\kappa_{s,p}(f;n)=\sup\left\{\sum_{k=1}^{n}(E_{s-1}(f)_{p,I_k})^\sigma\right\}^{1/\sigma},$$

where the sup is taken over all subdivisions of the interval $[-1,1]$ into n intervals $I_k=[x_{k-1},x_k]$, $k=1,\ldots,n$, $-1=x_0<x_1<\cdots<x_n=1$.

Lemma 6.10. *We have*

$$\kappa_{s,p}(f;n)\leqslant n^s\|f\|_p.$$

Proof. Since $E_{s-1}(f)_{p,I}\leqslant\|f\|_{p,I}$, the lemma follows from Hölder's inequality:

$$\left\{\sum_{k=1}^{n}(E_{s-1}(f)_{p,I_k})^\sigma\right\}^{1/\sigma}\leqslant\left\{\sum_{k-1}^{n}\|f\|_{p,I_k}^\sigma\right\}^{1/\sigma}$$
$$\leqslant\left\{\left(\sum_{k=1}^{n}1\right)^{1-\sigma/p}\left(\sum_{k=1}^{n}\|f\|_{p,I_k}^p\right)^{\sigma/p}\right\}^{1/\sigma}=n^s\|f\|_p. \quad □$$

The following lemma connects $\kappa_{s,p}(f;n)$ with the rational functions.

Lemma 6.11. *Let $r\in R_m$ be such that all poles of r are outside of the interval*

$[-1,1]$. *Then for* $1<p\leqslant\infty$, $1/\sigma=s+1/p$, *we have*

$$\kappa_{s,p}(r;n)\leqslant c(s,p)\,m^s\,\|r\|_p,$$

where the constant $c(s,p)$ *depends only on* s *and* p.

Proof. Let $\{I_k\}_1^n$, $I_k=[x_{k-1},x_k]$, $-1=x_0<\cdots<x_n=1$, be an arbitrary partition of the interval $[-1,1]$ into n subintervals. We shall separate the intervals I_k, $k=1,\ldots,n$, into two classes: M_1 and M_2. If $r^{(s)}(x)$ is a monotone function in the interval I_k, we set $I_k\in M_1$, otherwise we set $I_k\in M_2$. Obviously M_2 contains no more than $m(s+1)$ intervals. Therefore, using Hölder's inequality for $p'=p/\sigma$, $q'=(1-1/p')^{-1}$, we obtain

$$\left\{\sum_{I_k\in M_2}(E_{s-1}(r)_{p,I_k})^\sigma\right\}^{1/\sigma}\leqslant\left\{\sum_{I_k\in M_2}\|r\|^\sigma_{p,I_k}\right\}^{1/\sigma}$$

$$\leqslant\left\{\left(\sum_{I_k\in M_2}1\right)^{1-\sigma/p}\left(\sum_{I_k\in M_2}\|r\|^p_{p,I_k}\right)^{\sigma/p}\right\}^{1/\sigma}\leqslant c(s)m^s\|r\|_p. \tag{19}$$

If $I_k\in M_1$, then in view of the fact that $r^{(s)}$ is monotone in I_k, we can apply theorem 6.7 from the end of this chapter and we obtain that there exists algebraic polynomials q_k of degree $\leqslant s-1$ such that

$$\|r-q_k\|_{L_p(I_k)}\leqslant c(s)\,\|r^{(s)}\|_{\sigma,I_k}$$

or

$$E_{s-1}(r)_{p,I_k}\leqslant c(s)\,\|r^{(s)}\|_{\sigma,I_k}. \tag{20}$$

From (20), applying theorem 6.5, we obtain

$$\left\{\sum_{I_k\in M_1}(E_{s-1}(r)_{p,I_k})^\sigma\right\}^{1/\sigma}\leqslant c(s)\left\{\sum_{I_k\in M_1}\|r^{(s)}\|^\sigma_{\sigma,I_k}\right\}^{1/\sigma}$$

$$\leqslant c(s)\,\|r^{(s)}\|_\sigma\leqslant c'(s,p)m^s\,\|r\|_p. \tag{21}$$

The lemma follows from (19) and (21). □

In what follows in this section we shall denote the best rational approximation $R_n(f)_{L_p[-1,1]}$ by $R_n(f)_p$.

Theorem 6.6. *Let* $f\in L_p[-1,1]$, $p>1$. *For all natural numbers* $n>0$ *and* $s>0$ *we have*

$$\kappa_{s,p}(f;m)\leqslant c(s,p)\left\{\sum_{k=0}^n(2^{ks}R_{2^k}(f)_p)^\sigma\right\}^{1/\sigma}, \tag{22}$$

where $2^n\leqslant m<2^{n+1}$, $c(s,p)$ *depends only on* s *and* p *and* $1/\sigma=s+1/p$.

Proof. We shall apply Bernstein's method from Chapter 3. Let r_k be the rational function of order k of best $L_p[-1,1]$ approximation to f, i.e. $\| f - r_k \|_p = R_k(f)_p, k = 0, 1, \ldots$. If we set $q_k = r_{2^k} - r_{2^{k-1}}$, $r_{2^{-1}} = r_0$, we have

$$f - r_0 = f - r_{2^n} + \sum_{k=0}^{n} q_k.$$

Since $s \geqslant 1$, we have $\sigma \leqslant 1$ and therefore

$$\kappa_{s,p}^{\sigma}(f;m) = \kappa_{s,p}^{\sigma}(f - r_0;m) \leqslant \kappa_{s,p}^{\sigma}(f - r_{2^n};m) + \sum_{k=0}^{n} \kappa_{s,p}^{\sigma}(q_k;m)$$

$$((\textstyle\sum a_k)^{\sigma} \leqslant \sum |a_k|^{\sigma} \quad \text{for } \sigma \leqslant 1). \tag{23}$$

From lemma 6.10 we obtain

$$\kappa_{s,p}^{\sigma}(f - r_{2^n};m) \leqslant (m^s \| f - r_{2^n} \|_p)^{\sigma} = m^{s\sigma} R_{2^n}^{\sigma}(f)_p. \tag{24}$$

Since $q_k \in R_{2^{k+1}}$ and $\| q_k \|_p \leqslant 2 \| f - r_{2^{k-1}} \|_p = 2R_{2^{k-1}}(f)_p$, using lemma 6.11 we obtain

$$\kappa_{s,p}^{\sigma}(q_k;m) \leqslant (c(s,p) - 2^{(k+1)s} \cdot 2R_{2^{k-1}}(f)_p)^{\sigma}. \tag{25}$$

The inequalities (23)–(25) give us (22). □

Using the standard technique (see Chapter 3), we obtain from theorem 6.6 the following.

Corollary 6.3. *Let $f \in L_p[-1,1]$, $1 < p \leqslant \infty$, $s > 0$ and $n > 0$ be natural numbers and $1/\sigma = s + 1/p$. Then*

$$\kappa_{s,p}(f;n) \leqslant c(s,p) \left\{ \sum_{k=0}^{n} \frac{1}{k+1} ((k+1)^s R_k(f)_p)^{\sigma} \right\}^{1/\sigma}.$$

Theorem 6.7. *Let $f \in L_p(a,b)$ and let $f^{(k)}$ be a monotone function in the interval (a,b) and let $f^{(k)} \in L_{\sigma}(a,b)$ for $\sigma = (k + 1/p)^{-1}$. Then there exists an algebraic polynomial q_{k-1} of degree $k-1$ such that*

$$\| f - q_{k-1} \|_p := \| f - q_{k-1} \|_{L_p(a,b)} \leqslant \| f^{(k)} \|_{L_{\sigma}(a,b)}. \tag{26}$$

Proof. Obviously we can suppose that $f^{(k)}$ is a monotone increasing function in (a,b). Let $c \in (a,b)$ be such that $f^{(k)}(x) \leqslant 0$ for $x < c$, $f^{(k)}(x) \geqslant 0$ for $x \geqslant c$ (if such a c does not exist, we set $c = a$). By Taylor's formula there exists an algebraic polynomial q_{k-1} of degree $k-1$ such that

$$f(x) = q_{k-1}(x) + \frac{1}{(k-1)!} \int_c^x f^{(k)}(t)(t-c)^{k-1}\,dt, \quad x \in (c,b),$$

$$f(x) = q_{k-1}(x) + \frac{1}{(k-1)!} \int_x^c f^{(k)}(t)(t-c)^{k-1}\,dt, \quad x \in (a,c),$$

and therefore

$$\| f - q_{k-1} \|_p^p = \frac{1}{((k-1)!)^p} \left\{ \int_c^b \left| \int_c^x f^{(k)}(t)(t-c)^{k-1} \mathrm{d}t \right|^p \mathrm{d}x \right.$$

$$\left. + \int_a^c \left| \int_x^c f^{(k)}(t)(t-c)^{k-1} \mathrm{d}t \right|^p \mathrm{d}x \right\} = \frac{1}{((k-1)!)^p}(I_1 + I_2). \tag{27}$$

We shall prove that

$$I_1 \leqslant ((k-1)!)^p \| f^{(k)} \|_{L_\sigma(c,b)}^p, \quad I_2 \leqslant ((k-1)!)^p \| f^{(k)} \|_{L_\sigma(a,c)}^p. \tag{28}$$

Obviously from (27) and (28) the theorem follows.†

It is sufficient to estimate I_1, since I_2 can be estimated in a similar way.

We shall prove the inequality (28) for I_1 by induction with respect to k. Let first $k = 1$. Then we have $\sigma_1 = (1 + 1/p)^{-1}$. For $0 < p < \infty$, since f' is monotone nondecreasing and nonnegative in (c, b), we get:

$$I_1 = \int_c^b \left| \int_c^x f'(t) \mathrm{d}t \right|^p \mathrm{d}x = \int_c^b \left(\int_c^x (f'(t))^{p/(p+1)} (f'(t))^{1/(p+1)} \mathrm{d}t \right)^p \mathrm{d}x$$

$$\leqslant \int_c^b \left(\int_c^b (f'(t))^{p/(p+1)} \mathrm{d}t \right)^p (f'(x))^{p/(p+1)} \mathrm{d}x$$

$$= \left(\int_c^b (f'(x))^{p/(p+1)} \mathrm{d}x \right)^{p+1} = \| f' \|_{L_{\sigma_1}(c,b)}^p.$$

The case $p = \infty$ is trivial. Thus we have

$$\left\| \left(\int_c^x f'(t) \mathrm{d}t \right)^p \right\|_{L_p(c,b)} \leqslant \| f' \|_{L_{\sigma_1}(c,b)}, \quad 0 < p \leqslant \infty, \quad \sigma_1 = (1 + 1/p)^{-1}. \tag{29}$$

Let us set

$$\varphi^{(0)}(x) = \int_c^x f^{(k)}(t)(t-c)^{k-1} \mathrm{d}t,$$

$$\varphi^{(\nu)}(x) = \int_c^x \varphi^{(\nu+1)}(t) \mathrm{d}t, \quad \nu = 0, 1, \ldots, k-1,$$

$$x \in (c, b).$$

We have $\varphi^{(k)}(x) = (k-1)!\, f^{(k)}(x)$ a.e. on (c, b). Since evidently $\varphi^{(\nu)}$ is nonnegative and nondecreasing in (c, b), we can apply (29) for each derivative $\varphi^{(\nu)}$, $\nu = 0, \ldots, k-1$. We obtain consequently, using (29),

$$I_1^{1/p} = \| \varphi^{(0)} \|_{L_p(c,b)} \leqslant \| \varphi^{(1)} \|_{L_{\sigma_1}(c,b)} \leqslant \| \varphi^{(2)} \|_{L_{\sigma_2}(c,b)}$$

$$\leqslant \cdots \leqslant \| \varphi^{(k)} \|_{L_{\sigma_k}(c,b)} = (k-1)! \, \| f^{(k)} \|_{L_{\sigma_k}(c,b)},$$

† $(|a|^{p/\sigma} + |b|^{p/\sigma})^{1/p} \leqslant (|a| + |b|)^{1/\sigma}, \quad p/\sigma \geqslant 1.$

where

$$\sigma_1 = (1 + 1/p)^{-1},$$
$$\sigma_2 = (1 + 1/\sigma_1)^{-1}, \ldots, \sigma_k = (1 + 1/\sigma_{k-1})^{-1} = (k + 1/p)^{-1} = \sigma.$$

Thus the estimate (28) is proved and the theorem follows. □

6.4 Notes

We shall give first some interesting results of Dolženko and Sevastijanov concerning the connection between rational approximations and the convergence of Fourier series.

Let f be an integrable 2π-periodic function and let us consider the Fourier series for f,

$$S(f;x) \sim \frac{a_0}{2} + \sum_{k=1}^{\infty} (a_k \cos kx + b_k \sin kx),$$

where the Fourier coefficients are given by

$$a_k = \frac{1}{2\pi}\int_0^{2\pi} f(t)\cos kt\,dt, \quad b_k = \frac{1}{2\pi}\int_0^{2\pi} f(t)\sin kt\,dt. \tag{1}$$

Let $S_n(f;x)$ be the nth partial sum of the Fourier series for f:

$$S_n(f;x) = \frac{a_0}{2} + \sum_{k=1}^{n} (a_k \cos kx + b_k \sin kx).$$

We shall need also two more definitions.

Let ϕ be an increasing function, continuous and concave on $[0, \infty)$, $\phi(0) = 0$. We said that the function f has a bounded ϕ-variation if

$$V_\phi(f) = \sup \sum_{k=1}^{n} \phi(|f(x_k) - f(x_{k-1})|) < \infty,$$

where the sup is taken over all partitions $0 = x_0 < \cdots < x_n = 2\pi$ of the interval $[0, 2\pi]$ into n parts.

Let f be a bounded function on $[0, 2\pi]$. Let M_n be the set of all bounded functions on $[0, 2\pi]$ which are n times piecewise monotone on $[0, 2\pi]$, i.e. $\varphi \in M_n$ if there exist $n+1$ points y_i, $i = 0, \ldots, n$, $0 = y_0 < \cdots < y_n = 2\pi$ such that in every interval $[y_{i-1}, y_i]$, $i = 1, \ldots, n$, φ is a monotone function.

Following Dolženko and Sevastijanov (1976a), we consider the best uniform approximation to f by means of elements of M_n:

$$M_n(f)_{C[0,2\pi]} = \inf\{\|f - \varphi\|_{C[0,2\pi]} : \varphi \in M_n\}.$$

The following theorem was proved by E.A. Sevastijanov (1974a):

$$V_0^{2\pi} f = 2\sum_{n=0}^{\infty} M_n(f)_{C[0,2\pi]}.$$

E.P. Dolženko (1966b) gives the following result: *if*

$$\sum_{n=0}^{\infty} R_n(f)_{C[0,2\pi]}(n\phi^{-1}(1/n))^{-1} < \infty$$

then the function f has a bounded ϕ-variation, i.e. $V_\phi(f) < \infty$.

This result was improved by Sevastijanov (1974a, 1975) in the following way.

Let V_ϕ denote the class of all functions for which $V_\phi(f) < \infty$. The necessary and sufficient condition for every function f for which $R_n(f)_{C[0,2\pi]} \leqslant a_n$ to belong to V_ϕ is

$$\sum_{n=0}^{\infty} \phi(2a_n) < \infty, \quad a_0 \geqslant a_1 \geqslant \cdots \geqslant a_n \geqslant \cdots.$$

The first result for connection between rational uniform approximation of a function and the convergence for its Fourier series is due to Dolženko (1966b).

If

$$\sum_{n=1}^{\infty} \frac{1}{n} R_n^2(f)_{C[0,2\pi]} < \infty$$

then $S_n(f)$ converges to f almost everywhere on $[0, 2\pi]$.

If

$$\sum_{n=1}^{\infty} \frac{1}{n} \sqrt{R_n(f)_{C[0,2\pi]}} < \infty$$

then $S_n(f)$ converges to f uniformly on $[0, 2\pi]$.

E.A. Sevastijanov (1974a, 1975) improved the last result in the following way.

If

$$\sum_{n=1}^{\infty} \frac{1}{n} M_n(f)_{C[0,2\pi]} < \infty \tag{2}$$

then $S_n(f)$ converges uniformly to f on $[0, 2\pi]$.

Consequently if

$$\sum_{n=1}^{\infty} \frac{1}{n} R_n(f)_{C[0,2\pi]} < \infty \tag{3}$$

then $S_n(f)$ converges uniformly to f on $[0, 2\pi]$.

It is not possible to improve this result in the following sense.

Let $\{\alpha_n\}_1^\infty$ be a nonincreasing sequence of positive numbers such that

$$\sum_{n=1}^{\infty} \frac{1}{n} \alpha_n = \infty.$$

Then there exists a continuous 2π-periodic function f for which $R_n(f)_{C[0,2\pi]} \leqslant \alpha_n$ for all $n \geqslant 1$ and the Fourier series for f diverges at $x = 0$.

A very interesting result concerning absolute convergence of Fourier series was obtained by Sevastijanov (1978). Let R_n^{T} denote the set of all rational trigonometrical functions of order n, i.e. $t \in R_n^{\mathrm{T}}$ if

$$t = t_1/t_2, \quad t_i \in T_n, \quad i = 1, 2.$$

Let $R_n^{\mathrm{T}}(f)_{C[0,2\pi]}$ be the best uniform approximation of the 2π-periodic function f by means of trigonometric rational functions of order n:

$$R_n^{\mathrm{T}}(f)_{C[0,2\pi]} = \inf\{\|f - t\|_{C[0,2\pi]} : t \in R_n^{\mathrm{T}}\}.$$

Sevastijanov obtained the following estimate:

$$\sum_{n=1}^{\infty} (|a_n| + |b_n|) < 36 \sum_{n=0}^{\infty} R_n^{\mathrm{T}}(f)_{C[0,2\pi]},$$

where a_n, b_n are the Fourier coefficients of the function f, given by (1).

This fact that the condition $\sum_{n=0}^{\infty} R_n^{\mathrm{T}}(f)_C < \infty$ is a sufficient condition for absolute convergence of the Fourier series for f seems to be really connected with the rational approximation, as opposed to condition (3) for uniform convergence of the Fourier series, which is connected with piecewise monotone approximation (see (2)). The proof uses Russak's results from section 6.3.

Sevastijanov also gives the following inequalities for Fourier coefficients of rational trigonometric functions.

Let us set

$$\hat{t}(n) = \frac{1}{2\pi} \int_0^{2\pi} t(u) \mathrm{e}^{-\mathrm{i}nu} \, \mathrm{d}u, \quad n = 0, \pm 1, \ldots .$$

If $t \in R_n^{\mathrm{T}}$, then

$$\sum_{k=1}^{\infty} |\hat{t}(k)| \leqslant \frac{\pi n}{2} \|t\|_{C[0,2\pi]}.$$

If r is a rational complex function of nth degree without poles on $\Gamma = \{z: |z| = 1\}$, then

$$\sum_{k=-\infty}^{\infty} |c_k(r)| \leqslant \{2\pi n + 1\} \|r\|_{C(\Gamma)},$$

where $c_k(r) = \hat{t}(k)$, $t(u) = r(\mathrm{e}^{\mathrm{i}u})$.

The following results are given by Sevastijanov (1973).

(a) *Let $0 < p \leqslant \infty$, α a positive integer, $0 < q < 1/(\alpha + 1/p)$. Then there exists a constant $c(p, q, \alpha)$ such that if $r \in R_n$ then*

$$\|r^{(\alpha)}\|_q \leqslant c(p, q, \alpha) n^{\alpha} \|r\|_p. \tag{4}$$

(b) *Let $r \in R_n$. For every $\delta > 0$ there exists a set $E(\delta)$,* mes *$E(\delta) < \delta$, such that* *for every* $x \in [0,1] \backslash E(\delta)$, $x+h \in [0,1]$, *we have*

$$\left(\frac{1}{h} \int_x^{x+h} \left| r(t) - \sum_{i=0}^{[\lambda]} \frac{(t-x)^i}{i!} r^{(i)}(x) \right|^p \mathrm{d}t \right)^{1/p} \leqslant c(\lambda, p) n^{\lambda} \delta^{-\lambda - 1/p} |h|^{\lambda} \| r \|_p.$$

(c) *Let* $R_n(f)_{L_p[0,1]} = \mathrm{O}(n^{-\lambda-\varepsilon})$. *Then there exists a set G such that* mes $G = 0$ *and for every* $x \in [0,1] \backslash G$ *the function f has a local p-differential of order* λ. (We say that the function f has at the point x_0 a local p-differential of order λ if there exists a polynomial $p \in P_{[\lambda]}$ such that

$$\| f - p \|_{L_p(x_0 - h, x_0 + h)} = \mathrm{o}(h^{\lambda}), \quad h > 0.$$

See also Sevastijanov (1974a, 1980), Dolženko (1978)).

These results are generalized for the many-dimensional case by E.P. Dolženko and V.I. Danchenko (1977).

The situation is better if we consider complex rational approximation on the unit disk. A.A. Pekarskii (1984) obtained Bernstein type inequalities for the derivatives of rational functions on the unit disk, by means of which it is possible to get an exact converse theorem for rational approximation on the unit disk. We shall give here some of the results of Pekarskii (compare with section 6.3).

Let $\mathbb{C}$ be the complex plane, $D_+ = \{z: |z| < 1\}$, $D_- = \{z: |z| > 1\}$, $\Gamma = \{z: |z| = 1\}$.

We shall use the notation $\| f \|_p$ for

$$\| f \|_p = \left(\int_{\Gamma} |f(z)|^p |\mathrm{d}z| \right)^{1/p}, \quad 0 < p < \infty, \quad \| f \|_{\infty} = \operatorname*{ess\,sup}_{z \in \Gamma} |f(z)|, \quad p = \infty.$$

The Hardy space H_p, $0 < p \leqslant \infty$, is the set of all functions f, analytic in D_+ for which

$$\| f \|_{H_p} = \lim_{\rho \to 1-0} \| f(\cdot \rho) \|_p = \lim_{\rho \to 1-0} \left\{ \int_{\Gamma} |f(\rho e^{i\varphi})|^p \, \mathrm{d}\varphi \right\}^{1/p} < a.$$

We shall consider fractional derivatives of f. The α-derivative in the sense of Weil for the function f analytic in D_+ is given by

$$\mathfrak{I}^{\alpha} f(z) = \sum_{k=0}^{\infty} (k+1)^{\alpha} \hat{f}(k) z^k,$$

where $f(z) = \sum_{k=0}^{\infty} \hat{f}(k) z^k$ for $|z| < 1$.

By H_p^{α} we shall denote the Hardy–Sobolev space of all functions f for which $\mathfrak{I}^{\alpha} f \in H_p$. We set the seminorm in H_p^{α},

$$\| f \|_{H_p^{\alpha}} = \| \mathfrak{I}^{\alpha} f \|_p.$$

The Besov space B_{pq}^{α}, $\alpha \in (-\infty, \infty)$, $0 < p \leqslant \infty$, $0 < q \leqslant \infty$, is the set of all functions f analytic in D_+ for which the quasinorm

$$\|f\|_{B^\alpha_{pq}}=\left\{\int_0^1(1-\rho)^{q-1}\|\mathfrak{J}^{\alpha+1}f(\cdot\rho)\|_p^q\mathrm{d}\rho\right\}^{1/q},\quad 0<q<\infty,$$

$$\|f\|_{B^\alpha_{p\infty}}=\sup_{0<\rho<1}(1-\rho)\|\mathfrak{J}^{\alpha+1}f(\cdot\rho)\|_p,\quad q=\infty,$$

is finite.

We shall consider also the space BMOA (analytic functions with bounded mean oscillation): $f\in$BMOA if there exists $g\in L_\infty(\Gamma)$ such that

$$f(z)=\frac{1}{2\pi\mathrm{i}}\int_\Gamma\frac{g(t)\mathrm{d}t}{t-z},\quad z\in D_+. \tag{5}$$

The norm in BMOA is given by

$$\|f\|_{\mathrm{BMOA}}=\inf\|g\|_\infty,$$

where the inf is taken over all $g\in L_\infty(\Gamma)$ for which we have the representation (5).

Now we can formulate the Bernstein type inequalities of A.A. Pekarskii (1984).

Let r be a (complex) *rational function of nth degree with poles only in D_-. Let $\alpha>0$, $p\in[1,\infty]$, $\sigma=(\alpha+p^{-1})^{-1}$. Then the following inequalities hold*:

$$\|r\|_{H^\alpha_\sigma}\leqslant c_1(\alpha,p)n^\alpha\|r\|_p,$$

$$\|r\|_{B^\alpha_{\sigma\sigma}}\leqslant c_2(\alpha,p)n^\alpha\|r\|_p,$$

$$\|r\|_{H^\alpha_{1/\alpha}}\leqslant c_3(\alpha)n^\alpha\|r\|_{\mathrm{BMOA}},$$

$$\|r\|_{B^\alpha_{(1/\alpha)(1/\alpha)}}\leqslant c_4(\alpha)n^\alpha\|r\|_{\mathrm{BMOA}}.$$

The history of inequalities of this type is connected with the names of Gonchar (1966), Russak (1973), Sevastijanov (1973), Danchenko (1977), Peller (1980), Pekarskii (1980b) (see the paper of Pekarskii (1984)).

Using these inequalities it is possible to obtain converse theorems for the best rational approximation in Hardy spaces H_p and in BMOA.

We shall denote by $R_n(f)_{H_p}$ the best rational approximation in H_p of the function $f\in H_p$ by means of (complex) rational functions of nth degree, and by $R_n(f)_{\mathrm{BMOA}}$ the best approximation in BMOA of $f\in$BMOA by means of rational functions of nth degree. Let us set

$$\|f\|_{R^\alpha_{pq}}=\|f\|_{H_p}+\left\{\sum_{k=0}^\infty(2^{k\alpha}R_{2^k}(f)_{H_p})^q\right\}^{1/q},$$

$$\|f\|_{R^\alpha_{p\infty}}=\|f\|_{H_p}+\sup_{k=0,1,2,\ldots}2^{k\alpha}R_{2^k}(f)_{H^p},$$

$$\|f\|_{R^\alpha_{*q}} = \|f\|_{\text{BMOA}} + \left\{\sum_{k=0}^{\infty} (2^{k\alpha} R_{2^k}(f)_{\text{BMOA}})^q\right\}^{1/q},$$

$$\|f\|_{R^\alpha_{*\infty}} = \|f\|_{\text{BMOA}} + \sup_{k=0,1,2,\dots} 2^{k\alpha} R_{2^k}(f)_{\text{BMOA}}.$$

Let $\alpha > 0$, $1 < p \leqslant \infty$, $\sigma = (\alpha + p^{-1})^{-1}$. *The following inequalities hold* (Pekarskii, 1984):

$$\|f\|_{B^\alpha_{\sigma\sigma}} \leqslant c_1(\alpha, p) \|f\|_{R^\alpha_{p\sigma}},$$

$$\|f\|_{H^\alpha_\sigma} \leqslant c_2(\alpha, p) \|f\|_{R^\alpha_{p\min(2,\sigma)}},$$

$$\|f\|_{B^\alpha_{(1/\alpha)(1/\alpha)}} \leqslant c_3(\alpha) \|f\|_{R^\alpha_{*(1/2)}},$$

$$\|f\|_{H^\alpha_{1/\alpha}} \leqslant c_4(\alpha) \|f\|_{R^\alpha_{*\min(2,1/\alpha)}}.$$

These inequalities are exact.

For the history of inequalities of this type see the paper of Pekarskii.

Let $\omega_k(f;\delta)_p$ be the kth modulus of smoothness of the function $f \in L_p(\Gamma)$:

$$\omega_k(f;\delta)_p = \sup_{|h| \leqslant \delta} \left\| \sum_{\nu=0}^{k} (-1)^{k+\nu} \binom{k}{\nu} f(e^{i(\cdot + \nu h)}) \right\|_{L_p}.$$

Pekarskii (1984) obtained the following results.

Let $\alpha > 0$, $p \in (1, \infty]$, $\sigma = (\alpha + p^{-1})^{-1}$ *and let* k *be the smallest natural number such that* $k > \alpha$. *Then*

(a) *If* $f \in H_p$ *then*

$$\sum_{m=0}^{n} (2^{m\alpha} \omega_k(f; 2^{-m})_\sigma)^\sigma \leqslant c(\alpha, p) \sum_{m=0}^{n} (2^{m\alpha} R_{2^m}(f)_{H_p})^\sigma,$$

(b) *If* $f \in \text{BMOA}$ *then*

$$\sum_{m=0}^{n} (2^{m\alpha} \omega_k(f; 2^{-m})_{1/\alpha})^{1/\alpha} \leqslant c(\alpha) \sum_{m=0}^{n} (2^{m\alpha} R_{2^m}(f)_{\text{BMOA}})^{1/\alpha}$$

(compare with theorem 6.2 and (4)).

See also Y.A. Brudnyi (1979).

The first characterization of $R_n(f)_{\text{BMOA}}$ was given by Peller (1980):

$$\sum_{n=0}^{\infty} (R_n(f)_{\text{BMOA}})^p < \infty \textit{ if and only if } f \in B^{1/p}_{pp}$$

(see also Peller (1983), Semmes (1982), J. Peetre (1983)).

The modulus of variation $\kappa_{s,\infty}(f;n)$ for $s = 1$ was considered by René Lagrange (1965), Z.A. Chanturia (1974), V.A. Popov (1974b).

7

Spline approximation and Besov spaces

Besov spaces appear in a natural way in rational and spline approximations. In this chapter we shall use Besov spaces to obtain complete direct and converse theorems for spline approximation in the spaces L_p, C and BMO. In this direction we shall follow the plan from section 3.5 to find pairs of adjusted inequalities of Jackson and Bernstein type and then to characterize the spline approximation by the Peetre K-functional generated by the corresponding spaces. An essential fact here is the appearance of Besov spaces $B^{\alpha}_{\sigma,\sigma}$ with index $\sigma < 1$. This is the reason to begin in section 7.1 with some facts concerning the spaces L_p $(0 < p < 1)$. In section 7.2 we define Besov spaces and give some necessary equivalent quasi-norms. In section 7.3 are established direct and converse theorems for spline approximation in L_p, C and BMO.

The results proved in this chapter will be applied in Chapter 8 for the rational approximation.

7.1 L_p $(0 < p < 1)$ spaces

The spaces L_p $(0 < p < 1)$ appear in a natural way in the theory of rational and spline approximations in L_p $(1 \leqslant p \leqslant \infty)$ metric.

By definition $L_p[a,b]$, $0 < p < 1$ consists of all measurable functions f defined on $[a,b]$ such that $\int_a^b |f(x)|^p \mathrm{d}x < \infty$. The space L_p $(0 < p < 1)$ equipped with the distance

$$d(f,g) = \int_a^b |f(x) - g(x)|^p \, \mathrm{d}x$$

is a complete metric space. The completeness of $L_p[a,b]$ when $0 < p < 1$ can be proved exactly as in the case $p \geqslant 1$.

We shall denote

$$\| f \|_p = \| f \|_{L_p[a,b]} = \left(\int_a^b |f(x)|^p \, \mathrm{d}x \right)^{1/p}$$

which is a quasi-norm in L_p $(0 < p < 1)$:

$$\| f + g \|_p \leqslant C(p)(\| f \|_p + \| g \|_p), \quad C(p) = 2^{(1-p)/p}.$$

We shall frequently make use of the inequality

$$\left\| \sum_i f_i \right\|_p^p \leqslant \sum_i \| f_i \|_p^p \tag{1}$$

which replaces the Minkovski inequality. The inequality (1) follows immediately from the semiadditivity of the function x^p $(0 < p < 1)$:

$$\left(\sum_i |x_i| \right)^p \leqslant \sum_i |x_i|^p. \tag{2}$$

The spaces L_p $(0 < p < 1)$ have some exotic properties which make them unpleasant as function spaces. For instance, there is no nontrivial convex open set in L_p. Consequently, there is no linear continuous functional in L_p except the zero functional. However, for the purposes of approximation theory there is no substantial difference between the spaces L_p $(0 < p < 1)$ and L_p $(1 \leqslant p \leqslant \infty)$.

Next we give some well-known inequalities that will be useful later and will be used frequently even without any indication. Of course we shall use the Minkovski inequality in L_p and l_p $(1 \leqslant p \leqslant \infty)$, the inequalities (1) and (2), and Hölder's inequality for functions and sequences. Also, we shall use the following well-known inequality:

$$\left(\sum_i |x_i|^p \right)^{1/p} \leqslant \left(\sum_i |x_i|^q \right)^{1/q}, \quad 0 < q \leqslant p \leqslant \infty. \tag{3}$$

The well-known inequalities of Hardy will be of great importance for us: if f is measurable and $f(x) \geqslant 0$ for $x \in (0, \infty)$, $\alpha > 0$ and $p \geqslant 1$, then

$$\left(\int_0^\infty \left(t^{-\alpha} \int_0^t f(u) \frac{\mathrm{d}u}{u} \right)^p \frac{\mathrm{d}t}{t} \right)^{1/p} \leqslant \frac{1}{\alpha} \left(\int_0^\infty (t^{-\alpha} f(t))^p \frac{\mathrm{d}t}{t} \right)^{1/p}, \tag{4}$$

$$\left(\int_0^\infty \left(t^{+\alpha} \int_t^\infty f(u) \frac{\mathrm{d}u}{u} \right)^p \frac{\mathrm{d}t}{t} \right)^{1/p} \leqslant \frac{1}{\alpha} \left(\int_0^\infty (t^{\alpha} f(t))^p \frac{\mathrm{d}t}{t} \right)^{1/p}. \tag{5}$$

In order to prove the inequality (4) we first substitute $u = tx$ in the left-hand-side integral, then apply the Minkovski inequality and finally substitute $t = v/x$. We get

$$\left(\int_0^\infty \left(t^{-\alpha}\int_0^t f(u)\frac{du}{u}\right)^p \frac{dt}{t}\right)^{1/p} = \left(\int_0^\infty \left(\int_0^1 t^{-\alpha} f(tx)\frac{dx}{x}\right)^p \frac{dt}{t}\right)^{1/p}$$

$$\leqslant \int_0^1 \left(\int_0^\infty (t^{-\alpha} f(tx))^p \frac{dt}{t}\right)^{1/p} \frac{dx}{x}$$

$$= \int_0^1 x^\alpha \frac{dx}{x} \cdot \left(\int_0^\infty (v^{-\alpha} f(v))^p \frac{dv}{v}\right)^{1/p}$$

which implies (4). The inequality (5) can be proved in the same manner.

The modulus of smoothness $\omega_k(f;\delta)_p$ in $L_p[a,b]$ $(0<p<1)$ is defined exactly as in the case $1 \leqslant p \leqslant \infty$:

$$\omega_k(f;\delta)_p = \sup_{0\leqslant h\leqslant\delta}\left(\int_a^{b-kh} |\Delta_h^k f(x)|^p \, dx\right)^{1/p}.$$

Now, we get some properties of moduli of smoothness $\omega_k(f;\delta)_p$ for $f \in L_p[a,b]$, $0<p<1$, $k \geqslant 1$ (compare with section 3.1):

(i) $\lim_{\delta\to 0} \omega_k(f;\delta)_p = 0$,
(ii) $\omega_k(f;\delta)_p$ *is a nondecreasing function on* $[0,\infty)$,
(iii) $\omega_k(\alpha f + \beta g;\delta)_p^p \leqslant |\alpha|^p \omega_k(f;\delta)_p^p + |\beta|^p \omega_k(g;\delta)_p^p$,
(iv) $\omega_k(f;n\delta)_p \leqslant C(k,p) n^{k-1+1/p} \omega_k(f;\delta)_p$ *and therefore* $\omega_k(f;\lambda\delta)_p \leqslant C(k,p)(\lambda+1)^{k-1+1/p}\omega_k(f;\delta)_p$, $\lambda \geqslant 0$,
(v) *If* $f \in L_p[a,b]$ *and* $\omega_k(f;\delta)_p = o(\delta^{k-1+1/p})$, *then* f *is a polynomial of degree* $k-1$ *for almost all* $x \in (a,b)$.

Property (i) can be proved just as the same property is proved in the case $p \geqslant 1$; see A. Zygmund (1959) or A. Timan (1960). Properties (ii) and (iii) are trivial. In order to prove property (iv) we shall use the following equality (see (5) in section 3.1):

$$\Delta_{nh}^k f(x) = \sum_{v_1=0}^{n-1}\sum_{v_2=0}^{n-1}\cdots\sum_{v_k=0}^{n-1} \Delta_h^k f(x + v_1 h + v_2 h + \cdots + v_k h).$$

Hence

$$\Delta_{nh}^k f(x) = \sum_{v=0}^{(n-1)k} A_v^{(k)} \Delta_h^k f(x + vh), \tag{6}$$

where $A_v^{(k)}$, $v = 0, 1, \ldots, (n-1)k$ are given by the identity

$$(1 + t + t^2 + \cdots + t^{n-1})^k = \sum_{v=0}^{(n-1)k} A_v^{(k)} t^v.$$

Obviously $A_v^{(k)} > 0$. Now, we shall prove by induction that

$$A_v^{(k)} \leqslant n^{k-1}, \quad v = 0, 1, \ldots, (n-1)k. \tag{7}$$

Clearly $A_\nu^{(1)}=1$, $\nu=0,1,\ldots,n-1$. Suppose that $A_\nu^{(k)}\leqslant n^{k-1}$, $\nu=0,1,\ldots,(n-1)k$. Then by the equalities

$$(1+t+t^2+\cdots+t^{n-1})^{k+1}=\left(\sum_{\nu=0}^{(n-1)k}A_\nu^{(k)}t^\nu\right)(1+t+t^2+\cdots+t^{n-1})$$
$$=\sum_{\nu=0}^{(n-1)(k+1)}A_\nu^{(k+1)}t^\nu$$

we conclude that

$$A_\nu^{(k+1)}\leqslant n\max_\nu A_\nu^{(k)}\leqslant n^k,\quad \nu=0,1,\ldots,(n-1)(k+1).$$

Estimates (7) are proved.

Now by (6) and (7) in view of inequality (2) we get

$$\int_a^{b-knh}|\Delta_{nh}^k f(x)|^p\,dx\leqslant\sum_{\nu=0}^{(n-1)k}(A_\nu^{(k)})^p\int_a^{b-kh}|\Delta_h^k f(x)|^p\,dx$$
$$\leqslant C(k,p)n^{(k-1)p+1}\omega_k(f,\delta)_p^p,\quad 0\leqslant h\leqslant\delta,$$

which implies property (iv).

From property (iv) there follows immediately the following estimate:

$$\omega_k(f;\delta_2)_p/\delta_2^{k-1+1/p}\leqslant C(k,p)\omega_k(f;\delta_1)_p/\delta_1^{k-1+1/p},\quad 0<\delta_1<\delta_2.$$

Then, if $\omega_k(f;\delta)_p=o(\delta^{k-1+1/p})$, then $\omega_k(f;\delta)_p=0$, $\delta>0$, and according to lemma 7.8, which will be proved later on, we conclude that f coincides with a polynomial of degree $k-1$ a.e. in (a,b).

Note that we have for

$$f(x)=\begin{cases}0, & x\in(-1,0),\\ x^{k-1}, & x\in(0,1),\end{cases}$$

the estimate $\omega_k(f;\delta)_p=O(\delta^{k-1+1/p})$.

We remark that there is no upper estimate of $\omega_k(f;\delta)_p$ by $\omega_{k-1}(f';\delta)_p$ or $\|f^{(k)}\|_p$ when $f'\in L_p$ or $f^{(k)}\in L_p$, respectively, in the case $0<p<1$. Indeed, consider the function

$$\varphi_\varepsilon(x)=\begin{cases}0 & x\in[-1,0],\\ \varepsilon^{-1}x, & x\in(0,\varepsilon],\\ 1, & x\in(\varepsilon,1],\end{cases}$$

where $\varepsilon>0$ is sufficiently small. It is readily seen that $\omega_k(\varphi_\varepsilon;\delta)_p>C(\delta)>0$ and $\|\varphi'_\varepsilon\|_p=\varepsilon^{1/p-1}\to0$ as $\varepsilon\to0$. This fact is of central importance and will influence fundamentally our further discussion.

Next, we prove some more complicated properties of moduli of smoothness.

Lemma 7.1. *Let $f\in L_p[0,1]$, $0<p\leqslant\infty$, $k\geqslant1$ and $0\leqslant\delta\leqslant1/(k+1)$. Then we*

have:

(i) *If* $0<p<1$, *then*

$$\omega_k(f;\delta)_p^p \leqslant C\delta^{kp}\left\{\int_\delta^{1/(k+1)} t^{-kp}\omega_{k+1}(f;t)_p^p\frac{\mathrm{d}t}{t} + \|f\|_p^p\right\}. \tag{8}$$

(ii) *If* $1\leqslant p\leqslant\infty$, *then*

$$\omega_k(f;\delta)_p \leqslant C\delta^{k}\left(\int_\delta^{1/(k+1)} t^{-k}\omega_{k+1}(f;t)_p\frac{\mathrm{d}t}{t} + \|f\|_p\right). \tag{9}$$

In (8) and (9) $C=C(k,p)$.

Proof. We shall prove only the estimate (8). The estimate (9) can be proved similarly.

We shall use the following identity:

$$\Delta_h^k f(x) = 2^{-k}\left(\Delta_{2h}^k f(x) - \sum_{i=0}^{k-1}\sum_{j=i+1}^{k}\binom{k}{j}\Delta_h^{k+1}f(x+ih)\right). \tag{10}$$

Indeed, by induction it is readily seen that (see (5) in section 3.1)

$$\Delta_{nh}^k f(x) = \sum_{v_1=0}^{n-1}\sum_{v_2=0}^{n-1}\cdots\sum_{v_k=0}^{n-1}\Delta_h^k f(x+v_1h+v_2h+\cdots+v_kh).$$

Hence

$$\begin{aligned}\Delta_{2h}^k f(x) &= \sum_{v_1=0}^{1}\cdots\sum_{v_k=0}^{1}\Delta_h^k f(x+v_1h+\cdots+v_kh)\\ &= \Delta_h^k f(x) + \binom{k}{1}\Delta_h^k f(x+h)+\cdots+\binom{k}{k}\Delta_h^k f(x+kh).\end{aligned} \tag{11}$$

Obviously

$$\Delta_h^k f(x+jh) = \Delta_h^k f(x) + \sum_{i=0}^{j-1}\Delta_h^{k+1}f(x+ih).$$

Then by (11) we get

$$\begin{aligned}\Delta_{2h}^k f(x) &= \sum_{j=0}^{k}\binom{k}{j}\Delta_h^k f(x+jh)\\ &= 2^k\Delta_h^k f(x) + \sum_{j=1}^{k}\binom{k}{j}\sum_{i=0}^{j-1}\Delta_h^{k+1}f(x+ih)\\ &= 2^k\Delta_h^k f(x) + \sum_{i=0}^{k-1}\sum_{j=i+1}^{k}\binom{k}{j}\Delta_h^{k+1}f(x+ih)\end{aligned}$$

which implies (10).

Suppose $0<p<1$ and denote briefly

$$\Omega_k(\delta)=\sup_{0<h\leqslant\delta}\left(\int_0^{1/2}|\Delta_h^k f(x)|^p\,\mathrm{d}x\right)^{1/p}.$$

Let $0<h\leqslant 1/4k$. By (10) and semiadditivity of x^p $(0<p<1)$ (see inequality (2)), we get

$$|\Delta_h^k f(x)|^p\leqslant 2^{-kp}\left(|\Delta_{2h}^k f(x)|^p+\sum_{i=0}^{k-1}\sum_{j=i+1}^{k}\binom{k}{j}^p|\Delta_h^{k+1}f(x+ih)|^p\right)$$

Now we integrate over $[0,\frac{1}{2}]$ and take supremum. We obtain for $0<\delta\leqslant 1/4k$

$$\Omega_k(\delta)_p^p\leqslant 2^{-kp}\Omega_k(2\delta)+C\omega_{k+1}(f;\delta)_p^p,\quad C=C(p,k). \tag{12}$$

Let $r\geqslant 1$ and $0<\delta\leqslant 1/2^{r+1}k$. Then by (12) we have for $i=0,1,\ldots,r-1$

$$2^{-kpi}\Omega_k(2^i\delta)_p^p\leqslant 2^{-kp(i+1)}\Omega_k(2^{i+1}\delta)_p^p+C2^{-kpi}\omega_{k+1}(f,2^i\delta)_p^p.$$

Summing these inequalities over $i=0,1,\ldots,r-1$ we get

$$\begin{aligned}\Omega_k(\delta)_p^p&\leqslant 2^{-kpr}\Omega_k(2^r\delta)_p^p+C\sum_{i=0}^{r-1}2^{-kpi}\omega_{k+1}(f;2^i\delta)_p^p\\&\leqslant 2^{-kp(r-1)}\|f\|_p^p+C_1\delta^{kp}\sum_{i=0}^{r-1}\int_{2^i\delta}^{2^{i+1}\delta}t^{-kp}\omega_{k+1}(f;t)_p^p\frac{\mathrm{d}t}{t}.\end{aligned}$$

Consequently, for $0<\delta\leqslant 1/2^{r+1}k$,

$$\Omega_k(\delta)_p^p\leqslant 2^{-kp(r-1)}\|f\|_p^p+C_1\delta^{kp}\int_\delta^{2^r\delta}t^{-kp}\omega_{k+1}(f;t)_p^p\frac{\mathrm{d}t}{t}.$$

Clearly, from this estimate it follows that for $0<\delta\leqslant 1/2^{r+1}k$

$$\omega_k(f;\delta)_p^p\leqslant 2\cdot 2^{-kp(r-1)}\|f\|_p^p+2C_1\delta^{kp}\int_\delta^{2^r\delta}t^{-kp}\omega_{k+1}(f;t)_p^p\frac{\mathrm{d}t}{t}. \tag{13}$$

Let $0<\delta\leqslant 1/4k$. Choose $r\geqslant 1$ such that $1/2^{r+2}k<\delta\leqslant 1/2^{r+1}k$. Then by (13) we obtain

$$\omega_k(f;\delta)_p^p\leqslant C\delta^{kp}\left\{\int_\delta^{1/4k}t^{-kp}\omega_{k+1}(f;t)_p^p\frac{\mathrm{d}t}{t}+\|f\|_p^p\right\}$$

which implies (8). □

Corollary 7.1. *Let $f\in L_p[0,1]$, $0<p\leqslant\infty$, $m>k\geqslant 1$ and $0<\delta\leqslant 1$. Then we have*

(i) *If $0<p<1$, then*

$$\omega_k(f;\delta)_p^p\leqslant C\delta^{kp}\left\{\int_\delta^1 t^{-kp}\omega_m(f;t)_p^p\frac{\mathrm{d}t}{t}+\|f\|_p^p\right\}. \tag{14}$$

(ii) *If* $1 \leqslant p \leqslant \infty$, *then*

$$\omega_k(f;\delta)_p \leqslant C\delta^k\left\{\int_\delta^1 t^{-k}\omega_{k+1}(f;t)\frac{dt}{t} + \|f\|_p\right\}. \tag{15}$$

In (4) and (5) $C = C(k,p,m)$.

Proof. We shall only prove the inequality (14) by induction with respect to m. The inequality (14) holds with $m = k+1$ by lemma 7.1.

Suppose that (14) holds for some $m \geqslant k+1$. Then by lemma 7.1 with $k = m$ and the Hardy inequality (5) we get

$$\begin{aligned}\omega_k(f;\delta)_p^p \leqslant C\delta^{kp}&\left\{\int_\delta^1 \left(t^{mp-kp-1}\int_t^1 u^{-mp-1}\omega_{k+1}(f;u)_p^p\,du\right)dt\right.\\ &\left.+\int_\delta^1 t^{mp-kp}\|f\|_p^p\,dt + \|f\|_p^p\right\}\\ \leqslant C_1\delta^{kp}&\left\{\int_\delta^1 t^{-kp}\omega_{m+1}(f;t)_p^p\frac{dt}{t} + \|f\|_p^p\right\}.\end{aligned}$$

□

Lemma 7.2. *Let* $f \in L_p[a,b]$, $0<p\leqslant\infty$, $k \geqslant 1$ *and* $0<\delta\leqslant(b-a)/k$. *Then we have:*

(i) *If* $0<p<1$, *then*

$$\begin{aligned}\omega_k(f;\delta)_p &\leqslant c\left(\frac{1}{\delta}\int_0^\delta\int_a^{b-kt}|\Delta_t^k f(x)|^p\,dx\,dt\right)^{1/p}\\ &\leqslant \frac{c}{\delta}\int_0^\delta\left(\int_a^{b-kt}|\Delta_t^k f(x)|^p\,dx\right)^{1/p}dt.\end{aligned} \tag{16}$$

(ii) *If* $1 \leqslant p < \infty$, *then*

$$\begin{aligned}\omega_k(f;\delta)_p &\leqslant \frac{c}{\delta}\int_0^\delta\left(\int_a^{b-kt}|\Delta_t^k f(x)|^p\,dx\right)^{1/p}dt\\ &\leqslant c\left(\frac{1}{\delta}\int_0^\delta\int_a^{b-kt}|\Delta_t^k f(x)|^p\,dx\,dt\right)^{1/p}.\end{aligned} \tag{17}$$

(iii) *If* $p=\infty$, *then*

$$\omega_k(f;\delta)_\infty \leqslant \frac{c}{\delta}\int_0^\delta \sup_{x\in[a,b-kt]}|\Delta_t^k f(x)|\,dt. \tag{18}$$

In the estimates (16)–(18) c is a constant depending only on p and k: $c = c(p,k)$.

Proof. We shall prove only the estimate (16). Estimates (17) and (18) can be established similarly. The estimate (16) follows immediately from the following estimate:

If $s \geqslant 0$ and $a+s+2k\delta \leqslant b$, then

$$\sup_{0\leqslant h\leqslant\delta}\int_a^{a+s}|\Delta_h^k f(x)|^p\,\mathrm{d}x \leqslant \frac{c}{\delta}\int_0^{\delta}\int_a^{a+s+k\delta}|\Delta_h^k f(x)|^p\,\mathrm{d}x\,\mathrm{d}h, \quad 0<p<1, \tag{19}$$

where $c=c(p,k)$.

In order to prove the estimate (19) we shall use the following identity:

$$\Delta_h^k f(x) = \sum_{i=1}^{k}(-1)^i\binom{k}{i}\{\Delta_{i(t-h)/k}^k f(x+ih) - \Delta_{h+i(t-h)/k}^k f(x)\}. \tag{20}$$

We have:

$$\begin{aligned}
&\sum_{v=0}^{k}(-1)^{k-v}\binom{k}{v}\Delta_{h+v(t-h)/k}^k f(x)\\
&= \sum_{v=0}^{k}(-1)^{k-v}\binom{k}{v}\sum_{i=0}^{k}(-1)^{k-i}\binom{k}{i}f\left(x+i\left(h+\frac{v(t-h)}{k}\right)\right)\\
&= \sum_{i=0}^{k}(-1)^{k-i}\binom{k}{i}\sum_{v=0}^{k}(-1)^{k-v}\binom{k}{v}f\left(x+ih+\frac{vi(t-h)}{k}\right)\\
&= \sum_{i=1}^{k}(-1)^{k-i}\binom{k}{i}\Delta_{i(t-h)/k}^k f(x+ih)
\end{aligned}$$

which implies (20).

By (20) it follows that

$$|\Delta_h^k f(x)|^p \leqslant c(p,k)\sum_{i=1}^{k}(|\Delta_{i(t-h)/k}^k f(x+ih)|^p + |\Delta_{h+i(t-h)/k}^k f(x)|^p).$$

Integrating with respect to $t\in[0,\delta]$ and dividing by δ we get

$$\begin{aligned}
|\Delta_h^k f(x)|^p \leqslant \frac{c(k,p)}{\delta}\sum_{i=1}^{k}\Bigg\{&\int_0^{\delta}|\Delta_{i(t-h)/k}^k f(x+ih)|^p\,\mathrm{d}t\\
&+\int_0^{\delta}|\Delta_{h+i(t-h)/k}^k f(x)|^p\,\mathrm{d}t\Bigg\}, \quad x\in[a,a+s], \quad h\in[0,\delta].
\end{aligned}$$

Now we integrate with respect to $x\in[a,a+s]$ and we obtain for $h\in[0,\delta]$

$$\begin{aligned}
\int_a^{a+s}|\Delta_h^k f(x)|^p\,\mathrm{d}x \leqslant \frac{c(k,p)}{\delta}\sum_{i=1}^{k}\Bigg\{&\int_0^{\delta}\int_a^{a+s}|\Delta_{i(t-h)/k}^k f(x+ih)|^p\,\mathrm{d}x\,\mathrm{d}t\\
&+\int_0^{\delta}\int_a^{a+s}|\Delta_{h+i(t-h)/k}^k f(x)|^p\,\mathrm{d}x\,\mathrm{d}t\Bigg\}.
\end{aligned} \tag{21}$$

Next we shall estimate each term in the sum on the right side of (21). Substituting $x+ih=u$ and $i(t-h)/k=v$ we obtain

$$I_i = \int_0^{\delta}\int_a^{a+s}|\Delta_{i(t-h)/k}^k f(x+ih)|^p\,\mathrm{d}x\,\mathrm{d}t = \int_{-ih/k}^{i(\delta-h)/k}\int_{a+ih}^{a+s+ih}|\Delta_v^k f(u)|^p\,\mathrm{d}u\,\mathrm{d}v.$$

Using the fact that

$$\Delta_v^k f(u) = (-1)^k \Delta_{-v}^k (u + kv)$$

and the substitution $u - kv = x$ we get

$$\begin{aligned}\int_{-ih/k}^{0} \int_{a+ih}^{a+s+ih} |\Delta_v^k f(u)|^p \, du \, dv &= \int_0^{ih/k} \int_{a+ih}^{a+s+ih} |\Delta_v^k f(u - kv)|^p \, du \, dv \\ &= \int_0^{ih/k} \int_{a+ih-kv}^{a+s+ih-kv} |\Delta_v^k f(x)|^p \, dx \, dv \\ &\leqslant \int_0^{\delta} \int_a^{a+s+k\delta} |\Delta_t^k f(x)|^p \, dx \, dt.\end{aligned}$$

Consequently

$$I_i \leqslant 2 \int_0^{\delta} \int_a^{a+s+k\delta} |\Delta_t^k f(x)|^p \, dx \, dt.$$

The other integrals in (21) are estimated similarly. Thus the estimate (19) is proved.

If $0 \leqslant \delta \leqslant (b-a)/4k$, then by (19) we obtain

$$\begin{aligned}\sup_{0 \leqslant h \leqslant \delta} \int_a^{(a+b)/2} |\Delta_h^k f(x)|^p \, dx &\leqslant \frac{c}{\delta} \int_0^{\delta} \int_a^{(a+b)/2 + k\delta} |\Delta_t^k f(x)|^p \, dx \, dt \\ &\leqslant \frac{c}{\delta} \int_0^{\delta} \int_a^{b-kt} |\Delta_t^k f(x)|^p \, dx \, dt.\end{aligned}$$

By symmetry we have

$$\begin{aligned}\sup_{0 \leqslant h \leqslant \delta} \int_{(a+b)/2-kh}^{b-kh} |\Delta_h^k f(x)|^p \, dx &= \sup_{0 \leqslant h \leqslant \delta} \int_{(a+b)/2}^{b} |\Delta_{-h}^k f(x)|^p \, dx \\ &\leqslant \frac{c}{\delta} \int_0^{\delta} \int_a^{b-kt} |\Delta_t^k f(x)|^p \, dx \, dt.\end{aligned}$$

Consequently estimate (16) holds true when $0 \leqslant \delta \leqslant (b-a)/4k$.

In the case $(b-a)/4k < \delta \leqslant (b-a)/k$ estimate (16) follows from the case $\delta = (b-a)/4k$. □

Now we shall prove some technical inequalities for the polynomials.

Lemma 7.3. *Let $k \geqslant 0$, $0 < q \leqslant p \leqslant \infty$. Then for every polynomial $Q \in P_k$ and every finite interval Δ we have*

$$\left(\frac{1}{|\Delta|} \int_{\Delta} |Q(x)|^q \, dx\right)^{1/q} \leqslant \left(\frac{1}{|\Delta|} \int_{\Delta} |Q(x)|^p \, dx\right)^{1/p} \leqslant c \left(\frac{1}{|\Delta|} \int_{\Delta} |Q(x)|^q \, dx\right)^{1/q},$$

where $c = c(q, k)$ and when $q = \infty$ or $p = \infty$ the corresponding expression is replaced by $\| Q \|_{L_\infty(\Delta)}$, $|\Delta|$ the length of Δ.

Proof. The left hand inequality is an immediate consequence of Hölder's inequality. It is enough to prove the right hand inequality for $p=\infty$. To this end choose a point $x_0 \in \Delta$ such that $|Q(x_0)| = \|Q\|_{L_\infty(\Delta)}$. Using Markov's inequality

$$\|Q'\|_{L_\infty(\Delta)} \leqslant \frac{2k^2 \|Q\|_{L_\infty(\Delta)}}{|\Delta|}$$

(see theorem 3.12), we obtain that there exists a constant $c_0 = c_0(k) > 0$ such that for $x \in \Delta$ we have

$$|Q(x) - Q(x_0)| \leqslant |x - x_0| \, \|Q'\|_{L_\infty(\Delta)} \leqslant c_0 \|Q\|_{L_\infty(\Delta)} \frac{|x - x_0|}{|\Delta|}.$$

Thus if we set $\Delta_1 = \{x : x \in \Delta, \ |x - x_0| \leqslant |\Delta|/2c_0\}$, then $|\Delta_1| \geqslant |\Delta|/2c_0$ and for $x \in \Delta_1$ we have

$$|Q(x)| \geqslant |Q(x_0)| - |Q(x) - Q(x_0)| \geqslant \tfrac{1}{2} \|Q\|_{L_\infty(\Delta)}.$$

Integrating we find

$$\|Q\|_{L_\infty(\Delta)} \leqslant 2\left(\frac{1}{|\Delta_1|}\int_{\Delta_1} |Q(x)|^p \, dx\right)^{1/p} \leqslant c\left(\frac{1}{|\Delta|}\int_{\Delta} |Q(x)|^p \, dx\right)^{1/p}. \qquad \square$$

The next lemma estimates the coefficients of a polynomial.

Lemma 7.4. *Let $k \geqslant 0$, $p > 0$, and let Δ be an arbitrary finite interval. Then for every polynomial*

$$Q(x) = \sum_{\nu=0}^{k} a_\nu (x - x_0)^\nu, \quad x_0 \in \Delta,$$

we have

$$\sum_{\nu=0}^{k} |a_\nu| \, |\Delta|^\nu \leqslant c\left(\frac{1}{|\Delta|}\int_{\Delta} |Q(x)|^p \, dx\right)^{1/p}, \tag{22}$$

where $c = c(p,k)$.

Proof. By translating the interval we can assume that $x_0 = 0$ and $\Delta = [0,b]$. In view of lemma 7.3 we need to prove (22) for $p = \infty$. The case $\Delta = [0,b]$ and $p = \infty$ follows from the case $\Delta = [0,1]$, $p = \infty$, by a simple change of variables. Finally, the case $\Delta = [0,1]$, $p = \infty$, follows from the fact that any two norms in $(k+1)$th dimensional space P_k are equivalent. $\square$

Lemma 7.5. *Let $k \geqslant 0$ and $0 < p, q \leqslant \infty$. Then for every polynomial $Q \in P_k$ and for every finite interval Δ the following inequality holds true:*

$$\|Q'\|_{L_q(\Delta)} \leqslant c|\Delta|^{1/q - 1/p - 1} \|Q\|_{L_p(\Delta)},$$

where $c = c(k,p)$.

Proof. By lemma 7.3 and Markov's inequality

$$\|Q'\|_{L_\infty(\Delta)} \leqslant 2k^2|\Delta|^{-1}\|Q\|_{L_\infty(\Delta)}$$

we get

$$\begin{aligned}\|Q'\|_{L_q(\Delta)} &\leqslant |\Delta|^{1/q}\|Q'\|_{L_\infty(\Delta)} \leqslant 2k^2|\Delta|^{1/q-1}\|Q\|_{L_\infty(\Delta)}\\ &\leqslant c(k,p)|\Delta|^{1/q-1/p-1}\|Q\|_{L_p(\Delta)}.\end{aligned}$$ □

Theorem 7.1 (Whitney, 1957, p = ∞). *Let $0<p\leqslant\infty$, $f\in L_p[a,b]$ and $k\geqslant 1$. Then there exists a polynomial $Q\in P_{k-1}$ such that*

$$\|f-Q\|_{L_p[a,b]} \leqslant c\omega_k\left(f;\frac{b-a}{k}\right)_p, \tag{23}$$

where $c=c(k,p)$.

Proof of theorem 7.1 in the case $p=\infty$. Clearly it is enough to prove the theorem in the case $[a,b]=[0,1]$ since the general case follows from this case by a simple change of variables. Thus we shall assume that $f\in L_\infty(0,1)$. Next we shall use the following modified Steklov function as an intermediate approximation (compare with theorem 3.5):

$$\begin{aligned}f_{k,h}(x) = \frac{(-1)^{k-1}}{h^k}\int_0^h\int_0^h\cdots\int_0^h\Big\{&f(x+k\theta)-\binom{k}{1}f(x+(k-1)\theta)+\cdots\\ &+(-1)^{k-1}\binom{k}{k-1}f(x+\theta)\Big\}\mathrm{d}t_1\,\mathrm{d}t_2\cdots\mathrm{d}t_k,\end{aligned} \tag{24}$$

where

$$\theta=\theta(x)=\frac{t_1+t_2+\cdots+t_k}{k}-hx,\quad 0<h\leqslant 1/k.$$

Clearly if $0\leqslant x\leqslant 1$ and $0\leqslant t_i\leqslant h$, $i=1,2,\ldots,k$, then $0\leqslant x+i\theta\leqslant 1$, $i=1,2,\ldots,k$, and therefore $f_{k,h}(x)$ is defined correctly for each $x\in[0,1]$. For every $x\in[0,1]$ we have

$$f(x)-f_{k,h}(x)=\frac{(-1)^k}{h^k}\int_0^h\int_0^h\cdots\int_0^h\Delta_\theta^k f(x)\,\mathrm{d}t_1\,\mathrm{d}t_2\cdots\mathrm{d}t_k$$

and for almost all $x\in[0,1]$ we have

$$\begin{aligned}f_{k,h}^{(k)}(x)=\frac{(-1)^{k-1}}{h^k}\Big\{&(1-kh)^k\Delta_h^k f(x-khx)\\ &-\binom{k}{1}\left(\frac{k}{k-1}\right)^k(1-(k-1)h)^k\Delta_{(k-1)h/k}^k f(x-(k-1)hx)\\ &+\cdots+(-1)^k k^k(1-h)^k\Delta_{h/k}^k f(x-hx)\Big\}.\end{aligned}$$

Consequently

$$\| f - f_{k,h} \|_{L_\infty[0,1]} \leqslant \omega_k(f; h)_\infty, \tag{25}$$

$$\| f_{k,h}^{(k)} \|_{L_\infty[0,1]} \leqslant c(k) h^{-k} \omega_k(f; h)_\infty. \tag{26}$$

Let $x_0 \in [0, 1]$ and let us set

$$Q(x) = \sum_{\nu=0}^{k-1} f_{k,h}^{(\nu)}(x_0) \frac{(x - x_0)^\nu}{\nu!}.$$

Obviously $Q \in P_{k-1}$ and

$$\| f_{k,h} - Q \|_{L_\infty[0,1]} \leqslant \| f_{k,h}^{(k)} \|_{L_\infty[0,1]}.$$

From this estimate, (25) and (26), setting $h = 1/k$ we get

$$\| f - Q \|_\infty \leqslant \| f - f_{k,h} \|_\infty + \| f_{k,h} - Q \|_\infty \leqslant \omega_k\left(f; \frac{1}{k} \right)_\infty + \| f_{k,h}^{(k)} \|_\infty \leqslant c(k) \omega_k\left(f, \frac{1}{k} \right)_\infty.$$

Thus theorem 7.1 is proved in the case $p = \infty$.

Next, we shall prove theorem 7.1 in the case $0 < p < 1$. The case $1 \leqslant p < \infty$ is well known and can be proved similarly.

The following lemma proves theorem 7.1 in the case $k = 1$, $0 < p < \infty$.

Lemma 7.6. *Let $f \in L_p[a, b]$, $0 < p < \infty$. Then there exists a constant c such that*

$$\| f - c \|_{L_p[a,b]}^p \leqslant \frac{1}{b - a} \int_a^b \int_a^b | f(x) - f(y) |^p \, dx \, dy$$
$$= \frac{2}{b - a} \int_0^{b-a} \int_a^{b-t} | f(x + t) - f(x) |^p \, dx \, dt \leqslant 2 \omega_1(f; b - a)_p^p, \tag{27}$$

where the constant c can be taken as $c = (b - a)^{-1} \int_a^b f(t) dt$ when $p = 1$.

Proof. Consider the function

$$\phi(y) = \int_a^b | f(x) - f(y) |^p \, dx, \quad y \in [a, b].$$

Clearly, there exists $y_0 \in [a, b]$ such that

$$\phi(y_0) \leqslant \frac{1}{b - a} \int_a^b \phi(y) \, dy$$

and therefore setting $c = f(y_0)$ we obtain

$$\int_a^b | f(x) - c |^p \, dx \leqslant \frac{1}{b - a} \int_a^b \int_a^b | f(x) - f(y) |^p \, dx \, dy. \tag{28}$$

Also, we have

$$I=\int_a^b\int_a^b|f(x)-f(y)|^p\,dx\,dy=\int_a^b\int_x^b|f(x)-f(y)|^p\,dy\,dx$$
$$+\int_a^b\int_a^x|f(x)-f(y)|^p\,dy\,dx.$$

Substituting $y=x+t$ and $y=x-t$ respectively in the last two integrals and changing the order of integration we get

$$I=\int_a^b\int_0^{b-x}|f(x)-f(x+t)|^p\,dt\,dx+\int_a^b\int_0^{x-a}|f(x)-f(x-t)|^p\,dt\,dx$$
$$=\int_0^{b-a}\int_a^{b-t}|f(x)-f(x+t)|^p\,dx\,dt+\int_0^{b-a}\int_{a+t}^{b}|f(x)-f(x-t)|^p\,dt\,dx$$
$$=2\int_0^{b-a}\int_a^{b-t}|f(x+t)-f(x)|^p\,dx\,dt.$$

This equality together with (28) implies (27). □

Lemma 7.7. *Let $f\in L_p[0,1]$, $0<p\leqslant 1$. Then for every natural number $n\geqslant 1$ there exists a step-function φ_n with jumps at the points $i/n=x_i$, $i=1,\ldots,n-1$, such that*

$$\|f-\varphi_n\|^p_{L_p(0,1)}\leqslant 2n\int_0^{1/n}\int_0^{1-t}|f(x+t)-f(x)|^p\,dx\,dt\leqslant 2\omega_1(f;n^{-1})^p_p.$$

Proof. By lemma 7.6 there exist constants c_i, $i=1,\ldots,n$, such that

$$\int_{x_{i-1}}^{x_i}|f(x)-c|^p\,dx\leqslant 2n\int_0^{1/n}\int_{x_{i-1}}^{x_i-t}|f(x+t)-f(x)|^p\,dx\,dt,\quad i=1,\ldots,n.$$

Then the step function $\varphi_n(x)=c_i$ for $x\in(x_{i-1},x_i)$, $i=1,\ldots,n$, satisfies the assertions of lemma 7.7. □

Proof of theorem 7.1 in the case $0<p\leqslant 1$. A simple change of variables shows that we can consider only the case $[a,b]=[0,1]$.

Suppose that theorem 7.1 does not hold true. Then there exists a sequence of functions $\{f_m\}_{m=1}^{\infty}$, $f_m\in L_p(0,1)$, such that

$$\inf_{Q\in P_{k-1}}\|f_m-Q\|^p_{L_p(0,1)}>m\omega_k\left(f_m;\frac{1}{k}\right)^p_p,\quad m=1,2,\ldots.$$

Since the set of all polynomials $Q\in P_{k-1}$ such that $\|Q\|_{L_p(0,1)}\leqslant 1$ is a compact set in $L_p(0,1)$, then for each m there exists a polynomial $Q_m\in P_{k-1}$ such that

$$\|f_m-Q_m\|_{L_p(0,1)}=\inf_{Q\in P_{k-1}}\|f_m-Q\|_{L_p(0,1)}.$$

Consequently

$$\| f_m - Q_m \|_p^p > m\omega_k\left(f_m; \frac{1}{k}\right)_p^p, \quad m = 1, 2, \ldots. \tag{29}$$

Let us set

$$g_m = \lambda_m(f_m - Q_m), \quad \lambda_m = \| f_m - Q_m \|_p^{-1},$$

By (29) it follows that

$$\| g_m \|_p = \inf_{Q \in P_{k-1}} \| g_m - Q \|_p = 1 \tag{30}$$

and

$$\omega_k(g_m; k^{-1})_p^p < 1/m, \quad m = 1, 2, \ldots. \tag{31}$$

We shall prove that $\{g_m\}_1^\infty$ is a precompact set in $L_p(0, 1)$, i.e. there exist a function $g \in L_p$ and a subsequence $\{g_{m_i}\}_1^\infty$ such that $\| g - g_{m_i} \|_p \to 0$ as $i \to \infty$. To this end it is sufficient to prove that for each $\varepsilon > 0$ there exists a finite ε-net for $\{g_m\}_{m=1}^\infty$ in $L_p(0, 1)$ ($L_p(0, 1)$, $0 < p \leqslant 1$, is a complete metric space).

It follows by corollary 7.1 with $k = 1$, $m = k$ and (30), (31) that

$$\omega_1(g_m; \delta)_p^p \leqslant c\delta^p \left\{ \int_\delta^1 t^{-p} \frac{1}{m} \frac{dt}{t} + 1 \right\} \leqslant c_1 \left(\frac{1}{m} + \delta^p \right)$$

for $0 \leqslant \delta \leqslant 1$ and $m = 1, 2, \ldots$, and therefore for each $\varepsilon > 0$ there exist $m_0 > 0$ and $\delta_0 > 0$ such that

$$\omega_1(g_m; \delta)_p^p < \varepsilon \quad \text{for } 0 < \delta < \delta_0 \quad \text{and} \quad m > m_0. \tag{32}$$

Fix $n > 1/\delta_0$. It follows by lemma 7.7 and (32) that for each $m > m_0$ there exists a step function $\varphi_{m,n}$ with points of discontinuity i/n, $i = 1, \ldots, n-1$, such that

$$\| g_m - \varphi_{m,n} \|_p^p \leqslant 2\omega_1(g_m; n^{-1})_p^p < 2\varepsilon. \tag{33}$$

On the other hand by (30) and (33) we get

$$\| \varphi_{m,n} \|_p^p \leqslant \| g_m \|_p^p + \| g_m - \varphi_{m,n} \|_p^p < 1 + 2\varepsilon.$$

Since $\varphi_{m,n}(x)$ is a constant for $x \in ((i-1)/n, i/n)$, $i = 1, \ldots, n$, for $m > m_0$ we have

$$\| \varphi_{m,n} \|_{L_\infty[0,1]} \leqslant \left(n \int_0^1 |\varphi_{m,n}(x)|^p \, dx \right)^{1/p} \leqslant ((1 + 2\varepsilon)n)^{1/p} = M.$$

Consider the set Ψ of all step functions φ of the type

$$\varphi(x) = r\varepsilon^{1/p}, \quad x \in \left(\frac{i-1}{n}, \frac{i}{n} \right), \quad i = 1, \ldots, n, \quad r = 0, \pm 1, \ldots, \| \varphi \|_{L_\infty[0,1]} \leqslant M.$$

Clearly

$$\inf_{\varphi\in\Psi} \|\varphi_{m,n}-\varphi\|_p^p \leqslant \int_0^1 (\varepsilon^{1/p})^p\,dx=\varepsilon$$

and therefore Ψ is a finite ε-net for the set $\{\varphi_{m,n}\}_{m=m_0+1}^{\infty}$. From this and (33) it follows that Ψ is a finite 3ε-net for the set $\{g_m\}_{m=m_0+1}^{\infty}$.

Consequently the set $\{g_m\}_{m=1}^{\infty}$ is precompact in $L_p(0,1)$ and for an appropriate subsequence $\{g_{m_i}\}_{i=1}^{\infty}$ we have $\|g_{m_i}-g\|_p\to 0$ as $i\to\infty$ for some $g\in L_p(0,1)$. Hence, in view of (30), we have

$$\begin{aligned}\inf_{Q\in P_{k-1}} \|g-Q\|_p^p &\geqslant \inf_{Q\in P_{k-1}} \|g_{m_i}-Q\|_p^p - \|g-g_{m_i}\|_p^p\\ &=1-\|g-g_{m_i}\|_p^p \to 1 \quad \text{as} \quad i\to\infty,\end{aligned}$$

and therefore

$$\inf_{Q\in P_{k-1}} \|g-Q\|_p = 1. \tag{34}$$

On the other hand by (31) we get

$$\omega_k\left(g;\frac{1}{k}\right)_p^p \leqslant \omega_k\left(g_{m_i};\frac{1}{k}\right)_p^p + 2^{kp}\|g-g_{m_i}\|_p^p \to 0 \quad \text{as} \quad i\to\infty.$$

Thus $\omega_k(g;1/k)_p=0$. As we shall show below this equality implies that $g=Q$ a.e. for some $Q\in P_{k-1}$, and so we have a contradiction with (34). □

It remains to prove the following.

Lemma 7.8. *Let $f\in L_p(a,b)$, $0<p\leqslant 1$, and $\omega_k(f;(b-a)/k)_p=0$. Then there exists a polynomial $Q\in P_{k-1}$ such that $f=Q$ almost everywhere in $[a,b]$.*

Proof. We shall prove the lemma by induction with respect to k. In the case $k=1$ the lemma follows by lemma 7.6. Suppose that the lemma holds true for some $k\geqslant 1$. Without loss of generality we can assume that $[a,b]=[0,1]$. Suppose that

$$\omega_{k+1}\left(f;\frac{1}{k+1}\right)_p^p = \sup_{0\leqslant h\leqslant 1/(k+1)} \int_0^{1-(k+1)h} |\Delta_h^{k+1} f(x)|^p\,dx = 0. \tag{35}$$

First we shall prove that

$$\int_0^{1-kh_1-h} |\Delta_{h_1}^k \Delta_h^1 f(x)|^p\,dx = 0, \quad h_1,h\geqslant 0, \quad kh_1+h\leqslant 1. \tag{36}$$

Indeed, if $h_1=\alpha h$ and $\alpha=m/n$ with some integers m and n, then by the identity (5) from section 3.1 we get

$$|\Delta^k_{(m/n)h}\Delta^1_h f(x)|^p \leqslant \sum_{\nu_1=0}^{m-1}\sum_{\nu_2=0}^{m-1}\cdots\sum_{\nu_k=0}^{m-1}\left|\Delta^k_{h/n}\Delta^1_h f\left(x+\nu_1\frac{h}{n}+\cdots+\nu_k\frac{h}{n}\right)\right|^p$$

$$\leqslant \sum_{\nu_1=0}^{m-1}\sum_{\nu_2=0}^{m-1}\cdots\sum_{\nu_k=0}^{m-1}\sum_{\nu=0}^{n-1}\left|\Delta^{k+1}_{h/n} f\left(x+\frac{\nu_1 h}{n}+\cdots+\frac{\nu_k h}{n}+\frac{\nu h}{n}\right)\right|^p.$$

Integrating with respect to $x\in[0,1-(km/n+1)h]$ and using (35) we conclude that (36) holds true in the case considered.

Suppose that $h_1=\alpha h$, $\alpha>0$ an irrational number. Choose a sequence $\{\alpha_i\}_{i=1}^{\infty}$ of rational numbers such that $\alpha_i\to\alpha$ as $i\to\infty$ and $0<\alpha_i<\alpha$. We have

$$|\Delta^k_{\alpha h}\Delta^1_h f(x)| \leqslant |\Delta^k_{\alpha_i h}\Delta^1_h f(x)| + \sum_{\nu=0}^{k}\binom{k}{\nu}\{|f(x+\nu\alpha h+h)-f(x+\nu\alpha_i h+h)|$$
$$+|f(x+\nu\alpha h)-f(x+\nu\alpha_i h)|\},$$

and therefore

$$\begin{aligned}\int_0^{1-k\alpha h-h}|\Delta^k_{\alpha h}\Delta^1_h f(x)|^p\,\mathrm{d}x &\leqslant \int_0^{1-k\alpha_i h-h}|\Delta^k_{\alpha_i h}\Delta^1_h f(x)|^p\,\mathrm{d}x\\ &\quad + c(k,p)\omega_1(f;k(\alpha-\alpha_i)h)_p^p\\ &= c(k,p)\omega_1(f;k(\alpha-\alpha_i)h)_p^p,\end{aligned}\tag{37}$$

where we have used that (36) holds true for α_i a rational number. Since $\omega_1(f;\delta)_p\to 0$ as $\delta\to 0$, (37) implies (36).

We note, without proof, that $\Delta^k_{h_1}\Delta^1_h f(x)$ can be represented as a finite linear combination of differences of the type $\Delta^{k+1}_{h_2}f(x+\beta)$; see P. Binev, K. Ivanov (1985). This fact together with (35) implies (36) immediately.

In view of (36) our induction hypothesis gives that for each h, $0\leqslant h<1$, there exists a polynomial $Q_h\in P_{k-1}$ such that $\Delta^1_h f(x)=Q_h(x)$ for almost all $x\in[0,1-h]$, i.e.

$$f(x+h)-f(x)=\sum_{\nu=0}^{k-1}a_\nu(h)x^\nu \tag{38}$$

almost everywhere in $[0,1-h]$.

We shall prove that each coefficient $a_\nu(h)$ is a continuous function of $h\in[0,1)$. Let $0\leqslant h_1,h_2<1$. We apply lemma 7.4 to the polynomial

$$\sum_{\nu=0}^{k-1}(a_\nu(h_1)-a_\nu(h_2))x^\nu = f(x+h_1)-f(x+h_2)$$

for the interval $\Delta = [0, 1-h]$, $h = \max\{h_1, h_2\}$. We obtain

$$\sum_{\nu=0}^{k-1} |a_\nu(h_1) - a_\nu(h_2)||1-h|^\nu \leqslant c(k,p)\left(\frac{1}{1-h}\int_0^{1-h} |f(x+h_1) - f(x+h_2)|^p \,\mathrm{d}x\right)^{1/p}$$
$$\leqslant c_1(k,p)\left(\frac{1}{1-h}\omega_1(f; |h_1 - h_2|)_p^p\right)^{1/p}.$$

Since $\omega_1(f;\delta)_p \to 0$ as $\delta \to 0$, it follows that $a_\nu(h)$ is continuous function of $h \in [0, 1)$.

Applying now an arbitrary $(k+1)$th difference Δ_t^{k+1} to (38) as a function of h we obtain

$$\Delta_t^{k+1} f(x+h) = \sum_{\nu=0}^{k-1} (\Delta_t^{k+1} a_\nu(h)) x^\nu$$

for almost all $x \in (0, 1-h-(k+1)t)$ and t, $h \geqslant 0$, $h + (k+1)t < 1$. By (34) it follows that for almost all $x \in [0, 1-(k+1)h]$, $0 \leqslant h \leqslant 1/(k+1)$, we have $\Delta_h^{k+1} f(x) = 0$, and therefore since $a_\nu(h)$ is a continuous function of h we have

$$\Delta_t^{k+1} a_\nu(h) = 0, \quad 0 \leqslant h < 1-(k+1)t, \quad 0 \leqslant t < 1/(k+1), \quad \nu = 0, 1, \ldots, k-1.$$

Then by the case $p = \infty$ of theorem 7.1, which has already been proved, we conclude that for each interval $[0, \lambda]$, $0 < \lambda < 1$, $a_\nu(h)$ coincides with some polynomial $Q_\lambda \in P_k$. Consequently there exists exactly one polynomial $Q_\nu \in P_k$ such that $a_\nu(h) = Q_\nu(h)$ for $h \in [0, 1)$, $\nu = 0, 1, \ldots, k-1$. In view of (38) this fact implies the lemma. □

7.2 Besov spaces

Besov spaces occur in spline and rational approximations. Here we give only some needed notations and facts concerning Besov spaces. For more details one can see S.M. Nikol'skij (1969), J. Bergh, J. Löfström (1976), J. Peetre (1976), H. Triebel (1978).

Suppose $f \in L_p(0, 1)$, $0 < p, q \leqslant \infty$ and $\alpha > 0$. Taking $k = [\alpha] + 1$ we define

$$\|f\|_{B_{p,q}^\alpha} = \left(\int_0^1 (t^{-\alpha}\omega_k(f;t)_p)^q \frac{\mathrm{d}t}{t}\right)^{1/q}, \quad q < \infty, \tag{1}$$

and

$$\|f\|_{B_{p,\infty}^\alpha} = \sup_{t>0} t^{-\alpha}\omega_k(f;t)_p,$$

where $\omega_k(f;t)_p$ is the usual modulus of smoothness of f in $L_p(0, 1)$ (see section 7.1). The Besov space $B_{p,q}^\alpha$ is defined as the set of those functions $f \in L_p(0, 1)$ such that the quasi-norm $\|f\|_{B_{p,q}^\alpha}$ is finite. This is the homogeneous Besov space.

Let us indicate the properties of quasi-norm $\|\cdot\| = \|\cdot\|_{B^\alpha_{p,q}}$: (i) $\|f\| \geqslant 0$ for $f \in B^\alpha_{p,q}$, but it is possible $\|f\| = 0$ and $f \neq 0$, (ii) $\|\lambda f\| = |\lambda| \|f\|$ for $f \in B^\alpha_{p,q}$, λ a real number, (iii) $\|f+g\| \leqslant C(\|f\| + \|g\|)$ for $f, g \in B^\alpha_{p,q}$, where $C \geqslant 1$.

More frequently Besov spaces $B^\alpha_{p,q}$ are considered with the quasi-norm $\|f\| = \|f\|_p + \|f\|_{B^\alpha_{p,q}}$. It is well known that the Besov space $B^\alpha_{p,q}$ is a complete quasi-normed space (Banach space when $1 \leqslant p, q \leqslant \infty$).

Note that, since $\omega_k(f;t)_p = \omega_k(f;1/k)_p$ for $t \geqslant 1/k$, then the quasi-norm (1) is equivalent to the corresponding quasi-norm, if the integral in (1) is replaced by integral over $(0, 1/k)$ or $(0, \infty)$.

The choice $k = [\alpha] + 1$ in the definition of Besov space $B^\alpha_{p,q}$ is not essential. According to the properties of $\omega_k(f;t)_p$ (see section 7.1), it suffices to take $k > \alpha$ (more precisely $k > \alpha$ when $p \geqslant 1$ and $k > \alpha + 1 - 1/p$ when $0 < p < 1$).

In order to prove direct and converse theorems for spline approximation we shall make use of the following variant of Besov spaces $B^\alpha_{p,q}$. Denote by $B^\alpha_{p,q;k}$ the set of all functions $f \in L_p(0,1)$ such that the following quasi-norm is finite.

$$\|f\|_{B^\alpha_{p,q;k}} = \left(\int_0^1 (t^{-\alpha}\omega_k(f;t)_p)^q \frac{\mathrm{d}t}{t} \right)^{1/q}, \tag{2}$$

where $0 < p, q < \infty$, $\alpha > 0$, $k \geqslant 1$.

Obviously $B^\alpha_{p,q;k}$ coincides with $B^\alpha_{p,q}$ when $k = [\alpha] + 1$, but k in the definition of $B^\alpha_{p,q;k}$ is not directly connected with α.

In our considerations we shall mostly make use of spaces $B^\alpha_{\sigma,\sigma;k}$ with $\sigma < 1$ and we shall denote briefly $B^\alpha_{\sigma;k} = B^\alpha_{\sigma,\sigma;k}$ and $\|\cdot\|_{B^\alpha_{\sigma;k}} = \|\cdot\|_{B^\alpha_{\sigma,\sigma;k}}$. [†]We remark that the space $B^\alpha_{\sigma;k}$ with $0 < \sigma < 1$ is nontrivial when $0 < \alpha \leqslant k - 1 + 1/\sigma$, while, with $\sigma \geqslant 1$, it is nontrivial when $0 < \alpha \leqslant k$.

Next we give some other equivalent quasi-norms in $B^\alpha_{p,q;k}$. Denote

$$\|f\|^{(1)}_{B^\alpha_{p,q;k}} = \left(\sum_{\nu=0}^{\infty} \left(2^{\alpha\nu}\omega_k\left(f; \frac{1}{2^\nu}\right)_p \right)^q \right)^{1/q},$$

$$\|f\|^{(2)}_{B^\alpha_{p,q;k}} = \left(\int_0^{1/k} (t^{-\alpha}\|\Delta^k_t f(\cdot)\|_{L_p(0,1-kt)})^q \frac{\mathrm{d}t}{t} \right)^{1/q}.$$

Lemma 7.9. *Let $0 < p, q < \infty$, $\alpha > 0$ and $k \geqslant 1$. Then we have*

(i) $\|\cdot\|_{B^\alpha_{p,q;k}}$ *and* $\|\cdot\|^{(1)}_{B^\alpha_{p,q;k}}$ *are equivalent and*

(ii) $\|\cdot\|_{B^\alpha_{p,q;k}}$ *and* $\|\cdot\|^{(2)}_{B^\alpha_{p,q;k}}$ *are equivalent in the case* $p \leqslant q$.

Proof. The quasi-norm $\|\cdot\|^{(1)}_{B^\alpha_{p,q;k}}$ is the discrete variant of quasi-norm $\|\cdot\|_{B^\alpha_{p,q;k}}$. Clearly both quasi-norms are equivalent, since $\omega_k(f;t)_p$ is a monotone function.

[†] We set also $B^\alpha_\sigma = B^\alpha_{\sigma\sigma}$.

Obviously $\| f \|^{(2)}_{B^{\alpha}_{p,q;k}} \leqslant \| f \|_{B^{\alpha}_{p,q;k}}$ for $f \in B^{\alpha}_{p,q;k}$. In order to prove an estimate in the opposite direction we first observe that for $f \in B^{\alpha}_{p,q;k}$

$$\| f \|_{B^{\alpha}_{p,q;k}} \leqslant C\left(\int_0^{1/k} (t^{-\alpha}\omega_k(f;t)_p)^q \frac{dt}{t} \right)^{1/q}, \quad C = C(\alpha, p),$$

since $\omega_k(f;t)_p = \omega_k(f;1/k)_p$ for $t \geqslant 1/k$. Now, in view of lemma 7.2 in section 7.1 and Hardy's inequality (see (4) in section 7.1), we get for $p \leqslant q$

$$\| f \|_{B^{\alpha}_{p,q;k}} \leqslant C_1\left(\int_0^{1/k} \left(t^{-\alpha p - 1} \int_0^t u \| \Delta^k_u f(\cdot) \|^p_{L_p(0,1-ku)} \frac{du}{u} \right)^{q/p} \frac{dt}{t} \right)^{1/q}$$

$$\leqslant C_2\left(\int_0^{1/k} (t^{-\alpha} \| \Delta^k_t f(\cdot) \|_{L_p(0,1-kt)})^q \frac{dt}{t} \right)^{1/q} = C_2 \| f \|^{(2)}_{B^{(\alpha)}_{p,q;k}} \qquad \square$$

7.3 Spline approximation

The aim of this section is to prove direct and converse theorems for the spline approximation in the spaces L_p, C and BMO. Our point of view is that the natural way to obtain such theorems is to prove pairs of adjusted inequalities of Jackson and Bernstein type and then to characterize the spline approximation by the K-functional of Peetre between the corresponding spaces (see section 3.5). The spline approximation is a good illustration of this idea. Besov spaces appear in a natural way in this case.

7.3.1 Introduction

Spline functions were first introduced in approximation theory by I. Schoenberg (1946). They have a great number of applications for the following reasons:

(i) Splines are solutions of series of natural extremal problems;
(ii) Splines are convenient as a tool for interpolation and approximation;
(iii) Splines are useful in analysis, numerical methods, etc.

Denote by $\tilde{S}(k,n)$ the set of all spline functions of degree $k-1$ with $n-1$ free knots, i.e. $s \in \tilde{S}(k,n)$ if $s \in C^{k-2}(-\infty,\infty)$ and there exist points (knots) $-\infty = x_0 < x_1 \leqslant x_2 < \cdots \leqslant x_{n-1} < x_n = \infty$ such that s is some algebraic polynomial of degree $k-1$ in each interval (x_{i-1}, x_i).

It is readily seen that each spline $s \in \tilde{S}(k,n)$ can be represented in the form

$$s(x) = Q(x) + \sum_{\nu=1}^{n-1} a_\nu (x - x_\nu)^{k-1}_+,$$

where $Q \in P_{k-1}$,

$$x^{k-1}_+ = \begin{cases} 0, & x \leqslant 0, \\ x^{k-1}, & x > 0. \end{cases}$$

This and other basic facts related to splines one can find in C.de Boor (1978) and L. Schumaker (1981).

The splines, similarly to the rational functions of fixed degree, are a non-linear approximation tool. Moreover, as we shall see in Chapter 8, the rational approximation is closely connected to the spline approximation. This fact motivates the consideration of spline approximation in this book.

In this section we shall be concerned only with the order of the spline approximation of functions so that, as we shall see, the smoothness of splines will be not necessary.

Denote by $S(k,n)$ the set of all piecewise polynomial functions of degree $k-1$ with $n-1$ free knots, i.e. $s \in S(k,n)$ if there exist points (knots) $-\infty = x_0 < x_1 < \cdots < x_n = \infty$ and polynomials $Q_i \in P_{k-1}$ such that $s(x) = Q_i(x)$ for $x \in (x_{i-1}, x_i)$, $i = 1, 2, \ldots, n$. We shall suppose that $s(x_i) = s(x_i - 0)$ or $s(x_i) = s(x_i + 0)$ at each knot x_i. We shall call piecewise polynomials briefly also splines (they are splines with defect).

Also, we shall denote by $S(k, n, [a,b])$ the set of all splines $s \in S(k,n)$ restricted to $[a,b]$. In this case the end-points a and b of the interval will be also called knots of s.

Denote by $S_n^k(f)_p = S_n^k(f, [a,b])_p$, $S_n^k(f)_C = S_n^k(f, [a,b])_C$, $\tilde{S}_n^k(f)_p$ and $\tilde{S}_n^k(f)_C$ the best approximations of f in L_p and uniform metrics by means of the elements of $S(k,n)$ and $\tilde{S}(k,n)$ respectively. For instance

$$S_n^k(f)_p = \inf_{s \in S(k,n)} \| f - s \|_p.$$

We shall investigate the behavior of $\{S_n^k(f)_p\}$ instead of $\{\tilde{S}_n^k(f)_p\}$. The following lemma shows that there is no substantial difference between piecewise polynomials and splines with respect to the order of approximation in L_p and uniform metric that both classes produce.

Lemma 7.10. *Let $f \in L_p[a,b]$, $0 < p < \infty$ and $f \in C_{[a,b]}$ for $p = \infty$. Then we have for $k \geqslant 2$ and $n \geqslant 1$*

$$S_m^k(f)_p \leqslant \tilde{S}_m^k(f)_p \leqslant S_n^k(f)_p, \quad 0 < p \leqslant \infty,$$

where $m = (n-1)k + 1$.

In order to prove lemma 7.10 we need so-called B-splines. We shall use without proof the following lemma.

Lemma 7.11. *Let $-\infty < x_0 < x_1 < \cdots < x_k < \infty$, $k \geqslant 1$ and*

$$B(x) = \sum_{\nu=0}^{k} \frac{k(x - x_\nu)_+^{k-1}}{\omega'(x_\nu)}, \quad \omega(x) = \prod_{i=0}^{k} (x - x_i).$$

Then $B \in \tilde{S}(k, k+2)$, $B(x) = 0$ for $x \in (-\infty, x_0] \cup [x_k, \infty)$, $B(x) \geqslant 0$ for $x \in (-\infty, \infty)$ and $\int_{-\infty}^{\infty} B(x)\,dx = 1$. The spline B has knots $x_0, x_1, \ldots, x_k$. See C. de Boor (1978, p. 108), L. Schumaker (1981, p. 118).

Lemma 7.10 is an immediate consequence of the following.

Lemma 7.12. *Let $Q_1, Q_2 \in P_{k-1}$, $k \geqslant 2$. Then for each $\varepsilon > 0$ there exists $\delta_0 > 0$ such that for each δ, $0 < \delta \leqslant \delta_0$, there exists a spline $s_k \in \tilde{S}(k, k+1)$ with knots $0 = x_1 < x_2 < \cdots < x_k = \delta$ such that*

$$\left.\begin{aligned} s_k(x) &= Q_1(x), \quad x \leqslant 0, \\ s_k(x) &= Q_2(x), \quad x \geqslant \delta, \end{aligned}\right\} \tag{1}$$

$$\left.\begin{aligned} Q_1(0) - \varepsilon \leqslant s_k(x) \leqslant Q_2(\delta) + \varepsilon, \quad x \in (0, \delta), \quad Q_1(0) \leqslant Q_2(\delta), \\ Q_2(\delta) - \varepsilon \leqslant s_k(x) \leqslant Q_1(0) + \varepsilon, \quad x \in (0, \delta), \quad Q_1(0) > Q_2(\delta). \end{aligned}\right\} \tag{2}$$

Proof. We shall prove the lemma by induction with respect to k. The lemma is obvious for $k = 2$.

Suppose that it holds for some $k \geqslant 2$. Let $Q_1, Q_2 \in P_k$ and $\varepsilon > 0$.

Choose δ_1, $0 < \delta_1 \leqslant \delta_0$, such that for $0 < \delta \leqslant \delta_1$

$$\delta\{|Q'_1(0)| + |Q'_2(\delta)| + 2\varepsilon\} < \varepsilon. \tag{3}$$

Such a choice of δ_1 is possible since Q_1 and Q_2 are polynomials of fixed degree.

Let $0 < \delta \leqslant \delta_1 \leqslant \delta_0$. Because of $Q'_1, Q'_2 \in P_{k-1}$ and our assumption there exists a spline function $s_k \in \tilde{S}(k, k+1)$ with knots $0 = u_1 < u_2 < \cdots < u_k = \delta$ such that s satisfies (1), (2) with Q_1, Q_2 replaced by Q'_1, Q'_2.

Choose an arbitrary $u_0 \in (0, \delta)$ such that $u_0 \neq u_i$, $i = 1, 2, \ldots, k$. Let $\{u_i\}_{i=0}^k = \{x_i\}_{i=0}^k$ and $0 = x_0 < x_1 < \cdots < x_k = \delta$. Let $B \in \tilde{S}(k, k+2)$ be the B-spline from lemma 7.12 with knots $x_0, x_1, \ldots, x_k$. Set $\varphi(x) = \int_{-\infty}^x B(t)\mathrm{d}t$. It follows from lemma 7.12 that $\varphi(x) = 0$ for $x \leqslant 0$, $\varphi(x) = 1$ for $x \geqslant \delta$, $0 \leqslant \varphi(x) \leqslant 1$ for $x \in [0, \delta]$, φ is increasing and $\varphi \in \tilde{S}(k+1, k+2)$.

Now consider the spline

$$s_{k+1}(x) = Q_1(0) + \int_0^x s_k(t)\mathrm{d}t + \varphi(x)\left(Q_2(\delta) - Q_1(0) - \int_0^\delta s_k(t)\mathrm{d}t \right).$$

Clearly $s_{k+1} \in S(k+1, k+2)$ and s has knots $\{x_0, x_1, \ldots, x_k\} \subset [0, \delta]$. Also $s_{k+1}(x) = Q_1(x)$ for $x \leqslant 0$ and $s_{k+1}(x) = Q_2(\delta) + \int_\delta^x s_k(t)\mathrm{d}t = Q_2(x)$ for $x \geqslant \delta$. Finally, we have for $x \in (0, \delta)$

$$s_{k+1}(x) = (1 - \varphi(x))Q_1(0) + \varphi(x)Q_2(\delta) + \int_0^x s_k(t)\mathrm{d}t - \varphi(x)\int_0^\delta s_k(t)\mathrm{d}t.$$

Hence, in view of (2) and (3) with Q_1, Q_2 replaced by Q'_1, Q'_2,

$$\begin{aligned} |s_{k+1}(x) - \{(1 - \varphi(x))Q_1(0) + \varphi(x)Q_2(\delta)\}| &\leqslant \int_0^\delta |s_k(t)|\mathrm{d}t \\ &\leqslant \delta\{|Q'_1(0)| + |Q'_2(\delta)| + 2\varepsilon\} < \varepsilon. \end{aligned}$$

Consequently, s_{k+1} satisfies (2). □

Proof of lemma 7.10. Let $s \in S(k,n)$ and $0 < p < \infty$. Using lemma 7.12 on each knot of s in place of the origin we conclude that, for each $\varepsilon > 0$, $\tilde{S}_m^k(s)_p < S_n^k(f)_p + \varepsilon$ for $m = (n-1)k + 1$, which implies the assertion of lemma 7.10 in the case $0 < p < \infty$. The case when $f \in C_{[a,b]}$, $p = \infty$ is also clear in view of lemma 7.12. □

As already mentioned we are interested only in the order of spline approximation. Other questions connected with spline approximation will not be considered. The only fact we use is the existence of best approximating element in $S(k,n)$ in L_p and uniform metrics (compare J. Rice (1969, vol. 2, section 10–3)). Of course the existence is not necessary for our evaluations. We use this fact only to reduce some proofs.

Next we shall be concerned in some simple estimates for spline approximation that will be useful later.

Lemma 7.13. *Let f be defined on $[a,b]$, $f^{(r-1)}$ $(r \geqslant 1)$ be absolutely continuous and $f^{(r)} \in L_q[a,b]$ $(1 \leqslant q \leqslant \infty)$. Let $1 \leqslant p \leqslant \infty$ and $k \geqslant 1$. Then we have for $n = 1, 2, \ldots$*

$$S_{3n}^{k+r}(f)_p \leqslant \frac{(b-a)^{r+1/p-1/q} S_n^k(f^{(r)})_q}{n^r} \tag{4}$$

$$S_1^{k+r}(f)_p \leqslant (b-a)^{r+1/p-1/q} S_1^k(f^{(r)})_q. \tag{5}$$

Proof. Suppose $s \in S(k,n,[a,b])$ and

$$\| f^{(r)} - s \|_{L_q[a,b]} = S_n^k(f^{(r)})_q.$$

Then there exist points $a = x_0 < x_1 < \cdots < x_n = b$ such that s is an algebraic polynomial of degree $k-1$ in each interval (x_{i-1}, x_i).

Clearly, there exist points $a = u_0 < u_1 < \cdots < u_{3n} = b$ such that $\{x_i\}_{i=0}^n \subset \{u_i\}_{i=0}^{3n}$, $u_i - u_{i-1} \leqslant (b-a)/n$ and

$$\int_{u_{i-1}}^{u_i} |f^{(r)}(x) - s(x)|^q \,\mathrm{d}x \leqslant \frac{1}{n} S_n^k(f^{(r)})_q^q, \quad i = 1, 2, \ldots, 3n.$$

Set

$$\varphi(x) = \sum_{\nu=0}^{r-1} \frac{f^{(\nu)}(u_{i-1})(x-u_{i-1})}{\nu!} + \frac{1}{(r-1)!} \int_{u_{i-1}}^{x} (x-t)^{r-1} s(t)\,\mathrm{d}t$$

for $x \in [u_{i-1}, u_i)$, $i = 1, 2, \ldots, 3n$, and $\varphi(b) = \varphi(b-0)$. Clearly, $\varphi \in S(k+r, 3n, [a,b])$. Denote $\Delta_i = [u_{i-1}, u_i)$. By Taylor's formula we get

$$\begin{aligned}
\| f - \varphi \|_{C(\Delta_i)} &\leqslant \sup_{x \in \Delta_i} \frac{1}{(r-1)!} \left| \int_{u_{i-1}}^{x} (x-t)^{r-1} (f^{(r)}(t) - s(t))\,\mathrm{d}t \right| \\
&\leqslant |\Delta_i|^{r-1} \| f^{(r)} - s \|_{L_1(\Delta_i)} \leqslant |\Delta_i|^{r-1+1/q'} \| f^{(r)} - s \|_{L_q(\Delta_i)} \\
&\leqslant \frac{(b-a)^{r-1/q}}{n^r} S_n^k(f^{(r)})_q
\end{aligned}$$

and therefore

$$S_{3n}^{k+r}(f)_C \leqslant \frac{(b-a)^{r-1/q}}{n^r} S_n^k(f^{(r)})_q.$$

Hence

$$S_{3n}^{k+r}(f)_p \leqslant (b-a)^{1/p} S_{3n}^{k+r}(f)_C \leqslant \frac{(b-a)^{r+1/p-1/q} S_n^k(f^{(r)})_q}{n^r}.$$

Thus (4) is proved. The estimate (5) can be proved similarly. □

Theorem 7.2. (i) *If* $V_a^b f < \infty$ *and* $1 \leqslant p \leqslant \infty$ *then*

$$S_n^1(f)_p \leqslant \frac{V_a^b f (b-a)^{1/p}}{n}, \quad n = 1, 2, \ldots. \tag{6}$$

(ii) *If* $V_a^b f^{(r)} < \infty$, $r \geqslant 1$, *then*

$$S_n^{r+1}(f)_C \leqslant C(r) \frac{V_a^b f^{(r)} (b-a)^r}{n^{r+1}}, \quad n = 1, 2, \ldots. \tag{7}$$

Proof. Clearly, there exists a partition of $[a, b]$ into n subintervals $a = x_0 < x_1 < \cdots < x_n = b$ such that the variation of f in each open interval (x_{i-1}, x_i) does not exceed $V_a^b f/n$. Set $s(x) = f(x_{i-1} + 0)$ for $x \in [x_{i-1}, x_i)$, $i = 1, 2, \ldots, n$, and $s(b) = s(b - 0)$. Then $s \in S(1, n, [a, b])$ and

$$\| f - s \|_p \leqslant (b-a)^{1/p} \| f - s \|_\infty \leqslant \frac{V_a^b f (b-a)^{1/p}}{n}$$

which implies (6).†

The estimate (7) follows from lemma 7.13 and the estimate (6) applied to $f^{(r)}$ with $p = 1$. □

7.3.2 Direct and converse theorems in L_p $(0 < p < \infty)$

As already mentioned in section 3.5, in order to obtain complete direct and converse theorems in approximation theory it is enough to prove pairs of adjusted inequalities of Jackson and Bernstein type. Now we follow this plan approximating by splines in L_p $(0 < p < \infty)$ metric.

We begin with formulation of the main statements.

Theorem 7.3 (Jackson type inequality). *Let* $f \in B_{\sigma;k}^{\alpha}$, *where* $\alpha > 0$, $\sigma =$

† We suppose $f(a) = f(a + 0)$, $f(b) = f(b - 0)$.

$(\alpha + 1/p)^{-1}$, $0 < p < \infty$ *and* $k \geqslant 1$. *Then* $f \in L_p(0, 1)$ *and*

$$S_n^k(f)_p \leqslant C \frac{\| f \|_{B_{\sigma;k}^{\alpha}}}{n^{\alpha}}, \quad n = 1, 2, \ldots, \tag{8}$$

where $C = C(\alpha, p, k)$.

Theorem 7.4 (Bernstein type inequality). *Let*

$$s \in S(k, n, [0, 1]), \quad k, n \geqslant 1, \quad 0 < p < \infty, \quad \alpha > 0$$

and $\sigma = (\alpha + 1/p)^{-1}$. *Then*

$$\| s \| B_{\sigma;k}^{\alpha} \leqslant C n^{\alpha} \| s \|_p, \tag{9}$$

where $C = C(\alpha, p, k)$. *However, the estimate (9) does not hold in the case* $p = \infty$.

According to theorem 3.16 in section 3.5 theorems 7.3 and 7.4 imply the following direct and converse theorems:

Theorem 7.5 (direct theorem). *Let* $f \in L_p(0, 1)$, $0 < p < \infty$, $\alpha > 0$, $\sigma = (\alpha + 1/p)^{-1}$ *and* $k \geqslant 1$. *Then*

$$S_n^k(f)_p \leqslant C K(f, n^{-\alpha}; L_p, B_{\sigma,k}^{\alpha}), \quad n = 1, 2, \ldots,$$

where $C = C(\alpha, p, k)$.

Theorem 7.6 (converse theorem). *Let* $f \in L_p(0, 1)$, $0 < p < \infty$, $\alpha > 0$, $\sigma = (\alpha + 1/p)^{-1}$ *and* $k \geqslant 1$. *Then*

$$K(f, n^{-\alpha}; L_p, B_{\sigma;k}^{\alpha}) \leqslant C n^{-\alpha} \left(\sum_{\nu=1}^{n} \frac{1}{\nu} (\nu^{\alpha} S_{\nu}^k(f)_p)^{\lambda} \right)^{1/\lambda}, \quad n = 1, 2, \ldots,$$

where $\lambda = \min\{\sigma, 1\}$, $C = C(\alpha, p, k)$.

As a consequence of theorems 7.5 and 7.6 we obtain similarly as in section 3.5 the following.

Corollary 7.2. *Let* $f \in L_p(0, 1)$, $0 < p < \infty$, $k \geqslant 1$ *and let* ω *be nonnegative and nondecreasing function on* $[0, \infty)$ *such that* $\omega(2t) \leqslant 2^{\beta} \omega(t)$ *for* $t \geqslant 0$ $(\beta \geqslant 0)$. *Then we have*

$$S_n^k(f)_p = \mathrm{O}(n^{-\gamma} \omega(n^{-1})), \quad 0 < \beta + \gamma < \alpha \quad \textit{iff} \quad K(f, t; L_p, B_{\sigma,k}^{\alpha}) = \mathrm{O}(t^{\gamma/\alpha} \omega(t^{1/\alpha})).$$

In particular

$$S_n^k(f)_p = \mathrm{O}(n^{-\gamma}), \quad 0 < \gamma < \alpha, \quad \textit{iff} \quad K(f, t; L_p, B_{\sigma,k}^{\alpha}) = \mathrm{O}(t^{\gamma/\alpha}).$$

Denote

$$S_{q;k}^{\gamma}(L_p) = \left\{ f \in L_p[0, 1] \colon \| f \|_{S_{q;k}^{\gamma}(L_p)} = \left(\sum_{\nu=0}^{\infty} (2^{\nu\gamma} S_{2^{\nu}}^k(f)_p)^q \right)^{1/q} < \infty \right\}$$

and

$$S^{\gamma}_{\infty;k}(L_p)=\{f\in L_p[0,1]: \|f\|_{S^{\gamma}_{\infty;k}(L_p)}=\sup_n n^{\gamma}S^k_n(f)_p<\infty\},$$

where $L_p[0,1]=C_{[0,1]}$ when $p=\infty$.

As usually, we shall denote by $(X_0,X_1)_{\theta,q}$ the real interpolation space between quasi-normed spaces X_0 and X_1 with quasi-norm

$$\|f\|_{(X_0,X_1)\theta,q}=\left(\sum_{\nu=0}^{\infty}\left(2^{\nu\theta}K\left(f,\frac{1}{2^{\nu}};X_0,X_1\right)\right)^q\right)^{1/q},\quad 0<q<\infty,$$

and

$$\|f\|_{(X_0,X_1)_{\theta,\infty}}=\sup_{t>0}t^{-\theta}K(f,t;X_0,X_1).$$

***Corollary** 7.3. Let $0<p<\infty$, $\gamma>0$, $0<q\leqslant\infty$, $k\geqslant 1$, $\alpha>\gamma$ and $\sigma=(\alpha+1/p)^{-1}$. Then we have*

$$S^{\gamma}_{q;k}(L_p)=(L_p,B^{\alpha}_{\sigma;k})_{\gamma/\alpha,q}$$

with equivalent quasi-norms.

In order to prove theorem 7.3 we shall prove the following embedding result:

***Theorem** 7.7. Let $f\in B^{\alpha}_{\sigma,p;k}$, $\alpha>0$, $\sigma=(\alpha+1/p)^{-1}$, $0<p<\infty$, $k\geqslant 1$. Then $f\in L_p(0,1)$ and*

$$E_{k-1}(f)_p\leqslant C\|f\|_{B^{\alpha}_{\sigma,p;k}},\tag{10}$$

where $E_{k-1}(f)_p$ denotes the best approximation to f in $L_p(0,1)$ by means of all polynomials of degree at most $k-1$, $C=C(\alpha,p,k)$.

According to inequality (3) in section 7.1, if $f\in B^{\alpha}_{\sigma,\sigma;k}$ then $f\in B^{\alpha}_{\sigma,p;k}$ and $\|f\|_{B^{\alpha}_{\sigma,p;k}}\leqslant\|f\|_{B^{\alpha}_{\sigma,\sigma;k}}$, since $\sigma<p$. Then by (10) it follows that, if $f\in B^{\alpha}_{\sigma;k}$, then $f\in L_p$ and

$$E_{k-1}(f)_p\leqslant C\|f\|_{B^{\alpha}_{\sigma;k}},\quad C=C(\alpha,p,k).\tag{11}$$

In order to prove theorem 7.7 we need some auxiliary statements.

Denote by $E(k,n)$ the set of all piecewise polynomial functions of degree $k-1$ with fixed knots at the points i/n, $i=0,1,\dots,n$, i.e. $\varphi\in E(k,n)$ if φ is a polynomial of degree $k-1$ in each interval $((i-1)/n,i/n)$, $i=1,2,\dots,n$. Denote

$$E^k_n(f)_p=\inf_{\varphi\in E(k,n)}\|f-\varphi\|_p.$$

***Lemma** 7.14. If $f\in L_p(0,1)$, $0<p<\infty$ and $k\geqslant 1$, then*

$$E^k_n(f)_p\leqslant C\omega_k\left(f;\frac{1}{kn}\right)_p,\quad n=1,2,\dots,$$

where $C=C(p,k)$.

Proof. Denote $\Delta_i = ((i-1)/n, i/n)$. By theorem 7.1 in section 7.1 there exists a polynomial Q_i of degree $k-1$ such that

$$\| f - Q_i \|_{L_p(\Delta_i)} \leqslant C\omega_k\left(f; \frac{|\Delta_i|}{k}\right)_p, \quad C = C(k,p).$$

Hence, in view of lemma 7.2 in section 7.1 we have

$$\| f - Q_i \|^p_{L_p(\Delta_i)} \leqslant C_1 n \int_0^{1/kn} \int_{(i-1)/n}^{i/n - ku} |\Delta_u^k f(x)|^p \,\mathrm{d}x \,\mathrm{d}u. \tag{12}$$

Put $\varphi(x) = Q_i(x)$ for $x \in \Delta_i$, $i = 1, 2, \ldots, n$. Then by (12) we get

$$\begin{aligned} \| f - \varphi \|_{L_p} &= \left(\sum_{i=1}^n \| f - Q_i \|^p_{L_p(\Delta_i)} \right)^{1/p} \\ &\leqslant C_1 n \left(\int_0^{1/kn} \left(\sum_{i=1}^n \int_{(i-1)/n}^{(i/n) - ku} |\Delta_n^k f(x)|^p \,\mathrm{d}x \right) \mathrm{d}u \right)^{1/p} \\ &\leqslant C_1 n \left(\int_0^{1/kn} \int_0^{1-ku} |\Delta_u^k f(x)|^p \,\mathrm{d}x \,\mathrm{d}u \right)^{1/p} \leqslant C_1 \omega_k\left(f; \frac{1}{kn}\right)_p. \quad \square \end{aligned}$$

Lemma 7.14 shows that in order to prove theorem 7.7 it suffices to prove that $f \in L_p(0,1)$ and

$$E_{k-1}(f)_p \leqslant C\left(\sum_{\nu=0}^{\infty} (2^{\nu\alpha} E^k_{2^\nu}(f)_\sigma)^p \right)^{1/p}, \quad C = C(\alpha, p, k). \tag{13}$$

The estimate (13) will be proved by using the following.

Lemma 7.15. *Let $0 < \sigma < p < \infty$ and let there exist a sequence $\{u_n\}_{n=1}^{\infty}$ of nonnegative functions $u_n \in L_\infty(0,1)$ such that*

$$\| u_n \|_\sigma \leqslant \lambda_n, \quad \lambda_{n+1} \leqslant \beta\lambda_n, \quad n = 1, 2, \ldots, \quad \textit{where } 0 < \beta < 1, \tag{14}$$

and for some $\delta_n > 0$ and any r $(\sigma < r \leqslant \infty)$ we have

$$\| u_n \|_r \leqslant C(\sigma, r) \delta_n^{1/\sigma - 1/r} \lambda_n, \quad n = 1, 2, \ldots. \tag{15}$$

Then we have for the function $f = \sum_{n=1}^{\infty} u_n$ $f \in L_p(0,1)$ and

$$\| f \|_p \leqslant C(\sigma, p, \beta) \left(\sum_{n=1}^{\infty} \delta_n^{p/\sigma - 1} \lambda_n^p \right)^{1/p}. \tag{16}$$

This lemma we shall prove applying several times Hölder's inequality for sums and integrals, the inequality (3) from section 7.1 and Abel's transformation:

$$\sum_{n=1}^{N} u_n v_n = \sum_{n=1}^{N-1} U_n(v_n - v_{n+1}) + U_N v_N, \tag{17}$$

where $U_n = \sum_{k=1}^{n} u_k$ for $n = 1, 2, \ldots, N$ and also

$$\sum_{n=1}^{N} u_n v_n = \sum_{n=2}^{N} U_n^*(v_n - v_{n-1}) + U_1^* v_1 \tag{18}$$

where $U_n^* = \sum_{k=n}^{N} u_k$ for $n = 1, 2, \ldots, N$.

We shall use also the following.

Lemma 7.16. *Let $\alpha_1, \alpha_2, \ldots, \alpha_N$ be nonnegative numbers and $1 \leqslant p < \infty$. Then we have*

$$\sum_{n=1}^{N} \alpha_n \left(\sum_{\nu=n}^{N} \alpha_\nu \right)^p \leqslant p^p \sum_{n=1}^{N} \alpha_n \left(\sum_{\nu=1}^{n} \alpha_\nu \right)^p. \tag{19}$$

Proof. Obviously $b^p - a^p \leqslant pb^{p-1}(b-a)$ when $0 < a < b$, $p \geqslant 1$. Using this inequality and Abel's transformation (17) with $u_\nu = \alpha_\nu$ and $v_\nu = (\sum_{s=\nu}^{N} \alpha_s)^p$ we get

$$A = \sum_{n=1}^{N} \alpha_n \left(\sum_{\nu=n}^{N} \alpha_\nu \right)^p = \sum_{n=1}^{N-1} \left(\sum_{\nu=1}^{n} \alpha_\nu \right) \left(\left(\sum_{\nu=n}^{N} \alpha_\nu \right)^p - \left(\sum_{\nu=n+1}^{N} \alpha_\nu \right)^p \right) + \left(\sum_{\nu=1}^{N} \alpha_\nu \right) \alpha_N^p$$

$$\leqslant p \sum_{n=1}^{N-1} \left(\sum_{\nu=1}^{n} \alpha_\nu \right) \alpha_n \left(\sum_{\nu=n}^{N} \alpha_\nu \right)^{p-1} + \alpha_N^p \sum_{\nu=1}^{N} \alpha_\nu$$

$$\leqslant p \sum_{n=1}^{N} \alpha_n \left(\sum_{\nu=n}^{N} \alpha_\nu \right)^{p-1} \left(\sum_{\nu=1}^{n} \alpha_\nu \right).$$

Hence, if $p = 1$, then (19) is proved. Let $p > 1$. Then applying Hölder's inequality to the last sum we obtain

$$A \leqslant p \left\{ \sum_{n=1}^{N} \alpha_n \left(\sum_{\nu=n}^{N} \alpha_\nu \right)^p \right\}^{1-1/p} \left\{ \sum_{n=1}^{N} \alpha_n \left(\sum_{\nu=1}^{n} \alpha_\nu \right)^p \right\}^{1/p}$$

$$= pA^{1-1/p} \left\{ \sum_{n=1}^{N} \alpha_n \left(\sum_{\nu=1}^{n} \alpha_\nu \right)^p \right\}^{1/p},$$

which implies (19). □

Proof of lemma 7.15. Consider the case $p \geqslant 2$, $\sigma \geqslant 1$. Put $S_N = \sum_{n=1}^{N} u_n$. By lemma 7.17 and the inequality $(a+b)^r \leqslant 2^{r-1}(a^r + b^r)$, where $a, b \geqslant 0$, $r \geqslant 1$, we get

$$S_N^p = \sum_{n=1}^{N} u_n \left(\sum_{n=1}^{N} u_n \right)^{p-1} \leqslant 2^{p-2} \left\{ \sum_{n=1}^{N} u_n \left(\sum_{\nu=n}^{N} u_\nu \right)^{p-1} + \sum_{n=1}^{N} u_n \left(\sum_{\nu=1}^{n} u_\nu \right)^{p-1} \right\}$$

$$\leqslant C(p) \sum_{n=1}^{N} u_n \left(\sum_{\nu=1}^{n} u_\nu \right)^{p-1}.$$

Now, by Hölder's and Minkovski's inequalities we obtain

$$\begin{aligned}\int_0^1 S_N^p(x)\,dx &\leqslant C(p)\sum_{n=1}^N \int_0^1 u_n(x)\left(\sum_{\nu=1}^n u_\nu(x)\right)^{p-1} dx\\ &\leqslant C(p)\sum_{n=1}^N \|u_n\|_\sigma \left\|\left(\sum_{\nu=1}^n u_\nu\right)^{p-1}\right\|_{\sigma'}\\ &\leqslant C(p)\sum_{n=1}^N \|u_n\|_\sigma \left(\sum_{\nu=1}^n \|u_\nu\|_{\sigma'(p-1)}\right)^{p-1}\\ &\leqslant C(p)\sum_{n=1}^N \lambda_n \left(\sum_{\nu=1}^n \|u_\nu\|_{\sigma'(p-1)}\right)^{p-1} = C(p)A_N, \end{aligned} \tag{20}$$

where $1/\sigma + 1/\sigma' = 1$.

We continue applying Abel's transformation (18) and obtain

$$\begin{aligned}A_N &= \sum_{n=2}^N\left(\sum_{\nu=n}^N \lambda_\nu\right)\left\{\left(\sum_{\nu=1}^n \|u_\nu\|_{\sigma'(p-1)}\right)^{p-1} - \left(\sum_{\nu=1}^{n-1} \|u_\nu\|_{\sigma'(p-1)}\right)^{p-1}\right\}\\ &\quad + \left(\sum_{\nu=1}^N \lambda_\nu\right)(\|u_1\|_{\sigma'(p-1)})^{p-1}\\ &\leqslant (p-1)\sum_{n=1}^N\left(\sum_{\nu=n}^N \lambda_\nu\right)\|u_n\|_{\sigma'(p-1)}\left(\sum_{\nu=1}^n \|u_\nu\|_{\sigma'(p-1)}\right)^{p-2}.\end{aligned}$$

By (14) it follows that

$$\sum_{\nu=n}^N \lambda_\nu \leqslant \lambda_n \sum_{\nu=0}^\infty \beta^\nu = \frac{1}{1-\beta}\lambda_n.$$

Using the above inequalities and again Hölder's inequality we get

$$\begin{aligned}A_N &\leqslant C(p,\beta)\sum_{n=1}^N \lambda_n \|u_n\|_{\sigma'(p-1)}\left(\sum_{\nu=1}^n \|u_\nu\|_{\sigma'(p-1)}\right)^{p-2}\\ &\leqslant C(p,\beta)\left(\sum_{n=1}^N \lambda_n \|u_n\|^{p-1}_{\sigma'(p-1)}\right)^{1/(p-1)}\left(\sum_{n=1}^N \lambda_n\left(\sum_{\nu=1}^n \|u_\nu\|_{\sigma'(p-1)}\right)^{p-1}\right)^{(p-2)/(p-1)}\\ &= C(p,\beta)A_N^{(p-2)/(p-1)}\left(\sum_{n=1}^N \lambda_n \|u_n\|^{p-1}_{\sigma'(p-1)}\right)^{1/(p-1)}.\end{aligned}$$

Hence

$$A_N \leqslant C_1(p,\beta)\sum_{n=1}^N \lambda_n \|u_n\|^{p-1}_{\sigma'(p-1)}. \tag{21}$$

We have $\sigma'(p-1) = \sigma(p-1)/(\sigma-1) > \sigma$ and by (15)

$$\|u_n\|^{p-1}_{\sigma'(p-1)} \leqslant C(\sigma,p)\delta_n^{p/\sigma-1}\lambda_n^{p-1}. \tag{22}$$

The estimates (20), (21) and (22) entail

$$\|S_N\|_p^p \leqslant C(\sigma, p, \beta) \sum_{n=1}^{N} \delta_n^{p/\sigma - 1} \lambda_n^p,$$

which implies (16).

Now let $1 \leqslant \sigma < p < 2$. Put $v_n = (u_n)^{p/2}$. Then

$$\left(\sum_{n=1}^{N} u_n\right)^p = \left\{\left(\sum_{n=1}^{N} v_n^{2/p}\right)^{p/2}\right\}^2 \leqslant \left(\sum_{n=1}^{N} v_n\right)^2. \tag{23}$$

Write $q = 2\sigma/p$ $(q > 1)$. We have

$$\|v_n\|_q = \left(\int_0^1 u_n^\sigma(x)\mathrm{d}x\right)^{p/2\sigma} = \|u_n\|_\sigma^{p/2} \leqslant \lambda_n^{p/2} = \theta_n.$$

Thus $\{v_n\}_{n=1}^\infty$ satisfies (14). On the other hand, if $q < r < \infty$, then $s = \frac{1}{2}pr > \sigma$ and by (15) we have

$$\|v_n\|_r = \|u\|_s^{p/2} \leqslant C\delta_n^{p(1/\sigma - 1/s)/2} \lambda_n^{p/2} = C\delta_n^{1/q - 1/r}\theta_n.$$

Consequently, $\{v_n\}_{n=1}^\infty$ satisfies the assumptions of lemma 7.15 in the case already considered $(1 < q < 2)$ and we obtain using (23)

$$\|f\|_p^p \leqslant \left\|\sum_{n=1}^{\infty} v_n\right\|_2^2 \leqslant C(\sigma, p, \beta) \sum_{n=1}^{\infty} \delta_n^{2/q - 1}\theta_n^2 = C(\sigma, p, \beta) \sum_{n=1}^{\infty} \delta_n^{p/\sigma - 1}\lambda_n^p.$$

The case $0 < \sigma < 1$, $\sigma < p < \infty$ is considered similarly. Put $v_n = u_n^\sigma$, then

$$\left(\sum_{n=1}^{\infty} u_n\right)^p \leqslant \left(\sum_{n=1}^{\infty} v_n\right)^{p/\sigma}$$

and

$$\|v_n\|_1 = \int_0^1 u_n^\sigma(x)\mathrm{d}x \leqslant \lambda_n^\sigma = \eta_n.$$

We have for $1 < r < \infty$ $\|v_n\|_r \leqslant C\delta_n^{1 - 1/r}\eta_n$. Put $s = p/\sigma$ $(s > 1)$. Then we obtain as above

$$\|f\|_p^p \leqslant \left\|\sum_{n=1}^{\infty} v_n\right\|_s^s \leqslant C(\sigma, p, \beta) \sum_{n=1}^{\infty} \delta_n^{s-1}\eta_n^s = C(\sigma, p, \beta) \sum_{n=1}^{\infty} \delta_n^{p/\sigma - 1}\lambda_n^p. \qquad \square$$

Proof of theorem 7.7. As already mentioned, in order to prove theorem 7.7 it is sufficient to prove that $f \in L_p(0, 1)$ and the estimate (13).

Choose $\varphi_{2^n} \in E(k, 2^n)$ such that

$$\|f - \varphi_{2^n}\|_\sigma = E_{2^n}^k(f)_\sigma, \quad n = 0, 1, \ldots. \tag{24}$$

Now, we define by induction integers $n_0 = 0 < n_1 < n_2 < \cdots$ such that

$$E^k_{2^{n_i+1}}(f)_\sigma \leqslant \frac{1}{3^{1/\gamma}} E^k_{2^{n_i}}(f)_\sigma \leqslant E^k_{2^{n_{i+1}-1}}(f)_\sigma \tag{25}$$

for $i = 0, 1, \ldots,$ where $\gamma = \min\{1, \sigma\}$.

Put $u_i = |\varphi_{2^{n_i}} - \varphi_{2^{n_{i-1}}}|$. Then by Minkovski's inequality when $1 \leqslant \sigma < \infty$ and inequality (1) in section 7.1 when $0 < p < 1$, using (24) and (25), we get

$$\begin{aligned}\|u_{i+1}\|_\sigma^\gamma &= \|\varphi_{2^{n_{i+1}}} - \varphi_{2^{n_i}}\|_\sigma^\gamma \geqslant \|f - \varphi_{2^{n_i}}\|_\sigma^\gamma - \|f - \varphi_{2^{n_{i+1}}}\|_\sigma^\gamma \\ &= E^k_{2^{n_i}}(f)_\sigma^\gamma - E^k_{2^{n_i+1}}(f)_\sigma^\gamma \geqslant 2E^k_{2^{n_i+1}}(f)_\sigma^\gamma\end{aligned}$$

and

$$\|u_{i+1}\|_\sigma^\gamma \leqslant E^k_{2^{n_i}}(f)_\sigma^\gamma + E^k_{2^{n_i+1}}(f)_\sigma^\gamma \leqslant \tfrac{4}{3} E^k_{2^{n_i}}(f)_\sigma^\gamma.$$

Hence

$$\|u_{i+1}\|_\sigma \leqslant (\tfrac{2}{3})^{1/\gamma}\|u_i\|_\sigma, \quad i = 1, 2, \ldots. \tag{26}$$

The following inequality holds:

$$\|u_i\|_r \leqslant C \cdot 2^{n_i(1/\sigma - 1/r)}\|u_i\|_\sigma, \tag{27}$$

where $\sigma < r \leqslant \infty$, $C = C(k, p, r)$.

Indeed, by lemma 7.3 in section 7.1 we have

$$\|u_i\|_{L_r(\Delta_\nu)} \leqslant C|\Delta_\nu|^{1/r - 1/\sigma}\|u_i\|_{L_\sigma(\Delta_\nu)}, \quad C = C(k, r),$$

where $\Delta_\nu = ((\nu - 1)/2^{n_i}, \nu/2^{n_i})$, $\nu = 1, 2, \ldots, 2^{n_i}$. Then in view of inequality (3) in section 7.1 we get

$$\begin{aligned}\|u_i\|_r &= \left(\sum_{\nu=1}^{2^{n_i}} \|u_i\|^r_{L_r(\Delta_\nu)}\right)^{1/r} \\ &\leqslant C2^{n_i(1/\sigma - 1/r)}\left(\sum_{\nu=1}^{2^{n_i}} \|u_i\|^r_{L_\sigma(\Delta_\nu)}\right)^{1/r} \\ &\leqslant C2^{n_i(1/\sigma - 1/r)}\left(\sum_{\nu=1}^{2^{n_i}} \|u_i\|^\sigma_{L_\sigma(\Delta_\nu)}\right)^{1/\sigma} = C \cdot 2^{n_i(1/\sigma - 1/r)}\|u_i\|_\sigma.\end{aligned}$$

Inequality (27) is proved.

Put $\lambda_i = \|u_i\|_\sigma$, $\delta_i = 2^{n_i}$, $i = 1, 2, \ldots$. Inequalities (26) and (27) indicate that the requirements of lemma 7.15 are fulfilled. Therefore we have

$$\left\|\sum_{i=1}^\infty u_i\right\|_p \leqslant C(k, p, \sigma)\left(\sum_{i=1}^\infty 2^{n_i(p/\sigma - 1)}\|u_i\|_\sigma^p\right)^{1/p}. \tag{28}$$

By (24) and (25) we get as above

$$\begin{aligned}2^{n_i(p/\sigma - 1)}\|u_i\|_\sigma^p &\leqslant C(\sigma, p)2^{n_i(p/\sigma - 1)}(E^k_{2^{n_i}}(f)_\sigma^p + E^k_{2^{n_i-1}}(f)_\sigma^p) \\ &\leqslant C_1(\sigma, p)2^{(n_i - 1)(p/\sigma - 1)}E^k_{2^{n_i-1}}(f)_\sigma^p.\end{aligned}$$

Combining these estimates together with (28) we get

$$\left\| \sum_{i=1}^{\infty} u_i \right\|_p \leqslant C(k,\sigma,p)\left(\sum_{i=0}^{\infty} 2^{n_i(p/\sigma-1)} E^k_{2^{n_i}}(f)^p_\sigma \right)^{1/p}$$

$$\leqslant C(k,\sigma,p)\left(\sum_{n=0}^{\infty} (2^{n(1/\sigma-1/p)} E^k_{2^n}(f)_\sigma)^p \right)^{1/p}. \tag{29}$$

Since $f \in L_\sigma(0,1)$ and by the definition of φ_{2^n} it follows that the series

$$\sum_{i=1}^{\infty} (\varphi_{2^{n_i}} - \varphi_{2^{n_{i-1}}})$$

represents the function $f - \varphi_1$ in $L_\sigma(0,1)$. The inequality (29) shows that the same series converges in $L_p(0,1)$. Therefore $f \in L_p(0,1)$ and by (29)

$$E_{k-1}(f)_p \leqslant \| f - \varphi_1 \|_p \leqslant C(k,p,\alpha)\left(\sum_{n=0}^{\infty} (2^{n\alpha} E^k_{2^n}(f)_\sigma)^p \right)^{1/p}.$$

Theorem 7.7 is proved. □

Proof of theorem 7.3. In order to prove theorem 7.3 we shall use theorem 7.7, more precisely the estimate (11). Of course estimate (11) holds also with the following quasi-norm in the right-hand side (see section 7.2):

$$\| f \|^{(2)}_{B^\alpha_{\sigma;k}} = \left(\int_0^{1/k} (t^{-\alpha} \| \Delta^k_t f(\cdot) \|_{L_\sigma(0,1-kt)})^\sigma \frac{dt}{t} \right)^{1/\sigma}. \tag{30}$$

Now we transform the estimate (11) with quasi-norm (30) to an arbitrary interval $\Delta = (u,v) \subset (0,1)$. Consider the function $g(x) = f(u + (v-u)x)$, $x \in (0,1)$. By (11) we have

$$E_{k-1}(g)_{L_p(0,1)} \leqslant C \| g \|^{(2)}_{B^\alpha_{\sigma;k}}. \tag{31}$$

Clearly

$$E_{k-1}(g)_{L_p(0,1)} = \inf_{Q \in P_{k-1}} \left(\int_0^1 | f(u + (v-u)x) - Q(x)|^p dx \right)^{1/p}$$

$$= \frac{1}{(v-u)^{1/p}} \inf_{P \in P_{k-1}} \left(\int_u^v | f(t) - P(t)|^p dt \right)^{1/p}$$

$$= \frac{1}{|\Delta|^{1/p}} E_{k-1}(f)_{L_p(\Delta)}.$$

Thus we have

$$E_{k-1}(g)_{L_p(0,1)} = \frac{1}{|\Delta|^{1/p}} E_{k-1}(f)_{L_p(\Delta)}. \tag{32}$$

On the other hand we have

$$\|\Delta_t^k g(\cdot)\|_{L_\sigma(0,1-kt)} = \left(\int_0^{1-kt} |\Delta_{|\Delta|t}^k f(u+(v-u)x)|^\sigma \,\mathrm{d}x\right)^{1/\sigma}$$
$$= \frac{1}{|\Delta|^{1/\sigma}}\left(\int_u^{v-k|\Delta|t} |\Delta_{|\Delta|t}^k f(y)|^\sigma \,\mathrm{d}y\right)^{1/\sigma}$$
$$= \frac{1}{|\Delta|^{1/\sigma}} \|\Delta_{|\Delta|t}^k f(\cdot)\|_{L_\sigma(u,v-k|\Delta|t)},$$

where we have substituted $y = u + (v-u)x$. Now substituting $|\Delta|t = h$ we get

$$\|g\|_{B_{\sigma;k}^\alpha}^{(2)} = \left(\int_0^{1/k} (t^{-\alpha}\|\Delta_t^k g(\cdot)\|_{L_\sigma(0,1-kt)})^\sigma \frac{\mathrm{d}t}{t}\right)^{1/\sigma}$$
$$= \left(\int_0^{1/k} \left(\frac{\|\Delta_{|\Delta|t}^k f(\cdot)\|_{L_\sigma(u,v-k|\Delta|t)}}{|\Delta|^{1/\sigma}t^\alpha}\right)^\sigma \frac{\mathrm{d}t}{t}\right)^{1/\sigma}$$
$$= \frac{1}{|\Delta|^{1/p}}\left(\int_0^{|\Delta|/k} \left(h^{-\alpha}\|\Delta_h^k f(\cdot)\|_{L_\sigma(u,v-kh)}\right)^\sigma \frac{\mathrm{d}h}{h}\right)^{1/\sigma}.$$

Denote

$$\|f\|_{B(u,v)} = \left(\int_0^{(v-u)/k} (t^{-\alpha}\|\Delta_t^k f(\cdot)\|_{L_\sigma(u,v-kt)})^\sigma \frac{\mathrm{d}t}{t}\right)^{1/\sigma}.$$

The previous equalities together with (31) and (32) imply that for each interval $\Delta = (u,v) \subset (0,1)$

$$E_{k-1}(f)_{L_p(\Delta)} \leqslant C\|f\|_{B(u,v)}. \tag{33}$$

Let $n \geqslant 1$. We define by induction points $x_0, x_1, \ldots$. Set $x_0 = 0$. Let x_0, $x_1, \ldots, x_{i-1}$ be already defined such that $0 = x_0 < x_1 < \cdots < x_{i-1} < 1$. Now we define x_i as follows:

$$x_i = \sup\left\{y\colon x_{i-1} < y \leqslant 1,\ \|f\|_{B(x_{i-1},y)}^\sigma \leqslant \frac{1}{n}\|f\|_{B(0,1)}^\sigma\right\}.$$

Suppose $x_m \leqslant 1$ for some $m \geqslant 1$. Denote $\Delta_i = (x_{i-1}, x_i)$. Then we have

$$\sum_{i=1}^m \|f\|_{B(\Delta_i)}^\sigma = \sum_{i=1}^m \int_0^{|\Delta_i|/k} t^{-\alpha\sigma}\|\Delta_t^k f(\cdot)\|_{L_\sigma(x_{i-1},x_i-kt)}^\sigma \frac{\mathrm{d}t}{t}$$
$$= \int_0^{1/k} t^{-\alpha\sigma}\left(\sum_{|\Delta_i|\geqslant kt} \|\Delta_t^k f(\cdot)\|_{L_\sigma(x_{i-1},x_i-kt)}^\sigma\right)\frac{\mathrm{d}t}{t}$$
$$\leqslant \int_0^{1/k} t^{-\alpha\sigma}\|\Delta_t^k f(\cdot)\|_{L_\sigma(0,1-kt)}^\sigma \frac{\mathrm{d}t}{t} = \|f\|_{B(0,1)}.$$

Thus we have

$$\sum_{i=1}^{m} \| f \|_{B(\Delta_i)}^{\sigma} \leqslant \| f \|_{B(0,1)}^{\sigma}. \tag{34}$$

On the other hand the function $F(v) = \| f \|_{B(u,v)}$, $(u, v) \subset (0, 1)$, is obviously continuous and nondecreasing for $v \in (u, 1]$. Hence

$$\| f \|_{B(\Delta_i)}^{\sigma} = \frac{1}{n} \| f \|_{B(0,1)}^{\sigma}, \quad i = 1, 2, \ldots, m-1.$$

These equalities and (34) imply that there exists $m \leqslant n$ such that $x_m = 1$. Also, by the definition of $x_0, x_1, \ldots, x_m$ it follows that

$$\| f \|_{B(\Delta_i)}^{\sigma} \leqslant \frac{1}{n} \| f \|_{B(0,1)}^{\sigma}, \quad i = 1, 2, \ldots, m. \tag{35}$$

Now we apply the estimate (33) to the function f in each interval Δ_i. In view of (35) we get

$$E_{k-1}(f)_{L_p(\Delta_i)} \leqslant \frac{C}{n^{1/\sigma}} \| f \|_{B_{\sigma;k}^{\alpha}}^{(2)}, \quad i = 1, 2, \ldots, m, m \leqslant n.$$

Hence

$$S_n^k(f)_p \leqslant \left(\sum_{i=1}^{m} E_{k-1}(f)_{L_p(\Delta_i)}^p \right)^{1/p} \leqslant C \frac{\| f \|_{B_{\sigma;k}^{\alpha}}^{(2)}}{n^{\alpha}}. \qquad \square$$

Proof of theorem 7.4. Here we shall use the following Besov quasi-norm (see section 7.2):

$$\| f \|_{B_{\sigma;k}^{\alpha}} = \left(\int_0^1 (t^{-\alpha} \omega_k(f; t)_{\sigma})^{\sigma} \frac{dt}{t} \right)^{1/\sigma}.$$

Let $s \in S(k, n, [0, 1])$. Then there exists a partition of $[0, 1]$, $0 = x_0 < x_1 < \cdots < x_n = 1$ and polynomials $Q_i \in P_{k-1}$, $i = 1, 2, \ldots, n$, such that $s(x) = Q_i(x)$ for $x \in \Delta_i = (x_{i-1}, x_i)$, $i = 1, 2, \ldots, n$.

We need to estimate $\omega_k(s, t)_{\sigma}$ for fixed $t \in (0, 1)$. Observe that, since $s(x) = Q_i(x)$, $x \in \Delta_i$ and $Q_i \in P_{k-1}$, $\Delta_h^k s(x) = 0$ for x, $x + kh \in \Delta_i$. Hence

$$\begin{aligned} \omega_k(s; t)_{\sigma}^{\sigma} &= \sup_{0 < h \leqslant t} \int_0^{1-kh} |\Delta_h^k s(x)|^{\sigma} dx \\ &\leqslant C \left\{ \sum_{|\Delta_i| \leqslant kt} \int_{\Delta_i} |s(x)|^{\sigma} dx + \sum_{|\Delta_i| > kt} \left(\int_{\Delta_i'} |s(x)|^{\sigma} dx + \int_{\Delta_i''} |s(x)|^{\sigma} dx \right) \right\} \end{aligned} \tag{36}$$

where $\Delta_i' = (x_{i-1}, x_{i-1} + kt)$, $\Delta_i'' = (x_i - kt, x_i)$, the first sum is taken over all i such that $|\Delta_i| \leqslant kt$ and the second sum over all i such that $|\Delta_i| > kt$.

Next, we shall make use of the following inequalities:

$$\left(\int_{\Delta_i} |s(x)|^\sigma \mathrm{d}x\right)^{1/\sigma} = \|s\|_{L_\sigma(\Delta_i)} \leqslant |\Delta_i|^\alpha \|s\|_{L_p(\Delta_i)}, \tag{37}$$

$$\left(\int_{\Delta_i'} |s(x)|^\sigma \mathrm{d}x\right)^{1/\sigma} = \|s\|_{L_\sigma(\Delta_i')} \leqslant Ct^{1/\sigma}|\Delta_i|^{-1/p}\|s\|_{L_p(\Delta_i)}, \tag{38}$$

$$\left(\int_{\Delta_i''} |s(x)|^\sigma \mathrm{d}x\right)^{1/\sigma} \leqslant Ct^{1/\sigma}|\Delta_i|^{-1/p}\|s\|_{L_p(\Delta_i)}, \tag{39}$$

where $C = C(p, \alpha, k)$.

The inequality (37) follows from Hölder's inequality. In order to prove estimates (38) and (39) we shall apply again Hölder's inequality and the right-hand-side inequality from lemma 7.3 in section 7.1. We get

$$\|s\|_{L_\sigma(\Delta_i')} \leqslant |\Delta_i'|^{1/\sigma}\|s\|_{L_\infty(\Delta_i')} \leqslant (kt)^{1/\sigma}\|s\|_{L_\infty(\Delta_i)} \leqslant Ct^{1/\sigma}|\Delta_i|^{-1/p}\|s\|_{L_p(\Delta_i)}.$$

The estimate (39) can be proved similarly.

Combining (36) with (37)–(39) we obtain

$$\omega_k(s;t)_\sigma^\sigma \leqslant C\left\{\sum_{|\Delta_i|\leqslant kt} |\Delta_i|^{\alpha\sigma}\|s\|_{L_p(\Delta_i)}^\sigma + \sum_{|\Delta_i|>kt} t|\Delta_i|^{-\sigma/p}\|s\|_{L_p(\Delta_i)}^\sigma\right\}.$$

Now we are ready to estimate $\|s\|_{B^\alpha_{\sigma;k}}$. Applying the above estimate for $\omega_k(f;t)_\sigma$ we get

$$\begin{aligned}
\int_0^1 (t^{-\alpha}\omega_k(s;t)_\sigma)^\sigma \frac{\mathrm{d}t}{t} &\leqslant C\int_0^1 t^{-\alpha\sigma-1}\Bigg(\sum_{|\Delta_i|\leqslant kt} |\Delta_i|^{\alpha\sigma}\|s\|_{L_p(\Delta_i)}^\sigma \\
&\qquad + \sum_{|\Delta_i|>kt} t|\Delta_i|^{-\sigma/p}\|s\|_{L_p(\Delta_i)}^\sigma\Bigg)\mathrm{d}t \\
&\leqslant C\sum_{i=1}^n \Bigg(|\Delta_i|^{\alpha\sigma}\|s\|_{L_p(\Delta_i)}^\sigma \int_{|\Delta_i|/k}^\infty t^{-\alpha\sigma-1}\,\mathrm{d}t \\
&\qquad + |\Delta_i|^{-\sigma/p}\|s\|_{L_p(\Delta_i)}^\sigma \int_0^{|\Delta_i|/k} t^{-\alpha\sigma}\,\mathrm{d}t\Bigg) \\
&\leqslant C\sum_{i=1}^n \|s\|_{L_p(\Delta_i)}^\sigma \leqslant C\left(\sum_{i=1}^n \|s\|_{L_p(\Delta_i)}^p\right)^{\sigma/p} n^{1-\sigma/p} \\
&= Cn^{\alpha\sigma}\|s\|_{L_p(0,1)}^\sigma,
\end{aligned}$$

where we have applied the discrete variant of Hölder's inequality. These last estimates imply (9).

It remains to prove that the inequality (9) does not hold in the case $p = \infty$.

Consider the spline

$$s(x)=\begin{cases}0, & x\in[0,1/2],\\ \varepsilon^{-1}(x-1/2), & x\in(1/2,1/2+\varepsilon),\\ 1, & x\in(1/2+\varepsilon,1],\end{cases}$$

where $\varepsilon>0$ is a sufficiently small number. It is easily seen that $\|s\|_{B^{\alpha}_{1/\alpha;k}}> C\ln(1/\varepsilon)$, $C=C(\alpha,k)>0$ and $\|s\|_{L_\infty}=1$. Consequently, the estimate (9) does not hold when $p=\infty$. □

According to the arguments of section 3.5, theorems 7.5 and 7.6 and corollaries 7.2 and 7.3 are immediate consequences of theorems 7.3 and 7.4.

Finally, we give one simple inequality of Bernstein type which is weaker than inequality (9) in theorem 7.4. Unfortunately the corresponding Jackson type inequality does not hold.

Lemma 7.17. *Let $s\in S(k,n,[a,b])$, $k,n\geqslant 1$, $0<p\leqslant\infty$, $m\geqslant 1$ and $\sigma=(m+1/p)^{-1}$. Then we have*

$$\|s^{(m)}\|_{L_\sigma(a,b)}\leqslant C(p,k,m)n^m\|s\|_{L_p(a,b)}. \tag{40}$$

Proof. First we prove estimate (40) in the case $m=1$. Since $s\in S(k,n,[a,b])$, then there exist points $a=x_0<x_1<\cdots<x_n=b$ such that s is a polynomial of degree $k-1$ in each interval $\Delta_i=(x_{i-1},x_i)$. By lemma 7.5 in section 7.1 we have

$$\|s'\|_{L_\sigma(\Delta_i)}\leqslant C(p,k)\|s\|_{L_p(\Delta_i)}.$$

Hence, using the discrete variant of Hölder's inequality we get

$$\|s'\|_{L_\sigma(a,b)}=\left(\sum_{i=1}^n\|s'\|^\sigma_{L_\sigma(\Delta_i)}\right)^{1/\sigma}\leqslant C\left(\sum_{i=1}^n\|s\|^\sigma_{L_p(\Delta_i)}\right)^{1/\sigma}$$

$$\leqslant C\left\{\left(\sum_{i=1}^n\|s\|^p_{L_p(\Delta_i)}\right)^{\sigma/p}n^{1-\sigma/p}\right\}^{1/\sigma}=Cn\|s\|_{L_p(a,b)},$$

where we have supposed that $p<\infty$. The case $p=\infty$ is similar.

Let $m>1$. Write $\sigma_0=p$, $\sigma_1=1/(1+1/p)$, $\sigma_2=1/(2+1/p),\ldots,\sigma_m=1/(m+1/p)$. By the inequality (40) when $m=1$ it follows that

$$\|s^{(\nu)}\|_{L_{\sigma_\nu}(a,b)}\leqslant C_\nu n\|s^{(\nu-1)}\|_{L_{\sigma_{\nu-1}}(a,b)},\quad \nu=1,2,\ldots,m.$$

Multiplying the above inequalities we establish (40). □

7.3.3 Direct and converse theorems in uniform metric and in BMO

We have already explained in section 3.5 our point of view concerning direct and converse theorems in approximation theory. We need pairs of adjusted estimates of Jackson and Bernstein type.

In this section we prove estimates for uniform approximation by smooth splines instead of piecewise polynomials. In view of lemma 7.11 this is not essential for our discussion. The only formal reason for it is that the approximating splines are in the class of absolutely continuous functions.

We begin with the following trivial estimates.

Theorem 7.8. (i) *If* $f \in L_\infty(0,1)$ *and* $V_0^1 f < \infty$, *then*

$$S_n^1(f)_\infty \leqslant \frac{V_0^1 f}{n}, \quad n = 1, 2, \ldots. \tag{41}$$

(ii) *For each* $s \in S(k, n, [0,1])$, $n \geqslant 1$, $k \geqslant 1$, *we have*

$$V_0^1 s \leqslant C(k) n \| s \|_\infty . \tag{42}$$

Proof. The inequality (41) is established in theorem 7.2 in section 7.3.1. The estimate (42) is obvious since s is piecewise monotone with at most kn pieces. □

The estimates (41) and (42) form a pair of adjusted inequalities of Jackson and Bernstein type. Consequently, they imply just as in the method of section 3.5 complete direct and converse theorems. Those are theorems which characterize L_∞ spline approximation by the K-functional between spaces L_∞ and V. However, they can be used successfully for orders of approximation not better than $O(n^{-1})$. In order to characterize classes of functions with order of uniform spline approximation better than $O(n^{-1})$ we shall make use of the following inequalities of Jackson and Bernstein type.

Theorem 7.9 (Jackson type inequality). *If* f *is absolutely continuous on* $[0,1]$, $f' \in B^{\alpha-1}_{1/\alpha;k-1}$, $\alpha > 1$ *and* $k \geqslant 2$ *then*

$$\tilde{S}_n^k(f)_C \leqslant C(\alpha, k) \frac{\| f' \|_{B^{\alpha-1}_{1/\alpha;k-1}}}{n^\alpha}, \quad n = 1, 2, \ldots. \tag{43}$$

Proof. By lemma 7.10 and lemma 7.13 in subsection 7.3.1 we have for $n = 1, 2, \ldots$.

$$\left.\begin{aligned} &\tilde{S}_{3kn}^k(f)_C \leqslant S_{3n}^k(f)_C \leqslant \frac{S_n^{k-1}(f')_1}{n} \\ &\text{and} \\ &\tilde{S}_1^k(f)_C \leqslant S_1^{k-1}(f')_1 . \end{aligned}\right\} \tag{44}$$

On the other hand by theorem 7.3 in subsection 7.3.2 we have

$$S_n^{k-1}(f')_1 \leqslant C(\alpha) \frac{\| f' \|_{B^{\alpha-1}_{1/\alpha;k-1}}}{n^{\alpha-1}}, \quad n = 1, 2, \ldots. \tag{45}$$

The estimates (44) and (45) imply (43). □

Theorem 7.10 (Bernstein type inequality). *If* $s \in \tilde{S}(k, n, [0,1])$, $k \geqslant 2$, $n \geqslant 1$ *and* $\alpha > 1$, *then*

$$\| s' \|_{B^{\alpha-1}_{1/\alpha;k-1}} \leqslant C(\alpha, k) n^{\alpha} \| s \|_{C}. \tag{46}$$

Proof. By theorem 7.4 in subsection 7.3.2 we have

$$\| s' \|_{B^{\alpha-1}_{1/\alpha;k-1}} \leqslant C(\alpha, k) n^{\alpha-1} \| s' \|_{L_1} \tag{47}$$

and by estimate (42)

$$\| s' \|_{L_1} \leqslant C(k) n \| s \|_{C}. \tag{48}$$

The estimate (46) follows from (47) and (48). □

Denote by $\bar{B}^{\alpha}_{\sigma;k}$ the set of all functions f absolutely continuous on $[0,1]$ such that $f' \in B^{\alpha}_{\sigma;k}$ with quasi-norm $\| f \|_{\bar{B}^{\alpha}_{\sigma;k}} = \| f' \|_{B^{\alpha}_{\sigma;k}}$.

According to theorem 3.16 in section 3.5 theorems 7.9 and 7.10 imply the following direct and converse theorems.

Theorem 7.11 (direct theorem). *If* $f \in C_{[0,1]}$, $\alpha > 1$ *and* $k \geqslant 2$, *then we have for* $n = 1, 2, \ldots$

$$\tilde{S}^k_n(f)_C \leqslant CK(f, n^{-\alpha}; C, \bar{B}^{\alpha-1}_{1/\alpha;k-1}),$$

where $C = C(\alpha, k)$.

Theorem 7.12 (converse theorem). *If* $f \in C_{[0,1]}$, $\alpha > 1$ *and* $k \geqslant 2$, *then we have for* $n = 1, 2, \ldots$

$$K(f, n^{-\alpha}; C, \bar{B}^{\alpha-1}_{1/\alpha;k-1}) \leqslant Cn^{-\alpha} \left(\sum_{\nu=1}^{n} \frac{1}{\nu} (\nu^{\alpha} \tilde{S}^k_{\nu}(f)_C)^{1/\alpha} \right)^{\alpha}, \quad C = C(\alpha, k).$$

The following corollary follows from theorems 7.11 and 7.12 in view of corollary 3.6 in section 3.5.

Corollary 7.4. *If* $\alpha > 1$, $0 < \gamma < \alpha$, $0 < q \leqslant \infty$ *and* $k \geqslant 2$, *then*

$$S^{\gamma}_{q;k}(C) = (C, \bar{B}^{\alpha-1}_{1/\alpha;k-1})_{\gamma/\alpha, q}$$

with equivalent quasi-norms, where $C = C_{[0,1]}$ *and* $S^{\gamma}_{q;k}(C)$ *is defined in subsection 7.3.2.*

The BMO space appears in a natural way in many problems of analysis, see details in P. Koosis (1980), J. Garnett (1981). Next, we shall briefly establish direct and converse theorems for spline approximation in BMO which are similar to those in the uniform case.

The space BMO is defined as the set of all functions $f \in L_1(a, b)$ such that

$$\| f \|_{\mathrm{BMO}} = \| f \|_{\mathrm{BMO}(a,b)} = \sup_{\Delta \subset (a,b)} \frac{1}{|\Delta|} \int_{\Delta} |f(x) - f_{\Delta}| \mathrm{d}x < \infty,$$

where $f_\Delta = |\Delta|^{-1}\int_\Delta f(t)dt$ and supremum is taken over all intervals $\Delta \subset (a,b)$. Thus we have a semi-norm $\|\cdot\|_{\mathrm{BMO}}$ in BMO. Clearly $\|f\|_{\mathrm{BMO}} \leqslant 2\|f\|_{L_\infty}$.

The following lemma is basic in our discussion (compare with lemma 7.18 in subsection 7.3.2).

Lemma 7.19. *If* $s\in\tilde{S}(k,n,[0,1])$, $k\geqslant 2$, $n\geqslant 1$, *then*

$$\|s'\|_{L_1} \leqslant C(k)n\|s\|_{\mathrm{BMO}}. \tag{49}$$

Proof. First we shall prove that for each polynomial $Q\in P_{k-1}$ and for each interval Δ

$$\|Q'\|_{L_1(\Delta)} \leqslant C(k)\|Q\|_{\mathrm{BMO}(\Delta)}. \tag{50}$$

Indeed, the factor-space P_{k-1}/P_0 is finite dimensional and then $\|Q'\|^{(1)} = \|Q'\|_{L_1(0,1)}$ and $\|Q\|^{(2)} = \|Q\|_{\mathrm{BMO}(0,1)}$ are equivalent norms there. Consequently, for each polynomial $Q\in P_{k-1}$ the following inequality holds:

$$\|Q'\|_{L_1(0,1)} \leqslant C(k)\|Q\|_{\mathrm{BMO}(0,1)}. \tag{51}$$

The inequality (50) follows from (51) by simple change of variables.

If $s\in\tilde{S}(k,n,[0,1])$, then there exists a partition of $[0,1]$ into n subintervals $0=x_0<x_1<\cdots<x_n=1$ and polynomials $\{Q_i\}_{i=1}^n$, $Q_i\in P_{k-1}$ such that $s(x) = Q_i(x)$ for $x\in\Delta_i=(x_{i-1},x_i)$. Now applying (50) to s in each interval Δ_i we get

$$\|s'\|_{L_1(0,1)} = \sum_{i=1}^n \|Q_i'\|_{L_1(\Delta_i)} \leqslant C(k)\sum_{i=1}^n \|Q_i\|_{\mathrm{BMO}(\Delta_i)} \leqslant C(k)n\|s\|_{\mathrm{BMO}(0,1)}. \qquad \square$$

By theorem 7.4 in subsection 7.3.2 we have: if $s\in\tilde{S}(k,n,[0,1])$, $k\geqslant 2$, $n\geqslant 1$ and $\alpha>1$, then

$$\|s'\|_{B^{\alpha-1}_{1/\alpha;k-1}} \leqslant C(\alpha,k)n^{\alpha-1}\|s'\|_{L_1}.$$

This inequality together with (49) gives

$$\|s'\|_{B^{\alpha-1}_{1/\alpha;k-1}} \leqslant C(\alpha,k)n^{\alpha}\|s'\|_{\mathrm{BMO}}. \tag{52}$$

On the other hand by theorem 7.8 we have: if f is absolutely continuous on $[0,1]$ and $f'\in B^{\alpha-1}_{1/\alpha;k-1}$, $\alpha>1$, $k\geqslant 2$, then we have for $n=1,2,\ldots$

$$\tilde{S}_n^k(f)_{\mathrm{BMO}} \leqslant 2\tilde{S}_n^k(f)_C \leqslant C(\alpha,k)\frac{\|f'\|_{B^{\alpha-1}_{1/\alpha;k-1}}}{n^\alpha}, \tag{53}$$

where

$$\tilde{S}_n^k(f)_{\mathrm{BMO}} = \inf\{\|f-s\|_{\mathrm{BMO}}: s\in\tilde{S}(k,n,[0,1])\}.$$

The inequalities (52) and (53) form a pair of inequalities of Jackson and Bernstein type. They imply direct and converse theorems for spline approximation in BMO which are similar to theorems 7.10 and 7.11. The only difference is that uniform metric is replaced by BMO.

7.4 Notes

For the inequalities of Hardy ((4), (5) in section 7.1) see Hardy, Littlewood, Polya (1934).

Theorem 7.1 was proved by H. Whitney (1957, 1959) for the case $p = \infty$. For $1 \leqslant p < \infty$ see Yu.A. Brudnyi (1971), for $0 < p < 1$ see E.A. Storoženko (1977). Bl. Sendov (1985, 1987) proved that in the case $p = \infty$ the constant $c(k, \infty)$ is bounded, i.e. c does not depends on k. Bl. Sendov has the estimate $c \leqslant 6$.

The function $f_{k,h}$ in (24) of section 7.1 was introduced by Bl. Sendov.

Almost all of the lemmas in section 7.1 are well known, but we give their proofs for completeness.

The basic results of subsections 7.3.2 and 7.3.3 are due to P. Petrushev (1985). Let us mention that another connection between best spline approximations and Besov spaces was formulated without proofs by Yu.A. Brudnyi at the conference on approximation theory, Kiev, 1983.

The critical index $\sigma = (\alpha + 1/p)^{-1}$ appears, as far we know, for the first time in de Boor (1972). Lemma 7.10 is due to Brudnyi (1971). Theorem 7.2 appears in G. Freud, V.A. Popov (1969, 1970) and Yu.N. Subbotin, N.I. Chernyh (1970).

The problem of characterization of the best spline approximations (with free knots) was considered by many authors. We do not want to go here into details, but we want to mention the following works: D. Gaier (1970), Yu.A. Brudnyi (1971, 1974), J. Nitsche (1969a, b), J. Bergh, J. Peetre (1974), V.A. Popov (1973, 1976b), H.G. Burchard (1974, 1977), H.G. Burchard, D.F. Hall (1975), J. Peetre (1976), L. Schumaker (1981).

The embedding result – theorem 7.7 – was obtained by P. Oswald (1980).

We would like to remark that J. Peetre (1976) was the first who understood that Besov spaces B_σ^α with $\sigma < 1$ are very essential for spline approximation.

DeVore, Popov (1986) have obtained that

$$(L_p, B_\sigma^\alpha)_{\gamma/\alpha, q} = B_q^\gamma, \frac{1}{\sigma} = \alpha + \frac{1}{p}, \frac{1}{q} = \gamma + \frac{1}{p};$$

therefore corollary 7.3 becomes

$$S_{q;k}^\gamma(L_p) = B_{q;k}^\gamma, k > \gamma;$$

$$\frac{1}{q} = \gamma + \frac{1}{p}.$$

8

Relations between rational and spline approximations

It is well known that the rational and spline approximations are closely connected. Our point of view is that the splines are the most simple and well-known nonlinear tool for approximation and therefore it is very useful to investigate the connections between rational and spline approximations of functions.

In section 8.1 we prove that the rational functions are not worse than splines as a tool for approximation in L_p $(1 \leqslant p < \infty)$ metric. In section 8.2 we prove an estimate of spline approximation in L_p $(1 < p \leqslant \infty)$ by means of the rational approximation in L_p. In section 8.3 we establish some relations between the rational and spline approximations of functions and their derivatives in different L_p metrics.

8.1 The rational functions are not worse than spline functions as a tool for approximation in L_p $(1 \leqslant p < \infty)$ metric

The basic result in this chapter is the following.

Theorem 8.1. *If* $f \in L_p[a,b]$, $1 \leqslant p < \infty$, $k \geqslant 1$ *and* $\alpha > 0$, *then for* $n \geqslant \max\{1, k-1\}$

$$R_n(f)_p \leqslant Cn^{-\alpha} \sum_{\nu=1}^{n} \nu^{\alpha-1} S_\nu^k(f)_p. \tag{1}$$

Moreover, if we put $f(x) = 0$ *for* $x \in (-\infty, \infty) \setminus [a,b]$, *then for* $n \geqslant k-1$

$$R_n(f, (-\infty, \infty))_p \leqslant Cn^{-\alpha} \left\{ \| f \|_p + \sum_{\nu=1}^{n} \nu^{\alpha-1} S_\nu^k(f, [a,b])_p \right\}. \tag{2}$$

Also, if $f \in L_p(-\infty, \infty)$, *then*

$$R_n(f, (-\infty, \infty))_p \leqslant Cn^{-\alpha} \sum_{\nu=1}^{n} \nu^{\alpha-1} S_\nu^k(f, (-\infty, \infty))_p. \tag{3}$$

In the estimates (1)–(3) $C = C(p, k, \alpha)$ *depends only on* p, k *and* α.

Remark. Clearly, the estimates (1)–(3) do not hold true for $p=\infty$. The presence of spline approximations in the right-hand side of estimates (1)–(3) is essential. Spline approximations in these theorems cannot be replaced, for example, by piecewise monotone approximation.

Theorem 8.1 is full of corollaries which will be given after the proof. The proof is based on the following statement.

Theorem 8.2. *Let $\varphi\in S(k,m,[a,b])$, where $k\geqslant 1, m\geqslant 1$ and $[a,b]$ is an arbitrary compact interval, and let $1\leqslant p<\infty$. Put $\varphi(x)=0$ for $x\in(-\infty,\infty)\backslash[a,b]$. Then for each $\lambda>0$ there exists a rational function R such that*

$$\deg R\leqslant Dm\ln^2\left(\mathrm{e}+\frac{1}{\lambda}\right)$$

and

$$\|\varphi-R\|_{L_p(-\infty,\infty)}\leqslant\lambda\|\varphi\|_{L_p[a,b]},$$

where $D=D(p,k)>1$.

The same statement in another form: *Under the hypotheses above the following estimate holds true for $n\geqslant 1$:*

$$R_n(\varphi,(-\infty,\infty))_p\leqslant 2\exp\left\{-C\sqrt{\frac{n}{m}}\right\}\|\varphi\|_{L_p[a,b]},$$

where $C=C(p,k)>0$.

In order to prove theorem 8.2 we need the following lemma.

Lemma 8.1. *Let $Q\in P_k$, $k\geqslant 0$, $1\leqslant p<\infty$ and let $\Delta=[a,b]$ be an arbitrary compact interval. Then for each $\lambda>0$ there is a rational function R such that*

$$\deg R\leqslant D\ln^2\left(\mathrm{e}+\frac{1}{\lambda}\right),$$

$$\|Q-R\|_{L_p(\Delta)}\leqslant D\lambda\|Q\|_{L_p(\Delta)}$$

and

$$|R(x)|\leqslant D\left(\frac{|\Delta|}{|\Delta|+\rho(x,\Delta)}\right)^4\frac{\lambda\|Q\|_{L_p(\Delta)}}{|\Delta|^{1/p}},\quad x\in(-\infty,\infty)\backslash\Delta,$$

where $D=D(p,k)$,

$$\rho(x,\Delta)=\begin{cases}a-x, & x\leqslant a,\\ 0, & x\in\Delta,\\ x-b, & x\geqslant b,\end{cases}$$

is the distance from the point x to the interval Δ and $|\Delta|=b-a$.

The proof of lemma 8.1 is based on the following two lemmas.

Lemma 8.2. *If $Q\in P_k$, $k\geqslant 0$, $1\leqslant p<\infty$ and $\Delta=[-b,b]$, $b>0$, then for*

$x \in (-\infty, \infty) \setminus \Delta$

$$|Q(x)| \leqslant C \frac{\|Q\|_{L_p(\Delta)} |x|^k}{|\Delta|^{k+1/p}},$$

where $C = C(k)$.

Proof. The lemma follows immediately from the following well-known inequalities:

(i) If $Q \in P_k$, then for $|x| \geqslant 1$

$$|Q(x)| \leqslant \|Q\|_{L_\infty[-1,1]} |T_k(x)| \leqslant \|Q\|_{L_\infty[-1,1]} (|x| + \sqrt{(x^2-1)})^k$$
$$\leqslant 2^k \|Q\|_{L_\infty[-1,1]} |x|^k,$$

where $T_k(x) = \cos(k \arccos x)$, $|x| \leqslant 1$, is the Chebyshev polynomial (see Natanson (1949)). Consequently, if $Q \in P_k$, then for $|x| \geqslant b$

$$|Q(x)| \leqslant 2^k \|Q\|_{L_\infty[-b,b]} \frac{|x|^k}{b^k} = 2^{2k} \frac{\|Q\|_{L_\infty(\Delta)} |x|^k}{|\Delta|^k}.$$

(ii) If $Q \in P_k$, then

$$\|Q\|_{L_\infty(\Delta)} \leqslant C(k) \frac{\|Q\|_{L_p(\Delta)}}{|\Delta|^{1/p}}$$

(see lemma 7.3 in 7.1). □

Lemma 8.3 (The fundamental lemma). *If $d > 0$, $0 < \delta \leqslant 1$, $0 < \gamma \leqslant 1$ and $r \geqslant 0$, r integer, then there exists a rational function σ such that*

$$\deg \sigma \leqslant B \ln\left(e + \frac{1}{\delta}\right) \ln\left(e + \frac{1}{\gamma}\right) + 4r, \tag{4}$$

$$0 \leqslant 1 - \sigma(x) \leqslant \gamma, \quad |x| \leqslant d - \delta d, \tag{5}$$

$$0 \leqslant \sigma(x) \leqslant \left(\frac{2d}{d + |x|}\right)^{4r} \gamma, \quad |x| \geqslant d + \delta d \tag{6}$$

and

$$0 \leqslant \sigma(x) \leqslant 1, \quad x \in (-\infty, \infty), \tag{7}$$

where $B > 1$ is an absolute constant.

Proof. According to lemma 5.2 in section 5.1, if $0 < \varepsilon \leqslant 1/2$ and $n \geqslant 1$ then the rational function $S(x) = P(x)/P(-x)$, $P(x) = \prod_{i=1}^n (x - \varepsilon^{i/n})$, $S \in R_n$, satisfies the following inequalities

$$|S(x)| \geqslant \frac{1}{C_1} \exp\left\{\frac{C_2 n}{\ln(1/\varepsilon)}\right\}, \quad x \in [-1, -\varepsilon] \tag{8}$$

and

$$|S(x)| \leqslant C_1 \exp\left\{-\frac{C_2 n}{\ln(1/\varepsilon)}\right\}, \quad x \in [\varepsilon, 1], \tag{9}$$

where $C_1, C_2 > 0$ are absolute constants.

Put, using the notations above,

$$\varepsilon = \frac{1}{e + 2/\delta},$$

$$n = \left[\frac{1 + C_1^2}{2C_2} \ln\frac{1}{\varepsilon} \ln\left(e + \frac{1}{\gamma}\right) + 1\right], \quad \sigma_1(x) = \frac{1}{S^2(x)\left(\dfrac{1-x}{1+x}\right)^{2r} + 1},$$

where $[x]$ denotes the integer part of x.

It is readily seen that

$$\deg \sigma_1 = 2n + 2r \leqslant B_1 \ln\left(e + \frac{1}{\delta}\right) \ln\left(e + \frac{1}{\gamma}\right) + 2r, \tag{10}$$

where $B_1 > 0$ is an absolute constant. By (8) and the choice of ε and n we get for $x \in [-1, -\varepsilon] \supset [-1, -\delta/2]$

$$|\sigma_1(x)| \leqslant \frac{(1+x)^{2r}}{S^2(x)} \leqslant \frac{C_1^2 (1+x)^{2r}}{\exp\left\{\dfrac{2C_2 n}{\ln(1/\varepsilon)}\right\}} \leqslant (1+x)^{2r}\gamma \tag{11}$$

and by (9) we get for $x \in [\varepsilon, 1] \supset [\delta/2, 1]$

$$|1 - \sigma_1(x)| = \frac{S^2(x)\left(\dfrac{1-x}{1+x}\right)^{2r}}{S^2(x)\left(\dfrac{1-x}{1+x}\right)^{2r} + 1} \leqslant (1-x)^{2r} S^2(x) \leqslant (1-x)^{2r}\gamma. \tag{12}$$

Clearly

$$0 \leqslant \sigma_1(x) \leqslant 1 \quad \text{for } x \in (-\infty, \infty). \tag{13}$$

Consider the rational function

$$\sigma(x) = \sigma_1(\varphi(x)), \quad \varphi(x) = \frac{d^2 - x^2}{d^2 + x^2}.$$

We shall show that σ satisfies (4)–(7). Indeed, by (10) we obtain

$$\deg \sigma = 2 \deg \sigma_1 \leqslant 2B_1 \ln\left(e + \frac{1}{\delta}\right) \ln\left(e + \frac{1}{\gamma}\right) + 4r$$

$$= B \ln\left(e + \frac{1}{\delta}\right) \ln\left(e + \frac{1}{\gamma}\right) + 4r,$$

i.e. σ satisfies (4). Obviously, (13) implies (7). It remains to show that σ satisfies (5) and (6). Clearly, the function φ is even, strictly decreasing on $[0, \infty)$, $\varphi(0) = 1$, $\varphi(d) = 0$, $\lim_{x \to +\infty} \varphi(x) = -1$. Since $\varphi(d - \delta d) \geqslant \delta/2$ and $\varphi(d + \delta d) \leqslant -\delta/2$,

$$\delta/2 \leqslant \varphi(x) \leqslant 1, \quad |x| \leqslant d - \delta d, \tag{14}$$

$$-1 \leqslant \varphi(x) \leqslant -\delta/2, \quad |x| \geqslant d + \delta d. \tag{15}$$

By (12) and (14) we get $0 \leqslant 1 - \sigma(x) \leqslant \gamma$ for $|x| \leqslant d - \delta d$, i.e. σ satisfies (5). By (11) and (15) we obtain

$$0 \leqslant \sigma(x) \leqslant (1 + \varphi(x))^{2r}\gamma = \left(\frac{2d^2}{d^2 + x^2}\right)^{2r} \gamma \leqslant \left(\frac{2d}{d + x}\right)^{4r} \gamma$$

for $|x| \geqslant d + \delta d$, i.e. σ satisfies (6). □

Proof of lemma 8.1. Without loss of generality we shall assume that $\Delta = [-b, b]$. If $\lambda \geqslant 1$ then the rational function $R \equiv 0$ satisfies the requirments of lemma 8.1.

Let $0 < \lambda < 1$. Consider the rational function

$$R = \sigma Q,$$

where σ is the rational function from lemma 8.3 applied with

$$\delta = \frac{\lambda^p}{2}, \quad d = \frac{b}{1 + \delta}, \quad \gamma = \lambda, \quad r = k + 1.$$

By (4) we get

$$\deg R \leqslant \deg \sigma + \deg Q \leqslant B \ln\left(\mathrm{e} + \frac{1}{\delta}\right) \ln\left(\mathrm{e} + \frac{1}{\gamma}\right) + 4r + k$$

$$= B \ln\left(\mathrm{e} + \frac{2}{\lambda^p}\right) \ln\left(\mathrm{e} + \frac{1}{\lambda}\right) + 4(k + 1) + k$$

and hence

$$\deg R \leqslant B_1 \ln^2\left(\mathrm{e} + \frac{1}{\lambda}\right), \quad B_1 = B_1(p, k). \tag{16}$$

Now we estimate $\| f - R \|_{L_p(\Delta)}$. By (5) we obtain

$$|Q(x) - R(x)| = (1 - \sigma(x))|Q(x)| \leqslant \lambda |Q(x)|, \quad |x| \leqslant d - \delta d = \frac{1 - \delta}{1 + \delta} b. \tag{17}$$

If $d - \delta d \leqslant |x| \leqslant d + \delta d$, i.e.

$$\frac{1 - \delta}{1 + \delta} b \leqslant |x| \leqslant b,$$

then by (7) and lemma 8.2. we get

$$|Q(x)-R(x)| \leqslant |Q(x)| \leqslant C(k)\frac{\|Q\|_{L_p(\Delta)}|x|^k}{\left(2\frac{1-\delta}{1+\delta}b\right)^{k+1/p}} \leqslant C_1(k)\frac{\|Q\|_{L_p(\Delta)}}{|\Delta|^{1/p}}. \tag{18}$$

Using (17) and (18) we obtain

$$\begin{aligned}\|Q-R\|_{L_p(\Delta)} &= \left(\int_{-b}^{b}|Q(x)-R(x)|^p\,\mathrm{d}x\right)^{1/p}\\ &= \left(\int_{-b}^{-((1-\delta)/(1+\delta))b} + \int_{-((1-\delta)/(1+\delta))b}^{((1-\delta)/(1+\delta))b} + \int_{((1-\delta)/(1+\delta))b}^{b}\right)^{1/p}\\ &\leqslant \left\{2\left(b-\frac{1-\delta}{1+\delta}b\right)\left(C_1\frac{\|Q\|_{L_p(\Delta)}}{|\Delta|^{1/p}}\right)^p + \int_{-b}^{b}\lambda^p|Q(x)|^p\,\mathrm{d}x\right\}^{1/p}\\ &\leqslant C_2\lambda\|Q\|_{L_p(\Delta)}.\end{aligned}$$

Hence

$$\|Q-R\|_{L_p(\Delta)} \leqslant C_2\lambda\|Q\|_{L_p(\Delta)}, \quad C_2 = C_2(p,k). \tag{19}$$

If $|x|>b$, then by (6), lemma 8.2 and the fact that $4r \geqslant k+4$ we obtain

$$\begin{aligned}|R(x)| = |\sigma(x)\,Q(x)| &\leqslant C\frac{\|Q\|_{L_p(\Delta)}|x|^k}{|\Delta|^{k+1/p}}\left(\frac{2d}{d+|x|}\right)^{4r}\gamma\\ &\leqslant C\frac{\|Q\|_{L_p(\Delta)}|x|^k}{|\Delta|^{k+1/p}}\left(\frac{2b}{b+|x|}\right)^{k+4}\lambda \leqslant C\left(\frac{|\Delta|}{|\Delta|+|x|-b}\right)^4\frac{\lambda\|Q\|_{L_p(\Delta)}}{|\Delta|^{1/p}}.\end{aligned}$$

Hence

$$|R(x)| \leqslant C(k)\left(\frac{|\Delta|}{|\Delta|+\rho(x,\Delta)}\right)^4\frac{\lambda\|Q\|_{L_p(\Delta)}}{|\Delta|^{1/p}}, \quad |x|>b. \tag{20}$$

The estimates (16), (19) and (20) prove lemma 8.1. □

Proof of theorem 8.2. Suppose $\varphi\in S(k,m,[a,b])$, $k\geqslant 1$, $m\geqslant 1$. Then there exist a division $a=x_0<x_1<\cdots<x_m=b$ of $[a,b]$ and polynomials $Q_i\in P_{k-1}$, $i=1,2,\ldots,m$, such that $\varphi(x)=Q_i(x)$ for $x\in(x_{i-1},\ x_i)$. Put $\varphi(x)=0$ for $x\in(-\infty,\infty)\backslash[a,b]$. Let $\lambda>0$ and $1\leqslant p<\infty$.

In what follows we shall use the following notations:

$$\Delta=[a,b],\quad \Delta_0=(-\infty,x_0],\quad \Delta_i=[x_{i-1},x_i],\quad i=1,2,\ldots,m,$$

$$\Delta_{m+1}=[x_m,\infty),\quad A=\frac{\|\varphi\|_{L_p(\Delta)}}{m^{1/p}}.$$

Without loss of generality we shall assume that $\|\varphi\|_{L_p(\Delta_i)}\neq 0$ for $i=1,2,\ldots,m$.

Now we apply lemma 8.1 for the function φ in each interval Δ_i $(1 \leqslant i \leqslant m)$ with $\lambda_i = \lambda A / \|\varphi\|_{L_p(\Delta_i)}$. For each i $(1 \leqslant i \leqslant m)$ there exists a rational function R_i such that

$$\deg R_i \leqslant D \ln^2\left(e + \frac{1}{\lambda_i}\right) = D \ln^2\left(e + \frac{\|\varphi\|_{L_p(\Delta_i)}}{\lambda A}\right), \tag{21}$$

$$\|\varphi - R_i\|_{L_p(\Delta_i)} \leqslant D\lambda_i \|\varphi\|_{L_p(\Delta_i)} = D\lambda A \tag{22}$$

and

$$|R_i(x)| \leqslant D\left(\frac{|\Delta_i|}{|\Delta_i| + \rho(x, \Delta_i)}\right)^4 \frac{\lambda A}{|\Delta_i|^{1/p}}, \quad x \in (-\infty, \infty) \backslash \Delta_i, \tag{23}$$

where $D = D(p, k) > 0$.

We shall show that the rational function

$$R = \sum_{i=1}^{m} R_i$$

satisfies the requirements of theorem 8.2. First we estimate $\deg R$. To this end we use (21) and the facts that the function $F(x) = -\ln^2(e + x)$ is convex on $[0, \infty)$ and $\ln^2(e + x) \leqslant 4 \ln^2(e + x^p)$ for $x > 0$. We get

$$\begin{aligned} \deg R &\leqslant \sum_{i=1}^{m} \deg R_i \leqslant \sum_{i=1}^{m} D \ln^2\left(e + \frac{\|\varphi\|_{L_p(\Delta_i)}}{\lambda A}\right) \\ &\leqslant 4D \sum_{i=1}^{m} \ln^2\left(e + \frac{\|\varphi\|^p_{L_p(\Delta_i)}}{\lambda^p A^p}\right) \leqslant 4Dm \ln^2\left(e + \frac{\sum_{i=1}^{m} \|\varphi\|^p_{L_p(\Delta_i)}}{m\lambda^p A^p}\right) \\ &= 4Dm \ln^2\left(e + \frac{\|\varphi\|^p_{L_p(\Delta)}}{m\lambda^p A^p}\right) = 4Dm \ln^2\left(e + \frac{1}{\lambda^p}\right) \leqslant 4Dp^2 m \ln^2\left(e + \frac{1}{\lambda}\right). \end{aligned}$$

Thus we have

$$\deg R \leqslant D_1 m \ln^2\left(e + \frac{1}{\lambda}\right), \quad D_1 = D_1(p, k). \tag{24}$$

It remains to estimate $\|\varphi - R\|_{L_p}(-\infty, \infty)$. Note that if $p = 1$ then the required estimate follows from (22) and (23) immediately. But the case $p > 1$ is nontrivial. First we introduce some auxiliary notions.

Let $\{\Delta_i : i = 0, 1, \ldots, m + 1\}$ be the division of $(-\infty, \infty)$ considered above.

Definition 8.1. *We shall call the set of intervals $\{\Delta_\nu : i_0 \leqslant \nu \leqslant i_1\}$, $1 \leqslant i_0 \leqslant i_1 \leqslant m + 1$, a left class of intervals or briefly a left class, if $|\Delta_\nu| < |\Delta_{i_1}|$ for $\nu = i_0, i_0 + 1, \ldots, i_1 - 1$ and $|\Delta_{i_0 - 1}| \geqslant |\Delta_{i_1}|$.*

We shall suppose that $|\Delta_0| = |\Delta_{m+1}| > |\Delta_\nu|$, $\nu = 1, 2, \ldots, m$.

By Ω we shall denote the set of all left classes of intervals.

Some properties of the left classes of intervals:

(a) If $K, \tilde{K} \in \Omega$, then $K \cap \tilde{K} = \varnothing$ or $K \subset \tilde{K}$ or $\tilde{K} \subset K$. Therefore $K \subset \tilde{K}$ is an order-like relation in the set Ω with maximal element the left class $\{\Delta_\nu: 1 \leqslant \nu \leqslant m+1\}$.

(b) For each i $(1 \leqslant i \leqslant m+1)$ there is exactly one left class $K_i \in \Omega$ such that the interval Δ_i is the last interval in K_i, i.e. $K_i = \{\Delta_\nu: i_0 \leqslant \nu \leqslant i\}$ for some i_0. In the sequel K_i is always the left class corresponding to the interval Δ_i. Thus there is defined a one-to-one mapping of the set $\{\Delta_\nu: 1 \leqslant \nu \leqslant m+1\}$ onto the set Ω. Consequently, the number of the elements of Ω is $m+1$.

(c) If $K \in \Omega$ and $\Delta_i \in K$ $(1 \leqslant i \leqslant m+1)$, then $K_i \subset K$.

Definition 8.2. *We shall call the left class $\tilde{K}$ a left subclass of first order of the left class K, if $\tilde{K} \subset K$, $\tilde{K} \neq K$ and there is no class $K^* \in \Omega$, $K^* \neq K$ such that $\tilde{K} \subset K^* \subset K$, i.e. K follows $\tilde{K}$ immediately.*

By Ω_i $(1 \leqslant i \leqslant m+1)$ we shall denote the set of all subclasses of first order of the left class K_i and by μ_i the number of the elements of Ω_i.

(d) We have for $i = 1, 2, \ldots, m+1$

$$K_i = \bigcup_{K \in \Omega_i} K \cup \{\Delta_i\}.$$

More exactly, for each i $(1 \leqslant i \leqslant m+1)$

$$K_i = \{\Delta_\nu: j_0 + 1 \leqslant \nu \leqslant i\} = \bigcup_{\nu=1}^{\mu_i} K_{j_\nu} \cup \{\Delta_i\}$$

for appropriate indices $0 \leqslant j_0 < j_1 < \cdots < j_{\mu_i} = i - 1$,

$$\Omega_i = \{K_{j_\nu}: 1 \leqslant \nu \leqslant \mu_i\}, \quad K_{j_\nu} = \{\Delta_s: j_{\nu-1} + 1 \leqslant s \leqslant j_\nu\}.$$

Hence

$$|\Delta_{j_0}| \geqslant |\Delta_i| > |\Delta_{j_1}| \geqslant |\Delta_{j_2}| \geqslant \cdots \geqslant |\Delta_{j_{\mu_i}}|$$

and

$$|\Delta_s| < |\Delta_{j_\nu}|, \quad s = j_{\nu-1} + 1, j_{\nu-1} + 2, \ldots, j_\nu - 1, \quad \nu = 1, 2, \ldots, \mu_i.$$

(e) Each class $K \in \Omega$, $K \neq K_{m+1} = \{\Delta_\nu: 1 \leqslant \nu \leqslant m+1\}$, is a left subclass of first order of some left class and therefore

$$\Omega = \bigcup_{i=1}^{m+1} \Omega_i \cup \{K_{m+1}\}.$$

On the other hand $\Omega_i \cap \Omega_j = \varnothing$ for $i \neq j$. Consequently

$$\sum_{i=1}^{m+1} \mu_i = m. \tag{25}$$

The properties (a)–(e) of the left classes follow immediately from the definitions.

Analogously (more exactly by symmetry) we introduce the notion *right class of intervals* and *right subclass of first order of some right class.* We shall denote by Ω^* the set of all right classes, by K_i^* the right class in which Δ_i is the first interval, by Ω_i^* the set of all right subclasses of first order of K_i^* and by μ_i^* the number of elements in Ω_i^*. The right classes have properties symmetrical to the properties (a)–(e). We formulate only the following property:

$$\sum_{i=0}^{m} \mu_i^* = m. \tag{26}$$

The following lemma uses the notations introduced above.

Lemma 8.4. *The following estimate holds true for* $1 \leqslant i \leqslant m$:

$$\left|\sum_{\Delta_\nu \in K_i} R_\nu(x)\right| \leqslant C\left(\frac{|\Delta_i|}{|\Delta_i| + x - x_i}\right)^3 \frac{\lambda A}{|\Delta_i|^{1/p}}, \quad x > x_i, \tag{27}$$

where $C = C(p, k)$.

Proof. Let $K_i = \{\Delta_\nu : i_0 \leqslant \nu \leqslant i\}$. If $i_0 = i$, then the estimate (27) follows by (23) immediately.

Let $i_0 < i$ and $x \geqslant x_i$. By (23) we obtain

$$\left|\sum_{\Delta_\nu \in K_i} R_\nu(x)\right| \leqslant D\lambda A \sum_{\nu=i_0}^{i} \left(\frac{|\Delta_\nu|}{\sum_{s=\nu}^{i}|\Delta_s| + x - x_i}\right)^4 \frac{1}{|\Delta_\nu|^{1/p}}.$$

By the definition of left class it follows that $|\Delta_\nu| < |\Delta_i|$, $\nu = i_0, i_0 + 1, \ldots, i-1$. Denote for $r = 1, 2, \ldots$

$$G_{-r} = \{\nu : i_0 \leqslant \nu \leqslant i, 2^{-r}|\Delta_i| < |\Delta_\nu| \leqslant 2^{-r+1}|\Delta_i|\}.$$

Clearly, for $r \geqslant 1$

$$\sigma_r = \sum_{\nu \in G_{-r}} \left(\frac{|\Delta_\nu|}{\sum_{s=\nu}^{i}|\Delta_s| + x - x_i}\right)^4 \frac{1}{|\Delta_\nu|^{1/p}}$$

$$\leqslant \sum_{s=0}^{\infty} \left(\frac{2^{-r+1}|\Delta_i|}{s \cdot 2^{-r}|\Delta_i| + |\Delta_i| + x - x_i}\right)^4 \frac{1}{(2^{-r}|\Delta_i|)^{1/p}}.$$

Since the function under the last $\sum_{s=0}^{\infty}$ is decreasing on $s \in [0, \infty)$, the last series can be estimated by an appropriate integral. Thus we get

$$\sigma_r \leqslant \left(\frac{2^{-r+1}|\Delta_i|}{|\Delta_i| + x - x_i}\right)^4 \frac{1}{(2^{-r}|\Delta_i|)^{1/p}} + \int_0^\infty \left(\frac{2^{-r+1}|\Delta_i|}{t \cdot 2^{-r}|\Delta_i| + |\Delta_i| + x - x_i}\right)^4 \frac{dt}{(2^{-r}|\Delta_i|)^{1/p}}$$

$$\leqslant C \cdot 2^{-(3-1/p)r} \left(\frac{|\Delta_i|}{|\Delta_i| + x - x_i}\right)^3 \frac{1}{|\Delta_i|^{1/p}}.$$

Consequently

$$\left|\sum_{\Delta_\nu\in K_i} R_\nu(x)\right| \leqslant D\lambda A \sum_{r=1}^{\infty}\sum_{\nu\in G_{-r}}\left(\frac{|\Delta_\nu|}{\sum_{s=\nu}^{i}|\Delta_s|+x-x_i}\right)^4\frac{1}{|\Delta_\nu|^{1/p}}$$

$$\leqslant C_1\lambda A\left(\frac{|\Delta_i|}{|\Delta_i|+x-x_i}\right)^3\frac{1}{|\Delta_i|^{1/p}}\sum_{r=1}^{\infty}2^{-(3-1/p)r}$$

$$\leqslant C\left(\frac{|\Delta_i|}{|\Delta_i|+x-x_i}\right)^3\frac{1}{|\Delta_i|^{1/p}},\quad C=C(p,k). \qquad \square$$

The importance of the notations introduced above becomes clear by the following lemma.

Lemma 8.5. *The following estimate holds true for $i=2,3,\ldots,m+1$*

$$\int_{\Delta_i}\left|\sum_{\nu=1}^{i-1}R_\nu(x)\right|^p dx \leqslant C(\mu_i+1)\lambda^p A^p,$$

where $C=C(p,k)$.

Proof. Let $2\leqslant i\leqslant m+1$. By property (d) of the left classes of intervals it follows that there are indices $0\leqslant j_0<j_1<\cdots<j_{\mu_i}=i-1$ such that

$$K_i=\{\Delta_s: j_0+1\leqslant s\leqslant i\}=\bigcup_{\nu=1}^{\mu_i}K_{j_\nu}\cup\{\Delta_i\},$$

$$K_{j_\nu}=\{\Delta_s: j_{\nu-1}+1\leqslant s\leqslant j_\nu\},\quad \Omega_i=\{K_{j_\nu}: 1\leqslant\nu\leqslant\mu_i\}.$$

Hence

$$\left.\begin{array}{l}|\Delta_{j_0}|\geqslant|\Delta_i|>|\Delta_{j_1}|\geqslant|\Delta_{j_2}|\geqslant\cdots\geqslant|\Delta_{j_{\mu_i}}|\\ \text{and}\\ |\Delta_s|<|\Delta_{j_\nu}|,\quad s=j_{\nu-1}+1,j_{\nu-1}+2,\ldots,j_\nu-1,\quad \nu=1,2,\ldots,\mu_i.\end{array}\right\}\qquad(28)$$

We have

$$\int_{\Delta_i}\left|\sum_{\nu=1}^{i-1}R_\nu(x)\right|^p dx\leqslant 2^{p-1}\int_{\Delta_i}\left|\sum_{\nu=1}^{j_0}R_\nu(x)\right|^p dx$$

$$+2^{p-1}\int_{\Delta_i}\left|\sum_{\nu=j_0+1}^{i-1}R_\nu(x)\right|^p dx=I_1+I_2.$$

First we estimate I_1 using (23) and the fact that $|\Delta_{j_0}|\geqslant|\Delta_i|$ (see (28)). We obtain for $x\in\Delta_i$

$$\left|\sum_{\nu=1}^{j_0}R_\nu(x)\right|\leqslant\sum_{\nu=1}^{j_0}D\left(\frac{|\Delta_\nu|}{|\Delta_\nu|+x-x_\nu}\right)^4\frac{\lambda A}{|\Delta_\nu|^{1/p}}$$

$$\leqslant D\lambda A\sum_{\nu=1}^{j_0}\left(\frac{|\Delta_\nu|}{\sum_{s=\nu}^{j_0}|\Delta_s|+x-x_{i-1}}\right)^4\frac{1}{|\Delta_\nu|^{1/p}}$$

Denote for $r=0,\pm1,\pm2,\ldots$

$$G_r=\{\nu: 2^r|\Delta_{j_0}|<|\Delta_\nu|\leqslant 2^{r+1}|\Delta_{j_0}|,\quad 1\leqslant\nu\leqslant j_0\}.$$

Now, using arguments similar to the arguments of the proof of lemma 8.4 we get for $r \geqslant 0$, $x \in \Delta_i$

$$\sum_{\nu \in G_r} \left(\frac{|\Delta_\nu|}{\sum_{s=\nu}^{j_0} |\Delta_s| + x - x_{i-1}} \right)^4 \frac{1}{|\Delta_\nu|^{1/p}}$$

$$\leqslant \sum_{l=1}^{\infty} \left(\frac{2^{r+1}|\Delta_{j_0}|}{l \cdot 2^r |\Delta_{j_0}| + x - x_{i-1}} \right)^4 \frac{1}{(2^r|\Delta_{j_0}|)^{1/p}}$$

$$< \left(\frac{2^{r+1}|\Delta_{j_0}|}{2^r|\Delta_{j_0}| + x - x_{i-1}} \right)^4 \frac{1}{(2^r|\Delta_{j_0}|)^{1/p}}$$

$$+ \int_1^\infty \left(\frac{2^{r+1}|\Delta_{j_0}|}{t \cdot 2^r|\Delta_{j_0}| + x - x_{i-1}} \right)^4 \frac{1}{(2^r|\Delta_{j_0}|)^{1/p}} \, dt \leqslant \frac{C}{2^{r/p}|\Delta_{j_0}|^{1/p}}.$$

Also, if $r \leqslant -1$, then

$$\sum_{\nu \in G_r} \left(\frac{|\Delta_\nu|}{\sum_{s=\nu}^{j_0} |\Delta_s| + x - x_{i-1}} \right)^4 \frac{1}{|\Delta_\nu|^{1/p}}$$

$$\leqslant \sum_{s=0}^{\infty} \left(\frac{2^{r+1}|\Delta_{j_0}|}{s \cdot 2^r|\Delta_{j_0}| + |\Delta_{j_0}| + x - x_{i-1}} \right)^4 \frac{1}{(2^r|\Delta_{j_0}|)^{1/p}}$$

$$\leqslant \left(\frac{2^{r+1}|\Delta_{j_0}|}{|\Delta_{j_0}| + x - x_{i-1}} \right)^4 \frac{1}{(2^r|\Delta_{j_0}|)^{1/p}}$$

$$+ \int_0^\infty \left(\frac{2^{r+1}|\Delta_{j_0}|}{t \cdot 2^r|\Delta_{j_0}| + |\Delta_{j_0}| + x - x_{i-1}} \right)^4 \frac{dt}{(2^r|\Delta_{j_0}|)^{1/p}} \leqslant C \frac{2^{(3-1/p)r}}{|\Delta_{j_0}|^{1/p}}.$$

Consequently

$$\left| \sum_{\nu=1}^{j_0} R_\nu(x) \right| \leqslant D\lambda A \sum_{r=-\infty}^{\infty} \sum_{\nu \in G_r} \left(\frac{|\Delta_\nu|}{\sum_{s=\nu}^{j_0} |\Delta_s| + x - x_{i-1}} \right)^4 \frac{1}{|\Delta_\nu|^{1/p}}$$

$$\leqslant C_1 \lambda A \left\{ \sum_{r=0}^{\infty} \frac{1}{2^{r/p}|\Delta_{j_0}|^{1/p}} + \sum_{r=-\infty}^{1} \frac{2^{(3-1/p)r}}{|\Delta_{j_0}|^{1/p}} \right\}$$

$$\leqslant C\lambda A / |\Delta_{j_0}|^{1/p}, \quad C = C(p,k).$$

Integrating we obtain

$$I_1 = 2^{p-1} \int_{\Delta_i} \left| \sum_{\nu=1}^{j_0} R_\nu(x) \right|^p dx \leqslant C_1 \lambda^p A^p, \tag{29}$$

where $C_1 = C_1(p,k)$.

Now we estimate I_2. In view of lemma 8.4 we have

$$\left| \sum_{\nu=j_0+1}^{i-1} R_\nu(x) \right| \leqslant \sum_{s=1}^{\mu_i} \left| \sum_{\nu=j_{s-1}+1}^{j_s} R_\nu(x) \right| \leqslant C \sum_{s=1}^{\mu_i} \left(\frac{|\Delta_{j_s}|}{|\Delta_{j_s}| + x - x_{j_s}} \right)^3 \frac{\lambda A}{|\Delta_{j_s}|^{1/p}}. \tag{30}$$

Denote $y_l = x_{i-1} + \sum_{\nu=l}^{\mu_i} |\Delta_{j_\nu}|$ for $l = 1, 2, \ldots, \mu_i$ and $y_{\mu_i+1} = x_{i-1}$. Obviously $x_{i-1} = y_{\mu_i+1} < y_{\mu_i} < \cdots < y_1$.

If $x \in [y_1, \infty)$, then by (28) and (30) we get

$$\left|\sum_{\nu=j_0+1}^{i-1} R_\nu(x)\right| \leqslant C \sum_{s=1}^{\mu_i} \left(\frac{|\Delta_{j_s}|}{\sum_{r=1}^{\mu_i} |\Delta_{j_r}| + x - y_1}\right)^3 \frac{\lambda A}{|\Delta_{j_s}|^{1/p}}$$

$$\leqslant C\lambda A \sum_{s=1}^{\mu_i} \left(\frac{|\Delta_{j_s}|}{s|\Delta_{j_s}| + x - y_1}\right)^3 \frac{1}{|\Delta_{j_s}|^{1/p}}.$$

Hence

$$\left(\int_{y_1}^{\infty} \left|\sum_{\nu=j_0+1}^{i-1} R_\nu(x)\right|^p \mathrm{d}x\right)^{1/p}$$

$$\leqslant C\lambda A \left\{\int_{y_1}^{\infty} \left(\sum_{s=1}^{\mu_i} \left(\frac{|\Delta_{j_s}|}{s|\Delta_{j_s}| + x - y_1}\right)^3 \frac{1}{|\Delta_{j_s}|^{1/p}}\right)^p \mathrm{d}x\right\}^{1/p}$$

$$\leqslant C\lambda A \sum_{s=1}^{\mu_i} \left\{\int_{y_1}^{\infty} \left(\frac{|\Delta_{j_s}|}{s|\Delta_{j_s}| + x - y_1}\right)^{3p} \frac{1}{|\Delta_{j_s}|} \mathrm{d}x\right\}^{1/p}$$

$$\leqslant C\lambda A \sum_{s=1}^{\infty} \frac{1}{s^{3-1/p}}.$$

Consequently

$$\int_{y_1}^{\infty} \left|\sum_{\nu=j_0+1}^{i-1} R_\nu(x)\right|^p \mathrm{d}x \leqslant C_1 \lambda^p A^p, \quad C_1 = C_1(p,k). \tag{31}$$

Let $x \in [y_{l+1}, y_l]$, $1 \leqslant l \leqslant \mu_i$. By (30) we obtain

$$\left|\sum_{\nu=j_0+1}^{i-1} R_\nu(x)\right| \leqslant C\lambda A \left\{\sum_{s=1}^{l} \left(\frac{|\Delta_{j_s}|}{\sum_{\nu=s}^{l} |\Delta_{j_\nu}| + x - y_{l+1}}\right)^3 \frac{1}{|\Delta_{j_s}|^{1/p}}\right.$$

$$\left. + \sum_{s=l+1}^{\mu_i} \left(\frac{|\Delta_{j_s}|}{\sum_{\nu=l+1}^{\mu_i} |\Delta_{j_\nu}| + x - y_{l+1}}\right) \frac{1}{|\Delta_{j_s}|^{1/p}}\right\} = C\lambda A\{\sigma_1 + \sigma_2\}.$$

First we estimate σ_1. Denote for $r = 0, 1, \ldots$

$$G_r = \{s\colon 2^r |\Delta_{j_l}| \leqslant |\Delta_{j_s}| < 2^{r+1} |\Delta_{j_l}|, \quad 1 \leqslant s \leqslant l\}.$$

Using (28) we obtain

$$\sigma_1 \leqslant \sum_{r=0}^{\infty} \sum_{s \in G_r} \left(\frac{|\Delta_{j_s}|}{\sum_{\nu=s}^{l} |\Delta_{j_\nu}| + x - y_{l+1}}\right)^3 \frac{1}{|\Delta_{j_s}|^{1/p}}$$

$$\leqslant \sum_{r=0}^{\infty} \sum_{m=1}^{\infty} \left(\frac{2^{r+1}|\Delta_{j_l}|}{m \cdot 2^r |\Delta_{j_l}| + x - y_{l+1}}\right)^3 \frac{1}{(2^r |\Delta_{j_l}|)^{1/p}}$$

$$\leqslant \sum_{r=0}^{\infty} \frac{C}{2^{r/p} |\Delta_{j_l}|^{1/p}} \leqslant \frac{C_1}{|\Delta_{j_l}|^{1/p}}, \quad C_1 = C_1(p).$$

Using (28) again we get

$$\sigma_2 \leqslant \sum_{s=l+1}^{\mu_i} \left(\frac{|\Delta_{j_s}|}{(s-l)|\Delta_{j_s}| + x - y_{l+1}} \right)^3 \frac{1}{|\Delta_{j_s}|^{1/p}}.$$

Consequently

$$\left| \sum_{\nu=j_0+1}^{i-1} R_\nu(x) \right| \leqslant C\lambda A \left\{ \frac{1}{|\Delta_{j_l}|^{1/p}} + \sum_{s=l+1}^{\mu_i} \left(\frac{|\Delta_{j_s}|}{(s-l)|\Delta_{j_s}| + x - y_{l+1}} \right)^3 \frac{1}{|\Delta_{j_s}|^{1/p}} \right\}$$

where $C = C(p,k)$.

Now we take the L_p norm and obtain

$$\left(\int_{y_{l+1}}^{y_l} \left| \sum_{\nu=j_0+1}^{i-1} R_\nu(x) \right|^p \mathrm{d}x \right)^{1/p} \leqslant C\lambda A \left\{ \left(\int_{y_{l+1}}^{y_l} \frac{\mathrm{d}x}{|\Delta_{j_l}|} \right)^{1/p} \right.$$

$$\left. + \sum_{s=l+1}^{\mu_i} \left(\int_{y_{l+1}}^{y_l} \left(\frac{|\Delta_{j_s}|}{(s-l)|\Delta_{j_s}| + x - y_{l+1}} \right)^{3p} \frac{\mathrm{d}x}{|\Delta_{j_s}|} \right)^{1/p} \right\}$$

$$\leqslant C_1 \lambda A \left\{ 1 + \sum_{s=l+1}^{\infty} \frac{1}{(s-l)^{3-1/p}} \right\}.$$

Hence, for $l = 1, 2, \ldots, \mu_i$

$$\int_{y_{l+1}}^{y_l} \left| \sum_{\nu=j_0+1}^{i-1} R_\nu(x) \right|^p \mathrm{d}x \leqslant C\lambda^p A^p, \quad C = C(p,k).$$

Combining this estimate with (31) we get

$$I_2 = 2^{p-1} \int_{\Delta_i} \left| \sum_{\nu=j_0+1}^{i-1} R_\nu(x) \right|^p \mathrm{d}x \leqslant 2^{p-1} \int_{x_{i-1}}^{\infty} \left| \sum_{\nu=j_0+1}^{i-1} R_\nu(x) \right|^p \mathrm{d}x \leqslant C(\mu_i + 1)\lambda^p A^p,$$

where $C = C(p,k)$. From this and (29) follows the lemma. □

The following lemma can be proved in a similar (symmetrical) way.

Lemma 8.6. *The following estimate holds true for* $i = 0, 1, \ldots, m-1$:

$$\int_{\Delta_i} \left| \sum_{\nu=i+1}^{m} R_\nu(x) \right|^p \mathrm{d}x \leqslant C(\mu_i^* + 1)\lambda^p A^p,$$

where $C = C(p,k)$.

Completion of the proof of theorem 8.2. It follows by (22), (25), (26), lemma 8.5 and lemma 8.6 that

$$\|\varphi - R\|_{L_p(-\infty,\infty)} = \left\| \varphi - \sum_{\nu=1}^{m} R_\nu \right\|_{L_p(-\infty,\infty)}$$

$$= \left(\sum_{i=0}^{m+1} \left\| \varphi - \sum_{\nu=1}^{m} R_\nu \right\|_{L_p(\Delta_i)}^p \right)^{1/p}$$

$$\leqslant \left\{ \sum_{i=0}^{m+1} 3^{p-1} \left(\int_{\Delta_i} \left| \sum_{\nu<i} R_\nu(x) \right|^p dx + \int_{\Delta_i} |\varphi(x) - R_i(x)|^p dx \right.\right.$$

$$\left.\left. + \int_{\Delta_i} \left| \sum_{\nu>i} R_\nu(x) \right|^p dx \right) \right\}^{1/p}$$

$$\leqslant 3 \left\{ C(\mu_0^* + 1)\lambda^p A^p + \sum_{i=1}^{m} (C(\mu_i + 1)\lambda^p A^p + D^p \lambda^p A^p + C(\mu_i^* + 1)\lambda^p A^p) \right.$$

$$\left. + C(\mu_{m+1} + 1)\lambda^p A^p \right\}^{1/p}$$

$$\leqslant C_1 \lambda A \left(\sum_{i=1}^{m+1} \mu_i + m + \sum_{i=0}^{m} \mu_i^* \right)^{1/p}$$

$$\leqslant 3C_1 m^{1/p} \lambda A = 3C_1 \lambda \| \varphi \|_{L_p(\Delta)}.$$

Consequently

$$\| \varphi - R \|_{L_p(-\infty,\infty)} \leqslant C\lambda \| \varphi \|_{L_p[a,b]}, \quad C = C(p,k).$$

This estimate together with (24) establishes theorem 8.2. □

Proof of theorem 8.1. For each $m = 1, 2, \dots$ choose $\varphi_m \in S(k, m, [a,b])$ such that $\| f - \varphi_m \|_{L_p(\Delta)} = S_m^k(f)_p$, $\Delta = [a,b]$. Clearly, for $i \geqslant 1$ we have $\varphi_{2^i} - \varphi_{2^{i-1}} \in S(k, 2^{i+1}, \Delta)$ and

$$\| \varphi_{2^i} - \varphi_{2^{i-1}} \|_{L_p(\Delta)} \leqslant \| f - \varphi_{2^i} \|_{L_p(\Delta)} + \| f - \varphi_{2^{i-1}} \|_{L_p(\Delta)} \leqslant 2S_{2^{i-1}}^k(f)_p.$$

Let $s \geqslant 0$ be an integer. Applying theorem 8.2 for the function $\varphi_{2^i} - \varphi_{2^{i-1}}$ $(i \geqslant 1)$ with $\lambda_i = 2^{(i-s)\alpha}$ we find that there exists a rational function R_i such that

$$\deg R_i \leqslant D \cdot 2^{i+1} \ln^2 (e + 2^{(s-i)\alpha}), \quad D = D(p,k) > 1, \tag{32}$$

and

$$\| \varphi_{2^i} - \varphi_{2^{i-1}} - R_i \|_{L_p(\Delta)} \leqslant 2^{(i-s)\alpha} \| \varphi_{2^i} - \varphi_{2^{i-1}} \|_{L_p(\Delta)} \leqslant 2^{(i-s)\alpha+1} S_{2^{i-1}}^k(f)_p. \tag{33}$$

Consider the rational function $R = \sum_{i=0}^{s} R_i$ where $R_0 = \varphi_1 \in P_{k-1}$. First we estimate $\deg R$. By (32) we obtain

$$N = \deg R \leqslant \sum_{i=0}^{s} \deg R_i \leqslant k - 1 + \sum_{i=1}^{s} D \cdot 2^{i+1} \ln^2 (e + 2^{(s-i)\alpha})$$

$$\leqslant k + D_1(\alpha+1)^2 \sum_{i=1}^{s} 2^i (s-i)^2 \leqslant D(\alpha+1)^2 \cdot 2^s.$$

Consequently

$$N = \deg R \leqslant D(\alpha+1)^2 2^s, \quad D = D(p,k). \tag{34}$$

Now we estimate $\| f - R \|_{L_p(\Delta)}$. By (33) we get

$$\| f - R \|_{L_p(\Delta)} \leqslant \| f - \varphi_{2^s} \|_{L_p(\Delta)} + \sum_{i=1}^{s} \| \varphi_{2^i} - \varphi_{2^{i-1}} - R_i \|_{L_p(\Delta)}$$

$$+ \| \varphi_1 - R_0 \|_{L_p(\Delta)} \leqslant S_{2^s}^k(f)_p + \sum_{i=1}^{s} 2^{(i-s)\alpha+1} S_{2^{i-1}}^k(f)_p$$

$$\leqslant 2^{\alpha+1} 2^{-s\alpha} \sum_{i=0}^{s} 2^{i\alpha} S_{2^i}^k(f)_p \leqslant 2^{2\alpha+1} 2^{-s\alpha} \sum_{\nu=1}^{2^s} \nu^{\alpha-1} S_\nu^k(f)_p.$$

From this and (34) it follows that for each $s \geqslant 0$

$$R_N(f)_p \leqslant 2^{2\alpha+1} 2^{-s\alpha} \sum_{\nu=1}^{2^s} \nu^{\alpha-1} S_\nu^k(f)_p, \tag{35}$$

where $N \leqslant D(\alpha+1)^2 \cdot 2^s$, $\quad D = D(p,k) > 1$.

Now, let $n \geqslant \max\{1, k-1\}$. If $n \leqslant A = D(\alpha+1)^2$ then

$$R_n(f)_p \leqslant R_{k-1}(f)_p \leqslant S_1^k(f)_p \leqslant A^\alpha n^{-\alpha} \sum_{\nu=1}^{n} \nu^{\alpha-1} S_\nu^k(f)_p.$$

Consequently

$$R_n(f)_p \leqslant C n^{-\alpha} \sum_{\nu=1}^{n} \nu^{\alpha-1} S_\nu^k(f)_p, \quad \max\{1, k-1\} \leqslant n \leqslant A, \tag{36}$$

where $C = C(p, k, \alpha)$.

Let $n > A$. Choose $s \geqslant 0$ such that $A \cdot 2^s < n \leqslant A \cdot 2^{s+1}$. Then by (35)

$$R_n(f)_p \leqslant 2^{2\alpha+1} 2^{-s\alpha} \sum_{\nu=1}^{2^s} \nu^{\alpha-1} S_\nu^k(f)_p$$

$$\leqslant C_1(p, k, \alpha) n^{-\alpha} \sum_{\nu=1}^{n} \nu^{\alpha-1} S_\nu^k(f)_p.$$

This estimate together with (36) implies (1). The estimate (2) can be proved in a similar manner. It is readily seen that (2) implies (3). □

Next, we shall give some corollaries of theorem 8.1.

Corollary 8.1. *Let $f \in L_p(\Delta)$, $1 \leqslant p < \infty$ and $\Delta = [a, b]$ or $\Delta = (-\infty, \infty)$. Let $S_n^k(f, \Delta)_p = O(n^{-\gamma}\omega(n^{-1}))$, where $k \geqslant 1$, $\gamma > 0$ and ω is any increasing function on $[0, \infty)$ such that $\lim_{\delta \to 0} \omega(\delta) = 0$ and $\omega(2\delta) \leqslant 2\omega(\delta)$ for $\delta \geqslant 0$. Then*

$$R_n(f, \Delta)_p = O(n^{-\gamma}\omega(n^{-1})).$$

Of course theorem 8.1 can be used successfully in more general situations than in that of corollary 8.1.

Corollary 8.2. *Let $f \in L_p(\Delta)$, where $1 \leqslant p < \infty$, Δ is some interval (finite or infinite). Let ω be a non-decreasing and nonnegative function on $[0, \infty)$ such* *that* $\omega(2\delta) \leqslant 2^\beta \omega(\delta)$ *for* $\delta \geqslant 0$ ($\beta \geqslant 0$). *Then, if* $S_n^k(f)_{L_p(\Delta)} = O(n^{-\gamma}\omega(n^{-1}))$, $\gamma > 0$, $k \geqslant 1$, *then* $R_n(f)_{L_p(\Delta)} = O(n^{-\gamma}\omega(n^{-1}))$.

Proof of corollary 8.1 and corollary 8.2. Clearly corollary 8.1 is special case of corollary 8.2.

In order to prove corollary 8.2 we observe that, since $\omega(2\delta) \leqslant 2^\beta\omega(\delta)$ for $\delta \geqslant 0$, $\omega(2^\nu\delta) \leqslant 2^{\nu\beta}\omega(\delta)$ for $\delta \geqslant 0$, $\nu \geqslant 0$. Hence $\omega(n\delta) \leqslant (2n)^\beta\omega(\delta)$ for $\delta \geqslant 0$, $n \geqslant 1$ and therefore

$$\omega(\lambda\delta) \leqslant (2(\lambda + 1))^\beta \omega(\delta), \quad \delta, \lambda \geqslant 0. \tag{37}$$

The assertion of corollary 8.2 follows immediately from estimate (1) or estimate (2) in theorem 8.1 with some $\alpha > \gamma + \beta$. Indeed, since $S_n^k(f)_p = O(n^{-\gamma}\omega(n^{-1}))$, in view of (37), we get

$$\begin{aligned} R_n(f)_p &\leqslant Cn^{-\alpha} \sum_{\nu=1}^{n} \nu^{\alpha-1}\nu^{-\gamma}\omega(\nu^{-1}) \\ &\leqslant Cn^{-\alpha} \sum_{\nu=1}^{n} \nu^{\alpha-\gamma-1}\left(2\left(\frac{n}{\nu}+1\right)\right)^\beta \omega(n^{-1}) \\ &\leqslant C_1 n^{-\gamma}\omega(n^{-1}). \end{aligned}$$

□

Theorem 8.1 and theorem 7.3 in subsection 7.3.2 imply the following Jackson type estimate.

Theorem 8.3. *Let* $f \in B_\sigma^\alpha$, $\alpha > 0$, $\sigma = (\alpha + 1/p)^{-1}$ *and* $1 \leqslant p < \infty$; *then* $f \in L_p[0, 1]$ *and*

$$R_n(f)_p \leqslant C\frac{\|f\|_{B_\sigma^\alpha}}{n^\alpha}, \quad n > [\alpha],$$

where $C = C(\alpha, p)$.

Proof. By theorem 7.3 in subsection 7.3.2 we have $f \in L_p[0, 1]$ and

$$S_n^k(f)_p \leqslant C_1\frac{\|f\|_{B_\sigma^\alpha}}{n^\alpha}, \quad n = 1, 2, \ldots,$$

where $k = [\alpha] + 1$, $C_1 = C_1(\alpha, p)$.

Combining this estimate with estimate (1) in theorem 8.1 with α replaced by $\alpha + 1$ we get

$$\begin{aligned} R_n(f)_p &\leqslant C_2 n^{-\alpha-1} \sum_{\nu=1}^{n} \nu^\alpha S_\nu^k(f)_p \\ &\leqslant C_2 n^{-\alpha-1}\left(\sum_{\nu=1}^{n} 1\right)\|f\|_{B_\sigma^\alpha} = C_2 n^{-\alpha}\|f\|_{B_\sigma^\alpha}. \end{aligned}$$

□

For other applications of theorem 8.1 see section 8.3 and Chapter 10.

8.2 Estimate of spline approximation by means of rational in L_p, $1<p\leqslant\infty$

Here we shall obtain an estimation for the best spline approximation of functions by means of best rational approximation in L_p, $1<p\leqslant\infty$. This estimation, together with the results in section 8.1, gives a characterization of the best rational approximation by means of the best spline approximation in $L_p, 1<p<\infty$.

Theorem 8.4. *Let $f\in L_p[-1,1]$, $0<p\leqslant\infty$. Then for every natural number $k>0$ we have*

$$S_n^k(f)_p\leqslant n^{-k}\kappa_{k,p}(f;n),$$

where $\kappa_{k,p}(f;n)$ is the modulus of variation of f of order k in L_p, defined in section 6.3.

Proof. Let us consider the so called 'balanced' partition in L_p of the interval $[-1,1]$: $[-1,1]=\bigcup_{i=1}^n I_i$, $I_i=[x_{i-1},x_i]$, $-1=x_0<x_1<\cdots<x_n=1$ and†

$$(E_{k-1}(f)_{p,I_1})^p=(E_{k-1}(f)_{p,I_i})^p=\alpha_{n,p},\, i=2,\ldots,n.$$

(it is easy to see that such partition exists).

By the definition of the best spline approximation we have

$$S_n^k(f)_p\leqslant\left(\sum_{i=1}^n(E_{k-1}(f)_{p,I_i})^p\right)^{1/p}=(n\alpha_{n,p})^{1/p}. \tag{1}$$

On the other hand

$$\begin{aligned}\kappa_{k,p}(f;\ n)&\geqslant\left(\sum_{i=1}^n(E_{k-1}(f)_{p,I_i})^\sigma\right)^{1/\sigma}\\&=\left(\sum_{i=1}^n\alpha_{n,p}^{\sigma/p}\right)^{1/\sigma}=n^{1/\sigma}\alpha_{n,p}^{1/p}.\end{aligned} \tag{2}$$

From (1) and (2) the theorem follows:

$$\begin{aligned}S_n^k(f)_p&\leqslant n^{1/p}\alpha_{n,p}^{1/p}\leqslant n^{1/p-1/\sigma}\kappa_{k,p}(f;n)\\&=n^{-k}\kappa_{k,p}(f;n),\quad 1/\sigma=k+1/p.\end{aligned}$$

□

Theorem 8.5. *Let $f\in L_p(-1,1)$, $1<p\leqslant\infty$. Then:*

(a) *for $2^n\leqslant m<2^{n+1}$, $1/\sigma=k+1/p$, we have*

$$S_m^k(f)_p\leqslant c(k,p)m^{-k}\left\{\sum_{\nu=0}^n(2^{\nu k}R_{2^\nu}(f)_p)^\sigma\right\}^{1/\sigma}; \tag{3}$$

† we use the notations from section 6.3.

(b) *for every* α, $0<\alpha<k$, *we have*

$$S_m^k(f)_p \leqslant c(k,p,\alpha)m^{-\alpha}\sum_{\nu=0}^{m}(\nu+1)^{\alpha-1}R_\nu(f)_p; \tag{4}$$

(c) *for every* $q>0$ *and* $0<\alpha<k$ *we have*

$$\sum_{\nu=0}^{\infty}(2^{\nu\alpha}S_{2^\nu}(f)_p)^q \leqslant c(\alpha,k,p,q)\sum_{\nu=0}^{\infty}(2^{\nu\alpha}R_{2^\nu}(f)_p)^q. \tag{5}$$

Proof. The inequality (3) follows direct from theorems 6.6 and 8.4. To prove (4), we obtain from (3), using Hölder's inequality for $p'=1/\sigma>1$, $q'=(1-\sigma)^{-1}$,

$$S_m^k(f)_p \leqslant c(k,p)m^{-k}\left(\sum_{\nu=0}^{n}2^{\nu(k-\alpha)\sigma q'}\right)^{1/\sigma q'}\sum_{\nu=0}^{n}2^{\nu\alpha}R_{2^\nu}(f)_p$$

$$\leqslant c(\alpha,k,p)m^{-\alpha}\sum_{\nu=0}^{m}(\nu+1)^{\alpha-1}R_\nu(f)_p.$$

The inequality (5) follows now from (4), if we change the order of summation; see lemma 3.10. □

Corollary 8.3. *If we have* $R_n(f)_p=\mathrm{O}(n^{-\alpha})$, $1<p\leqslant\infty$, $0<\alpha<k$, *then*

$$S_n^k(f)_p=\mathrm{O}(n^{-\alpha}).$$

The corollary follows from the inequality (3).

The following result can be considered as an inverse to theorem 6.6 and corollary 6.3.

Lemma 8.7. *We have for* $1\leqslant p<\infty$

$$R_n(f)_p\leqslant c(p,k)n^{-k}\kappa_{k,p}(f;n),$$

where the constant $c(p,k)$ *depends only on* p *and* k.

Proof. The lemma follows from theorem 8.1 and theorem 8.4. Indeed, let us set in theorem 8.1 $\alpha=k+2$. Since the modulus of variation $\kappa_{k,p}(f;\nu)$ of f of order k in L_p is a monotone increasing function, we obtain from theorems 8.1 and 8.4

$$R_n(f)_p\leqslant c_1(p,k)n^{-k-2}\sum_{\nu=1}^{n}\nu^{k+1}S_\nu^k(f)_p$$

$$\leqslant c_1(p,k)n^{-k-2}\sum_{\nu=1}^{n}\nu\kappa_{k,p}(f;\nu)$$

$$\leqslant c_2(p,k)n^{-k-2}\kappa_{k,p}(f;n)\sum_{\nu=1}^{n}\nu\leqslant c(p,k)n^{-k}\kappa_{k,p}(f;n). \qquad \square$$

The inequality (5) together with (1) from theorem 8.1 gives us the following impressive theorem.

Theorem 8.6. *Let $1 < p < \infty$, $\alpha < k, q > 0$. There exist constants $c_1(\alpha, k, p, q)$ and $c_2(\alpha, k, p, q)$ depending only on α, k, p and q such that*

$$c_1(\alpha, k, p, q)\left\{\sum_{\nu=0}^{\infty}(2^{\nu\alpha}S_{2^\nu}^k(f)_p)^q\right\}^{1/q} \leqslant \left\{\sum_{\nu=0}^{\infty}(2^{\nu\alpha}R_{2^\nu}(f)_p)^q\right\}^{1/q}$$
$$\leqslant c_2(\alpha, k, p, q)\left\{\sum_{\nu=0}^{\infty}(2^{\nu\alpha}S_{2^\nu}^k(f)_p)^q\right\}^{1/q}.$$

The set of all functions, for which we have

$$\left\{\sum_{\nu=0}^{\infty}(2^{\nu\alpha}E_{2^\nu}(f)_X)^q\right\}^{1/q} < \infty$$

for some best approximation E_{2^ν} in some space X, is called the approximation space for the corresponding best approximation in X. Following DeVore and Popov (1986) we shall denote these spaces by $A_q^\alpha(E_\nu; X)$:

$$A_q^\alpha(E_\nu; X) = \left\{f: \left\{\sum_{\nu=0}^{\infty}(2^{\nu\alpha}E_{2^\nu}(f)_X)^q\right\}^{1/q} < \infty\right\}.$$

So theorem 8.6 gives us the remarkable fact that the approximation spaces for the best free-knots spline approximation in L_p and for the best rational approximation in L_p, $1 < p < \infty$, coincide.

A special case of theorem 8.6 is the case $q = \infty$, which follows from corollaries 8.1 and 8.3. We shall formulate this case separately because of its importance.

Theorem 8.7. *For $1 < p < \infty$, $0 < \alpha < k$, we have*

$$R_n(f)_p = O(n^{-\alpha})$$

if and only if

$$S_n^k(f)_p = O(n^{-\alpha}).$$

The connection between approximation spaces and interpolation spaces is studied in detail in DeVore and Popov (1986), where historical remarks are given also (compare with Chapter 3, 3.5, and Chapter 7). In particular, on the basis of interpolation of Besov spaces the following theorem is proved in DeVore, Popov (1986):

$$A_\sigma^\alpha(S_\nu^k; L_p) = B_{\sigma\sigma}^\alpha, \quad \sigma = (\alpha + 1/p)^{-1}, \quad p < \infty, \quad k > \alpha, \tag{6}$$

The equality (6) together with theorem 8.6 gives us the following basic characterization of the best rational and best free-knots spline approximation in L_p by means of Besov spaces.

Theorem 8.8. *For $1 < p < \infty$, $\sigma = (\alpha + 1/p)^{-1}$, $k > \alpha$, we have*

$$A_\sigma^\alpha(S_\nu^k; L_p) = A_\sigma^\alpha(R_\nu; L_p) = B_{\sigma\sigma}^\alpha.$$

Remark. The equality (6) was formulated without any indication of proof by Yu.A. Brudnyi in 1983 at the conference on approximation theory in Kiev.

We shall end this section with one useful counter-example.

Lemma 8.8. *Let $f_p(x) = x^{-1/p}$, $0 < p < \infty$ and $k \geqslant 1$. There exists a constant $c = c(p,k) > 0$ such that*

$$S_n^k(f_p)_{L_p[1/2^{2n},1]} > c \, \| f_p \|_{L_p[1/2^{2n},1]}, \quad n = 1, 2, \ldots. \tag{7}$$

Proof. We have

$$\inf_{Q \in P_{k-1}} \int_{2^{-(i+1)}}^{2^{-i}} |x^{-1/p} - Q(x)|^p \, \mathrm{d}x = \inf_{Q \in P_{k-1}} \int_{1/2}^{1} \left| t^{-1/p} - \frac{Q(2^{-i}t)}{2^{i/p}} \right|^p \mathrm{d}t$$

$$= \inf_{Q \in P_{k-1}} \int_{1/2}^{1} |t^{-1/p} - Q(t)|^p \, \mathrm{d}t = c_1(k,p) > 0. \tag{8}$$

Let $s \in S(k, n, [2^{-2n}, 1])$. It is readily seen that there exist intervals $(1/2^{m_\nu + 1}, 1/2^{m_\nu})$, $0 \leqslant m_1 < m_2 < \cdots m_n \leqslant 2n - 1$, such that each of them does not contain a knot of the spline s. Therefore using (8) we get

$$\left(\int_{2^{-2n}}^{1} |x^{-1/p} - s(x)|^p \, \mathrm{d}x \right)^{1/p} \geqslant \left(\sum_{\nu=1}^{n} \inf_{Q \in P_{k-1}} \int_{2^{-(m_\nu+1)}}^{2^{-m_\nu}} |x^{-1/p} - Q(x)|^p \, \mathrm{d}x \right)^{1/p}$$

$$\geqslant (c_1(p,k) n)^{1/p}.$$

Hence

$$S_n^k(f_p)_{L_p[1/2^{2n},1]} \geqslant (c_1(p,k))^{1/p} n^{1/p}. \tag{9}$$

On the other hand

$$\| f_p \|_{L_p[1/2^{2n},1]} = \left(\int_{2^{-2n}}^{1} \frac{\mathrm{d}x}{x} \right)^{1/p} = (2 \ln 2)^{1/p} n^{1/p}. \tag{10}$$

The estimates (9) and (10) imply (7). □

Lemma 8.8 shows that theorem 8.5 does not hold in the case $p = 1$, since $f_1(x) = 1/x \in R_1$.

8.3 Relations between rational and spline approximations of functions and rational and spline approximations of their derivatives

We begin with the basic results. As usually we denote by $W_p^r[a,b]$ the Sobolev class of all functions f such that $f^{(r-1)}$ is absolutely continuous on $[a,b]$ and $f^{(r)} \in L_p[a,b]$.

Theorem 8.9. *If $f \in W_p^r[a,b]$, $\alpha > 0$ and $m \geqslant 1$, then the estimates*

$$R_n(f)_q \leqslant c_1 n^{-r-\alpha} \sum_{\nu=0}^{n} (\nu+1)^{\alpha-1} R_\nu(f^{(r)})_p, \quad n \geqslant r, \tag{1}$$

and

$$R_n(f)_q \leqslant c_2 n^{-r-\alpha} \sum_{\nu=1}^{n} \nu^{\alpha-1} S_\nu^m(f^{(r)})_p, \quad n \geqslant r+m-1, \tag{2}$$

hold in the following situations:

(i) $r=1$, $1 \leqslant q < \infty$, $p=1$,
(ii) $r=1$, $q=\infty$, $1 < p \leqslant \infty$,
(iii) $r \geqslant 2$, $q=\infty$, $p=1$,

where $c_1 = c(p,q,r,\alpha)(b-a)^{r+(1/q)-(1/p)}$, $c_2 = c(p,q,r,\alpha,m)(b-a)^{r+(1/q)-(1/p)}$.

Remark. The estimates (1) and (2) do not hold in the case $r=1$, $q=\infty$, $p=1$, since $\{R_n(f)_C\}_{n=1}^{\infty}$ may tend to zero as slowly as we want in the class of all absolutely continuous functions f (see theorem 11.3 in subsection 11.1.2).

Theorem 8.10. *If $f \in W_p^r[a,b]$, $\alpha > 0$, $k > r+\alpha$, then the estimate*

$$S_n^k(f)_q \leqslant c n^{-r-\alpha} \sum_{\nu=0}^{n} (\nu+1)^{\alpha-1} R_\nu(f^{(r)})_p, \quad n \geqslant 1, \tag{3}$$

holds in the following situations:

(i) $r=1$, $1 \leqslant q < \infty$, $p=1$,
(ii) $r=1$, $q=\infty$, $1 < p \leqslant \infty$,
(iii) $r \geqslant 2$, $q=\infty$, $p=1$,

where $c = c_1(p,q,r,\alpha)(b-a)^{r+(1/q)-(1/p)}$.

Remark. The right-hand-sides of estimates (1)–(3) involve rational or spline functions as tools for approximation. From our proofs it will become clear that these tools can be replaced by more general ones. Namely, it is sufficient that the approximating functions and their derivatives up to a certain order are monotone in O(n) subintervals. However, such a general formulation is not used here, since other approximating tools that produce a better order of approximation are not known to us.

Next, we shall prove theorem 8.9. We need some auxiliary results.

Theorem 8.11. *Let $f \in W_p^1[a,b]$, $1 < p \leqslant \infty$, and assume that there exist a partition of $[a,b]$ into m ($m \geqslant 1$) subintervals $a = x_0 < x_1 < \cdots < x_m = b$ and rational functions $\{r_i\}_{i=0}^{m-1}$ such that for $i = 0, 1, \ldots, m-1$*

$$\| f - r_i \|_{C[x_i, x_{i+1}]} \leqslant \varepsilon_1 \tag{4}$$

and

$$\deg r_i \leqslant k_i, \tag{5}$$

where $\varepsilon_1 > 0$ *and* $k_i \geqslant 0$ *are given numbers.*

Then for each $\varepsilon_2 > 0$ *there exists a rational function* r *such that*

$$\| f - r \|_{C[a,b]} \leqslant \varepsilon_1 + \varepsilon_2 \tag{6}$$

and

$$\deg r \leqslant 2 \sum_{i=0}^{m-1} k_i + Dp'm \ln^2 \left(\mathrm{e} + \frac{\| f' \|_{L_p}(b-a)^{1/p'}}{m\varepsilon_2} \right), \tag{7}$$

where $D > 1$ *is an absolute constant,* $p' = (1 - 1/p)^{-1}$.

Proof. We shall proceed similarly as in the proof of theorem 5.2 in section 5.2.

Since the function $F(x) = x \ln^2(\mathrm{e} + 1/x)$ is increasing in $(0, \infty)$, it is readily seen that, if the theorem holds for $m = 2^s$, $s = 0, 1, \ldots$, then it holds with some other constant D for all $m = 1, 2, \ldots$. Thus we shall suppose that $m = 2^s$, s integer, and also without loss of generality $1 < p < \infty$.

Next, we shall make use of the following notations: $f_\Delta(x) = f(x) - f(u)$ for each interval $\Delta = [u, v] \subset [a, b]$;

$$N(\mu, \Delta) = 6B_1 p' \sum_{\nu=0}^{\mu} 2^\nu \ln^2(\mathrm{e} + 2^\mu 4^{-\nu} \| f' \|_{L_p(\Delta)} |\Delta|^{1/p'} \varepsilon_2^{-1}), \tag{8}$$

where $B_1 > 1$ is the constant from lemma 5.3 in section 5.1, f, p and ε_2 are from the assumptions of theorem 8.11, μ and the interval Δ are parameters.

To avoid more complicated indexation we shall denote

$$\Omega_\mu = \{[x_i, x_{i+2^\mu}] : i = 0, 1, \ldots, m - 2^\mu\}, \quad 0 \leqslant \mu \leqslant s, \tag{9}$$

and

$$\Omega_\Delta = \{\Delta^* : \Delta^* \in \Omega_0, \Delta^* \subset \Delta\}. \tag{10}$$

Note that $\Omega_0 = \{[x_i, x_{i+1}] : i = 0, 1, \ldots, m-1\}$. Also, we set $r_\Delta = r_i$ and $k_\Delta = k_i$ for each interval $\Delta = [x_i, x_{i+1}]$, $\Delta \in \Omega_0$.

Briefly we shall denote $\| \cdot \|_\Delta = \| \cdot \|_{C(\Delta)}$.

We need some lemmas, where we shall use the assumptions and notations introduced above.

Lemma 8.9. *For each interval* $\Delta \in \Omega_0$ *there exists a rational function* Q_Δ *such that*

$$\| f_\Delta - Q_\Delta \|_\Delta \leqslant \varepsilon_1 + \varepsilon_2/2, \tag{11}$$

$$\| Q_\Delta \|_{(-\infty, \infty)} \leqslant 2(V_\Delta f)^{3/2} \varepsilon_2^{-1/2} \tag{12}$$

and

$$\deg Q_\Delta \leqslant 2k_\Delta. \tag{13}$$

Proof. Let $\Delta \in \Omega_0$ and $\Delta = [u, v]$. If $V_\Delta f = 0$, then, obviously, the rational function $Q_\Delta \equiv 0$ satisfies (11)–(13).

If $\| r_\Delta - f(u) \|_\Delta > 2 \| f \|_\Delta$, then by (4) we get $\| f_\Delta \|_\Delta \leqslant \| r_\Delta - f(u) \|_\Delta - \| f_\Delta \|_\Delta \leqslant \| f - r_\Delta \|_\Delta \leqslant \varepsilon_1$ and therefore the rational function $Q_\Delta \equiv 0$ satisfies (11)–(13).

Let $V_\Delta f > 0$ and $\| r_\Delta - f(u) \|_\Delta \leqslant 2 \| f_\Delta \|_\Delta$. Consider the rational function

$$Q_\Delta(x) = \frac{r_\Delta(x) - f(u)}{1 + \eta_\Delta (r_\Delta(x) - f(u))^2}, \quad \eta_\Delta = \tfrac{1}{8}(V_\Delta f)^{-3} \varepsilon_2.$$

It is readily seen, in view of (5), that

$$\deg Q_\Delta \leqslant 2 \deg r_\Delta \leqslant 2k_\Delta$$

and

$$\| Q_\Delta \|_{(-\infty,\infty)} \leqslant \frac{1}{2\sqrt{\eta_\Delta}} < 2(V_\Delta f)^{3/2} \varepsilon_2^{-1/2},$$

i.e. Q_Δ satisfies (12) and (13). It remains to estimate $\| f_\Delta - Q_\Delta \|_\Delta$. By our assumptions and (4) we get $\| f_\Delta - Q_\Delta \|_\Delta \leqslant \| f - r_\Delta \|_\Delta + \eta_\Delta \| r_\Delta - f(u) \|_\Delta^2 \| f_\Delta \|_\Delta \leqslant \varepsilon_1 + 4\eta_\Delta \| f_\Delta \|_\Delta^3 \leqslant \varepsilon_1 + 4\eta (V_\Delta f)^3 = \varepsilon_1 + \varepsilon_2/2$. Hence Q_Δ satisfies (11). □

We shall prove theorem 8.11 starting from lemma 8.9 and applying several times the following lemma.

Lemma 8.10. *Let $0 \leqslant \mu \leqslant s-1$. Assume that for each $\Delta \in \Omega_\mu$ there exists a rational function Q_Δ such that*

$$\| f_\Delta - Q_\Delta \|_\Delta \leqslant \varepsilon_1 + \varphi(\mu) \varepsilon_2, \tag{14}$$

where $\varphi(\mu) \geqslant \frac{1}{2}$ depends only on μ,

$$\| Q_\Delta \|_{(-\infty,\infty)} \leqslant 2(V_\Delta f)^{3/2} \varepsilon_2^{-1/2} \tag{15}$$

and

$$\deg Q_\Delta \leqslant 2 \sum_{\Delta^* \in \Omega_\Delta} k_{\Delta^*} + N(\mu, \Delta). \tag{16}$$

Then for each $\Delta \in \Omega_{\mu+1}$ there exists a rational function Q_Δ such that

$$\| f_\Delta - Q_\Delta \|_\Delta \leqslant \varepsilon_1 + \left(\varphi(\mu) + \frac{1}{2^{\mu+1}} \right) \varepsilon_2, \tag{17}$$

$$\| Q_\Delta \|_{(-\infty,\infty)} \leqslant 2(V_\Delta f)^{3/2} \varepsilon_2^{-1/2} \tag{18}$$

$$\deg Q_\Delta \leqslant 2 \sum_{\Delta^* \in \Omega_\Delta} k_{\Delta^*} + N(\mu+1, \Delta), \tag{19}$$

where the last sum is taken over all intervals $\Delta^* \in \Omega_\Delta$, $N(\mu, \Delta)$, Ω_μ and Ω_Δ are defined in (8), (9) and (10) respectively.

Proof. Let $\Delta \in \Omega_{\mu+1}$ and $\Delta = [z_1, z_3]$. If $V_\Delta f \leqslant \varepsilon_2/2$, then, since $\| f_\Delta \|_\Delta \leqslant V_\Delta f$, the rational function $Q_\Delta \equiv 0$ satisfies (17)–(19).

Let $V_\Delta f > \varepsilon_2/2$. Clearly, there exists a point $z_2 \in \Delta$ such that the intervals

$\Delta_1 = [z_1, z_2]$ and $\Delta_2 = [z_2, z_3]$ belong to Ω_μ. Next, we shall denote for brevity

$$M_1 = V_{\Delta_1} f, \quad M_2 = V_{\Delta_2} f, \quad M = V_\Delta f = M_1 + M_2,$$

$$\eta = \left(\frac{\varepsilon_2}{2^{\mu+2} \| f' \|_{L_p(\Delta)}} \right)^{p'}, \quad d_1 = [z_1, z_2 + \eta], \quad d_2 = [z_2 - \eta, z_3].$$

Note that $\| f' \|_{L_p(\Delta)} > 0$, since $V_\Delta f = \| f' \|_{L_1(\Delta)} > 0$.

By (14)–(16) there exist rational functions Q_{Δ_1}, Q_{Δ_2} such that for $i = 1, 2$

$$\| f_{\Delta_i} - Q_{\Delta_i} \|_{\Delta_i} \leqslant \varepsilon_1 + \varphi(\mu) \varepsilon_2, \tag{20}$$

$$\| Q_{\Delta_i} \|_{(-\infty, \infty)} \leqslant 2 M_i^{3/2} \varepsilon_2^{-1/2}, \tag{21}$$

$$\deg Q_{\Delta_i} \leqslant 2 \sum_{\Delta^* \in \Omega_{\Delta_i}} k_{\Delta^*} + N(\mu, \Delta_i). \tag{22}$$

We need an estimate of the modulus of continuity of the function f on Δ:

$$\omega(f, \Delta; \delta)_C = \sup \{ | f(x') - f(x'') | : x', x'' \in \Delta, | x' - x'' | \leqslant \delta \}.$$

By Hölder's inequality we get, for $x', x'' \in \Delta$, $x' \leqslant x''$,

$$| f(x') - f(x'') | \leqslant \int_{x'}^{x''} | f'(x) | \mathrm{d}x \leqslant \| f' \|_{L_p[x', x'']} | x' - x'' |^{1/p'}.$$

Consequently

$$\omega(f, \Delta; \delta)_C \leqslant \| f' \|_{L_p(\Delta)} \delta^{1/p'}, \quad \delta \geqslant 0. \tag{23}$$

Let α_i $(i = 1, 2)$ be the linear increasing function which maps d_i onto Δ_i. Clearly

$$\| \alpha_i(x) - x \|_{d_i} = \eta, \quad i = 1, 2. \tag{24}$$

Put $r_1(x) = Q_{\Delta_1}(\alpha_1(x))$ and $r_2(x) = Q_{\Delta_2}(\alpha_2(x)) - f(z_1) + f(z_2)$. By (20), (23) and (24) we get

$$\begin{aligned} \| f_\Delta - r_1 \|_{d_1 \cap \Delta} &\leqslant \| f_{\Delta_1} - f_{\Delta_1}(\alpha_1) \|_{d_1 \cap \Delta} + \| f_{\Delta_1}(\alpha_1) - Q_{\Delta_1}(\alpha_1) \|_{d_1} \leqslant \omega(f_\Delta, \Delta; \eta) \\ &+ \| f_{\Delta_1} - Q_{\Delta_1} \|_{\Delta_1} \leqslant \varepsilon_1 + \left(\varphi(\mu) + \frac{1}{2^{\mu+2}} \right) \varepsilon_2. \end{aligned}$$

Thus we have

$$\| f_\Delta - r_1 \|_{d_1 \cap \Delta} \leqslant \varepsilon_1 + \left(\varphi(\mu) + \frac{1}{2^{\mu+2}} \right) \varepsilon_2. \tag{25}$$

By (21), (22) we obtain

$$\| r_1 \|_{(-\infty, \infty)} \leqslant 2 M_1^{3/2} \varepsilon_2^{-1/2} \leqslant 2 M^{3/2} \varepsilon_2^{-1/2} \tag{26}$$

$$\deg r_1 \leqslant 2 \sum_{\Delta^* \in \Omega_{\Delta_1}} k_{\Delta^*} + N(\mu, \Delta_1). \tag{27}$$

Similarly, by (20), (23) and (24) we get

$$\| f_\Delta - r_2 \|_{d_2 \cap \Delta} \leqslant \varepsilon_1 + \left(\varphi(\mu) + \frac{1}{2^{\mu+2}} \right) \varepsilon_2. \tag{28}$$

By (21) and the fact that $M > \varepsilon_2/2$ we get

$$\begin{aligned} \| r_2 \|_{(-\infty,\infty)} &\leqslant |f(z_1) - f(z_2)| + \| Q_{\Delta_2} \|_{(-\infty,\infty)} \leqslant M_1 + 2M_2^{3/2}\varepsilon^{-1/2} \\ &\leqslant 2M_1 M^{1/2}\varepsilon_2^{-1/2} + 2M_2 M^{1/2}\varepsilon_2^{-1/2} = 2M^{3/2}\varepsilon_2^{-1/2}. \end{aligned} \tag{29}$$

By (22) we have

$$\deg r_2 \leqslant 2 \sum_{\Delta^* \in \Omega_\Delta} k^* + N(\mu, \Delta_2). \tag{30}$$

If $\Delta \subset d_1$ or $\Delta \subset d_2$, then it follows by (25)–(30) that the rational function r_1 or r_2, respectively, satisfies (17)–(19).

In the opposite case we have $|d_1 \cap d_2 \cap \Delta| > \eta$. Now we are able to apply lemma 5.3 in section 5.1 to the function f_Δ with the corresponding values of parameters Δ_1, Δ_2, ε_1, A and k_i from (25)–(30). Setting $\varepsilon_2/2^{\mu+2}$ in place of ε_2 we conclude that there exists a rational function Q_Δ such that

$$\| f_\Delta - Q_\Delta \|_\Delta \leqslant \varepsilon_1 + \left(\varphi(\mu) + \frac{1}{2^{\mu+1}} \right) \varepsilon_2, \quad \| Q_\Delta \|_{(-\infty,\infty)} \leqslant 2M^{3/2}\varepsilon_2^{-1/2},$$

$$\begin{aligned} \deg Q_\Delta \leqslant 2 \sum_{\Delta^* \in \Omega_\Delta} k_{\Delta^*} + N(\mu, \Delta_1) + N(\mu, \Delta_2) \\ + B_1 \ln\left(e + \frac{|\Delta|}{|d_1 \cap d_2 \cap \Delta|} \right) \ln\left(e + \frac{2M^{3/2}\varepsilon_2^{-1/2}}{\varepsilon_2/2^{\mu+2}} \right). \end{aligned}$$

Consequently Q_Δ satisfies (17) and (18).

It remains to prove that Q_Δ satisfies (19). Since

$$|d_1 \cap d_2 \cap \Delta| > \eta, \quad M = V_\Delta f = \| f' \|_{L_1(\Delta)} \leqslant \| f' \|_{L_p(\Delta)} |\Delta|^{1/p'} \quad \text{and} \quad M > \varepsilon_2/2$$

we have

$$\begin{aligned} & B_1 \ln\left(e + \frac{|\Delta|}{|d_1 \cap d_2 \cap \Delta|} \right) \ln\left(e + \frac{2M^{3/2}\varepsilon_2^{-1/2}}{\varepsilon_2/2^{\mu+2}} \right) \\ &= B_1 \ln(e + (2^{\mu+2} \| f' \|_{L_p(\Delta)} |\Delta|^{1/p'} \varepsilon_2^{-1})^{p'}) \ln(e + 2^{\mu+3}(M\varepsilon_2^{-1})^{3/2}) \\ &\leqslant 6B_1 p' \ln^2(e + 2^{\mu+1} \| f' \|_{L_p(\Delta)} |\Delta|^{1/p'} \varepsilon_2^{-1}). \end{aligned}$$

Hence, it suffices to prove that

$$N(\mu, \Delta_1) + N(\mu, \Delta_2) + 6B_1 p' \ln^2(e + 2^{\mu+1} \| f' \|_{L_p(\Delta)} |\Delta|^{1/p'} \varepsilon_2^{-1}) \leqslant N(\Delta, \mu + 1). \tag{31}$$

To this end we shall make use of the following inequalities:

$$\ln^2(\mathrm{e}+x_1)+\ln^2(\mathrm{e}+x_2)\leqslant 2\ln^2\left(\mathrm{e}+\frac{x_1+x_2}{2}\right),\quad x_1,x_2\geqslant 0, \tag{32}$$

$$\|f'\|_{L_p(\Delta_1)}|\Delta_1|^{1/p'}+\|f'\|_{L_p(\Delta_2)}|\Delta_2|^{1/p'}\leqslant\|f'\|_{L_p(\Delta)}|\Delta|^{1/p'}. \tag{33}$$

The inequality (32) follows from the fact that the function $F_1(x)=-\ln^2(\mathrm{e}+x)$ is convex on $[0,\infty)$ and (33) from the fact that the function $F_2(x,y)=-x^{1/p}y^{1/p'}$ is convex on the set $\{(x,y): x,y\geqslant 0\}$ $(1/p+1/p'=1)$ and also from the discrete Hölder inequality.

Using the definition of $N(\mu,\Delta)$ in (8) we get

$$\begin{aligned}
&N(\mu,\Delta_1)+N(\mu,\Delta_2)+6B_1p'\ln^2(\mathrm{e}+2^{\mu+1}\|f'\|_{L_p(\Delta)}|\Delta|^{1/p'}\varepsilon_2^{-1})\\
&\quad=6B_1p'\sum_{\nu=0}^{\mu}2^{\nu}\{\ln^2(\mathrm{e}+2^{\mu}4^{-\nu}\|f'\|_{L_p(\Delta_1)}|\Delta_1|^{1/p'}\varepsilon_2^{-1})\\
&\qquad+\ln^2(\mathrm{e}+2^{\mu}4^{-\nu}\|f'\|_{L_p(\Delta_2)}|\Delta_2|^{1/p'}\varepsilon_2^{-1/2})\}\\
&\qquad+6B_1p'\ln^2(\mathrm{e}+2^{\mu+1}\|f'\|_{L_p(\Delta)}|\Delta|^{1/p'}\varepsilon_2^{-1})\\
&\quad\leqslant 6B_1p'\sum_{\nu=0}^{\mu}2^{\nu+1}\ln^2(\mathrm{e}+\tfrac{1}{2}2^{\mu}4^{-\nu}(\|f'\|_{L_p(\Delta_1)}|\Delta_1|^{1/p'}+\|f'\|_{L_p(\Delta_2)}|\Delta_2|^{1/p'})\varepsilon_2^{-1})\\
&\qquad+6B_1p'\ln^2(\mathrm{e}+2^{\mu+1}\|f'\|_{L_p(\Delta)}|\Delta|^{1/p'}\varepsilon_2^{-1})\\
&\quad\leqslant 6B_1p'\sum_{\nu=0}^{\mu}2^{\nu+1}\ln^2(\mathrm{e}+2^{\mu+1}4^{-(\nu+1)}\|f'\|_{L_p(\Delta)}|\Delta|^{1/p'}\varepsilon_2^{-1})\\
&\qquad+6B_1p'\ln^2(\mathrm{e}+2^{\mu+1}\|f'\|_{L_p(\Delta)}|\Delta|^{1/p'}\varepsilon_2^{-1})=N(\mu+1,\Delta).
\end{aligned}$$

Thus the inequality (31) is established. □

Completion of the proof of theorem 8.11. Starting from lemma 8.9 and applying lemma 8.10 s times we obtain that there exists a rational function Q_Δ, $\Delta=[a,b]$ such that

$$\|f_\Delta-Q_\Delta\|_\Delta\leqslant\varepsilon_1+\left(\sum_{\mu=0}^{s}\frac{1}{2^{\mu+1}}\right)\varepsilon_2<\varepsilon_1+\varepsilon_2$$

and

$$\deg Q_\Delta\leqslant 2\sum_{\Delta\in\Omega_0}k_\Delta+N(s,[a,b]).$$

Putting $r=Q_\Delta+f(a)$ we have

$$\|f-r\|_{[a,b]}\leqslant\varepsilon_1+\varepsilon_2$$

and

$$\deg r\leqslant 2\sum_{\Delta\in\Omega_0}k_\Delta+N(s,[a,b]).$$

It remains to prove that

$$N(s,\Delta) \leqslant Dp'm \ln^2\left(\mathrm{e} + \frac{\| f' \|_{L_p} |\Delta|^{1/p'}}{m\varepsilon_2}\right). \tag{34}$$

By (8) it follows that

$$N(s,\Delta) \leqslant 6B_1 p' \sum_{\nu=0}^{s} 2^\nu \ln^2\left(\left(\mathrm{e} + \frac{\| f' \|_{L_p(\Delta)} |\Delta|^{1/p'}}{2^s \varepsilon_2}\right) 4^{s-\nu}\right)$$

Arguments similar to that of the proof of theorem 5.2 (see p. 121) show that the last estimate implies (34). □

Theorem 8.11 provides a new proof of the well-known theorem 5.4 in sections 5.3.2 and 5.4.2:

Theorem 5.4. *If $f \in W_p^1[a,b]$, $p > 1$, then*

$$R_n(f)_C \leqslant C \frac{\| f' \|_{L_p}}{n}, \quad n = 1, 2, \ldots, \tag{35}$$

where $C = C_1 p'(b-a)^{1/p'}$, $C_1 =$ constant.

Proof. Clearly (see theorem 7.2 in section 7.3), for each $m \geqslant 1$ there exist points $a = x_0 < x_1 < \cdots < x_m = b$ such that for $i = 0, 1, \ldots, m-1$

$$\| f - f(x_i) \|_{C[x_i, x_{i+1}]} \leqslant \frac{V_a^b f}{m} = \frac{\| f' \|_{L_1}}{m} \leqslant \frac{\| f' \|_{L_p} (b-a)^{1/p'}}{m}.$$

Then applying theorem 8.11 for the function f with $\varepsilon_1 = \varepsilon_2 = \| f' \|_{L_p}(b-a)^{1/p'}/m$ when $\| f' \|_{L_p} > 0$ (the case $\| f' \|_{L_p} = 0$ is trivial) we conclude that there exists a rational function r such that

$$\| f - r \|_{C[a,b]} \leqslant \frac{2 \| f' \|_{L_p} (b-a)^{1/p'}}{m}$$

and

$$\deg r \leqslant Dp'm \ln^2\left(\mathrm{e} + \frac{\| f' \|_{L_p} (b-a)^{1/p'}}{m\varepsilon_2}\right) \leqslant Cp'm.$$

These estimates imply (35) immediately. □

Theorem 8.12. *Let f be defined on $[a,b]$, $f' \in L_p[a,b]$, $1 < p \leqslant \infty$, and f', f'', $\ldots$, $f^{(k)}$ $(k \geqslant 1)$ be monotone but possible unbounded on (a,b). Then*

$$R_n(f)_C \leqslant C \frac{\| f' \|_{L_p} (b-a)^{1/p'}}{n^{k+1}}, \quad n = 1, 2, \ldots, \tag{36}$$

where $C = C_1(k)(p')^{4(k+1)}$.

Proof. Suppose $[a,b]=[0,1]$, $\|f'\|_{L_p}=1$, $1<p<\infty$, and $f^{(k)}\in C(0,1)$. Clearly, if theorem 8.12 holds in this case then it holds also in the general case.

Denote $x_i=1/2^i$ and $y_i=1-1/2^i$, $i=1,2,\ldots$. First we estimate $|f^{(k)}(x_i)|$ and $|f^{(k)}(y_i)|$ for $i=1,2,\ldots$.

Since f' is monotone in $(0,1)$ and $0<x_i\leqslant\frac{1}{2}$, $1=\int_0^1|f'(x)|^p\,dx\geqslant|f'(x_i)|^p x_i$ and therefore $|f'(x_i)|\leqslant 2^{i/p}$. Similarly $|f'(y_i)|\leqslant 2^{i/q}$, $i=1,2,\ldots$. Since f'' is monotone, f' is either convex or concave and therefore, for $i=2,3,\ldots$,

$$|f''(x_i)|\leqslant\max\left\{\left|\frac{f'(x_i)-f'(x_{i+1})}{x_i-x_{i+1}}\right|,\left|\frac{f'(x_i)-f'(x_{i-1})}{x_i-x_{i-1}}\right|\right\}$$

and

$$|f''(x_1)|\leqslant\max\left\{\left|\frac{f'(x_1)-f'(x_2)}{x_1-x_2}\right|,\left|\frac{f'(y_1)-f'(y_2)}{y_1-y_2}\right|\right\}.$$

Hence

$$|f''(x_i)|\leqslant 2^3\cdot 2^{(1+1/p)i},\ |f''(y_i)|\leqslant 2^3\cdot 2^{(1+1/p)i},\quad i=1,2,\ldots.$$

Just in the same manner we estimate $|f^{(v)}(x_i)|$ and $|f^{(v)}(y_i)|$, $i=1,2,\ldots$. Finally we get

$$\left.\begin{aligned}|f^{(k)}(x_i)|&\leqslant 2^{(k+1)^2}2^{i(k-1+1/p)}=2^{(k+1)^2}2^{i(k-1/p')}\\ |f^{(k)}(y_i)|&\leqslant 2^{(k+1)^2}2^{i(k-1/p')},\quad i=1,2,\ldots.\end{aligned}\right\}\tag{37}$$

Denote $A=e^{18}D^2(k+1)^6(p')^4$, where $D>1$ is the constant from theorem 8.11.

Clearly

$$R_n(f)_C\leqslant V_0^1 f=\|f'\|_{L_1}\leqslant\|f'\|_{L_p}\leqslant\frac{A^{k+1}}{n^{k+1}},\quad 1\leqslant n\leqslant A.\tag{38}$$

Let $n>A$. Choose s to be an integer and $s\geqslant 2$ such that

$$2^s\leqslant\left(\frac{n\ln 2}{16k(k+1)p'}\right)^{(k+1)p'}\leqslant 2^{s+1}.\tag{39}$$

Denote $\Delta_i=[x_{i+1},x_i]$ and $\Delta_i^*=[y_i,y_{i+1}]$ for $i=1,2,\ldots,s$, $\Delta_{s+1}=[0,x_{s+1}]$ and $\Delta_{s+1}^*=[y_{s+1},1]$. Put

$$n_i=\left[\frac{n\ln 2}{16(k+1)p'\cdot 2^{i/(k+1)p'}}\right],\quad n_i^*=n_i,\quad i=1,2,\ldots,s,$$

where $[x]$ denotes the integer part of x.

From the choice of n and s it follows that $n_i=n_i^*\geqslant k$, $i=1,2,\ldots,s$. Then by theorem 5.1 in section 5.2 and (37) we get for $i = 1, 2, \ldots, s$

$$R_{n_i}(f)_{C(\Delta_i)} \leqslant C(k)\frac{V_{\Delta_i} f^{(k)} |\Delta_i|^k}{n_i^{k+1}} \leqslant C(k)\frac{(|f^{(k)}(x_i)| + |f^{(k)}(x_{i+1})|)|\Delta_i|^k}{n_i^{k+1}}$$

$$\leqslant C(k)\frac{2^{(k+1)^2+k+1}2^{i(k-1/p')}2^{-(i+1)k}}{\left(\dfrac{n \ln 2}{32(k+1)p' \cdot 2^{i/(k+1)p'}}\right)^{k+1}} \leqslant \frac{C_1(k)(p')^{k+1}}{n^{k+1}}.$$

Hence, for each $i = 1, 2, \ldots, s$ there exists a rational function $r_i \in R_{n_i}$ such that

$$\| f - r_i \|_{C(\Delta_i)} \leqslant \frac{C_1(k)(p')^{k+1}}{n^{k+1}} \tag{40}$$

Put $r_{s+1} = f(x_{s+1})$. Clearly, by (39) we get

$$\| f - r_{s+1} \|_{C(\Delta_{s+1})} \leqslant V_{\Delta_{s+1}} f = \| f' \|_{L_1(\Delta_{s+1})} \leqslant \| f' \|_{L_p(\Delta_{s+1})} |\Delta_{s+1}|^{1/p'}$$

$$\leqslant \left(\frac{1}{2^{s+1}}\right)^{1/p'} \leqslant \left(\frac{16k(k+1)p'}{n \ln 2}\right)^{k+1}.$$

Hence

$$\| f - r_{s+1} \|_{C(\Delta_{s+1})} \leqslant \frac{C_2(k)(p')^{k+1}}{n^{k+1}}, \quad r_{s+1} \in R_0. \tag{41}$$

Similarly, for each $i = 1, 2, \ldots, s$ there exists a rational function $r_i^* \in R_{n_i^*}$ such that

$$\| f - r_i^* \|_{C(\Delta_i^*)} \leqslant \frac{C_1(k)(p')^{k+1}}{n^{k+1}}. \tag{42}$$

Put $r_{s+1}^* \equiv f(y_{s+1})$. As above we have

$$\| f - r_{s+1}^* \|_{C(\Delta_{s+1}^*)} \leqslant \frac{C_2(k)(p')^{k+1}}{n^{k+1}}, \quad r_{s+1}^* \in R_0. \tag{43}$$

In addition by the choice of n_i and n_i^*

$$\sum_{i=1}^{s} n_i + \sum_{i=1}^{s} n_i^* \leqslant 2 \sum_{i=1}^{s} \frac{n \ln 2}{8(k+1)p' \cdot 2^{i/(k+1)p'}} \leqslant \frac{n \ln 2}{4(k+1)p'} \sum_{i=1}^{\infty} \left(\frac{1}{2^{1/(k+1)p'}}\right)^i$$

$$= \frac{n \ln 2}{4(k+1)p'(e^{\ln 2/(k+1)p'} - 1)} \leqslant \frac{n}{4}.$$

Thus we have

$$\sum_{i=1}^{s} n_i + \sum_{i=1}^{s} n_i^* \leqslant \frac{n}{4}. \tag{44}$$

Now, in view of (40)–(43) we are able to apply theorem 8.11. Setting

$\varepsilon_2 = 1/n^{k+1}$ we conclude that there exists a rational function r such that

$$\| f - r \|_{C[0,1]} \leqslant \frac{C(k)(p')^{k+1}}{n^{k+1}}$$

and

$$\deg r \leqslant 2 \sum_{i=1}^{s} n_i + 2 \sum_{i=1}^{s} n_i^* + Dp' \cdot 2(s+1) \ln^2 \left(\mathrm{e} + \frac{\| f' \|_{L_p}}{2(s+1)n^{-k-1}} \right).$$

By (39), (44) and because of $\| f' \|_{L_p} = 1$, $n > A = \mathrm{e}^{18} D^2 (k+1)^6 (p')^4$ and $\ln^3 x \leqslant \sqrt{x}$ for $x > \mathrm{e}^{18}$ we get

$$\deg r \leqslant \frac{n}{2} + 4Dp's \ln^2 (n^{k+1}) \leqslant \frac{n}{2} + \frac{4D(k+1)^3 (p')^2 \ln^3 n}{\ln 2}$$

$$\leqslant \frac{n}{2} + 8D(k+1)^3 (p')^2 \sqrt{n} \leqslant n.$$

Consequently

$$R_n(f)_C \leqslant \frac{C(k)(p')^{k+1}}{n^{k+1}}, \quad n > A. \tag{45}$$

The estimates (38) and (45) prove theorem 8.12. □

Theorem 8.13. *Let $f \in W_p^1[a,b]$, $1 < p \leqslant \infty$, and assume that there exists a partition of $[a,b]$ into m ($m \geqslant 1$) subintervals $a = x_0 < x_1 < \cdots < x_m = b$ such that $f', f'', \ldots, f^{(k)}$ ($k \geqslant 1$) are monotone but possibly unbounded in each interval (x_{i-1}, x_i). Then*

$$R_n(f)_C \leqslant C \frac{\| f' \|_{L_p[a,b]} (b-a)^{1/p'} m^k}{n^{k+1}}, \quad n = 1, 2, \ldots, \tag{46}$$

where $C = C_1(k)(p')^{4(k+1)}$.

Proof. Denote $A = \mathrm{e}^{10} D^2 (k+1)^4 (p')^2 m$, where $D > 1$ is the constant from theorem 8.11. If $1 \leqslant n \leqslant A$, then by theorem 5.4 in subsection 5.3.2 we have

$$R_n(f)_C \leqslant C \frac{\| f' \|_{L_p} (b-a)^{1/p'}}{n} \leqslant C A^k \frac{\| f' \|_{L_p} (b-a)^{1/p'}}{n^{k+1}}.$$

Hence

$$R_n(f)_C \leqslant C_1 \frac{\| f' \|_{L_p} (b-a)^{1/p'} m^k}{n^{k+1}}, \quad 1 \leqslant n \leqslant A. \tag{47}$$

Now consider the case $n > A$. Suppose $\| f' \|_{L_p} > 0$ (the case $\| f' \|_{L_p} = 0$ is trivial) and put

$$n_i = \left[\frac{n}{8} \left(\frac{\| f' \|_{L_p(\Delta_i)} |\Delta_i|^{1/p'}}{\| f' \|_{L_p[a,b]} (b-a)^{1/p'} m^k} \right)^{1/(k+1)} + 1 \right], \quad \Delta_i = [x_{i-1}, x_i].$$

Then by theorem 8.12 for f in the interval Δ_i we obtain

$$R_{n_i}(f,\Delta_i)_C \leqslant C\frac{\|f'\|_{L_p(\Delta_i)}|\Delta_i|^{1/p'}}{n_i^{k+1}} \leqslant C_1\frac{\|f'\|_{L_p[a,b]}(b-a)^{1/p'}m^k}{n^{k+1}}, \quad C_1 = C_2(k)(p')^{k+1}.$$

Hence, for each $i = 1, 2, \ldots, n$ there exists a rational function $r_i \in R_{n_i}$ such that

$$\|f - r_i\|_{C(\Delta_i)} \leqslant C_1\frac{\|f'\|_{L_p}(b-a)^{1/p'}m^k}{n^{k+1}}.$$

From this estimate using theorem 8.11 with

$$\varepsilon_1 = C_1\frac{\|f'\|_{L_p}(b-a)^{1/p'}m^k}{n^{k+1}}, \quad \varepsilon_2 = \frac{\|f'\|_{L_p}(b-a)^{1/p'}m^k}{n^{k+1}}$$

and $k_i = n_i$ we conclude that there exists a rational function r such that

$$\|f - r\|_{C[a,b]} \leqslant (C_1 + 1)\frac{\|f'\|_{L_p}(b-a)^{1/p'}m^k}{n^{k+1}} \tag{48}$$

and

$$\deg r \leqslant 2\sum_{i=1}^{m} n_i + Dp'm\ln^2\left(\mathrm{e} + \left(\frac{n}{m}\right)^{k+1}\right).$$

It remains to estimate $\deg r$. Using twice discrete variants of Hölder's inequality (see (33)) we get

$$\begin{aligned}
2\sum_{i=1}^{m} n_i &\leqslant 2\sum_{i=1}^{m}\left\{\frac{n}{8}\left(\frac{\|f'\|_{L_p(\Delta_i)}|\Delta_i|^{1/p'}}{\|f'\|_{L_p}(b-a)^{1/p'}m^k}\right)^{1/(k+1)} + 1\right\} \\
&\leqslant 2m + \frac{n}{4(\|f'\|_{L_p}(b-a)^{1/p'})^{1/(k+1)}m^{k/(k+1)}} \\
&\quad \times \sum_{i=1}^{m}(\|f'\|_{L_p(\Delta_i)}|\Delta_i|^{1/p'})^{1/(k+1)}1^{k/(k+1)} \\
&\leqslant 2m + \frac{n}{4(\|f'\|_{L_p}(b-a)^{1/p'})^{1/(k+1)}m^{k/(k+1)}} \\
&\quad \times \left(\sum_{i=1}^{m}\|f'\|_{L_p(\Delta_i)}|\Delta_i|^{1/p'}\right)^{1/(k+1)} m^{k/(k+1)} \\
&\leqslant 2m + \frac{n}{4} \leqslant \frac{n}{2}.
\end{aligned}$$

On the other hand, since $\ln^2 x \leqslant \sqrt{x}$ for $x > \mathrm{e}^{10}$ and $n > A$ we have

$$Dp'm\ln^2\left(\mathrm{e} + \left(\frac{n}{m}\right)^{k+1}\right) \leqslant 4D(k+1)^2p'm\ln^2\left(\frac{n}{m}\right) \leqslant 4D(k+1)^2p'\sqrt{(mn)} \leqslant \frac{n}{2}.$$

Thus $r \in R_n$ and by (48) we obtain

$$R_n(f)_C \leqslant C \frac{\| f' \|_{L_p} (b-a)^{1/p'} m^k}{n^{k+1}}, \quad n > A.$$

This estimate and (47) imply (46). □

Proof of theorem 8.9. First we consider the case $r = 1$, $p > 1$, $q = \infty$. We shall prove only the estimate (2). The estimate (1) can be proved similarly. We shall proceed similarly as in the proof of theorem 8.1.

Choose $\varphi_{2^\nu} \in S(m, 2^\nu, [a,b])$ such that

$$\| f' - \varphi_{2^\nu} \|_{L_p} = S^m_{2^\nu}(f')_p. \tag{49}$$

Clearly

$$\| \varphi_{2^\nu} - \varphi_{2^{\nu-1}} \|_{L_p} \leqslant 2 S^m_{2^{\nu-1}}(f')_p. \tag{50}$$

Set $\phi_{2^\nu}(x) = \int \varphi_{2^\nu}(x) \mathrm{d}x$ and $k = [\alpha] + 2$.

Let s be an arbitrary non-negative integer. For each ν $(1 \leqslant \nu \leqslant s)$ choose N_ν to be a positive integer and

$$\tfrac{1}{2} N_\nu \leqslant 2^{(s(\alpha+1) + \nu(k-\alpha))/(k+1)} \leqslant N_\nu. \tag{51}$$

Put $N_0 = m + 1$, $N_{s+1} = 2^s$, $N = \sum_{\nu=0}^{s+1} N_\nu$. Clearly, we have

$$\begin{aligned} R_N(f)_C \leqslant R_{N_{s+1}}(f - \phi_{2^s})_C + R_{N_s}(\phi_{2^s} - \phi_{2^{s-1}})_C + R_{N_{s-1}}(\phi_{2^{s-1}} - \phi_{2^{s-2}})_C \\ + \cdots + R_{N_1}(\phi_{2^1} - \phi_{2^0})_C + R_{N_0}(\phi_{2^0})_C. \end{aligned} \tag{52}$$

Consider the function $\psi_{2^\nu} = \phi_{2^\nu} - \phi_{2^{\nu-1}}$. Obviously $\psi'_{2^\nu} = \varphi_{2^\nu} - \varphi_{2^{\nu-1}} \in S(m, 3\cdot 2^{\nu-1}, [a,b])$ and therefore there exist at most $m_\nu + 1 \leqslant 3\cdot 2^{\nu-1}(1 + 1 + 2 + \cdots + m) + 1 < 3\cdot 2^{\nu-1}(m+1)^2 + 1$ points $a = x_0^{(\nu)} < x_1^{(\nu)} < \cdots < x_{m_\nu}^{(\nu)} = b$ such that the functions $\psi'_{2^\nu}, \psi''_{2^\nu}, \ldots, \psi^{(k)}_{2^\nu}$ are monotone in each interval $(x_{i-1}^{(\nu)}, x_i^{(\nu)})$. Then by theorem 8.13

$$R_{N_\nu}(\phi_{2^\nu} - \phi_{2^{\nu-1}})_C \leqslant C \frac{\| \varphi_{2^\nu} - \varphi_{2^{\nu-1}} \|_{L_p} (b-a)^{1/p'} m_\nu^k}{N_\nu^{k+1}},$$

where $C = C_1(k)(p')^{4(k+1)}$. From this estimate and (50) we obtain

$$R_{N_\nu}(\phi_{2^\nu} - \phi_{2^{\nu-1}})_C \leqslant C_2 (b-a)^{1/p'} \frac{2^{(\nu-1)\alpha} S^m_{2^{\nu-1}}(f')_p}{2^{s(\alpha+1)}}. \tag{53}$$

Also, by theorem 5.4 in subsection 5.3.2 and (49) we get

$$R_{N_{s+1}}(f - \phi_{2^s})_C \leqslant C p' (b-a)^{1/p'} \frac{\| f' - \varphi_{2^s} \|_{L_p}}{2^s} \leqslant C p' (b-a)^{1/p'} \frac{2^{s\alpha} S^m_{2^s}(f')_p}{2^{s(\alpha+1)}}. \tag{54}$$

Obviously

$$R_{N_0}(\phi_0)_C = 0. \tag{55}$$

Combining (52) with (53)–(55) and (51) we get

$$R_N(f)_C \leqslant C_3(b-a)^{1/p'}2^{-s(\alpha+1)}\sum_{\nu=0}^{s} 2^{\nu\alpha}S_{2^\nu}^m(f')_p,$$

where

$$N=\sum_{\nu=0}^{s+1} N_\nu \leqslant m+1+\sum_{\nu=1}^{s} 2\cdot 2^{(s(\alpha+1)+\nu(k-\alpha))/(k+1)}+2^s$$

$$\leqslant \frac{9(k+1)(m+1)}{k-\alpha}2^s \leqslant 9(\alpha+3)(m+1)2^s.$$

Consequently, for each $s \geqslant 0$

$$R_N(f)_C \leqslant C(b-a)^{1/p'}\cdot 2^{-s(\alpha+1)}\sum_{\nu=0}^{s} 2^{\nu\alpha}S_{2^\nu}^m(f')_p \tag{56}$$

for some $N \leqslant 9(\alpha+3)(m+1)2^s$.

If $m \leqslant n \leqslant A = 9(\alpha+3)(m+1)$, then

$$\begin{aligned} R_n(f)_C &\leqslant R_0(f-\phi_{2^0})_C + R_m(\phi_1)_C = R_0(f-\phi_1)_C \leqslant V_a^b(f-\phi_1) \\ &= \| f' - \phi_1' \|_{L_1} \leqslant (b-a)^{1/p'} \| f' - \phi_1' \|_{L_p} = (b-a)^{1/p'} \| f' - \varphi_1 \|_{L_p} \\ &= (b-a)^{1/p'} S_1^m(f')_p \\ &\leqslant (9(\alpha+3)(m+1))^{\alpha+1}(b-a)^{1/p'}n^{-(\alpha+1)}\sum_{\nu=1}^{n} \nu^{\alpha-1}S_\nu^m(f')_p. \end{aligned}$$

Thus we have

$$R_n(f)_C \leqslant C(b-a)^{1/p'}n^{-(\alpha+1)}\sum_{\nu=1}^{n} \nu^{\alpha-1}S_\nu^m(f')_p, \quad m \leqslant n \leqslant A. \tag{57}$$

Let $n > A$. Choose s to be integer and $A\cdot 2^s \leqslant n < A\cdot 2^{s+1}$. By (56) we obtain

$$\begin{aligned} R_n(f)_C \leqslant R_N(f)_C &\leqslant C_1(b-a)^{1/p'}\cdot 2^{-s(\alpha+1)}\left\{ S_1^m(f')_p + \sum_{\nu=1}^{s} 2^{(\nu-1)\alpha}S_{2^\nu}^m(f')_p \right\} \\ &\leqslant C_2(b-a)^{1/p'}n^{-(\alpha+1)}\left\{ S_1^m(f')_p + \sum_{\nu=1}^{s}\sum_{\mu=2^{\nu-1}+1}^{2^\nu} \mu^{\alpha-1}S_\mu^m(f')_p \right\}. \end{aligned}$$

Hence

$$R_n(f)_C \leqslant C(b-a)^{1/p'}n^{-(\alpha+1)}\sum_{\nu=1}^{n} \nu^{\alpha-1}S_\nu^m(f')_p, \quad n > A.$$

This estimate and (57) imply (2) in the case $r=1$, $p>1$, $q=\infty$.

Now we prove theorem 8.9 in the case $r=1$, $1 \leqslant q < \infty$, $p=1$. We shall prove only the estimate (1). The proof of estimate (2) is similar.

If $1 \leqslant n \leqslant [\alpha]+1$, then

$$\begin{aligned} R_n(f)_q &\leqslant (b-a)^{1/q}R_1(f)_C \leqslant (b-a)^{1/q}R_0(f')_1 \\ &\leqslant (\alpha+1)^{\alpha+1}(b-a)^{1/q}n^{-\alpha-1}\sum_{\nu=0}^{n} (\nu+1)^{\alpha-1}R_\nu(f')_1. \end{aligned}$$

Thus we have

$$R_n(f)_q \leqslant C(\alpha)(b-a)^{1/q} n^{-\alpha-1} \sum_{\nu=0}^{n} (\nu+1)^{\alpha-1} R_\nu(f')_1, \quad 1 \leqslant n \leqslant [\alpha]+1. \tag{58}$$

Let $n > [\alpha]+1$. By theorem 8.1 with α replaced by $\alpha+2$ and $k=[\alpha]+2$ we have for $n \geqslant k-1 = [\alpha]+1$

$$R_n(f)_q \leqslant C n^{-\alpha-2} \sum_{\nu=1}^{n} \nu^{\alpha+1} S_\nu^k(f)_q, \quad C = C(q,\alpha). \tag{59}$$

On the other hand by theorem 8.10 with $k=[\alpha]+2>\alpha+1$ we have for $n=1,2,\ldots$

$$S_n^k(f)_q \leqslant C n^{-\alpha-1} \sum_{\mu=0}^{n} (\mu+1)^{\alpha-1} R_\mu(f')_1 \tag{60}$$

where $C = C_1(q,\alpha)(b-a)^{1/q}$.

Combining (59) and (60) we obtain

$$R_n(f)_q \leqslant C_2 n^{-\alpha-2} \sum_{\nu=1}^{n} \sum_{\mu=0}^{\nu} (\mu+1)^{\alpha-1} R_\mu(f')_1$$

$$= C_2 n^{-\alpha-2} \left\{ \sum_{\mu=1}^{n} (\mu+1)^{\alpha-1} \left(\sum_{\nu=\mu}^{n} 1 \right) R_\mu(f')_1 + \left(\sum_{\nu=1}^{n} 1 \right) R_0(f')_1 \right\}.$$

Hence

$$R_n(f)_q \leqslant C n^{-\alpha-1} \sum_{\nu=0}^{n} (\nu+1)^{\alpha-1} R_\nu(f')_1, \quad n > [\alpha]+1, \quad C = C_3(q,\alpha)(b-a)^{1/q}. \tag{61}$$

The estimates (58) and (61) imply estimate (1) in the case $r=1$, $1 \leqslant q < \infty$, $p=1$.

Theorem 8.9 in the case $r \geqslant 2$, $q=\infty$, $p=1$ can be proved in the same manner as above using inequality (2) in the case $r=1$, $q=\infty$, $p>1$ and the estimate (3) in theorem 8.10 in an appropriate situation. The details are omitted. □

In order to prove theorem 8.10 we need the following.

Theorem 8.14. *Let f be absolutely continuous on $[a,b]$ and assume that there exists a partition of $[a,b]$ into m ($m \geqslant 1$) subintervals $a = x_0 < x_1 < \cdots < x_m = b$ such that $f, f', \ldots, f^{(k-1)}$ ($k \geqslant 1$) are monotone but possibly unbounded in each interval (x_{i-1}, x_i). Let $1 \leqslant q < \infty$. Then for $n=1,2,\ldots$*

$$S_n^k(f)_q \leqslant C \frac{m^{k-1} \| f' \|_{L_1[a,b]}}{n^k}, \tag{62}$$

where $C = C(q,k)(b-a)^{1/q}$.

Proof of theorem 8.14 in the case $m=1$. Without loss of generality we shall suppose that $[a,b]=[0,1]$, $\|f'\|_{L_1}=1$ and $f^{(k-1)}\in C_{(0,1)}$, $k\geqslant 2$.

Denote $A=16^2k^2q/(\ln 2)^2$.

If $1\leqslant n\leqslant A$, then obviously

$$S_n^k(f)_q\leqslant S_n^k(f)_C\leqslant\frac{V_0^1 f}{n}=\frac{\|f'\|_{L_1}}{n}\leqslant\frac{A^{k-1}}{n^k}.$$

Thus we have

$$S_n^k(f)_q\leqslant\frac{C(q,k)}{n^k},\quad 1\leqslant n\leqslant A. \tag{63}$$

Let $n>A$. Denote $x_i=1/2^i$, $y_i=1-1/2^i$, $i=1,2,\ldots$. Exactly as in the proof of theorem 8.12 we get

$$\left.\begin{aligned}&|f^{(k-1)}(x_i)|\leqslant C_1(k)\cdot 2^{i(k-1)},\\&|f^{(k-1)}(y_i)|\leqslant C_1(k)\cdot 2^{i(k-1)},\quad i=1,2,\ldots\end{aligned}\right\} \tag{64}$$

Choose $s\geqslant 0$ to be an integer such that

$$\frac{1}{2^{s+1}}\leqslant\left(\frac{q}{n}\right)^{kq}\leqslant\frac{1}{2^s}. \tag{65}$$

Denote $\Delta_i=[x_{i+1},x_i]$, $\Delta_i^*=[y_i,y_{i+1}]$, $i=1,2,\ldots,s$, $\Delta_{s+1}=[0,1/2^{s+1}]$, $\Delta_{s+1}^*=[1-1/2^{s+1},1]$. Put $n_i=[n\ln 2/(16kq\cdot 2^{i/2kq})]$ and $n_i^*=n_i$ for $i=1,2,\ldots,s$. From the condition $n>A$ and (65) it follows that $n_i=n_i^*\geqslant 1$, $i=1,2,\ldots,s$.

Using theorem 7.2 in subsection 7.3.1, (64) and the choice of n_i we obtain

$$\begin{aligned}S_{n_i}^k(f,\Delta_i)_q&\leqslant|\Delta_i|^{1/q}S_{n_i}^k(f,\Delta_i)_C\leqslant C(k)\frac{(V_{\Delta_i}f^{(k-1)})|\Delta_i|^{k-1+1/q}}{n_i^k}\\&\leqslant C(k)\frac{(|f^{(k-1)}(x_i)|+|f^{(k-1)}(x_{i+1})|)|\Delta_i|^{k-1+1/q}}{n_i^k}\\&\leqslant C_1(q,k)\frac{2^{i(k-1)}2^{-i(k-1+1/q)}}{(n/2^{i/2kq})^k}.\end{aligned}$$

Hence

$$S_{n_i}^k(f,\Delta_i)_q\leqslant\frac{C(q,k)}{n^k\cdot 2^{i/2q}},\quad i=1,2,\ldots,s. \tag{66}$$

Similarly

$$S_{n_i^*}^k(f,\Delta_i)_q\leqslant\frac{C(q,k)}{n^k\cdot 2^{i/2q}},\quad i=1,2,\ldots,s. \tag{67}$$

Obviously in view of (65)

$$S_1^k(f,\Delta_{s+1})_q\leqslant|\Delta_{s+1}|^{1/q}\|f'\|_{L_1}\leqslant|\Delta_{s+1}|^{1/q}=\left(\frac{1}{2^{s+1}}\right)^{1/q}\leqslant\frac{q^k}{n^k}. \tag{68}$$

Similarly

$$S_1^k(f, \Delta_{s+1}^*)_q \leqslant \frac{q^k}{n^k}. \tag{69}$$

In addition, by the choice of n_i and n_i^* we have

$$\sum_{i=1}^{s} n_i + \sum_{i=1}^{s} n_i^* + 2 \leqslant 2 \sum_{i=1}^{s} \frac{n \ln 2}{16kq \cdot 2^{i/2kq}} + 2 \leqslant 2 + \frac{2n \ln 2}{16kq(1 - 2^{-1/2kq})} \leqslant n.$$

Hence

$$\sum_{i=1}^{s} n_i + \sum_{i=1}^{s} n_i^* + 2 \leqslant n. \tag{70}$$

Finally, using (66)–(70) we obtain

$$S_n^k(f, [0, 1])_q \leqslant \left\{ S_1^k(f, \Delta_{s+1})_q^q + \sum_{i=1}^{s} S_{n_i}^k(f, \Delta_i)_q^q + \sum_{i=1}^{s} S_{n_i^*}^k(f, \Delta_i^*)_q^q + S_1^k(f, \Delta_{s+1}^*)_q^q \right\}^{1/q}$$

$$\leqslant C_1(q, k) \left\{ \frac{1}{n^{kq}} + \sum_{i=1}^{s} \left(\frac{1}{n^k \cdot 2^{i/2q}} \right)^q \right\}^{1/q}.$$

Thus we have

$$S_n^k(f)_q \leqslant \frac{C(q, k)}{n^k}, \quad n > A. \tag{71}$$

The estimates (63) and (71) imply (62) in the case $m = 1$.

Proof of theorem 8.14 in the general case. If $1 \leqslant n \leqslant 2m$, then

$$S_n^k(f)_q \leqslant (b - a)^{1/q} S_n^k(f)_C \leqslant \frac{(b - a)^{1/q} \| f' \|_{L_1}}{n} \leqslant 2^{k-1}(b - a)^{1/q} \frac{m^{k-1} \| f' \|_{L_1}}{n^k}$$

as required.

Let $n > 2m$. Set

$$n_i = \left[\frac{n}{2} \left(\frac{\| f' \|_{L_1(\Delta_i)}}{m^{k-1} \| f' \|_{L_1[a,b]}} \right)^{1/k} + 1 \right], \quad \Delta_i = [x_{i-1}, x_i],$$

when $\| f' \|_{L_1} > 0$ (the case $\| f' \|_{L_1} = 0$ is trivial). Then by theorem 8.14 with $m = 1$ we have

$$S_{n_i}^k(f, \Delta_i)_q \leqslant C_1 \frac{|\Delta_i|^{1/q} \| f' \|_{L_1(\Delta_i)}}{n_i^k} \leqslant C_2(q, k) \frac{\| f' \|_{L_1[a,b]} |\Delta_i|^{1/q} m^{k-1}}{n^k}. \tag{72}$$

Using a discrete variant of the Hölder's inequality we obtain

$$\sum_{i=1}^{m} n_i \leqslant m + \frac{n}{2 \| f' \|_{L_1} m^{(k-1)/k}} \cdot \sum_{i=1}^{m} \| f' \|_{L_1(\Delta_i)}^{1/k} \cdot 1^{(k-1)/k} \leqslant n. \tag{73}$$

Finally, by (72) and (73) we get

$$S_n^k(f)_q \leqslant \left(\sum_{i=1}^m S_{n_i}^k(f, \Delta_i)_q^q \right)^{1/q} \leqslant C(q,k) \left\{ \sum_{i=1}^m \left(\frac{\| f' \|_{L_1} m^{k-1} |\Delta_i|^{1/q}}{n^k} \right)^q \right\}^{1/q}$$
$$\leqslant C_1(q,k)(b-a)^{1/q} \frac{m^{k-1} \| f' \|_{L_1}}{n^k}.$$

Theorem 8.14 is proved. □

Proof of theorem 8.10. One can prove theorem 8.10 in the case $r = 1$, $1 \leqslant q < \infty$, $p = 1$ as a consequence of theorem 8.14 just as theorem 8.1 was proved in section 8.1 using theorem 8.2 and also as theorem 8.9 was proved by means of theorem 8.13. The details are omitted.

Theorem 8.10 in the case $r = 1$, $q = \infty$, $p > 1$ can be proved using theorem 8.5 in section 8.2 and lemma 7.13 in subsection 7.3.1 just as theorem 8.9 was proved in the case $r = 1$, $1 \leqslant q < \infty$, $p = 1$. In the same manner theorem 8.10 in the case $r \geqslant 2$, $q = \infty$, $p = 1$ can be proved using theorem 8.10, $r = 1$, $q < \infty$, $p = 1$, and lemma 7.13 in subsection 7.3.1. The details are omitted. □

Remark. Corollaries similar to corollaries 8.1–8.3 follow from theorems 8.9 and 8.10. For some other consequences of theorem 8.9 see Chapter 10.

8.4 Notes

Theorem 8.2, and as a consequence theorem 8.1, is based on lemma 8.3, which gives a 'good' rational approximation of the function

$$\varphi(x) = \begin{cases} 0, & |x| > d, \\ 1, & |x| \leqslant d. \end{cases}$$

In our opinion this lemma is the L_p analog of Newman's theorem 4.1 for uniform rational approximation of $|x|$:

$$R_n(|x|)_{C[-1,1]} = \mathrm{O}(\mathrm{e}^{-c\sqrt{n}}). \tag{1}$$

Lemma 8.3 can be used successfully also for rational uniform approximation on the whole real line of functions with finite support. It is easy to see that lemma 8.3 implies for example the following estimate:

$$\sup_{\alpha > 0} R_n(\psi_\alpha)_{C(-\infty,\infty)} = \mathrm{O}(\mathrm{e}^{-c\sqrt{n}}), \quad c > 0, \tag{2}$$

where

$$\psi_\alpha(x) = \begin{cases} 1 - |x|/\alpha, & |x| \leqslant \alpha, \\ 0, & |x| > \alpha. \end{cases} \tag{3}$$

The estimate (2) is a generalization of Newman's estimate (1).

The functions ψ_α, given by (3), are a typical example of integrals of 'atoms'. Let us remember that a function φ, defined on $(-\infty, \infty)$, is called an 'atom' if its support is a finite interval Δ and

$$\int_\Delta \varphi(x)\,\mathrm{d}x = 0, \quad |\varphi(x)| \leqslant 1/|\Delta| \text{ for every } x \in (-\infty, \infty).$$

The most essential fact connected with the 'atoms' is the following description of the Hardy space $H_1(-\infty, \infty)$:

$f \in H_1(-\infty, \infty)$ if and only if $f = \sum_{i=1}^{\infty} \lambda_i \varphi_i$, where $\sum_{i=1}^{\infty} |\lambda_i| < \infty$ and φ_i, $i = 1, 2, \ldots,$ are 'atoms' (see Coifman (1974), Latter (1978), Kashin and Saakjan (1984)).

Let us note that from here follows the famous Fefferman theorem (Fefferman, 1971) that H_1 and BMO are dual spaces.

E. Moskona proved that if $f(x) = \int_{-\infty}^{x} \varphi(t)\mathrm{d}t$, where φ is an 'atom', then

$$R_n(f)_{C(-\infty,\infty)} = \mathrm{o}(n^{-1}).$$

Our conjecture, which in our opinion is very important for rational approximations, is that if $f' \in H_1$, then

$$R_n(f)_{C(-\infty,\infty)} \leqslant C \frac{\|f'\|_{H_1}}{n}, \quad C = \text{constant}.$$

If this conjecture is true, then together with the Bernstein type inequality of Russak from section 6.2, theorem 6.3, we can obtain a complete characterization of the best uniform rational approximations of order $\mathrm{O}(n^{-\alpha})$, $0 < \alpha < 1$.

The theorems 8.1, 8.2, 8.3, 8.9 and 8.10 are due to P. Petrushev (1984a, b, 1987), see also P. Petrushev (1981, 1983a). Theorem 8.5 is proved by A. Pekarskii (1986).

The first who remarked that the spaces L_p, $0 < p < 1$, are important for rational approximation, was Yu.A. Brudnyi (1979); see also Brudnyi (1980). Let us mention that at the conference on approximation theory, Kiev, 1983, Brudnyi announced without proofs some connections between best rational approximations and Besov spaces.

In Brudnyi (1979, 1980) the following estimates are given without proof:

For every $n > \lambda$ we have

$$R_n(f)_{L_p(0,1)} \leqslant \frac{c}{n^\lambda} \left\{ \sum_{j=n}^{\infty} \frac{1}{j} \left(j^\lambda \omega_k \left(f; \frac{1}{j} \right)_q \right)^q \right\}^{1/q},$$

where $q > q(\lambda, p) = (\lambda + 1/p)^{-1}$.

For every $t > 0$ we have

$$\omega_k(f;t)_q \leqslant ct^\lambda \left\{ \sum_{0 \leqslant j \leqslant t^{-1}} \frac{1}{j} (j^\lambda R_j(f)_p)^{q^*} \right\}^{1/q^*},$$

where $q^* = \min[1, q]$, $q = q(\lambda, p) = (\lambda + 1/p)^{-1}$.

The Jackson type theorem for best rational approximations to analytic functions on the unit disk as well as a characterization of the corresponding approximation spaces as Besov spaces are proved by Pekarskii (1985).

9

Approximation with respect to Hausdorff distance

In this chapter we shall consider rational approximation of functions with respect to the Hausdorff distance. The Hausdorff distance in the space $C[a, b]$ of the continuous functions in the interval $[a, b]$ was introduced by Bl. Sendov and B. Penkov (1962). After this Bl. Sendov developed the theory of approximation of bounded functions by means of algebraic polynomials with respect to the Hausdorff distance. Many mathematicians have obtained results in the theory of approximation of functions with respect to the Hausdorff distance – the results are collected in the book of Bl. Sendov (1979).

In section 9.1 we give the definition of Hausdorff distance in the set of all bounded functions in a given interval and we consider some of its properties.

In section 9.2 we consider the most interesting examples of rational approximation in Hausdorff distance – rational approximation of sign x. In our opinion this result is basic in the theory of rational approximation – from here follows the most essential results for uniform and L_p rational approximation – for example Newman's result for $|x|$. The Hausdorff distance is the natural distance by means of which we can explain the fact that sign x can be approximated to order $O(e^{-c\sqrt{n}})$ by means of rational functions.

In section 9.3 we consider the general case of rational approximation of bounded functions with respect to the Hausdorff distance. From the estimate obtained it follows for example that functions with bounded variation can be approximated by rational functions in order $O(n^{-1})$ in Hausdorff distance, while the order of the best polynomial Hausdorff approximation of this class of functions is $O(\ln n/n)$.

In the notes at the end of the chapter we give some other results connected with rational Hausdorff approximation.

9.1 Hausdorff distance and its properties

Let F_Δ be the set of all closed and bounded sets in the plane which are convex with respect to the y-axis and the projection of which on the x-axis coincides with the interval $\Delta = [a, b]$.

The Hausdorff distance with a parameter α, $\alpha > 0$, between two sets F and G, $F \in F_\Delta$, $G \in F_\Delta$, is defined as follows:

$$r(F, G; \alpha) = \max \left\{ \max_{A \in F} \min_{B \in G} d_\alpha(A, B), \max_{A \in G} \min_{B \in F} d_\alpha(A, B) \right\}, \tag{1}$$

where

$$d_\alpha(A, B) = d_\alpha(A(a_1, a_2), B(b_1, b_2)) = \max \{\alpha^{-1} |a_1 - b_1| |a_2 - b_2|\}. \tag{2}$$

Here $A(a_1, a_2)$, $B(b_1, b_2)$ denote the point A, respectively B, in the plane with coordinates (a_1, a_2), respectively (b_1, b_2).

It is easily seen that $r(F, G; \alpha)$ is a real distance in F_Δ, i.e. r satisfies the three axioms for distance.

The distance $d_\alpha(A, B)$ between the points A and B in the plane may be defined in another way as well, not only by (2), for example in the case $\alpha = 1$ we can use the usual Euclidean distance $e(A, B) = \sqrt{((a_1 - b_1)^2 + (a_2 - b_2)^2)}$. The choice of the distance d_α is convenient for some calculations which appear in the theory of Hausdorff approximation.

Now we shall define the notion of the complemented graph $\bar{f}$ of the function f which is bounded on the interval $\Delta = [a, b]$. The complemented graph $\bar{f}$ of the function f is the following set of points in the plane:

$$\bar{f} = \{\bigcap F : F \in F_\Delta, f \subset F\},$$

where f denotes the usual graph of the function f, i.e. the set of points $f = \{(x, y): x \in \Delta, y = f(x)\}$.

Obviously $\bar{f} \in F_\Delta$.

We define the Hausdorff distance with a parameter α in the set of all functions bounded in the interval $[a, b]$ as the Hausdorff distance with parameter α between the complemented graphs of the functions:

$$r(f, g; \alpha) = r(\bar{f}, \bar{g}; \alpha).$$

The so defined distance $r(f, g; \alpha)$ is not a real distance in the set of all bounded functions in the interval $[a, b]$, since evidently there exist different functions f and g for which $\bar{f} = \bar{g}$ and consequently $r(f, g; \alpha) = 0$, for example for the functions f and g, given in $[-1, 1]$ by

$$f(x) = \begin{cases} 0, & x \in [-1, 0] \\ 1, & x \in (0, 1] \end{cases} \quad g(x) = \begin{cases} 0, & x \in [-1, 0), \\ \frac{1}{2}, & x = 0, \\ 1, & x \in (0, 1], \end{cases}$$

This fact will not be essential for us.

The Hausdorff distance with a parameter α may be considered as a generalization of the uniform distance in the space $C[a,b]$. For $f\in C[a,b]$ we have obviously $\bar{f}=f$, i.e. the complemented graph of f coincides with the graph of f. It is easily seen that in this case the Hausdorff distance between $f\in C[a,b]$ and $g\in C[a,b]$ can be defined as follows:

$$r(f,g;\alpha)=\max\left\{\max_{x\in[a,b]}\min_{y\in[a,b]}\max\{\alpha^{-1}|x-y|,|f(x)-g(y)|\},\right.$$
$$\left.\max_{x\in[a,b]}\min_{y\in[a,b]}\max\{\alpha^{-1}|x-y|,|f(y)-g(x)|\}\right\}. \tag{3}$$

From (3) it follows immediately that

$$r(f,g;\alpha)\leqslant\|f-g\|_{C[a,b]} \tag{4}$$

(take in (3) $y=x$ instead of $\min_{y\in[a,b]}$).

It is not difficult also to obtain an estimate for $\|f-g\|_{C[a,b]}$ by means of $r(f,g;\alpha)$ and the modulus of continuity of one of the functions f or g. The following lemma holds.

Lemma 9.1. *Let $f\in C[a,b]$, $g\in C[a,b]$. Then*

$$\|f-g\|_{C[a,b]}\leqslant r(f,g;\alpha)+\omega(g;\alpha r(f,g;\alpha)). \tag{5}$$

Proof. From the continuity of f and g and the definition (3) it follows that for every $x\in[a,b]$ there exists a point $y_x\in[a,b]$ such that

$$r(f,g;\alpha)\geqslant\max\{\alpha^{-1}|x-y_x|,|f(x)-g(y_x)|\},$$

i.e.

$$\alpha^{-1}|x-y_x|\leqslant r(f,g;\alpha),$$
$$|f(x)-g(y_x)|\leqslant r(f,g;\alpha).$$

On the other hand for every $x\in[a,b]$ we have

$$|f(x)-g(x)|\leqslant|f(x)-g(y_x)|+|g(y_x)-g(x)|$$
$$\leqslant r(f,g;\alpha)+\omega(g;|x-y_x|)\leqslant r(f,g;\alpha)+\omega(g;\alpha r(f,g;\alpha)). \quad \square$$

From lemma 9.1 we immediately obtain the following.

Corollary 9.1. *Let $f\in C[a,b]$ and $\{f_n\}_{n=1}^{\infty}$, $f_n\in C[a,b]$, be such that $r(f,f_n;\alpha)\underset{n\to\infty}{\longrightarrow}0$. Then $\|f-f_n\|_{C[a,b]}\underset{n\to\infty}{\longrightarrow}0$.*

In other words, the topologies in the space $C[a,b]$ generated by the Hausdorff distance and the uniform distance coincide.

Corollary 9.2. *If $f \in C[a,b]$, $g \in C[a,b]$, then*

$$\lim_{\alpha \to 0} r(f, g; \alpha) = \| f - g \|_{C[a,b]}.$$

Corollary 9.2 follows directly from (4) and (5), setting in (5) $\alpha \to 0$.

This corollary shows that the Hausdorff distance with a parameter α can be considered as a generalization of the uniform distance – if we have some result for the Hausdorff distance with a parameter α we can obtain the corresponding result for the uniform distance setting $\alpha \to 0$.

To conclude this section we shall give one working lemma, as follows.

Lemma 9.2. *Let f and g be bounded functions in the interval $\Delta = [a,b]$ and $\delta > 0$. If for every $x \in \Delta$ there are intervals $\Delta_x, \Delta'_x, x \in \Delta_x, x \in \Delta'_x, |\Delta_x| \leqslant \delta, |\Delta'_x| \leqslant \delta$, such that*

$$\inf\{g(t): t \in \Delta_x\} \leqslant y \leqslant \sup\{g(t): t \in \Delta_x\}, \quad \forall (x,y) \in \bar{f},$$

$$\inf\{f(t): t \in \Delta'_x\} \leqslant y \leqslant \sup\{f(t): t \in \Delta'_x\}, \quad \forall (x,y) \in \bar{g},$$

then

$$r(f, g; \alpha) \leqslant \alpha^{-1}\delta.$$

Proof. Since $\bar{f}$ and $\bar{g}$ are connected, from the conditions of the lemma it follows that for every $\varepsilon > 0$ and every $(x,y) \in \bar{f}$ there is $(x', z) \in \bar{g}$ such that

$$\left.\begin{aligned} &|x' - x| \leqslant |\Delta_x| \leqslant \delta, \\ &|y - z| \leqslant \varepsilon, \end{aligned}\right\} \tag{6}$$

and for every $(x, y') \in \bar{g}$ there is $(x'', z') \in \bar{f}$ such that

$$\begin{aligned} &|x'' - x| \leqslant |\Delta'_x| \leqslant \delta, \\ &|y' - z'| \leqslant \varepsilon. \end{aligned} \tag{7}$$

From (6), (7) and the definition (1), (2) it follows that

$$r(f, g; \alpha) = r(\bar{f}, \bar{g}; \alpha) \leqslant \max\{\alpha^{-1}\delta, \varepsilon\}.$$

Since $\varepsilon > 0$ is arbitrary, the lemma follows. □

9.2 Hausdorff approximation of the jump

We shall consider best rational and polynomial approximation of bounded functions on the interval $[a,b]$ in respect to the Hausdorff distance with a parameter $\alpha, \alpha > 0$.

The best Hausdorff approximation with a parameter α to the function f bounded on the interval $[a,b]$ by means of algebraic polynomials of nth degree is given by

$$E_n(f;\alpha) = E_n(f;\alpha;[a,b]) = \inf\{r(f,p;\alpha): p \in P_n\},$$

and the best Hausdorff approximation to f with a parameter α by means of rational functions of nth degree is given by

$$R_n(f;\alpha) = R_n(f;\alpha;[a,b]) = \inf\{r(f,q;\alpha): q \in R_n\}.$$

The basic result in the theory of approximation of function with respect to the Hausdorff distance is the following result of Bl. Sendov (1962), see also Bl. Sendov (1979).

Let f be a bounded function in the interval $[a,b]$. Then

$$E_n(f;\alpha) \leqslant c(\alpha,(b-a))\frac{\ln n}{n}, \tag{1}$$

where the constant $c(\alpha, b-a)$ depends only on α and the length $b-a$ of the interval $[a,b]$.

We shall not give here the proof of this result.

Typical functions for which the order $\ln n/n$ is obtained are

$$\sigma(x) = \operatorname{sign} x = \begin{cases} -1, & x \in [-1,0), \\ 0, & x = 0, \\ 1, & x \in (0,1], \end{cases} \qquad \delta(x) = \begin{cases} 0, & x \in [-1,0), \\ 1, & x = 0, \\ 0, & x \in (0,1]. \end{cases}$$

In Bl. Sendov, V.A. Popov (1972) (see also Bl. Sendov (1979, pp. 127, 135)), it is proved that

$$\left.\begin{aligned} \lim_{n\to\infty} \frac{n}{\ln n} E_n(\sigma; 1; [-1,1]) = 1, \\ \lim_{n\to\infty} \frac{n}{\ln n} E_n(\delta; 1; [-1,1]) = 1. \end{aligned}\right\} \tag{2}$$

These exact asymptotics can be obtained also for the whole class of uniformly bounded functions, see Bl. Sendov, V.A. Popov (1972), Bl. Sendov (1979), p. 135.

In this section we shall consider the best Hausdorff rational approximation to the functions σ and δ.

Theorem 9.1. *We have for $n \geqslant 2$*

$$R_n(\delta; \alpha; [-1,1]) = 0.$$

Proof. Obviously for every $\alpha > 0$ we have

$$r\left(\delta, \frac{1}{1+Ax^2}; \alpha\right) \xrightarrow[A\to\infty]{} 0. \qquad \square$$

To obtain an estimate for $R_n(\operatorname{sign} x; 1; [-1, 1])$ we shall need two lemmas. The first lemma is a very useful one in obtaining lower bounds in rational approximation (not only for the Hausdorff rational approximation – see Chapter 11).

Lemma 9.3. *Let* $0 < \varepsilon < 1$ *and* $n \geqslant 1$. *Then the rational function* $r \in R_n$,

$$r(x) = \frac{p(x) - p(-x)}{p(x) + p(-x)}, \quad p(x) = \prod_{\nu=1}^{n} (x + \varepsilon^{(2\nu-1)/2n}),$$

satisfies the conditions

$$|\operatorname{sign} x - r(x)| > \exp\left\{-\frac{\pi^2 n}{\ln(1/\varepsilon)}\right\} \tag{3}$$

for

$$x \in \{\varepsilon^{\nu/n}\}_{\nu=0}^{n} \bigcup \{-\varepsilon^{\nu/n}\}_{\nu=0}^{n}$$

and $\operatorname{sign} x - r(x)$ *has alternate signs at the points* $-1, -\varepsilon^{1/n}, \ldots, -\varepsilon, \varepsilon, \ldots, \varepsilon^{1/n}, 1$.

Proof. Let $x = \varepsilon^{i/n}$, $0 \leqslant i \leqslant n$. Obviously $|p(-x)|/|p(x)| \leqslant 1$. Then we have

$$|\operatorname{sign} x - r(x)| = \frac{2|p(-x)|}{|p(x) + p(-x)|} \geqslant \frac{2|p(-x)|/|p(x)|}{1 + |p(-x)|/|p(x)|}$$

$$\geqslant \left|\frac{p(-x)}{p(x)}\right| = \prod_{\nu=1}^{n} \frac{|\varepsilon^{i/n} - \varepsilon^{(2\nu-1)/2n}|}{\varepsilon^{i/n} + \varepsilon^{(2\nu-1)/2n}} \geqslant \left(\prod_{j=1}^{n} \frac{1 - \varepsilon^{(2j-1)/2n}}{1 + \varepsilon^{(2j-1)/2n}}\right)^2$$

$$> \left(\prod_{j=1}^{n} \frac{1 - \varepsilon^{j/2n}}{1 + \varepsilon^{j/2n}}\right)^2 = \exp\left\{2 \sum_{j=1}^{n} \ln \frac{1 - \varepsilon^{j/2n}}{1 + \varepsilon^{j/2n}}\right\}.$$

For $|x| < 1$ we have

$$\ln \frac{1 - x}{1 + x} = -2 \sum_{k=1}^{\infty} \frac{x^{2k-1}}{2k - 1}.$$

Therefore

$$|\operatorname{sign} x - r(x)| > \exp\left\{-4 \sum_{j=1}^{n} \sum_{k=1}^{\infty} \frac{(\varepsilon^{j/2n})^{2k-1}}{2k - 1}\right\}$$

$$= \exp\left\{-4 \sum_{k=1}^{\infty} \frac{1}{2k - 1} \sum_{j=1}^{n} (\varepsilon^{(2k-1)/2n})^j\right\}$$

$$= \exp\left\{-4 \sum_{k=1}^{\infty} \frac{1}{2k - 1} \frac{\varepsilon^{(2k-1)/2n}(1 - \varepsilon^{(2k-1)/2})}{1 - \varepsilon^{(2k-1)/2n}}\right\}$$

$$> \exp\left\{-4 \sum_{k=1}^{\infty} \frac{1}{2k - 1} \frac{\varepsilon^{(2k-1)/2n}}{1 - \varepsilon^{(2k-1)/2n}}\right\}.$$

We have $\ln(1/x) \leqslant (1-x)/x$ for $0 < x \leqslant 1$. Using this inequality we obtain

$$|\operatorname{sign} x - r(x)| > \exp\left\{-4\sum_{k=1}^{\infty}\frac{1}{2k-1}\frac{1}{\ln(1/\varepsilon)^{(2k-1)/2n}}\right\}$$
$$= \exp\left\{-\frac{8n}{\ln(1/\varepsilon)}\sum_{k=1}^{\infty}\frac{1}{(2k-1)^2}\right\} = \exp\left\{-\pi^2\frac{n}{\ln(1/\varepsilon)}\right\}.$$

Therefore the estimate (3) is proved for $x = \varepsilon^{i/n}$, $i = 0,\ldots,n$. The case $x = -\varepsilon^{i/n}$, $i = 0,\ldots,n$, is similar.

Since

$$r(x) = \frac{p(x) - p(-x)}{p(x) + p(-x)}, \quad p(x) = \prod_{v=1}^{n}(x + \varepsilon^{(2v-1)/2n})$$

it is evident that $r(-\varepsilon^{(2v-1)/2n}) = -1$, $v = 1,\ldots,n$, and $r(\varepsilon^{(2v-1)/2n}) = 1$, $v = 1,\ldots,n$. Since the points $-\varepsilon^{(2v-1)/2n}$, $v = 1,\ldots,n$, $\varepsilon^{(2v-1)/2n}$, $v = 1,\ldots,n$, separate the points $-\varepsilon^{v/n}$, $v = 0,\ldots,n$, $\varepsilon^{v/n}$, $v = 0,\ldots,n$, condition (3) gives us that $\operatorname{sign} x - r(x)$ has alternate signs at the points $-1, -\varepsilon^{1/n},\ldots,-\varepsilon, \varepsilon,\ldots,\varepsilon^{1/n}, 1$.

□

We shall need also one particular application of a theorem of the type 2.3 (de la Vallée–Poussin theorem) to the function $\operatorname{sign} x$. By the same method as in theorem 2.3 it is possible to obtain the following result.

Lemma 9.4. *Let $\Delta(\varepsilon) = [-1, -\varepsilon] \cup [\varepsilon, 1]$. Let there exist $2n+2$ points $x_1 < x_2 < \cdots < x_{2n+2}$, $x_i \in \Delta(\varepsilon)$, $i = 1,\ldots,2n+2$, and a rational function $r \in R_n$ such that*

$$\operatorname{sign} x_i - r(x_i) = \mu(-1)^i\lambda_i, \quad \mu = \pm 1, \quad \lambda_i > 0, \quad i = 1,\ldots,2n+2.$$

Then

$$R_n(\operatorname{sign} x)_{C[\Delta(\varepsilon)]} \geqslant \min\{\lambda_i : i = 1,\ldots,2n+2\}.$$

Lemmas 9.3 and 9.4 give us

Corollary 9.3. *For every ε, $0 < \varepsilon < 1$, we have*

$$R_n(\operatorname{sign} x)_{C[\Delta(\varepsilon)]} \geqslant \exp\left\{-\pi^2\frac{n}{\ln(1/\varepsilon)}\right\}.$$

Theorem 9.2. *There exist absolute constants c_1 and c_2 such that*

$$e^{-\pi\sqrt{n}} \leqslant R_n(\operatorname{sign} x; 1; [-1, 1]) \leqslant c_1 e^{-c_2\sqrt{n}}.$$

Proof. The upper bound for $R_n(\operatorname{sign} x; 1; [-1, 1])$ we shall obtain using lemma 5.1. Let us set in lemma 5.1 $\beta = 1$, $\alpha = e^{-\sqrt{n}}$, $\gamma = e^{-\sqrt{n}}$. We obtain that there exists a rational function $r \in R_N$, $N \leqslant B\ln(e + e^{\sqrt{n}})\ln(e + e^{\sqrt{n}}) \leqslant B_1 n$, B_1 an absolute constant, such that

$$|r(x)| \leqslant e^{-\sqrt{n}}, \quad x \in [-1, -e^{-\sqrt{n}}],$$
$$|1 - r(x)| \leqslant e^{-\sqrt{n}}, \quad x \in [e^{-\sqrt{n}}, 1],$$
$$0 \leqslant r(x) \leqslant 1, \quad \forall x.$$

Therefore the rational function $q = 2r - 1 \in R_N$, $N \leqslant B_1 n$, satisfies

$$|-1 - q(x)| \leqslant 2e^{-\sqrt{n}}, \quad x \in [-1, -e^{-\sqrt{n}}],$$
$$|1 - q(x)| \leqslant 2e^{-\sqrt{n}}, \quad x \in [e^{-\sqrt{n}}, 1],$$
$$0 \leqslant r(x) \leqslant 1, \quad \forall x.$$

Therefore the graph of q belongs to the domain

$$G = \{(x, y): e^{-\sqrt{n}} \leqslant |x| \leqslant 1, |y - \operatorname{sign} x| \leqslant 2e^{-\sqrt{n}}; |x| \leqslant e^{-\sqrt{n}}, |y| \leqslant 1\}.$$

Using the definition of the Hausdorff distance with a parameter 1 we obtain that

$$r(\operatorname{sign} x, q; 1; [-1, 1]) \leqslant 2e^{-\sqrt{n}}.$$

Since $q \in R_N$, $N \leqslant B_1 n$, B_1 an absolute constant, we obtain the upper bound

$$R_n(\operatorname{sign} x; 1; [-1, 1]) \leqslant c_1 e^{-c_2\sqrt{n}}.$$

To prove the lower bound let us assume the converse, that

$$R_n(\operatorname{sign} x; 1; [-1, 1]) < e^{-\pi\sqrt{n}}.$$

Using the definition of the Hausdorff distance we obtain that there exists a rational function $r \in R_n$ such that

$$|\operatorname{sign} x - r(x)| < e^{-\pi\sqrt{n}}, \quad \varepsilon \leqslant |x| \leqslant 1, \tag{4}$$

where $\varepsilon = e^{-\pi\sqrt{n}}$,

$$-1 - \varepsilon \leqslant r(x) \leqslant 1 + \varepsilon, \quad |x| \leqslant \varepsilon.$$

But (4) gives us that

$$R_n(\operatorname{sign} x)_{C(\Delta(\varepsilon))} \leqslant \exp\left\{-\pi^2 \frac{n}{\ln(1/\varepsilon)}\right\} = e^{-\pi\sqrt{n}}, \tag{5}$$

since $\varepsilon = e^{-\pi\sqrt{n}}$.

The inequality (5) contradicts corollary 9.3. □

Remark. Using the same method it is possible to prove that

$$c_3(\alpha)e^{-c_4(\alpha)\sqrt{n}} \leqslant R_n(\operatorname{sign} x; \alpha; [-1, 1]) \leqslant c_1(\alpha)e^{-c_2(\alpha)\sqrt{n}}$$

where the constants $c_i(\alpha)$, $i = 1, \ldots, 4$, depend only on α.

9.3 Bounded functions

In this section we shall consider the general case of rational Hausdorff approximation of bounded functions. We shall obtain an analog of Sendov's theorem (1) in section 9.2, but in the logarithm will appear the so-called averaged modulus $\tau(f;\delta)$. Therefore we shall give first the definition and some properties of the averaged modulus.

We define the local modulus of continuity of the bounded on $[0,1]$ function f at the point $x\in[0,1]$ by

$$\omega(f,x;\delta)=\sup\{|f(x')-f(x'')|:x',x''\in[x-\delta/2,x+\delta/2]\cap[0,1]\}.$$

Then the averaged modulus of f is the following function of δ, $\delta>0$:

$$\tau(f;\delta)=\|\omega(f,\cdot;\delta)\|_{L_1(0,1)}=\int_0^1\omega(f,x;\delta)\,\mathrm{d}x.$$

For the history of the averaged modulus see Bl. Sendov (1979).

The following properties of $\tau(f;\delta)$ are evident.

(i) $\tau(f;\delta)\leqslant\tau(f;\delta')$, $\delta\leqslant\delta'$.

(ii) $\tau(f+g;\delta)\leqslant\tau(f;\delta)+\tau(g;\delta)$.

We shall need also the following two properties.

(iii) $\tau(f;n\delta)\leqslant n\tau(f;\delta)$, $n>0$, *n integer.*

Proof. We have

$$\begin{aligned}\omega(f,x;n\delta)&=\sup\left\{|f(x')-f(x'')|:x',x''\in\left[x-\frac{n\delta}{2},x+\frac{n\delta}{2}\right]\cap[0,1]\right\}\\&\leqslant\sum_{i=0}^{n-1}\sup\left\{|f(x')-f(x'')|:x',x''\right.\\&\left.\in\left[x-\frac{n-1-2i}{2}\delta-\frac{\delta}{2},x-\frac{n-1-2i}{2}\delta+\frac{\delta}{2}\right]\cap[0,1]\right\}\\&\leqslant\sum_{i=0}^{n-1}\omega\left(f,x-\frac{n-1-2i}{2}\delta;\delta\right),\end{aligned}$$

where we set $f(x)=f(0)$ for $x<0$, $f(x)=f(1)$ for $x>1$.

From here we get

$$\tau(f;n\delta)=\int_0^1\omega(f,x;\delta)\,\mathrm{d}x\leqslant\sum_{i=0}^{n-1}\int_0^1\omega\left(f,x-\frac{n-1-2i}{2}\delta;\delta\right)\mathrm{d}x\leqslant n\tau(f;\delta).\ \square$$

(iv) *If f is a function with bounded variation in* $[0,1]$ *then*

$$\tau(f;\delta)\leqslant\delta V_0^1 f.$$

Proof. Let us set again $f(x)=f(0)$ for $x<0$, $f(x)=f(1)$ for $x>1$. Then

$$\omega(f,x;\delta)\leqslant V_{x-\delta/2}^{x+\delta/2}f,$$

therefore

$$\tau(f;\delta)\leqslant\int_0^1 V_{x-\delta/2}^{x+\delta/2}f\,\mathrm{d}x=\int_0^1 V_0^{x+\delta/2}f\,\mathrm{d}x-\int_0^1 V_0^{x-\delta/2}f\,\mathrm{d}x$$

$$=\int_{\delta/2}^{1+\delta/2} V_0^x f\,\mathrm{d}x-\int_{-\delta/2}^{1-\delta/2} V_0^x f\,\mathrm{d}x\leqslant\int_{1-\delta/2}^{1} V_0^x\,\mathrm{d}x+\delta/2V_0^1 f\leqslant\delta V_0^1 f.$$

□

We shall need also the following lemma.

Lemma 9.5. *Let f be a bounded function in the interval $[0,1]$. For every natural number $m>0$ there exists a step function φ_m with jumps at the points $x_i=i/2m$, $i=1,\dots,2m-1$, such that*

(a) $r(f,\varphi_m)\leqslant 1/m$,[†]
(b) $\|f-\varphi_m\|_{L_1(0,1)}\leqslant 2\tau(f;m^{-1})$,
(c) $V_0^1\varphi_m\leqslant 6m\tau(f;m^{-1})$.

Proof. Let us denote $x_i=i/2m$, $i=0,\dots,2m$, $x_{-1}=x_0=0$, $x_{2m+1}=x_{2m}=1$,

$$m_i=\inf\{f(x):x\in[x_{i-1},x_{i+1}]\},$$
$$M_i=\sup\{f(x):x\in[x_{i-1},x_{i+1}]\}.$$

We set

$$\varphi_m(x)=\begin{cases} m_{2i-1}, & x\in[x_{2i-2},x_{2i-1}),\quad i=1,\dots,m,\\ M_{2i-1}, & x\in[x_{2i-1},x_{2i}),\quad i=1,\dots,m,\\ \varphi_m(x_{2m-1}), & x=1.\end{cases}$$

From the construction of φ_m and lemma 9.2 it follows that $r(f,\varphi_m)\leqslant m^{-1}$, i.e. we have (a).

For the difference $f(x)-\varphi_m(x)$ we have, using again the definition of φ_m,

$$|f(x)-\varphi_m(x)|\leqslant\omega(f,x;2m^{-1});$$

therefore, using property (iii) of $\tau(f;\delta)$ we obtain (b):

$$\|f-\varphi_m\|_1=\int_0^1|f(x)-\varphi_m(x)|\,\mathrm{d}x\leqslant\int_0^1\omega(f,x;2m^{-1})\mathrm{d}x\leqslant 2\tau(f;m^{-1}).$$

[†] In this section we set $r(f,g;1)=r(f,g)$.

Finally let us estimate the variation of φ_m. We have

$$V_0^1\varphi_m = \sum_{i=1}^{2m-1} |\varphi_m(x_i) - \varphi_m(x_{i-1})| \leqslant \sum_{i=1}^{2m-1} \omega(f, x_i; 2m^{-1})$$

$$= 2m \sum_{i=1}^{2m-1} \int_{x_{i-1}}^{x_i} \omega(f, x_i; 2m^{-1})\mathrm{d}x \leqslant 2m \sum_{i=1}^{2m-1} \int_{x_{i-1}}^{x_i} \omega(f, x; 3m^{-1})\mathrm{d}x$$

$$\leqslant 2m \int_0^1 \omega(f, x; 3m^{-1})\mathrm{d}x = 2m\tau(f; 3m^{-1}) \leqslant 6m\tau(f; \delta),$$

since evidently for $x \in [x_{i-1}, x_i]$ we have

$$\omega(f, x_i; 2m^{-1}) \leqslant \omega(f, x; 3m^{-1}).$$

□

Theorem 9.3. *For every function f bounded in the interval $[0,1]$ we have*

$$R_n(f, 1; [0,1]) \leqslant c\frac{\ln(\mathrm{e} + n\tau(f; n^{-1}))}{n}, \quad n \geqslant 1,$$

where c is an absolute constant.

Proof. Let f be a bounded function in the interval $[0,1]$. Let us consider first the case when

$$\alpha_n \equiv 8\mathrm{e}D\frac{\ln(\mathrm{e} + n\tau(f; n^{-1}))}{n} > 1,$$

where D is the constant from theorem 5.1.

Since obviously $R_n(f, 1; [0,1]) \leqslant 1$, we have

$$R_n(f, 1; [0,1]) \leqslant c_1\frac{\ln(\mathrm{e} + n\tau(f; n^{-1}))}{n} \tag{1}$$

with a constant $c_1 = 8\mathrm{e}D$.

Now let $\alpha_n \leqslant 1$. We set

$$r = [\ln(\mathrm{e} + n\tau(f; n^{-1}))], \quad h = 2\mathrm{e}D/n, \quad m = \left[\frac{1}{4rh}\right]. \tag{2}$$

Since $rh \leqslant \frac{1}{4}, m \geqslant 1, D > 1$, we have

$$\frac{1}{8m} \leqslant rh \leqslant \frac{1}{4m}, \tag{3}$$

$$m \leqslant (4rh)^{-1} \leqslant n/8. \tag{4}$$

For m so defined let us consider the step function φ_m from lemma 9.5. We set $\varphi_m(x) = \varphi(0)$ for $x < 0$ and $\varphi_m(x) = \varphi_m(1)$ for $x > 1$. Let us consider the function

$$\psi(x)=\frac{1}{2H}\int_{-H}^{H}\varphi_m(x+t)\,\mathrm{d}t,\quad H=\frac{1}{16m}.$$

For the function ψ we have $\varphi_m(x)=\psi(x)$ if

$$x\in[x_{i-1}+H,\,x_i-H]=\left[x_{i-1}+\frac{1}{16m},x_i-\frac{1}{16m}\right],\quad i=1,\dots,2m.$$

Moreover ψ is continuous and linear in the intervals $[x_i-H,x_i+H]$, $i=0,\dots,2m$.

From the properties of ψ and lemma 9.5 we obtain

$$V_0^1\psi=V_0^1\varphi_m\leqslant 6m\tau(f;m^{-1}).\tag{5}$$

Let us consider the function $\psi_{r,h}$ given by

$$\psi_{r,h}(x)=h^{-r}\int_{-h/2}^{h/2}\cdots\int_{-h/2}^{h/2}\psi(x+t_1+\cdots+t_r)\mathrm{d}t_1\cdots\mathrm{d}t_r.$$

Since $rh/2\leqslant 1/8m$, we obtain, using the properties of the function ψ given above, that

$$\psi_{r,h}(x_i+1/4m)=\psi(x_i+1/4m)=\varphi_m(x_i+1/4m),\quad i=0,\dots,2m-1,$$

$\min\{\varphi_m(x_{i-1}),\varphi_m(x_i)\}\leqslant\psi_{r,h}(x)\leqslant\max\{\varphi_m(x_{i-1}),\varphi_m(x_i)\}$ for $x\in[x_i-1/4m, x_i+1/4m]$, $i=1,\dots,2m-1$.

Therefore lemma 9.2 gives us

$$r(\varphi_m,\psi_{r,h})\leqslant\frac{1}{2m}.\tag{6}$$

From here and lemma 9.5 we get

$$r(f,\psi_{r,h})\leqslant r(f,\varphi_m)+r(\varphi_m,\psi_{r,h})\leqslant 2m^{-1}.\tag{7}$$

On the other hand we have

$$\psi_{r,h}^{(r)}(x)=h^{-r}\sum_{\nu=0}^{r}\binom{r}{\nu}(-1)^{r+\nu}\psi(x+\nu h-rh/2).$$

From here and (5) we obtain

$$V_0^1\psi_{r,h}^{(r)}\leqslant 2^r h^{-r}\cdot 6m\tau(f;m^{-1}).\tag{8}$$

Using theorem 5.1, (8), (2)–(4) and property (iii) of $\tau(f;\delta)$ we get.

$$R_n(\psi_{r,h})_{C[0,1]}\leqslant D^r\frac{V_0^1\psi_{r,h}^{(r)}}{n^{r+1}}$$

$$\leqslant D^r\left(\frac{2}{hn}\right)^r\frac{6m\tau(f;m^{-1})}{n}\leqslant\frac{6m\tau(f;m^{-1})}{\mathrm{e}^r n}$$

$$\leqslant \frac{6me\tau(f;8rh)}{(e+n\tau(f;n^{-1}))n} \leqslant \frac{6en([2eD\ln(e+n\tau(f;n^{-1}))]+1)\tau(f;n^{-1})}{(e+n\tau(f;n^{-1}))n}$$

$$\leqslant 18e^2D\frac{\ln(e+n\tau(f;n^{-1}))}{n}. \tag{9}$$

From (7), (2)–(4) and (9) we obtain

$$\begin{aligned} R_n(f,1;[0,1]) &\leqslant r(f,\psi_{r,h}) + R_n(\psi_{r,h},1;[0,1]) \\ &\leqslant 2m^{-1} + R_n(\psi_{r,h})_{C[0,1]} \leqslant 16rh \\ &\quad + 18e^2D\frac{\ln(e+n\tau(f;n^{-1}))}{n} \leqslant 50e^2D\frac{\ln(e+n\tau(f;n^{-1}))}{n}, \end{aligned} \tag{10}$$

i.e. we have the statement of the theorem with $c = 50e^2D$.

From (1) and (10) the theorem follows with $c = 50e^2D$. □

Corollary 9.4. *Let f be a function of bounded variation on the interval* $[0,1]$. *Then*

$$R_n(f;1;[0,1]) = O(n^{-1}).$$

The corollary follows directly from theorem 9.3 and property (iv) of $\tau(f;\delta)$.

9.4 Notes

As we mentioned in the introduction to this chapter, the Hausdorff distance between bounded functions was introduced by Bl. Sendov and B. Penkov (1962). The results for approximation of functions with respect to the Hausdorff distance by means of polynomials, splines, rational functions, linear operators, are collected in the book of Bl. Sendov (1979). We shall restrict ourselves to rational Hausdorff approximations.

In connection with theorem 9.2 we want to mention the following upper bound given by S.A. Agahanov and N.Sh. Zagirov (1978) (compare with A.P. Bulanov (1975a)):

$$R_n(\operatorname{sign} x)_{C(\Delta(\varepsilon))} \leqslant c\exp\left(-\frac{\pi^2}{2}\frac{n}{\ln(1/\varepsilon)}\right), \tag{1}$$

where $\Delta(\varepsilon) = [-1,-\varepsilon]\cup[\varepsilon,1]$, $\varepsilon\in(0,e^{-\sqrt{n}})$, c a constant.

The exact lower bound is given by A.A. Gonchar (1967b), (see also A.P. Bulanov (1975a)):

$$\exp\left(-\frac{\pi^2}{2}\frac{n}{\ln(1/\varepsilon)}\right) \leqslant R_n(\operatorname{sign} x)_{C(\Delta(\varepsilon))}. \tag{2}$$

It is not very difficult to obtain from the estimates (1) and (2) the following asymptotics for the best Hausdorff rational approximation to sign x, which improves theorem 9.2:

$$R_n(\operatorname{sign} x; 1; [-1, 1]) \asymp \exp\left(-\frac{\pi}{\sqrt{2}}\sqrt{n}\right)$$

(compare with theorem 4.2).

Theorem 9.3 is proved by P. Petrushev (1980b). This theorem is exact in the following sense (P. Petrushev, 1980c):

For every function $\tau(\delta)$, $\delta \geqslant 0$, *such that* $\tau(\delta) = \tau(g; \delta)$ *for* $\delta \geqslant 0$, *where* g *is a bounded function on the interval* $[0, 1]$ *with unbounded variation* (i.e. $\tau(\delta)/\delta \underset{\delta \to 0}{\longrightarrow} \infty$), *there exists a bounded function on the interval* $[0, 1]$ *such that* $\tau(f; \delta) \leqslant \tau(\delta)$, $\delta \geqslant 0$, *and*

$$\limsup_{n \to \infty} (R_n(f; 1; [0, 1])n/\ln(e + n\tau(1/n))) > 0.$$

Corollary 9.4 also can be improved (P. Petrushev, 1980d): for every function f with bounded variation on $[0, 1]$ we have $R_n(f; 1; [0, 1]) = o(1/n)$.

Let us remark that there exists an absolutely continuous function f on $[0, 1]$ for which the order of the best polynomial Hausdorff approximation is exactly $\ln n/n$ (see Bl. Sendov (1979)).

Many interesting results concerning rational Hausdorff approximation are given by E.P. Dolženko and E.A. Sevastijanov (1976a, b) and E.P. Dolženko (1976). These results are connected with the so-called piecewise monotone approximation.

For example if $\sum_{n=0}^{\infty}(R_n(f; 1; \Delta))^{1/s}$ converges for $s \geqslant 1$ then the function f has almost everywhere on the interval Δ a differential of order s, and this result is exact

If $R_n(f; 1; \Delta) = o(1/n)$, then f is univalent and continuous almost everywhere on Δ.

Finally, we want to give the generalizations of Gonchar's results from section 5.6 given by Bl. Sendov (see Bl. Sendov (1979)).

Let the function f be analytic in the unit disk $D = \{z: |z| < 1\}$. We say that f belongs to the class A if

$$\lim_{\rho \to 1-0} \frac{1}{2\pi} \int_0^{2\pi} \ln_+ |f(\rho e^{i\theta})| \, dt = A(f) < \infty,$$

where $\ln_+ a = \max\{0, \ln a\}$, and that f belongs to the class H_p (Hardy spaces) if

$$\lim_{\rho \to 1-0} \frac{1}{2\pi} \int_0^{2\pi} |f(\rho e^{i\theta})|^p \, d\theta = H_p(f) < \infty.$$

The result of Bl. Sendov is the following.

Let f *be a continuous real valued function on the interval* $(0, 1]$, *which is bounded from above or from below in* $(0, 1]$. *Let there exist an analytic function* $F(z)$ *in* $\{z: |z-1| < 1\}$ *which coincides with* f *on* $(0, 1]$. *Then if* $F(z-1) \in A$, *then*

$$R_n(f; 1; (0, 1]) = O\left(\frac{\ln n}{n}\right).$$

If $F(z-1) \in H_p$, *then*

$$R_n(f; 1; (0, 1]) = O(e^{-c\sqrt{n}}),$$

where c *is a positive constant.*

10

The o-effect

As we have noted in section 5.1 a characteristic property of the rational approximation is the appearance of the o-effect in the order of approximation of individual functions of some functional classes. The chapter is devoted to the study of this phenomenon.

In section 10.1 is established and characterized the o-effect for uniform rational approximation of the functions from the class V_r and for the rational L_1 approximation of functions of bounded variation. Section 10.2 investigates the o-effect for the rational uniform approximation of functions in some classes of absolutely continuous functions. The same effect for the rational uniform approximation of convex functions is considered in section 10.3. The o-effect for the rational L_p approximation of functions of bounded variation is investigated in section 10.4.

10.1 Existence and characterization for uniform approximation of individual functions of the class V_r and for L_1 approximation of functions of bounded variation

The class $V_r = V_r(M, [a, b])$ of all functions f for which $V_a^b f^{(r)} \leqslant M < \infty$ is basic for rational approximation. In theorem 5.1 in section 5.2 we established the exact order for the rational uniform approximation of the class $V_r (r \geqslant 1)$:

$$\sup_{f \in V_r} R_n(f)_C = O\left(\frac{1}{n^{r+1}}\right).$$

In this section we prove that for each function $f \in V_r$ there exists a sequence $\{\varepsilon_n(f)\}_{n=1}^{\infty}$, $\varepsilon_n(f) \to 0$ as $n \to \infty$, such that

$$R_n(f)_C \leqslant \frac{\varepsilon_n(f)}{n^{r+1}}, \quad n = 1, 2, \ldots,$$

i.e. the o-effect appears. In addition, we characterize this effect. More precisely we replace $\varepsilon_n(f)$ in the last estimate by one new functional characteristic. Also, we consider the same problem for the spline approximation. We investigate the o-effect for the rational L_1 approximation of functions of bounded variation. Finally we present some generalizations.

10.1.1 One new functional characteristic

Definition of the function $\theta(f)$. Suppose that f is a function of bounded variation on $[a, b]$. The general notion of complemented graph $\bar{f}$ for a given function f was defined in section 9.1. In our case ($V_a^b f < \infty$) $\bar{f}$ consists of the graph f of the function f and all closed line-segments in the plane that joint the points $(x, f(x-0))$, $(x, f(x))$ and $(x, f(x+0))$ for each point $x \in [a, b]$ of discontinuity for f. Note that the complemented graph $\bar{f}$ coincides with the graph f when the function f is continuous.

We consider the complemented graph $\bar{f}$ of f as a curve in the plane. Since $V_a^b f < \infty$, the curve $\bar{f}$ is rectifiable, i.e. $\bar{f}$ is of finite length. Further we shall denote always the length of $\bar{f}$ by $l = l(f)$.

Let s be the natural parameter (the arc length) of $\bar{f}$ so that

$$\bar{f}: x = x(s), \quad y = y(s), \quad s \in [0, l], \quad x(0) = a, \quad x(l) = b. \tag{1}$$

Naturally, the points $(x, y) \in \bar{f}$ for which y is not between $f(x-0)$ and $f(x+0)$ are obtained from (1) for two different values of s. For example, the complemented graph $\bar{f}$ of the function

$$f(x) = \begin{cases} 0, & 0 < |x| \leqslant 1, \\ 1, & x = 0, \end{cases}$$

is the set $\bar{f} = \{(x, 0): -1 \leqslant x \leqslant 1\} \cup \{(0, y): 0 \leqslant y \leqslant 1\}$ and $\bar{f}$ has the following parametric equations:

$$x(s) = \begin{cases} s-1, & s \in [0, 1], \\ 0, & s \in (1, 3], \\ s-3, & s \in (3, 4], \end{cases} \quad y(s) = \begin{cases} 0, & s \in [0, 1], \\ s-1, & s \in (1, 2], \\ 3-s, & s \in (2, 3], \\ 0, & s \in (3, 4], \end{cases}$$

$$l(f) = 4.$$

Lemma 10.1. *For the parametric equations* (1) *of $\bar{f}$ ($V_a^b f < \infty$) we have $x(s)$, $y(s) \in \mathrm{Lip}_1 1$*[†] *and*

$$(x'(s))^2 + (y'(s))^2 = 1 \quad \textit{for almost all } s \in [0, l]. \tag{2}$$

[†] $x(s) \in \mathrm{Lip}_1 1$ means that $|x(s') - x(s'')| \leqslant |s' - s''|$, for $s', s'' \in [0, l]$.

Proof. From the definition of length of arc it follows that

$$\{(x(s')-x(s''))^2+(y(s')-y(s''))^2\}^{1/2}\leqslant|s'-s''|,\quad s',s''\in[0,l],$$

and therefore $x(s), y(s)\in\mathrm{Lip}_1\, 1$.

Now we prove (2). First we observe that

$$(x'(s))^2+(y'(s))^2\leqslant 1\quad\text{almost everywhere (a.e.) in } [0,l]. \tag{3}$$

Indeed, by the definition of length of arc we have

$$((x(s+h)-x(s))^2+(y(s+h)-y(s))^2)^{1/2}\leqslant|h|,\quad s,s+h\in[0,l]$$

and hence

$$\left(\frac{x(s+h)-x(s)}{h}\right)^2+\left(\frac{y(s+h)-y(s)}{h}\right)^2\leqslant 1.$$

Taking the limit in this inequality with $h\to 0$ we obtain (3) a.e. in $[0,l]$.

Also, we shall prove that

$$\int_0^l((x'(s))^2+(y'(s))^2)^{1/2}\,\mathrm{d}s\geqslant l. \tag{4}$$

To this end it suffices to prove that

$$((x(s')-x(s''))^2+(y(s')-y(s''))^2)^{1/2}\leqslant\int_{s'}^{s''}((x'(s))^2+(y'(s))^2)^{1/2}\,\mathrm{d}s \tag{5}$$

for $0\leqslant s'\leqslant s''\leqslant l$.

Set $\int_{s'}^{s''}x'(s)\,\mathrm{d}s=\rho\cos\alpha$ and $\int_{s'}^{s''}y'(s)\,\mathrm{d}s=\rho\sin\alpha,\rho\geqslant 0$. Then we have

$$((x(s')-x(s''))^2+(y(s')-y(s''))^2)^{1/2}=\left\{\left(\int_{s'}^{s''}x'(s)\,\mathrm{d}s\right)^2+\left(\int_{s'}^{s''}y'(s)\,\mathrm{d}s\right)^2\right\}^{1/2}$$

$$=\rho=\rho\cos^2\alpha+\rho\sin^2\alpha=\int_{s'}^{s''}(x'(s)\cos\alpha+y'(s)\sin\alpha)\,\mathrm{d}s.$$

This and the obvious inequality

$$|x'(s)\cos\alpha+y'(s)\sin\alpha|\leqslant((x'(s))^2+(y'(s))^2)^{1/2}$$

give (5) and (5) implies (4). The equality (2) follows from (3) and (4) immediately.

□

Denote

$$E=E(f)=\{s\in[0,l]:(x'(s))^2+(y'(s))^2=1\}.$$

By lemma 10.1 we have $\operatorname{mes}E=l$.

Now we are able to define the function $\theta=\theta(f)$. For each $s\in E$ we define $\theta(s)=\theta(f,s)$ as the oriented angle between the real axes and the tangent vector $(x'(s), y'(s))$ to $\bar{f}$ at the point $(x(s), y(s))$, i.e. we define

$$\theta(s) = \arctan \frac{y'(s)}{x'(s)}, \quad s \in E, \quad x'(s) \neq 0,$$

and

$$\theta(s) = \frac{\pi}{2} \operatorname{sign} y'(s), \qquad s \in E, \quad x'(s) = 0.$$

Some properties of the function $\theta = \theta(f)$. By the definition of θ it follows that θ is defined a.e. in $[0, l]$, θ is measurable and $|\theta(s)| \leqslant \pi/2$ a.e. in $[0, l]$.

The essential difference between the functions θ and f' is that $(a, f(a))$ and θ a.e. in $[0, l]$ determine uniquely the complemented graph $\bar{f}$ of f, while $f(a)$ and f' a.e. in $[a, b]$ determine completely f and hence $\bar{f}$ only when f is absolutely continuous.

The complemented graph $\bar{f}$ of f has the following representation which uses θ:

$$\bar{f}: \left\{ \begin{aligned} x &= x(s) = a + \int_0^s \cos \theta(s_1)\, ds_1, \\ y &= y(s) = f(a) + \int_0^s \sin \theta(s_1)\, ds_1, \end{aligned} \right\} s \in [0, l]. \tag{6}$$

Indeed, by the definition of θ it follows that $x'(s) = \cos \theta(s)$ and $y'(s) = \sin \theta(s)$ a.e. in $[0, l]$ which implies (6).

Let the function f be absolutely continuous on $[a, b]$. Then the function $x(s)$ is strictly increasing on $[0, l]$. Denote by $s(x), x \in [a, b]$ the converse function to $x(s)$. It is well known that $s(x) = \int_a^x \sqrt{(1 + (f'(t))^2)}\, dt$ for $x \in [a, b]$. Clearly $\theta(s) = \arctan f'(x(s))$ a.e. in $[0, l]$. It is readily seen that, if $f \in C^1_{[a,b]}$, then $\theta(s) = \arctan f'(x(s))$ for each $s \in [0, l]$.

Further we shall apply the following functional characteristic

$$\omega(\theta, \delta)_L = \sup_{0 \leqslant h \leqslant \delta} \int_0^{l-h} |\theta(s+h) - \theta(s)|\, ds, \quad \delta \geqslant 0,$$

to describe the o-effect in some situations.

Since $\theta(f)$ is measurable and bounded a.e. in $[0, l]$, then $\omega(\theta; \delta)_L \to 0$ as $\delta \to 0$. If f is convex and bounded on $[a, b]$, then θ is monotone and bounded. Hence $\omega(\theta; \delta)_L = O(\delta)$. Also, $\omega(\theta; \delta)_L$ satisfies the usual properties of the integral moduli of continuity; see section 3.1. A very essential property of $\omega(\theta; \delta)_L$ is provided by the following lemma.

Lemma 10.2. *If f is absolutely continuous on $[a, b]$ and $\theta = \theta(f)$, then*

$$\omega(\theta; \delta)_L \leqslant C\omega(f'; \delta)_L, \quad \delta \geqslant 0, \tag{7}$$

where $C > 0$ is an absolute constant.

Proof. Our arguments are based on the following well-known inequality (see lemma 7.6 in section 7.1): if $g \in L_{[u,v]}$, then

$$\int_u^v |g(x) - \frac{1}{v-u}\int_u^v g(t)\,\mathrm{d}t|\,\mathrm{d}x \leqslant \frac{2}{v-u}\int_0^{v-u}\int_u^{v-t} |g(x+t) - g(x)|\,\mathrm{d}x\,\mathrm{d}t. \quad (8)$$

Consider the case $0 < h \leqslant \delta \leqslant (b-a)/4 < l/3$, where $l = l(f)$ is the length of the graph f of the function f. Note that $\bar{f} = f$, since f is continuous. Set $n = [l/2h]$. Then we have

$$\int_0^{l-h} |\theta(s+h) - \theta(s)|\,\mathrm{d}s$$

$$\leqslant \sum_{\nu=0}^{n-1} \int_{2\nu h}^{(2\nu+1)h} |\theta(s+h) - \theta(s)|\,\mathrm{d}s$$

$$+ \sum_{\nu=0}^{n-2} \int_{(2\nu+1)h}^{(2\nu+2)h} |\theta(s+h) - \theta(s)|\,\mathrm{d}s + \int_{l-3h}^{l-2h} |\theta(s+h) - \theta(s)|\,\mathrm{d}s$$

$$+ \int_{l-2h}^{l-h} |\theta(s+h) - \theta(s)|\,\mathrm{d}s = \sigma_1 + \sigma_2 + \sigma_3 + \sigma_4.$$

Now we estimate each integral in the sum σ_1. Let $f: x = x(s)$, $y = y(s)$, $s \in [0, l]$, $x(0) = a$ be the parametric equations of the graph f of the function f with respect to its natural parameter s. By $s(x)$, $x \in [a, b]$ we denote the converse function to $x(s)$, $s \in [0, l]$.

Denote $x_\nu = x(\nu h)$, $\Delta_\nu = [x_\nu, x_{\nu+2}]$, $K_\nu = |\Delta_\nu|^{-1} \int_{\Delta_\nu} f'(u)\,\mathrm{d}u$.

Clearly, we have

$$I_\nu = \int_{2\nu h}^{(2\nu+1)h} |\theta(s+h) - \theta(s)|\,\mathrm{d}s \leqslant 2\int_{2\nu h}^{(2\nu+2)h} |\theta(s) - \arctan K_{2\nu}|\,\mathrm{d}s.$$

Since f is absolutely continuous, $s'(x) = \sqrt{(1 + (f'(x))^2)}$ a.e. in $[a, b]$ and

$$I_\nu \leqslant 2\int_{2\nu h}^{(2\nu+2)h} |\arctan f'(x(s)) - \arctan K_{2\nu}|\,\mathrm{d}s$$

$$= 2\int_{x_{2\nu}}^{x_{2\nu+2}} |\arctan f'(t) - \arctan K_{2\nu}| \sqrt{(1 + (f'(t))^2)}\,\mathrm{d}t,$$

where we have made the substitution $t = x(s)$.

Because of

$$(\sqrt{(1+x^2)}\arctan x)' = \frac{x \arctan x}{\sqrt{(1+x^2)}} + \frac{1}{\sqrt{(1+x^2)}}$$

and

$$(\sqrt{(1+x^2)})' = \frac{x}{\sqrt{(1+x^2)}}$$

we have for $x_1, x_2 \in (-\infty, \infty)$

$$|\sqrt{(1+x_1^2)}\arctan x_1 - \sqrt{(1+x_2^2)}\arctan x_2|$$
$$\leqslant \sup_{x\in(-\infty,\infty)}\left(\frac{x\arctan x}{\sqrt{(1+x^2)}} + \frac{x}{\sqrt{(1+x^2)}}\right)|x_1 - x_2| \leqslant \left(\frac{\pi}{2}+1\right)|x_1 - x_2|$$

and

$$|(\sqrt{(1+x_1^2)} - \sqrt{(1+x_2^2)})\arctan x_2|$$
$$\leqslant \frac{\pi}{2}\sup_{x\in(-\infty,\infty)}\frac{x}{\sqrt{(1+x^2)}}|x_1 - x_2| \leqslant \frac{\pi}{2}|x_1 - x_2|.$$

The last inequalities imply that

$$I_\nu \leqslant 2\int_{\Delta_{2\nu}} |\sqrt{(1+(f'(t))^2)}\arctan f'(t) - \sqrt{(1+K_{2\nu}^2)}\arctan K_{2\nu}|\,\mathrm{d}t$$
$$+2\int_{\Delta_{2\nu}} |(\sqrt{(1+K_{2\nu}^2)} - \sqrt{(1+(f'(t))^2)})\arctan K_{2\nu}|\,\mathrm{d}t$$
$$\leqslant 2(\pi+1)\int_{\Delta_{2\nu}} |f'(t) - K_{2\nu}|\,\mathrm{d}t.$$

Thus we have for $\nu = 0, 1, \ldots, n-1$

$$\int_{2\nu h}^{(2\nu+1)h} |\theta(s+h) - \theta(s)|\,\mathrm{d}s \leqslant 2(\pi+1)\int_{\Delta_{2\nu}} |f'(t) - K_{2\nu}|\,\mathrm{d}t. \tag{9}$$

Choose η such that $2\delta \leqslant \eta \leqslant 4\delta$ $(0 < \delta \leqslant (b-a)/4)$ and $\eta = (b-a)/2m$ for some positive integer m.

Denote $d_i = [a + i\eta, a + (i+2)\eta]$. Also, denote by Ω_1 the set of all intervals $\Delta_\nu = [x_\nu, x_{\nu+2}]$ such that $\Delta_\nu \subset d_{2i}$ for some i $(0 \leqslant i \leqslant m-1)$ and by Ω_2 the set of all intervals Δ_ν such that $\Delta_\nu \subset d_{2i+1}$ for some i $(0 \leqslant i \leqslant m-1)$. Clearly $\{\Delta_\nu : \nu = 0, 1, \ldots, n-1\} = \Omega_1 \cup \Omega_2$. Set

$$C_i = \frac{1}{2\eta}\int_{d_i} f'(u)\,\mathrm{d}u.$$

If $\Delta_\nu \subset d_i$, then

$$\int_{\Delta_\nu} |f'(t) - K_\nu|\,\mathrm{d}t \leqslant \int_{\Delta_\nu} |f'(t) - C_i|\,\mathrm{d}t$$
$$+\int_{\Delta_\nu}\left|C_i - \frac{1}{|\Delta_\nu|}\int_{\Delta_\nu} f'(u)\,\mathrm{d}u\right|\mathrm{d}t$$
$$\leqslant 2\int_{\Delta_\nu} |f'(t) - C_i|\,\mathrm{d}t.$$

Consequently, we obtain for fixed i $(0 \leqslant i \leqslant 2m-2)$

$$\sum_{\Delta_\nu \subset d_i} \int_{\Delta_\nu} |f'(x) - K_\nu| \mathrm{d}x \leqslant 2 \sum_{\Delta_\nu \subset d_i} \int_{\Delta_\nu} |f'(x) - C_i| \mathrm{d}x$$

$$\leqslant 2 \int_{d_i} |f'(x) - C_i| \mathrm{d}x.$$

Hence, in view of the inequality (8) we obtain

$$\sum_{\Delta_\nu \in \Omega_1} \int_{\Delta_\nu} |f'(x) - K_\nu| \mathrm{d}x \leqslant 2 \sum_{i=0}^{m-1} \int_{d_{2i}} |f'(x) - C_{2i}| \mathrm{d}x$$

$$\leqslant 2 \sum_{i=0}^{m-1} \frac{1}{\eta} \int_0^{2\eta} \int_{a+2i\eta}^{a+(2i+2)\eta - t} |f'(x+t) - f'(x)| \mathrm{d}x \, \mathrm{d}t$$

$$\leqslant \frac{2}{\eta} \int_0^{2\eta} \int_a^{b-t} |f'(x+t) - f'(x)| \mathrm{d}x \, \mathrm{d}t \leqslant 4\omega(f', 2\eta)_L.$$

Thus we have

$$\sum_{\Delta_\nu \in \Omega_1} \int_{\Delta_\nu} |f'(x) - K_\nu| \mathrm{d}x \leqslant 32\omega(f'; \delta)_L. \tag{10}$$

Similarly we find

$$\sum_{\Delta_\nu \in \Omega_2} \int_{\Delta_\nu} |f'(x) - K_\nu| \mathrm{d}x \leqslant 32\omega(f'; \delta)_L. \tag{11}$$

The estimates (9)–(11) imply that for $0 \leqslant \delta \leqslant (b-a)/4$

$$\sigma_1 \leqslant C\omega(f'; \delta)_L, \quad C = \text{constant}. \tag{12}$$

Similarly we obtain for $i = 2, 3, 4$

$$\sigma_i \leqslant C\omega(f'; \delta)_L, \quad 0 \leqslant \delta \leqslant \frac{b-a}{4}.$$

Consequently, we have in the case $0 \leqslant \delta \leqslant (b-a)/4$

$$\int_0^{l-h} |\theta(s+h) - \theta(s)| \mathrm{d}s \leqslant 4C\omega(f'; \delta)_L, \quad 0 \leqslant h \leqslant \delta. \tag{13}$$

Hence

$$\omega(\theta; \delta)_L \leqslant 4C\omega(f'; \delta)_L, \quad 0 \leqslant \delta \leqslant (b-a)/4. \tag{14}$$

Since $\omega(\theta; \delta)_L = \omega(\theta; l)_L$ for $\delta \geqslant l$ $(l \geqslant b-a)$, the case $\delta = l$ for the inequality

(14) contains the case $\delta > l$. On the other hand $\omega(f'; b-a)_L \leqslant 4\omega(f'; \delta)_L$ for $(b-a)/4 < \delta \leqslant b-a$ and therefore the case $(b-a)/4 < \delta \leqslant b-a$ for (14) is contained in the case $\delta = b-a$. Consequently, to prove the inequality (14) when $\delta > (b-a)/4$ it suffices to establish the inequality (13) with $0 < h \leqslant l$ and $\delta = b-a$.

Now, suppose $0 < h \leqslant l$. Then we obtain similarly as in the proof of estimate (12)

$$\begin{aligned}
&\int_0^{l-h} |\theta(s+h) - \theta(s)|\,\mathrm{d}s \\
&\leqslant 2\int_0^l \left|\theta(s) - \arctan\left(\frac{1}{b-a}\int_a^b f'(u)\,\mathrm{d}u\right)\right|\mathrm{d}s \\
&\leqslant 2(\pi+1)\int_a^b \left|f'(x) - \frac{1}{b-a}\int_a^b f'(u)\,\mathrm{d}u\right|\mathrm{d}x \\
&\leqslant 4(\pi+1)\frac{1}{b-a}\int_0^{b-a}\int_a^{b-t} |f'(x+t) - f'(x)|\,\mathrm{d}x\,\mathrm{d}t \\
&\leqslant 4(\pi+1)\omega(f'; b-a)_L
\end{aligned}$$

as required. Thus lemma 10.2 is proved. □

Remark. A lower estimate of $\omega(\theta(f); \delta)_L$ by $\omega(f'; \delta)$ is not true in general. Indeed, for instance, if f is convex and bounded on $[0, 1]$, then $\theta = \theta(f)$ is monotone and bounded on $[0, l(f)]$ and therefore $\omega(\theta; \delta)_L = \mathrm{O}(\delta)$. On the other hand, it is readily seen that $\omega(f'; \delta)_L \geqslant \omega(f; \delta)_C$ for $0 \leqslant \delta \leqslant \frac{1}{2}$ where $\omega(f; \delta)_C$ is the modulus of continuity of f, which may tend to zero as slow as we want with $\delta \to 0$. However, if $f \in \mathrm{Lip}\,1$, then one can easily prove that $\omega(f'; \delta)_L = \mathrm{O}(\omega(\theta; \delta)_L)$, i.e. in this case $\omega(\theta; \delta)_L \asymp \mathrm{O}(\omega(f'; \delta)_L)$.

10.1.2 The o-effect in some spline approximations

We start with two lemmas concerning the intermediate approximation by means of polygons (broken lines). The first lemma is trivial, but the second one not and plays the main role in this chapter.

Lemma 10.3. *Let f be absolutely continuous on $[a, b]$ and φ be the polygon with knots $x_i = a + i(b-a)/n$, $i = 0, 1, \ldots, n$, $(n \geqslant 1)$ which interpolates f at these knots, i.e. $\varphi(x_i) = f(x_i)$ and φ is a linear function on each interval $[x_{i-1}, x_i]$.*

Then we have for $i = 1, 2, \ldots, n$

$$\|f - \varphi\|_{C[x_{i-1}, x_i]} \leqslant \frac{2n}{b-a}\int_0^{(b-a)/n}\int_{x_{i-1}}^{x_i - t} |f'(x+t) - f'(x)|\,\mathrm{d}x\,\mathrm{d}t$$

and therefore

$$\| f-\varphi \|_{L[a,b]} \leqslant \frac{2(b-a)}{n}\,\omega\left(f',\frac{b-a}{n}\right)_L.$$

Moreover

$$\| f'-\varphi' \|_{L[a,b]} \leqslant 2\omega\left(f';\frac{b-a}{n}\right)_L$$

and

$$V_a^b\varphi' \leqslant \frac{n}{b-a}\,\omega\left(f';\frac{b-a}{n}\right)_L,$$

where we take $\varphi'(x_i)=\frac{1}{2}(\varphi'(x_i-0)+\varphi'(x_i+0))$, $i=1,2,\dots,n-1$, *when we calculate* $V_a^b\varphi'$.

Proof. By our assumptions we have for $x\in[x_{i-1},x_i]$

$$\begin{aligned}\varphi(x)&=f(x_{i-1})+\frac{f(x_i)-f(x_{i-1})}{x_i-x_{i-1}}(x-x_{i-1})\\&=f(x_{i-1})+\int_{x_{i-1}}^{x}\left(\frac{1}{x_i-x_{i-1}}\int_{x_{i-1}}^{x_i}f'(u)\,\mathrm{d}u\right)\mathrm{d}t.\end{aligned}$$

Using the estimate (8) in subsection 10.1.1, see also lemma 7.6 in section (7.1), we get for $x\in[x_{i-1},x_i]$

$$\begin{aligned}|f(x)-\varphi(x)|&=\left|\int_{x_{i-1}}^{x}\left(f'(t)-\frac{1}{x_i-x_{i-1}}\int_{x_{i-1}}^{x_i}f'(u)\,\mathrm{d}u\right)\mathrm{d}t\right|\\&\leqslant\int_{x_{i-1}}^{x_i}\left|f'(t)-\frac{1}{x_i-x_{i-1}}\int_{x_{i-1}}^{x_i}f'(u)\,\mathrm{d}u\right|\mathrm{d}t\\&\leqslant\frac{2n}{b-a}\int_0^{(b-a)/n}\int_{x_{i-1}}^{x_i-t}|f'(x+t)-f'(x)|\,\mathrm{d}x\,\mathrm{d}t.\end{aligned}$$

In the same manner we obtain

$$\begin{aligned}&\int_a^b|f'(x)-\varphi'(x)|\,\mathrm{d}x\\&=\sum_{i=1}^{n}\int_{x_{i-1}}^{x_i}\left|f'(x)-\frac{1}{x_i-x_{i-1}}\int_{x_{i-1}}^{x_i}f'(u)\,\mathrm{d}u\right|\mathrm{d}x\\&\leqslant\frac{2n}{b-a}\sum_{i=1}^{n}\int_0^{(b-a)/n}\int_{x_{i-1}}^{x_i-t}|f'(x+t)-f'(x)|\,\mathrm{d}x\,\mathrm{d}t\\&\leqslant\frac{2n}{b-a}\int_0^{(b-a)/n}\int_a^{b-t}|f'(x+t)-f'(x)|\,\mathrm{d}x\,\mathrm{d}t\leqslant 2\omega\left(f';\frac{b-a}{n}\right)_L\end{aligned}$$

and also

$$V_a^b\varphi' = \sum_{i=1}^{n-1}\left|\frac{n}{b-a}\int_{x_i}^{x_{i+1}} f'(u)\,du - \frac{n}{b-a}\int_{x_{i-1}}^{x_i} f'(u)\,du\right|$$

$$\leqslant \frac{n}{b-a}\sum_{i=1}^{n-1}\int_{x_{i-1}}^{x_i}\left|f'\left(x+\frac{b-a}{n}\right)-f'(x)\right|dx$$

$$\leqslant \frac{n}{b-a}\int_a^{b-(b-a)/n}\left|f'\left(x+\frac{b-a}{n}\right)-f'(x)\right|dx \leqslant \frac{n}{b-a}\omega\left(f';\frac{b-a}{n}\right)_L. \quad \square$$

Lemma 10.4. *Let f be a function of bounded variation on $[a,b]$ and the complemented graph of f have the parametric equations*

$$\bar{f}: x = x(s), \quad y = y(s), \quad s \in [0,l], \quad l = l(f), \quad x(0) = a, \tag{15}$$

with respect to the natural parameter s. Denote $\theta = \theta(f)$. Let $n \geqslant 1$ and φ be the polygon that interpolates $\bar{f}$ at the points $(x(s_i), y(s_i))$, $s_i = il/n$, $i = 0, 1, \ldots, n$, i.e. *φ is the polygon which connects consecutively with line-segments the points $(x(s_i), y(s_i))$.*

Then we have

$$\| f - \varphi \|_{L[a,b]} \leqslant C\frac{l}{n}\omega\left(\theta;\frac{l}{n}\right)_L \tag{16}$$

and

$$V_0^{l(\varphi)}\theta(\varphi) \leqslant C\frac{n}{l}\omega\left(\theta;\frac{l}{n}\right)_L, \tag{17}$$

where we have taken $\theta(\varphi, s_i) = \frac{1}{2}(\theta(\varphi, s_i - 0) + \theta(\varphi, s_i + 0))$ at the knots s_i when $V_0^{l(\varphi)}\theta(\varphi)$ is calculated, $C > 0$ is an absolute constant.

To prove lemma 10.4 we need some auxiliary statements.

The Hausdorff distance between the functions will play an essential role in our evaluations. The Hausdorff distance with a parameter α was defined and applied in Chapter 9. Here we shall consider only the Hausdorff distance $r(f,g) = r(f,g;1)$ with parameter $\alpha = 1$, i.e. the Hausdorff distance generated by the following distance in the plane:

$$d(A,B) = d(A(a_1,a_2), B(b_1,b_2)) = \max\{|a_1 - b_1|, |a_2 - b_2|\}. \tag{18}$$

We shall need the Hausdorff distance $r(f_1, f_2)$ between functions f_1 and f_2 that may be defined on different intervals Δ_1 and Δ_2 respectively. By definition

$$r(f_1, f_2) = r(f_1, \Delta_1; f_2, \Delta_2) = r(\bar{f}_1, \bar{f}_2),$$

where $\bar{f}_i$ $(i = 1, 2)$ is the complemented graph of the function f_i defined on Δ_i.

Also, the Hausdorff distance can be defined as follows. For fixed $\varepsilon > 0$ denote by $D_\varepsilon = \{(x,y): d((x,y),(0,0)) \leqslant \varepsilon\}$ the ε-neighborhood of the origin with respect to the distance (18). Define for each two sets F and G in the plane

$$F+G=\{(x,y): x=x_1+x_2, y=y_1+y_2, (x_1,y_1)\in F, (x_2,y_2)\in G\}. \quad (19)$$

Then the ε-neighborhood $\bar{f}^{\varepsilon}$ of $\bar{f}$ is defined by $\bar{f}^{\varepsilon}=\bar{f}+D_{\varepsilon}$. Clearly, we have

$$r(f_1,f_2)=\inf\{\varepsilon: \bar{f}_1\subset \bar{f}_2^{\varepsilon}, \bar{f}_2\subset \bar{f}_1^{\varepsilon}\}. \quad (20)$$

We shall denote by $\rho(A,B)$ the Euclidean distance between the points A and B in the plane, i.e.

$$\rho(A,B)=\rho(A(a_1,a_2),B(b_1,b_2))=((a_1-b_1)^2+(a_2-b_2)^2)^{1/2}.$$

Note that $d(A,B)\leqslant \rho(A,B)$.

Lemma 10.5. *Let f be a function of bounded variation on $[a,b]$ and the complemented graph of f have the following parametric equations:*

$$\bar{f}: x=x(s), \quad y=y(s), \quad s\in[0,l], \quad l=l(f), \quad x(0)=a, \quad (21)$$

where s is the natural parameter of $\bar{f}$. Suppose $0\leqslant s_1\leqslant s_2\leqslant l$ and $\tilde{f}$ is the arc of $\bar{f}$ which is obtained from (21) for $s\in[s_1,s_2]$. Denote $x_i=x(s_i)$, $y_i=y(s_i)$, $i=1,2$. Let φ be the closed line-segment with endpoints (x_1,y_1) and (x_2,y_2).

Then we have

$$r(\tilde{f},\varphi)\leqslant \frac{8}{s_2-s_1}\int_0^{s_2-s_1}\int_{s_1}^{s_2-t}|\theta(s+t)-\theta(s)|\,\mathrm{d}s\,\mathrm{d}t \quad (22)$$

and

$$\left|\theta(\varphi)-\frac{1}{s_2-s_1}\int_{s_1}^{s_2}\theta(s)\,\mathrm{d}s\right|\leqslant \frac{2\pi}{(s_2-s_1)^2}\int_0^{s_2-s_1}\int_{s_1}^{s_2-t}|\theta(s+t)-\theta(s)|\,\mathrm{d}s\,\mathrm{d}t, \quad (23)$$

where

$$\theta(\varphi)=\begin{cases}\arctan\left(\dfrac{y_2-y_1}{x_2-x_1}\right), & x_1<x_2,\\[2ex] \dfrac{\pi}{2}\operatorname{sign}(y_2-y_1), & x_1=x_2, \theta=\theta(f).\end{cases}$$

Proof. The case $x_1=x_2$ is trivial.

Let $x_1<x_2$. Consider the linear function

$$\psi(x)=y_1+\tan\theta(\psi)\cdot(x-x_1), \quad x\in[x_1,x_3],$$

where

$$\theta(\psi)=\frac{1}{s_2-s_1}\int_{s_1}^{s_2}\theta(s)\,\mathrm{d}s, \quad x_3=x_1+(s_2-s_1)\cos\theta(\psi).$$

Clearly, $l(\psi)=l(\tilde{f})=s_2-s_1$. Denote $y_3=\psi(x_3)$.

The arcs $\tilde{f}$ and ψ (the graph of the function ψ) have the following parametric equations (see (6) in subsection 10.1.1):

$$\tilde{f}: \begin{cases} x = x_f(s) = x_1 + \displaystyle\int_0^s \cos\theta(s_1 + t)\,dt, \\ y = y_f(s) = y_1 + \displaystyle\int_0^s \sin\theta(s_1 + t)\,dt, \quad s \in [0, s_2 - s_1], \end{cases}$$

and

$$\psi: \begin{cases} x = x_\psi(s) = x_1 + \displaystyle\int_0^s \cos\theta(\psi)\,dt, \\ y = y_\psi(s) = y_1 + \displaystyle\int_0^s \sin\theta(\psi)\,dt, \quad s \in [0, s_2 - s_1]. \end{cases}$$

From the above representations of $\tilde{f}$ and ψ we get

$$\begin{aligned} &\rho((x_f(s), y_f(s)), (x_\psi(s), y_\psi(s))) \\ &\quad = ((x_f(s) - x_\psi(s))^2 + (y_f(s) - y_\psi(s))^2)^{1/2} \\ &\quad \leqslant \left| \int_0^s (\cos\theta(s_1 + t) - \cos\theta(\psi))\,ds \right| + \left| \int_0^s (\sin\theta(s_1 + t) - \sin\theta(\psi))\,ds \right| \\ &\quad \leqslant 2 \int_0^{s_2 - s_1} |\theta(s_1 + t) - \theta(\psi)|\,ds = 2 \int_{s_1}^{s_2} |\theta(s) - \theta(\psi)|\,ds. \end{aligned}$$

Set

$$T = \frac{4}{s_2 - s_1} \int_0^{s_2 - s_1} \int_{s_1}^{s_2 - t} |\theta(s + t) - \theta(s)|\,ds\,dt.$$

Then by the last estimates and the estimate (8) in subsection 10.1.1 (see also lemma 7.6 in section 7.1), it follows that

$$\rho((x_f(s), y_f(s)), (x_\psi(s), y_\psi(s))) \leqslant T, \quad s \in [0, s_2 - s_1]. \tag{24}$$

Hence

$$r(\tilde{f}, \psi) \leqslant T. \tag{25}$$

Since φ and ψ are line-segments with common endpoint (x_1, y_1) and the other endpoint of φ is on $\tilde{f}$, we have from (24) with $s = s_2 - s_1$

$$r(\varphi, \psi) \leqslant \rho((x_2, y_2), (x_3, y_3)) \leqslant T.$$

From this and (25) we get

$$r(\tilde{f}, \varphi) \leqslant r(\tilde{f}, \psi) + r(\psi, \varphi) \leqslant 2T,$$

i.e. the estimate (22) holds.

It remains to estimate $|\theta(\varphi) - \theta(\psi)|$. Consider the case $|\theta(\varphi) - \theta(\psi)| < \pi/2$. Let (x_4, y_4) be the orthogonal projection of the point (x_3, y_3) on the line continuation of φ. By (24) with $s = s_2 - s_1$ we have $\rho((x_2, y_2), (x_3, y_3)) \leqslant T$

and hence $\rho((x_4, y_4), (x_3, y_3)) \leqslant T$. Then we have

$$\frac{2}{\pi}|\theta(\varphi) - \theta(\psi)| \leqslant \sin|\theta(\varphi) - \theta(\psi)| \leqslant \frac{\rho((x_4, y_4), (x_3, y_3))}{l(\psi)} \leqslant \frac{T}{s_2 - s_1}$$

which implies (23).

Now, let $|\theta(\varphi) - \theta(\psi)| \geqslant \pi/2$. Because of $\rho((x_2, y_2), (x_3, y_3)) \leqslant T$ we have

$$s_2 - s_1 \leqslant \rho((x_2, y_2), (x_3, y_3)) \leqslant T.$$

Hence $|\theta(\varphi) - \theta(\psi)| \leqslant \pi \leqslant \pi T/(s_2 - s_1)$ which implies (23). □

The next lemma gives a relation between L_1-distance and the Hausdorff distance between functions of bounded variation.

Lemma 10.6. *If $V_a^b f < \infty$ and $V_a^b g < \infty$, then*

$$\| f - g \|_{L_1[a,b]} \leqslant 9 \min \{l(f), l(g)\} r(f, g), \tag{26}$$

where $l(f)$ and $l(g)$ are the lengths of $\bar{f}$ and $\bar{g}$ respectively.

Proof. Suppose $l(f) \leqslant l(g)$. Set $\varepsilon = r(f, g)$. Obviously $\varepsilon \leqslant b - a$. Denote by $\bar{f}^\varepsilon$ the ε-neighborhood of $\bar{f}$ with respect to the distance $d(A, B)$ from (18), i.e. $\bar{f}^\varepsilon = \bar{f} + D_\varepsilon$, where $D_\varepsilon = \{(x, y): d((x, y), (0, 0)) \leqslant \varepsilon\}$. Then in view of (20) we have $\bar{g} \subset \bar{f}^\varepsilon$ and therefore

$$\| f - g \|_{L[a,b]} \leqslant \operatorname{mes}_2 \bar{f}^\varepsilon, \tag{27}$$

where $\operatorname{mes}_2 \bar{f}^\varepsilon$ is the two-dimensional Lebesgue measure of the set $\bar{f}^\varepsilon$.

Denote

$$E'_\varepsilon = \{(x, y): x \in \{0, \varepsilon, -\varepsilon\}, |y| \leqslant \varepsilon\},$$
$$E''_\varepsilon = \{(x, y): y \in \{0, \varepsilon, -\varepsilon\}, |x| \leqslant \varepsilon\},$$

and $E_\varepsilon = E'_\varepsilon \cup E''_\varepsilon$.

Clearly, since $\varepsilon \leqslant b - a$, it follows that (see (19))

$$\bar{f}^\varepsilon = \bar{f} + D_\varepsilon = \bar{f} + E_\varepsilon = (\bar{f} + E'_\varepsilon) \cup (\bar{f} + E''_\varepsilon)$$

and therefore by (27)

$$\| f - g \|_{L[a,b]} \leqslant \operatorname{mes}_2 (\bar{f} + E'_\varepsilon) + \operatorname{mes}_2 (\bar{f} + E''_\varepsilon).$$

It is readily seen that all the sets in consideration are measurable and

$$\operatorname{mes}_2(\bar{f} + E'_\varepsilon) = 6\varepsilon(b - a), \quad \operatorname{mes}_2(\bar{f} + E''_\varepsilon) \leqslant 6\varepsilon V_a^b f.$$

Hence

$$\| f - g \|_{L[a,b]} \leqslant 6\varepsilon(b - a + V_a^b f) \leqslant 9 l(f) r(f, g).$$ □

Proof of lemma 10.4. Denote by f_i the arc of the curve $\bar{f}$ which is obtained from (15) for $s \in [s_{i-1}, s_i]$, $s_i = il/n$, and by φ_i the line-segment of φ which connects the points $(x(s_{i-1}), y(s_{i-1}))$ and $(x(s_i), y(s_i))$. Also denote

$$\theta_i = \begin{cases} \arctan \dfrac{y(s_i) - y(s_{i-1})}{x(s_i) - x(s_{i-1})}, & x(s_{i-1}) < x(s_i), \\ \dfrac{\pi}{2} \operatorname{sign}(y(s_i) - y(s_{i-1})), & x(s_{i-1}) = x(s_i), \end{cases}$$

$$\theta_i^* = \frac{1}{s_i - s_{i-1}} \int_{s_{i-1}}^{s_i} \theta(s)\, \mathrm{d}s$$

and

$$\Delta_i = [x(s_{i-1}), x(s_i)].$$

By lemma 10.5 we have for $i = 1, 2, \ldots, n$

$$r(f_i, \varphi_i) \leqslant \frac{8n}{l} \int_0^{l/n} \int_{s_{i-1}}^{s_i - t} |\theta(s+t) - \theta(s)|\, \mathrm{d}s\, \mathrm{d}t \tag{28}$$

and

$$|\theta_i - \theta_i^*| \leqslant \frac{2\pi n^2}{l^2} \int_0^{l/n} \int_{s_{i-1}}^{s_i - t} |\theta(s+t) - \theta(s)|\, \mathrm{d}s\, \mathrm{d}t. \tag{29}$$

The estimate (28) and lemma 10.6 imply that for $i = 1, 2, \ldots, n$

$$\| f - \varphi \|_{L(\Delta_i)} \leqslant \frac{9l}{n} r(f_i, \varphi_i) \leqslant 72 \int_0^{l/n} \int_{s_{i-1}}^{s_i - t} |\theta(s+t) - \theta(s)|\, \mathrm{d}s\, \mathrm{d}t.$$

Summing these inequalities we get

$$\begin{aligned} \| f - \varphi \|_{L[a,b]} &= \sum_{i=1}^{n} \| f - \varphi \|_{L(\Delta_i)} \\ &\leqslant 72 \sum_{i=1}^{n} \int_0^{l/n} \int_{s_{i-1}}^{s_i - t} |\theta(s+t) - \theta(s)|\, \mathrm{d}s\, \mathrm{d}t \\ &\leqslant 72 \int_0^{l/n} \int_0^{l-t} |\theta(s+t) - \theta(s)|\, \mathrm{d}s\, \mathrm{d}t \\ &\leqslant \frac{72l}{n} \omega\left(\theta; \frac{l}{n}\right)_L, \end{aligned}$$

i.e. the estimate (16) holds.

Now we estimate $V_0^{l(\varphi)}\theta(\varphi)$. By (29) we obtain

$$V_0^{l(\varphi)}\theta(\varphi) = \sum_{i=2}^{n} |\theta_i - \theta_{i-1}| \leqslant \sum_{i=2}^{n} |\theta_i^* - \theta_{i-1}^*| + 2 \sum_{i=1}^{n} |\theta_i - \theta_i^*|$$

$$\leqslant \frac{n}{l} \sum_{i=1}^{n-1} \int_{s_{i-1}}^{s_i} \left| \theta\left(s + \frac{l}{n}\right) - \theta(s) \right| \mathrm{d}s$$
$$+ \frac{4\pi n^2}{l^2} \int_0^{l/n} \left(\sum_{i=1}^{n} \int_{s_{i-1}}^{s_i - t} |\theta(s+t) - \theta(s)| \mathrm{d}s \right) \mathrm{d}t$$
$$\leqslant (1 + 4\pi) \frac{n}{l} \omega\left(\theta, \frac{l}{n}\right)_L.$$

Thus the estimate (17) is established. □

Now we shall apply the functional characteristic $\omega(\theta, \delta)_L$ for describing the o-effect in some spline approximations. As in the previous chapters (see section 7.3), we shall denote by $S_n^k(f)_p$ and naturally by $S_n^k(f)_r$ the best approximations of the function f by means of all piecewise polynomial functions of degree $k-1$ with $n+1$ (free) knots on $[a, b]$ with respect to L_p and Hausdorff metric respectively.

Lemma 10.4 directly implies the following theorem.

Theorem 10.1. *Let $V_a^b f < \infty$. Then we have*

$$S_n^2(f)_1 = \mathrm{o}\left(\frac{1}{n}\right). \tag{30}$$

Moreover

$$S_n^2(f)_1 \leqslant C \frac{l}{n} \omega\left(\theta(f); \frac{l}{n}\right)_L, \quad n = 1, 2, \ldots, \tag{31}$$

where $l = l(f)$, $C =$ constant.

Theorem 10.2. *Let $f \in V_r$, $r \geqslant 1$. Then we have*

$$S_n^{r+2}(f)_C = \mathrm{o}\left(\frac{1}{n^{r+1}}\right). \tag{32}$$

Moreover

$$S_n^{r+2}(f)_C \leqslant C \frac{\omega(\theta(f^{(r)}), l/n)_L}{n^{r+1}}, \quad n = 1, 2, \ldots, \tag{33}$$

where $l = l(f^{(r)})$, $C = C_1(r) l (b-a)^{r-1}$.

Proof. By lemma 7.13 in section 7.3 we have

$$S_{3n}^{r+2}(f)_C \leqslant \frac{(b-a)^{r-1}}{n^r} S_n^2(f^{(r)})_1, \quad n = 1, 2, \ldots$$

This estimate and (31) imply (33). □

Theorem 10.3. *Let $V_a^b f < \infty$. Then we have*

$$S_n^2(f)_r = \mathrm{o}\left(\frac{1}{n}\right). \tag{34}$$

Moreover

$$S_n^2(f)_r \leqslant C\frac{\omega(\theta(f); l/n)_L}{n}, \quad n=1,2,\ldots, \tag{35}$$

where $l = l(f)$, $C =$ constant.

Proof. Let

$$\bar{f}: x = x(s), \quad y = y(s), \quad s \in [0, l], \quad x(0) = a, \tag{36}$$

be the parametric representation of the complemented graph $\bar{f}$ of the function f with respect to the natural parameter s. Suppose $n \geqslant 1$. Set

$$\theta_1(s) = \frac{n}{l}\int_{\Delta_i} \theta(s)\,\mathrm{d}s, \quad s \in \Delta_i = \left[\frac{(i-1)l}{n}, \frac{il}{n}\right],$$

$i = 1, 2, \ldots, n$, where $\theta = \theta(f)$. By lemma 7.6 in section 7.1 it follows that

$$\|\theta - \theta_1\|_{L[0,l]} = \sum_{i=1}^{n} \|\theta - \theta_1\|_{L(\Delta_i)}$$
$$\leqslant \sum_{i=1}^{n} \frac{2n}{l}\int_0^{l/n}\int_{(i-1)l/n}^{il/n - t} |\theta(s+t) - \theta(s)|\,\mathrm{d}s\,\mathrm{d}t \leqslant 2\omega\left(\theta; \frac{l}{n}\right)_L.$$

Hence, there exist points $0 = t_0 < t_1 < \cdots < t_n = l$ such that

$$\|\theta - \theta_1\|_{L[t_{i-1}, t_i]} \leqslant \frac{2}{n}\omega\left(\theta; \frac{l}{n}\right)_L, \quad i = 1, 2, \ldots, n. \tag{37}$$

Let $\{s_i\}_{i=0}^{2n} = \{il/n\}_{i=0}^{n} \cup \{t_i\}_{i=0}^{n}$ and $0 = s_0 \leqslant s_1 \leqslant \cdots \leqslant s_{2n} = l$. Then (37) implies that

$$\|\theta - \theta_1\|_{L[s_{i-1}, s_i]} \leqslant \frac{2}{n}\omega\left(\theta; \frac{l}{n}\right)_L \tag{38}$$

and θ_1 equals some constant in each interval (s_{i-1}, s_i).

Denote by f_i the arc of $\bar{f}$ which is obtained from (36) for $s \in [s_{i-1}, s_i]$. Let φ be the polygon that is defined connecting the consecutive points $(x(s_i), y(s_i))$, $i = 0, 1, \ldots, 2n$, with line-segments. By φ_i we denote the line-segment of φ which connects the points $(x(s_{i-1}), y(s_{i-1}))$ and $(x(s_i), y(s_i))$.

By lemma 10.5 and (38) we get

$$r(f_i, \varphi_i) \leqslant \frac{8}{s_i - s_{i-1}}\int_0^{s_i - s_{i-1}}\int_{s_{i-1}}^{s_i - t} |\theta(s+t) - \theta(s)|\,\mathrm{d}s\,\mathrm{d}t$$
$$= \frac{8}{s_i - s_{i-1}}\int_0^{s_i - s_{i-1}}\int_{s_{i-1}}^{s_i - t} |\theta(s+t) - \theta_1(s+t) - (\theta(s) - \theta_1(s))|\,\mathrm{d}s\,\mathrm{d}t$$
$$\leqslant 16\|\theta - \theta_1\|_{L[s_{i-1}, s_i]} \leqslant \frac{32}{n}\omega\left(\theta; \frac{l}{n}\right)_L.$$

Consequently

$$r(f,\varphi)\leqslant\frac{32}{n}\omega\left(\theta;\frac{l}{n}\right)_L$$

and

$$\varphi\in S(2,2n,[a,b]).$$

Hence

$$S_{2n}^2(f)_r\leqslant\frac{32}{n}\omega\left(\theta;\frac{l}{n}\right)_L,\quad n=1,2,\ldots$$

This estimate implies (35) for $n\geqslant 2$. The estimate (35) for $n=1$ follows from lemma 10.5. □

Remark. The estimates (30), (32) and (34) are exact in the corresponding function classes. The exactness of these estimates can be established similarly as for the rational approximation; see section 11.1.

10.1.3 Uniform approximation of individual functions of the class V_r and L_1 approximation of functions of bounded variation

In theorem 5.1 in section 5.2 we proved that

$$\sup_{f\in V_r} R_n(f)_C=\mathrm{O}\left(\frac{1}{n^{r+1}}\right),\quad r\geqslant 1.$$

The following theorem establishes the existence and character of the o-effect for the rational uniform approximation of every individual function $f\in V_r$.

Theorem 10.4. *Let $f\in V_r$ $(r\geqslant 1)$. Then we have*

$$R_n(f)_C=\mathrm{o}\left(\frac{1}{n^{r+1}}\right).\tag{39}$$

Moreover

$$R_n(f)_C\leqslant C\frac{\omega(\theta(f^{(r)});l/n)_L}{n^{r+1}},\quad n\geqslant r+1,\tag{40}$$

where $l=l(f^{(r)})$ and $C=C_1(r)l(b-a)^{r-1}$.

Remark. In theorem 11.4 in subsection 11.1.3 we prove that the estimate (39) is exact with respect to the order in the class.

Corollary 10.1. (i) *If f is defined on $[a,b]$ and $f^{(r)}$ $(r\geqslant 1)$ is absolutely continuous, then*

$$R_n(f)_C=\mathrm{O}\left(\frac{\omega(f^{(r+1)};1/n)_L}{n^{r+1}}\right)\tag{41}$$

(compare with theorem 10.6 in section 10.2).

(ii) *If f is defined on $[a,b]$ and $f^{(r)}$ $(r \geqslant 1)$ is convex and bounded, then*

$$R_n(f)_C = \mathrm{O}\left(\frac{1}{n^{r+2}}\right). \tag{42}$$

Proof of theorem 10.4. We shall prove the estimate (40) only in the case $r \geqslant 2$. The proof of (40) in the most essential case $r = 1$ is complicated and needs more precise techniques which are unfortunately too long to be included in this book.

By theorem 8.9 in section 8.3 we have for $n \geqslant r+1$ and $r \geqslant 2$

$$R_n(f)_C \leqslant C \frac{\sum_{\nu=1}^{n} \nu^{\alpha-1} S_\nu^2(f^{(r)})_1}{n^{r+\alpha}}, \tag{43}$$

where $C = C_1(r,\alpha)(b-a)^{r-1}$ and $\alpha > 0$. On the other hand, by theorem 10.1 in subsection 10.1.2 we have for $n = 1, 2, \ldots$

$$S_n^2(f^{(r)})_1 \leqslant C \frac{l}{n} \omega\left(\theta(f^{(r)}); \frac{l}{n}\right)_L \tag{44}$$

where $l = l(f^{(r)})$ and $C =$ constant.

Combining (43) with $\alpha = 3$ and (44) we get

$$\begin{aligned} R_n(f)_C &\leqslant C(r)(b-a)^{r-1} l \frac{\sum_{\nu=1}^{n} \nu\omega(\theta(f^{(r)}); l/\nu)_L}{n^{r+3}} \\ &\leqslant C(r)(b-a)^{r-1} l \frac{\sum_{\nu=1}^{n} \nu(n/\nu + 1)\omega(\theta(f^{(r)}); l/n)_L}{n^{r+3}} \\ &\leqslant C_1(r)(b-a)^{r-1} l \frac{\omega(\theta(f^{(r)}); l/n)_L}{n^{r+1}}. \end{aligned}$$

Thus the estimate (40) is proved when $r \geqslant 2$ and (40) implies (39) in this case.

Now we prove (39) in the case $r = 1$. By theorem 8.9 in section 8.3 we have for $n = 1, 2, \ldots$

$$R_n(f)_C \leqslant C \frac{\sum_{\nu=0}^{n} (\nu+1)^{\alpha-1} R_\nu(f')_p}{n^{\alpha+1}}, \tag{45}$$

where $C = C_1(p,\alpha)(b-a)^{1-1/p}$, $\alpha > 0$, $p > 1$. Theorem 10.11 in section 10.4 implies that

$$R_n(f')_p = \mathrm{o}\left(\frac{1}{n}\right). \tag{46}$$

The estimates (45) and (46) imply (39) in the case $r = 1$. □

Proof of corollary 10.1. If $f^{(r)}$ $(r \geqslant 1)$ is absolutely continuous on $[a,b]$, then by lemma 10.2 in subsection 10.1.1 we have $\omega(\theta(f^{(r)});\delta)_L \leqslant C\omega(f^{(r+1)};\delta)_L$ for $\delta \geqslant 0$. Then (41) follows from (40) immediately.

Suppose $f^{(r)}$ $(r \geqslant 1)$ is convex and bounded on $[a,b]$. Then $\theta(f^{(r)})$ is monotone and bounded on $[0, l(f^{(r)})]$ and hence $\omega(\theta(f^{(r)});\delta)_L = \mathrm{O}(\delta)$. Thus (40) implies (42) directly.

Theorem 10.1 in subsection 10.1.2 establishes the o-effect for the L_1 spline approximation of functions of bounded variations: if $V_a^b f < \infty$, then

$$S_n^2(f)_1 \leqslant C\frac{l}{n}\omega\left(\theta(f);\frac{l}{n}\right)_L, \quad n = 1, 2, \ldots, \tag{47}$$

where $l = l(f)$, $C =$ constant.

A similar estimate holds also for the rational approximation.

Theorem 10.5. *Let $V_a^b f < \infty$. Then we have*

$$R_n(f)_1 = \mathrm{o}\left(\frac{1}{n}\right). \tag{48}$$

Moreover

$$R_n(f)_1 \leqslant C\frac{l}{n}\omega\left(\theta(f);\frac{l}{n}\right)_L, \quad n = 1, 2, \ldots, \tag{49}$$

where $l = l(f)$, $C =$ constant.

Remark. The exactness of the estimate (48) is proved in theorem 11.6 in subsection 11.1.5.

Proof of theorem 10.5. By theorem 8.1 in section 8.1 we have for each $\alpha > 0$

$$R_n(f)_1 \leqslant C(\alpha)\frac{\sum_{v=1}^n v^{\alpha-1} S_v^2(f)_1}{n^\alpha}, \quad n = 1, 2, \ldots. \tag{50}$$

Then by (47) and (50) with $\alpha = 3$ we get

$$R_n(f)_1 \leqslant C_1 l \frac{\sum_{v=1}^n v\omega(\theta(f); l/v)_L}{n^3}$$

$$\leqslant C_1 l \frac{\sum_{v=1}^n v(n/v+1)\omega(\theta(f); l/n)_L}{n^3} \leqslant C_2 \frac{l}{n}\omega\left(\theta(f);\frac{l}{n}\right)_L.$$

Thus the estimate (49) is proved and obviously (49) implies (48). □

10.2 Uniform approximation of absolutely continuous functions

In section 5.3 we considered the rational uniform approximation of some classes of absolutely continuous functions. Here we investigate the o-effect for these approximations.

Theorem 10.6. *If* $f \in W_p^r[0,1]$, *then the estimate*

$$R_n(f)_q = \mathrm{o}\left(\frac{1}{n^r}\right) \tag{1}$$

holds in the following situations:

(i) $r = 1, p = 1, 1 \leqslant q < \infty$,
(ii) $r = 1, p > 1, q = \infty$,
(iii) $r \geqslant 2, p = 1, q = \infty$.

Moreover for an arbitrary positive integer k the estimate

$$R_n(f)_q \leqslant C\frac{\omega_k(f^{(r)}; 1/n)_p}{n^r}, \quad n \geqslant r + k - 1, \quad C = C(p, q, r, k), \tag{2}$$

holds in the above situations (i)–(iii).

Proof. If $f \in W_p^r[0,1]$, then by theorem 8.9 in section 8.3 the estimate

$$R_n(f)_q \leqslant C\frac{\sum_{v=1}^n v^{\alpha-1} S_v^k(f^{(r)})_p}{n^{r+\alpha}}, \quad n \geqslant r + k - 1, \quad C = C(p, q, r, k, \alpha), \tag{3}$$

holds in the situations (i)–(iii) from the suppositions of the theorem.

On the other hand by lemma 7.14 it follows that for each $f \in L_p[0,1]$, $1 \leqslant p \leqslant \infty$, and $k \geqslant 1$

$$S_n^k(f)_p \leqslant C(p,k)\omega_k\left(f; \frac{1}{n}\right)_p, \quad n = 1, 2, \ldots. \tag{4}$$

The estimates (3) with $\alpha = k + 1$ and (4) imply that

$$R_n(f)_q \leqslant C_1 \frac{\sum_{v=1}^n v^k \omega_k(f^{(r)}; 1/v)_p}{n^{r+k+1}}$$

$$\leqslant C_1 \frac{\sum_{v=1}^n v^k (n/v + 1)^k \omega_k(f^{(r)}; 1/n)_p}{n^{r+k+1}} \leqslant C\frac{\omega_k(f^{(r)}; 1/n)_p}{n^r}.$$

The estimate (2) implies (1). □

Theorem 10.7. *If f is absolutely continuous on* $[0,1]$ *and* $f' \in L\log L$, *then*

$$R_n(f)_C = \mathrm{o}\left(\frac{1}{n}\right). \tag{5}$$

Proof. By theorem 5.6 (see also theorem 5.5) we obtain

$$R_{2n+1}(f)_C \leqslant C\frac{E_n(f')_{L\log L}}{n}, \quad n = 1, 2, \ldots, \tag{6}$$

where $E_n(f')_{L\log L}$ is the best approximation to f' in the Orlicz space $L\log L$

by means of algebraic polynomials of degree no greater than n. It is well known that for $g \in L \log L$ $E_n(g)_{L \log L} \to 0$ as $n \to \infty$ and (6) implies (5). □

Now we establish the exact order of the rational uniform approximation of individual absolutely continuous function with a given modulus of continuity. Consider the class $V(\omega) = V(M, [a,b], \omega)$ of all functions f continuous on $[a,b]$ such that $V_a^b f \leqslant M$ and $\omega(f;\delta)_C \leqslant \omega(\delta)$, $\delta \geqslant 0$ where ω is a given modulus of continuity. In theorem 5.7 in subsection 5.3.4 we proved that for each

$$R_n(f)_C \leqslant C \min_{1 \leqslant t \leqslant n} \left\{ \frac{M}{t} + \omega\left(\frac{b-a}{te^{n/t}}\right) \right\}, \quad n = 1, 2, \ldots . \tag{7}$$

In theorem 11.7 in section 11.2 we establish the exactness of this estimate in the class $V(\omega)$ when $\omega(\delta)/\delta \to \infty$ as $\delta \to 0$. Of course the estimate (7) holds also for all absolutely continuous functions in $V(\omega)$ and it is exact for this class. However, the o-effect appears for 'good' moduli of continuity ω.

Theorem 10.8. *Let f be absolutely continuous on $[a,b]$ and $\omega(f;\delta)_C \leqslant \omega(\delta)$ for $\delta \geqslant 0$, where ω is a given modulus of continuity. Then there exists a sequence $\{\varepsilon_n(f)\}_{n=1}^{\infty}$, $\varepsilon_n(f) \to 0$ as $n \to \infty$, such that*

$$R_n(f)_C \leqslant C \min_{1 \leqslant t \leqslant n} \left\{ \frac{\varepsilon_n(f)}{t} + \omega\left(\frac{b-a}{te^{n/t}}\right) \right\}, \quad n = 1, 2, \ldots, \tag{8}$$

where $C > 0$ is an absolute constant. Moreover

$$R_n(f)_C \leqslant C \min_{1 \leqslant t \leqslant n} \left\{ \frac{\omega(f'; (b-a)t/n^2)_L}{t} + \omega\left(\frac{b-a}{te^{n/t}}\right) \right\}, \quad n = 1, 2, \ldots . \tag{9}$$

The estimates (8) and (9) imply the following estimates.

Corollary 10.2 (Newman's conjecture). *If $f \in$* Lip 1, then

$$R_n(f)_C = \mathrm{o}\left(\frac{1}{n}\right). \tag{10}$$

Corollary 10.3. (i) *If f is absolutely continuous on $[a,b]$ and $\omega(f;\delta)_C = \mathrm{O}(\delta^\gamma)$, $0 < \gamma < 1$, then*

$$R_n(f)_C = \mathrm{o}(\ln n/n).$$

(ii) *If f is absolutely continuous on $[a,b]$ and $\omega(f;\delta)_C = \mathrm{O}((\ln(1/\delta))^{-\gamma})$, $\gamma > 0$, then*

$$R_n(f)_C = \mathrm{o}(n^{-\gamma/(1+\gamma)}).$$

(iii) *If f is absolutely continuous on $[a,b]$ and $\omega(f;\delta)_C = \mathrm{O}((\underbrace{\ln \cdots \ln}_{k}(1/\delta))^{-\gamma})$, $\gamma > 0$, $k \geqslant 2$, then*

$$R_n(f)_C = \mathrm{O}((\underbrace{\ln \cdots \ln}_{k-1} n)^{-\gamma}),$$

Remark. The estimates (8), (10) and those of corollary 10.3 are exact with respect to the order in the corresponding classes. This fact can be proved in a similar way to the exactness of estimate (7) in $V(\omega)$ (see theorem 11.7 in section 11.2). The precise proof of the exactness of the above estimates is omitted.

Proof of theorem 10.8. We shall prove the estimate (9) which implies (8) with $\varepsilon_n(f)=\omega(f';(b-a)/n)_L$.

Let $n\geqslant 1$ and $1\leqslant t\leqslant n$. Set $m=[n^2/t]$. Let φ be the polygon with knots $x_i=a+i(b-a)/m$, $i=0,1,\dots,m$, which interpolate f at these knots. Set $\varphi'(x_i)=\frac{1}{2}(\varphi'(x_i-0)+\varphi'(x_i+0))$ for $i=1,2,\dots,m-1$. Then by lemma 10.3 in subsection 10.1.2 we have

$$V_a^b(f-\varphi)=\|f'-\varphi'\|_{L[a,b]}\leqslant 2\omega\left(f',\frac{b-a}{m}\right)_L \tag{11}$$

and

$$V_a^b\varphi'\leqslant\frac{m}{b-a}\omega\left(f',\frac{b-a}{m}\right)_L. \tag{12}$$

It is not difficult to see that

$$\omega(f-\varphi;\delta)_C\leqslant 2\omega(f;\delta)_C\leqslant 2\omega(\delta),\quad \delta\geqslant 0. \tag{13}$$

By (7), (11) and (13) we get

$$\begin{aligned}R_n(f-\varphi)_C&\leqslant C\left\{\frac{V_a^b(f-\varphi)}{t}+\omega\left(f-\varphi;\frac{b-a}{te^{n/t}}\right)_C\right\}\\&\leqslant 4C\left\{\frac{\omega(f';(b-a)t/n^2)_L}{t}+\omega\left(\frac{b-a}{te^{n/t}}\right)\right\}.\end{aligned}$$

On the other hand by theorem 5.1 in section 5.2 and (12) we obtain

$$\begin{aligned}R_n(\varphi)_C&\leqslant C_1\frac{V_a^b\varphi'(b-a)}{n^2}\leqslant C_1\frac{m\omega(f';(b-a)/m)_L}{n^2}\\&\leqslant 2C_1\frac{\omega(f';(b-a)t/n^2)_L}{t}.\end{aligned}$$

Consequently, for each $t\in[1,n]$

$$\begin{aligned}R_{2n}(f)_C&\leqslant R_n(f-\varphi)_C+R_n(\varphi)_C\\&\leqslant(4C+2C_1)\left\{\frac{\omega(f';(b-a)t/n^2)_L}{t}+\omega\left(\frac{b-a}{te^{n/t}}\right)\right\}\end{aligned}$$

which implies the estimate (9). □

Proof of corollary 10.2. The estimate (10) follows immediately from (9) setting $t = \max\{1, n/\ln(e + 1/\omega(f', 1/n)_L)\}$. □

Proof of corollary 10.3. The assertions (i)–(iii) in corollary 10.3 follow immediately setting for sufficiently large n consecutively $t = \alpha n/(2 \ln n)$, $t = \sqrt{\varepsilon_n(f)} \cdot n^{\gamma/(1+\gamma)} + 2$ and $t = (\underbrace{\ln \cdots \ln}_{k-1} n)^\gamma$ in the estimate (8). □

10.3 Uniform approximation of convex functions

In section 5.5 we found the exact order of the rational uniform approximation of the class $\mathrm{Conv}_M[a, b]$ of all functions f convex and continuous on $[a, b]$ such that $\| f \|_{C[a,b]} \leqslant M$. The same problem was solved also for the class $\mathrm{Conv}_M(\alpha, [a, b])$ of all functions f convex on $[a, b]$ such that $\omega(f, \delta)_C \leqslant M\delta^\alpha$ for $\delta \geqslant 0$. Here we shall prove the existence of the o-effect for the rational uniform approximation of each individual function of these classes.

Theorem 10.9. *Let the function f be convex and continuous on $[a, b]$. Then*

$$R_n(f)_C = o\left(\frac{1}{n}\right). \tag{1}$$

Moreover

$$R_n(f)_C \leqslant C \frac{\omega(f; (b-a)/n)_C}{n}, \quad n = 1, 2, \ldots, \tag{2}$$

where $C =$ constant.

Remark. The estimate (1) is exact in the class $\mathrm{Conv}_M[a, b]$ (see theorem 11.5 in subsection 11.1.4). However, the estimate (2) is not exact with respect to the order of $\omega(f; \delta)_C$. In this book we do not consider the rational approximation of the class of all convex functions with a given modulus of continuity except the class of all convex and $\mathrm{Lip}\,\alpha$ functions.

Proof of theorem 10.9. Set

$$\tilde{f}'(x) = \tfrac{1}{2}(f'(x-0) + f'(x+0)), \quad x \in (a, b),$$

$$f_1(x) = \begin{cases} f(a+\delta) + \tilde{f}'(a+\delta)(x - a - \delta), & x \in [a, a+\delta], \\ f(x), & x \in (a+\delta, b-\delta), \\ f(b-\delta) + \tilde{f}'(b-\delta)(x - b + \delta), & x \in [b-\delta, b], \end{cases}$$

where $\delta = (b-a)/2n$ and $f_2 = f - f_1$.

Clearly, f_1 is a primitive of the function $\tilde{f}'_1(x) = \frac{1}{2}(f'_1(x-0) + f'_1(x+0))$, $x \in (a, b)$, and since f is convex

$$V_a^b \tilde{f}'_1 \leqslant \tilde{f}'(b-\delta) - \tilde{f}'(a+\delta) \leqslant \frac{2\omega(f; \delta)_C}{\delta} = \frac{4n}{b-a}\omega\left(f; \frac{b-a}{2n}\right)_C.$$

Then by theorem 5.1 in section 5.2 we obtain

$$R_n(f_1)_C \leqslant C\frac{V_a^b\tilde{f}'(b-a)}{n^2} \leqslant 4C\frac{\omega(f;(b-a)/2n)_C}{n}. \tag{3}$$

On the other hand, obviously f_2 is convex and continuous on $[a, b]$ and

$$\| f_2 \|_{C[a,b]} \leqslant \omega(f;\delta)_C = \omega\left(f;\frac{b-a}{2n}\right)_C.$$

Hence, by theorem 5.11 in section 5.5

$$R_n(f_2)_C \leqslant C_1\frac{\| f_2 \|_{C[a,b]}}{n} \leqslant C_1\frac{\omega(f;(b-a)/2n)_C}{n}.$$

This estimate together with (3) gives

$$R_{2n}(f)_C \leqslant R_n(f_1)_C + R_n(f_2)_C \leqslant (4C + C_1)\frac{\omega(f;(b-a)/2n)_C}{n}$$

which implies (2). □

Theorem 10.10. *If f is convex on $[a, b]$ and $\omega(f;\delta)_C \leqslant M\delta^\alpha$ for $\delta \geqslant 0$, where $M > 0$, $0 < \alpha \leqslant 1$, then*

$$R_n(f)_C = \mathrm{o}\left(\frac{1}{n^2}\right).$$

Proof. It is readily seen that it suffices to prove the theorem only in the case $[a, b] = [0, 1]$, $M = 1$ and f nonincreasing on $[0, 1]$.

Let $\varepsilon > 0$. Choose $d \in (0, 1)$ such that $12^2 C(\alpha) d^\alpha \leqslant \varepsilon/2$, where $C(\alpha) > 0$ is the constant from theorem 5.13 in section 5.5. Denote $\Delta_1 = [0, d]$ and $\Delta_2 = [d, 1]$. By theorem 5.13 it follows that for each $n \geqslant 6$ there exists a rational function $r_1 \in R_{[n/6]}$ such that

$$\| f - r_1 \|_{C(\Delta_1)} \leqslant C(\alpha)\frac{|\Delta_1|^\alpha}{[n/6]^2} \leqslant \frac{12^2 C(\alpha) d^\alpha}{n^2} < \frac{\varepsilon}{2n^2}. \tag{4}$$

Since f is convex and nonincreasing on $[0, 1]$ and $\omega(f;\delta)_C \leqslant \delta^\alpha$, $\delta \geqslant 0$, f is a primitive of the function $\tilde{f}'(x) = \frac{1}{2}(f'(x-0) + f'(x+0))$, $x \in (0, 1)$ and $V_{\Delta_2}\tilde{f}' \leqslant \tilde{f}'(d) < \infty$. Then by theorem 10.4 in subsection 10.1.3 it follows that there exists a sequence $\{\varepsilon_n\}_{n=1}^\infty$, $\varepsilon_n \to 0$ as $n \to \infty$ such that

$$R_n(f, \Delta_2)_C \leqslant \frac{\varepsilon_n}{n^2}, \quad n = 1, 2, \ldots. \tag{5}$$

Choose n_0 such that for $n \geqslant n_0$ we have $\varepsilon_{[n/6]}/[n/6]^2 \leqslant \varepsilon/2n^2$, $2/n < \varepsilon/2$ and $B_2 \ln(\mathrm{e} + n^{3/\alpha}) \ln(\mathrm{e} + n^3) \leqslant n/3$, where $B_2 > 1$ is the constant from lemma 5.4 in section 5.1.

By (5) it follows that for each $n \geqslant n_0$ there exists a rational function $r_2 \in R_{[n/6]}$ such that

$$\| f - r_2 \|_{C(\Delta_2)} \leqslant \frac{\varepsilon_{[n/6]}}{[n/6]^2} \leqslant \frac{\varepsilon}{2n^2}. \tag{6}$$

Now we apply lemma 5.4 in section 5.1 for 'joining' of rational functions with $\varepsilon_2 = 1/n^3$ and $\delta = (1/n^3)^{1/\alpha}$. We conclude that for each $n \geqslant n_0$ there exists a rational function r such that

$$\| f - r \|_{C[0,1]} \leqslant \frac{\varepsilon}{2n^2} + \frac{2}{n^3} < \frac{\varepsilon}{n^2}$$

and

$$\deg r \leqslant 2 \deg r_1 + 2 \deg r_2 + B_2 \ln\left(\mathrm{e} + \frac{1}{\delta}\right) \ln\left(\mathrm{e} + \frac{1}{\varepsilon_2}\right)$$

$$\leqslant \frac{2n}{3} + B_2 \ln(\mathrm{e} + n^{3/\alpha}) \ln(\mathrm{e} + n^3) \leqslant n.$$

Hence for $n \geqslant n_0$ we have $R_n(f)_C < \varepsilon/n^2$ and therefore $n^2 R_n(f)_C \to 0$ as $n \to \infty$. □

10.4 L_p approximation of functions of bounded variation

In theorem 10.5 in subsection 10.1.3 we have proved that, if $V_a^b f < \infty$, then $R_n(f)_1 = \mathrm{o}(1/n)$. Here we extend this result for L_p rational approximation.

Theorem 10.11. *If $V_a^b f < \infty$ and $1 \leqslant p < \infty$, then*

$$R_n(f)_p = \mathrm{o}\left(\frac{1}{n}\right). \tag{1}$$

Remark. In theorem 11.6 in subsection 11.1.5 we prove that the estimate (1) is exact with respect to the order in the class under consideration.

To prove theorem 10.11 we shall make use of the relation between rational and spline approximations in L_p metric from theorem 8.1 in section 8.1.

It is well known that each function f of bounded variation on $[a, b]$ can be represented in the form $f = f_1 + f_2$, where f_1 is absolutely continuous and f_2 is a singular function. We have established in theorem 10.6 in section 10.2 that the estimate (1) holds for absolutely continuous functions f. It remains to prove it for singular functions.

We shall make use of the following functional characteristic for singular functions f. Let $\Omega_n = \{\Delta_i\}_{i=1}^n$ be a partition of $[a, b]$ into n compact intervals $\Delta_i = [x_{i-1}, x_i]$ such that $a = x_0 < x_1 < \cdots < x_n = b$. Denote

$$\rho_p(f, \Omega_n) = \left\{ \sum_{i=1}^{n} |\Delta_i|^{1/(p+1)} (V_{\Delta_i} f)^{p/(p+1)} \right\}^{(p+1)/p}, \quad p > 0.$$

We define

$$\rho_p(f, n) = \inf_{\Omega_n} \rho_p(f, \Omega_n) \tag{2}$$

where inf is taken over all partitions of $[a, b]$ into n compact subintervals disjoint except for the endpoints.

Lemma 10.7. *If f is a singular function on $[a, b]$ and $p > 0$, then $\rho_p(f, n) \to 0$ as $n \to \infty$.*

Proof. Since f is singular, then for each $\varepsilon > 0$ there exists a partition $\Omega_{2n} = \{\Delta_i\}_{i=1}^{2n}$ of $[a, b]$ such that $\sum_{k=1}^{n} |\Delta_{2k}| \leqslant \varepsilon$ and $\sum_{k=1}^{n} V_{\Delta_{2k-1}} f \leqslant \varepsilon$ (see S. Saks (1937)). Hence, using Hölder's inequality we get

$$\begin{aligned}
\rho_p(f, \Omega_{2n}) &= \left\{ \sum_{k=1}^{n} |\Delta_{2k}|^{1/(p+1)} (V_{\Delta_{2k}} f)^{p/(p+1)} \right. \\
&\quad \left. + \sum_{k=1}^{n} |\Delta_{2k-1}|^{1/(p+1)} (V_{\Delta_{2k-1}} f)^{p/(p+1)} \right\}^{(p+1)/p} \\
&\leqslant \left\{ \left(\sum_{k=1}^{n} |\Delta_{2k}| \right)^{1/(p+1)} \left(\sum_{k=1}^{n} V_{\Delta_{2k}} f \right)^{p/(p+1)} \right. \\
&\quad \left. + \left(\sum_{k=1}^{n} |\Delta_{2k-1}| \right)^{1/(p+1)} \left(\sum_{k=1}^{n} V_{\Delta_{2k-1}} f \right)^{p/(p+1)} \right\}^{(p+1)/p} \\
&\leqslant \{ \varepsilon^{1/(p+1)} (V_a^b f)^{p/(p+1)} + (b-a)^{1/(p+1)} \varepsilon^{p/(p+1)} \}^{(p+1)/p}.
\end{aligned}$$

The lemma is proved. □

Lemma 10.8. *Let f be a singular function on $[a, b]$ and $1 \leqslant p < \infty$. Then we have*

$$S_n^1(f)_p = o\left(\frac{1}{n}\right). \tag{3}$$

Moreover

$$S_{2n}^1(f)_p \leqslant \frac{\rho_p(f, n)}{n}, \quad n = 1, 2, \ldots, \tag{4}$$

where $\rho_p(f, n)$ is defined in (2).

Proof. We shall apply the following trivial estimate: if $V_\Delta f < \infty$, Δ an interval, then

$$S_n^1(f, \Delta)_p \leqslant |\Delta|^{1/p} S_n^1(f, \Delta)_\infty \leqslant \frac{|\Delta|^{1/p} V_\Delta f}{n}, \quad n = 1, 2, \ldots. \tag{5}$$

Indeed, it is sufficient to divide $\Delta = [u, v]$ into n subintervals $u = x_0 < x_1 < \cdots < x_n = v$ such that the variation of f in each open interval (x_{i-1}, x_i) does not exceed $V_\Delta f/n$ and to approximate f by means of the step-function $\varphi(x)=f(x_{i-1}+0)$ for $x\in(x_{i-1},x_i)$, $i=1,2,\dots,n$.

Let $\Omega_n=\{\Delta_i\}_{i=1}^n$ be an arbitrary partition of $[a,b]$ into n subintervals disjoint except for the endpoints. Choose

$$n_i=\left[\frac{n|\Delta_i|^{1/(p+1)}(V_{\Delta_i}f)^{p/(p+1)}}{\sum_{k=1}^n|\Delta_k|^{1/(p+1)}(V_{\Delta_k}f)^{p/(p+1)}}+1\right].$$

Clearly, we have $\sum_{i=1}^n n_i\leqslant 2n$ and by (5)

$$S_{n_i}^1(f,\Delta_i)_p\leqslant\frac{|\Delta_i|^{1/p}V_{\Delta_i}f}{n_i}.$$

Hence

$$S_{2n}^1(f,[a,b])_p\leqslant\left(\sum_{i=1}^n S_{n_i}^1(f,\Delta_i)_p^p\right)^{1/p}\leqslant\left(\sum_{i=1}^n\frac{|\Delta_i|(V_{\Delta_i}f)^p}{n_i^p}\right)^{1/p}$$
$$\leqslant\frac{(\sum_{i=1}^n|\Delta_i|^{1/(p+1)}(V_{\Delta_i}f)^{p/(p+1)})^{(p+1)/p}}{n}=\frac{\rho_p(f,\Omega_n)}{n}$$

which implies (4). In view of lemma 10.7 (4) implies (3). □

Proof of theorem 10.11. If $V_a^bf<\infty$, then f can be represented in the form $f=f_1+f_2$, where f_1 is absolutely continuous and f_2 is singular. By theorem 10.6 in section 10.2 we have

$$R_n(f_1)_p=\mathrm{o}\left(\frac{1}{n}\right). \tag{6}$$

By theorem 8.1 in section 8.1. we have for each $n\geqslant 1$

$$R_n(f_2)_p\leqslant C(p,\alpha)\frac{\sum_{\nu=1}^n\nu^{\alpha-1}S_\nu^1(f_2)_p}{n^\alpha},\quad n=1,2,\dots. \tag{7}$$

On the other hand, by lemma 10.8 we have

$$S_n^1(f_2)_p=\mathrm{o}\left(\frac{1}{n}\right). \tag{8}$$

It follows from (7) with $\alpha=2$ and (8) that

$$R_n(f_2)_p=\mathrm{o}\left(\frac{1}{n}\right).$$

This estimate and (6) imply (1). □

10.5 Notes

D. Newman (1964b) asked whether the estimate

$$R_n(f)_C=\mathrm{o}\left(\frac{1}{n}\right)\quad\text{for each } f\in\mathrm{Lip}\,1 \tag{1}$$

holds. G. Freud remarked that, if

$$R_n(f)_C = O\left(\frac{1}{n^{r+1}}\right) \quad \text{for each } f\in V_r \quad (r\geqslant 1) \tag{2}$$

(see section 5.2), then the estimate (1) holds. V.A. Popov (1977) proved the estimate (2) and as a consequence proved Newman's conjecture (the estimate (1)); see corollary 10.2.

The o-effect for the rational uniform approximation of each individual function in the class V_r $(r\geqslant 1)$ is proved by P. Petrushev (1979); see the estimate (39) in theorem 10.4. The results in section 10.1 are due to P. Petrushev (1980a, 1983b) except the estimate (41) in corollary 10.1 which is due to Yu.A. Brudnyi (1979). Note that A. Abdulgaparov (1974), A. Hatamov (1975b) and P. Petrushev (1976a) have estimated the rational uniform approximation of functions with convex rth derivative. The final estimate is found by P. Petrushev (1976b, c); see the estimate (42) in corollary 10.1.

Theorem 10.5 is proved in P. Petrushev (1980d).

A natural generalization of the function $\theta(f)$ is the following function $\Theta_\gamma(f)$. Suppose $V_a^b f < \infty$ and $\bar{f}: x = x(s),\ y = y(s),\ s\in[0,l],\ l = l(f),\ x(0) = a$, are the parametric equations of the complemented graph $\bar{f}$ of f with respect to its natural parameter s. Set

$$F_\gamma(t) = \int_0^t \frac{dt}{1+|t|^\gamma}, \quad t\in(-\infty,\infty),\ \gamma\geqslant 0.$$

Denote as in subsection 10.1.1

$$E = \{s: s\in[0,l], (x'(s))^2 + (y'(s))^2 = 1\}, \quad \text{mes } E = l.$$

We define

$$\Theta_\gamma(f,s) = \begin{cases} F_\gamma\left(\dfrac{y'(s)}{x'(s)}\right), & s\in E, \quad x'(s)\neq 0, \\ \lim\limits_{t\to\infty} F_\gamma(t)\,\text{sign}\, y'(s) & s\in E, \quad x'(s) = 0. \end{cases}$$

Clearly $\Theta_2(f) = \theta(f)$. Note that $\Theta_\gamma(f,s) = F_\gamma(\tan\theta(f,s))$ for $s\in E$, where we take $\tan(\pm\pi/2) = \pm\infty$ and $F_\gamma(\pm\infty) = \lim_{t\to\pm\infty} F_\gamma(t)$.

If necessary we define in a suitable way $\Theta_\gamma(f,s)$ also for $s\in[0,l]\setminus E$. Thus when we calculate the variation $V_0^l\Theta_\gamma(f)$ we always take (if possible)

$$\Theta_\gamma(f,s) = \tfrac{1}{2}(\Theta_\gamma(f,s-0) + \Theta_\gamma(f,s+0)), \quad s\in(0,l),$$

$$\Theta_\gamma(0) = \Theta_\gamma(0+0)$$

and

$$\Theta_\gamma(l) = \Theta_\gamma(l-0).$$

Note that, if $V_0^l\Theta_{\gamma_1}(f)<\infty$ and $0\leqslant\gamma_1\leqslant\gamma_2$ then

$$V_0^l\Theta_{\gamma_2}(f)\leqslant 2V_0^l\Theta_{\gamma_1}(f). \tag{3}$$

Indeed, we have for $s_1, s_2\in E$

$$\begin{aligned}|\Theta_{\gamma_2}(f,s_1)-\Theta_{\gamma_2}(f,s_2)| &= \left|\int_{\tan\theta(f,s_2)}^{\tan\theta(f,s_1)}\frac{dt}{1+|t|^{\gamma_2}}\right| \\ &\leqslant \left|2\int_{\tan\theta(f,s_2)}^{\tan\theta(f,s_1)}\frac{dt}{1+|t|^{\gamma_1}}\right| \\ &= 2|\Theta_{\gamma_1}(f,s_1)-\Theta_{\gamma_1}(f,s_2)|\end{aligned}$$

which implies (3).

On the other hand it is not difficult to see that, if $1\leqslant\gamma_1<\gamma_2$, then there exists a function f or bounded variation such that

$$V_0^l\Theta_{\gamma_2}(f)<\infty$$

but

$$V_0^l\Theta_{\gamma_1}(f)=\infty.$$

The following theorems involve Θ_γ in the rational approximation.

Theorem 10.12. *Let $f\in V_r$ $(r\geqslant 1)$ and $V_0^l\Theta_\gamma(f^{(r)})<\infty$, $0\leqslant\gamma\leqslant r+1$, $l=l(f)$. Then we have*

$$R_n(f)_C\leqslant C\frac{V_0^l\Theta_\gamma(f^{(r)})}{n^{r+2}},\quad n\geqslant r+1, \tag{4}$$

where $C=C_1(r)(b-a)^{r+1-\gamma}l^\gamma$.

Moreover we have

$$R_n(f)_C=\mathrm{o}\left(\frac{1}{n^{r+2}}\right). \tag{5}$$

In particular, if $f^{(r)}$ $(r\geqslant 1)$ is convex and bounded, then the estimate (5) holds, compare with estimate (42) in corollary 10.1.

Theorem 10.13. *If $V_a^bf<\infty$ and $V_0^l\theta(f)<\infty$, $l=l(f)$, then*

$$R_n(f)_{L_1}=\mathrm{o}\left(\frac{1}{n^2}\right). \tag{6}$$

In particular, if f is convex and continuous, then the estimate (6) holds.

Note that estimates similar to estimates (4)–(6) hold also for the corresponding spline approximations.

The proofs of theorem 10.12 and theorem 10.13 are much more difficult and long. Therefore we omit them.

Theorem 10.6 was announced by Yu.A. Brudnyi (1979). Theorem 10.7 was obtained by A. Pekarskii (1982). Theorem 10.8 and corollary 10.2 are proved independently by P. Petrushev (1976c, 1977) and A. Pekarskii (1977, 1978a). The estimate (9) in theorem 10.8 is due to A. Pekarskii (1978a). Theorem 10.9 was obtained by V. Popov, P. Petrushev (1977). Theorem 10.10 was announced by P. Petrushev (1980a). Theorem 10.11 in the case $1 < p < \infty$ is proved by A. Pekarskii (1980a).

11

Lower bounds

In the previous chapters a number of estimates for rational approximation were established. Here we shall be concerned with the exactness of these estimates in the sense of definitions 5.1–5.3 from section 5.1. We use alternance techniques based on some variants of the well-known Chebyshev theorem and Vallée-Poussin theorem for rational approximation.

In section 11.1 there will be given some relatively simple lower bounds, almost all of which are not purely rational in scope. That is, almost all of them are valid for approximation by piecewise monotone functions or piecewise convex functions, particularly for spline approximation. In section 11.2 a non-trivial lower bound is obtained for the rational uniform approximation of functions of bounded variation and given modulus of continuity. Other lower bounds which can be analogously obtained will be omitted.

11.1 Some simple lower bounds

In this section we give some relatively elementary lower bounds for rational approximations which are not intrinsically dependent on the nature of the rational functions as an approximating tool. These bounds are based on some more general properties of the rational functions such as piecewise monotony and piecewise convexity.

11.1.1 Negative results for uniform approximation of continuous functions with given modulus of smoothness

In the preceding chapters classes of functions have been found which can be approximated by rational functions better than by polynomials. In this section we show that in the class of all continuous functions with a given modulus of smoothness the rational functions are in general not better than the polynomials as an approximation tool in the uniform metric. However, the o-effect appears when Lipschitz functions are approximated by rational functions.

Theorem 11.1. *Let ω_k be a given modulus of smoothness of order $k \geqslant 1$, i.e. $\omega_k(\delta) = \omega_k(g; \delta)_C$, $\delta \geqslant 0$, for some $g \in C_{[-1,1]}$ and let*

$$\lim_{\delta \to 0} \frac{\omega_k(\delta)}{\delta^k} = \infty. \tag{1}$$

Then there exists a function f continuous on $[-1, 1]$ such that

$$\omega_k(f; \delta)_C \leqslant \omega_k(\delta), \quad \delta \geqslant 0 \tag{2}$$

and

$$\limsup_{n \to \infty} \frac{R_n(f, [-1, 1])_C}{\omega_k(1/n)} > 0. \tag{3}$$

Proof. Select indices $\{n_\nu\}_{\nu=1}^{\infty}$ such that

$$n_\nu = 9^{m_\nu},\ m_\nu \text{ integer},\ 1 \leqslant m_1 < m_2 < \cdots, \tag{4}$$

$$2\omega_k\left(\frac{1}{n_{\nu+1}}\right) \leqslant \omega_k\left(\frac{1}{n_\nu}\right) \tag{5}$$

and

$$2n_\nu^k \omega_k\left(\frac{1}{n_\nu}\right) \leqslant n_{\nu+1}^k \omega_k\left(\frac{1}{n_{\nu+1}}\right). \tag{6}$$

Condition (5) is possible because $\lim_{\delta \to 0} \omega_k(\delta) = 0$ and (6) because of (1).

Consider the function

$$f(x) = \sum_{\nu=1}^{\infty} g_\nu(x), \quad x \in [-1, 1],$$

where

$$g_\nu(x) = 40^{-k} \omega_k\left(\frac{1}{n_\nu}\right) T_{9n_\nu}\left(\frac{x}{2}\right), \quad T_n(x) = \cos(n \arccos x)$$

is the Chebyshev polynomial. Here the right-hand series converges uniformly in $[-1, 1]$ because of (5).

First we shall prove that $\omega_k(f; \delta)_C \leqslant \omega_k(\delta)$ for $\delta \geqslant 0$. To this end we shall make use of the following inequalities.

(i) $\omega_k(g; \delta)_C \leqslant 2^k \| g \|_C$, $\quad \omega_k(g; \delta)_C \leqslant \delta^k \| g^{(k)} \|_C$, $\quad$ when $\quad g \in C^k[-1, 1]$, $\omega_k(g_1 + g_2; \delta)_C \leqslant \omega_k(g_1; \delta)_C + \omega_k(g_2; \delta)_C$ and $\omega_k(g; \delta_2)/\delta_2^k \leqslant 2^k \omega_k(g; \delta_1)/\delta_1^k$ for $0 < \delta_1 \leqslant \delta_2$ which follows directly from the inequality $\omega_k(g; \lambda\delta)_C \leqslant (\lambda + 1)^k \omega_k(g; \delta)_C$; see the properties of moduli of smoothness in section 3.1.

(ii) Bernstein's inequality (see theorem 3.11 in section 3.4) provides for $P \in P_n$

$$\| P^{(k)} \|_{C[-\frac{1}{2}, \frac{1}{2}]} \leqslant \left(\frac{2}{\sqrt{3}}\right)^k n(n-1) \cdots (n-k+1) \| P \|_{C[-1,1]} \leqslant 2^k n^k \| P \|_{C[-1,1]}.$$

Now let $1/n_{j+1} \leqslant \delta \leqslant 1/n_j$, $j \geqslant 1$. Then by (5), (6), (i) and (ii) we get

$$\begin{aligned}
\omega_k(f;\delta)_C &\leqslant \sum_{\nu=1}^{j} \omega_k(g_\nu;\delta)_C + 2^k \left\| \sum_{\nu=j+1}^{\infty} g_\nu \right\|_C \leqslant \delta^k \sum_{\nu=1}^{j} \| g_\nu^{(k)} \|_C + 2^k \sum_{\nu=j+1}^{\infty} \| g_\nu \|_C \\
&\leqslant 40^{-k} \left\{ \frac{\delta^k}{2^k} \sum_{\nu=1}^{j} \omega_k\left(\frac{1}{n_\nu}\right) \| T_{9n_\nu}^{(k)} \|_{C[-\frac{1}{2},\frac{1}{2}]} + 2^k \sum_{\nu=j+1}^{\infty} \omega_k\left(\frac{1}{n_\nu}\right) \right\} \\
&\leqslant 40^{-k} \left\{ 9^k \delta^k \sum_{\nu=1}^{j} n_\nu^k \omega_k\left(\frac{1}{n_\nu}\right) + 2^k \sum_{\nu=j+1}^{\infty} \omega_k\left(\frac{1}{n_\nu}\right) \right\} \\
&\leqslant 40^{-k} \left\{ 9^k \delta^k n_j^k \omega_k\left(\frac{1}{n_j}\right) \sum_{\nu=1}^{j} \frac{1}{2^{j-\nu}} + 2^k \omega_k\left(\frac{1}{n_{j+1}}\right) \cdot \sum_{\nu=j+1}^{\infty} \frac{1}{2^{\nu-j-1}} \right\} \\
&\leqslant 40^{-k} \left\{ 2 \cdot 9^k \delta^k \frac{\omega_k(1/n_j)}{(1/n_j)^k} + 2^{k+1} \omega_k\left(\frac{1}{n_{j+1}}\right) \right\} \\
&\leqslant 40^{-k} \{ 2^{k+1} \cdot 9^k \omega_k(\delta) + 2^{k+1} \omega_k(\delta) \} \leqslant \omega_k(\delta).
\end{aligned}$$

Similarly it can be proved that $\omega_k(f;\delta)_C \leqslant \omega_k(\delta)$ for $\delta \geqslant 1/n_1$. Consequently f satisfies (2).

Consider the polynomial $P_i = \sum_{\nu=1}^{i-1} g_\nu$ of degree $9n_{i-1} \leqslant n_i$. Let $x_j = 2\cos(j\pi/9n_i)$, $3n_i \leqslant j \leqslant 6n_i$. Clearly $x_j \in [-1, 1]$ and in view of (4)

$$T_{9n_\nu}\left(\frac{x_j}{2}\right) = \cos(9^{m_\nu - m_i} j\pi) = (-1)^j, \quad \nu \geqslant i, \quad 3n_i \leqslant j \leqslant 6n_i.$$

Hence

$$f(x_j) - P_i(x_j) = \sum_{\nu=i}^{\infty} g_\nu(x_j) = (-1)^j 40^{-k} \sum_{\nu=i}^{\infty} \omega_k\left(\frac{1}{n_\nu}\right), \tag{7}$$

i.e. $f - P_i$ attains the maximum of its absolute value in $[-1, 1]$ with alternate signs at the points x_j. The number of these points is $3n_i + 1 \geqslant 2n_i + 2$. Then by Chebyshev's theorem (see theorem 2.2 in section 2.2) it follows that P_i is the best approximating function to f in R_{n_i} and by (7)

$$R_{n_i}(f)_C = 40^{-k} \sum_{\nu=i}^{\infty} \omega_k\left(\frac{1}{n_\nu}\right) > 40^{-k} \omega_k\left(\frac{1}{n_i}\right)$$

which implies (3). □

Theorem 11.2. *Let $k \geqslant 1$, $C > 0$, $\varepsilon_1 \geqslant \varepsilon_2 \geqslant \cdots > 0$ and $\varepsilon_n \to 0$ with $n \to \infty$ arbitrarily slowly. Then there exists a function f continuous on $[-1, 1]$ such that*

$$\omega_k(f;\delta)_C \leqslant C\delta^k, \quad \delta \geqslant 0, \tag{8}$$

and

$$\limsup_{n \to \infty} R_n(f)_C \left(\frac{\varepsilon_n}{n^k}\right)^{-1} \geqslant 1. \tag{9}$$

Proof. Choose indices $\{n_\nu\}_{\nu=1}^{\infty}$ such that $n_\nu = 9^{m_\nu}$, $1 \leqslant m_1 < m_2 < \cdots$, and $9^k \sum_{\nu=1}^{\infty} \varepsilon_{n_\nu} \leqslant C$, where $C > 0$ is from the hypothesis of the theorem.

Consider the function

$$f(x) = \sum_{\nu=1}^{\infty} \frac{\varepsilon_{n_\nu}}{n_\nu^k} T_{9n_\nu}\left(\frac{x}{2}\right), \quad x \in [-1, 1].$$

Obviously the right-hand series converges uniformly in $[-1, 1]$. It is readily seen that for $\delta \geqslant 0$

$$\omega_k(f; \delta)_C \leqslant \delta^k 9^k \sum_{\nu=1}^{\infty} \varepsilon_{n_\nu} \leqslant C\delta^k,$$

i.e. f satisfies (8).

Consider the polynomial

$$P_i(x) = \sum_{\nu=1}^{i-1} \frac{\varepsilon_{n_\nu}}{n_\nu^k} T_{9n_\nu}\left(\frac{x}{2}\right)$$

of degree $9n_{i-1} \leqslant n_i$. As above in the proof of theorem 11.1 P_i is the best approximating function to f in R_{n_i} and

$$R_{n_i}(f)_C = \sum_{\nu=i}^{\infty} \frac{\varepsilon_{n_\nu}}{n_\nu^k} > \frac{\varepsilon_{n_i}}{n_i^k},$$

which implies (9). □

11.1.2 One negative result for uniform approximation of absolutely continuous functions

It was mentioned in section 5.3 that for absolutely continuous functions f $R_n(f)_C$ may tend to zero with $n \to \infty$ as slow as we want. More precisely the following statement holds.

Theorem 11.3. *For each sequence $\{\varepsilon_n\}_{n=1}^{\infty}$, $\varepsilon_1 \geqslant \varepsilon_2 \geqslant \cdots > 0$, $\lim_{n\to\infty} \varepsilon_n = 0$, there exist an index n_0 and a function f which is absolutely continuous in $[0, 1]$ (also f is non-decreasing in $[0, 1]$, $f(0) = 0$ and $f(1) = 1$) such that*

$$R_n(f, [0, 1])_C > \varepsilon_n, \quad n \geqslant n_0.$$

Proof. Consider the function

$$g_\delta(x) = \begin{cases} 0, & -1 \leqslant x \leqslant 0, \\ x/\delta, & 0 < x \leqslant \delta, \\ 1, & \delta < x \leqslant 1, \end{cases}$$

where $0 < \delta < 1$.

Corollary 9.3 in section 9.2 implies that for each $n \geqslant 1$ there exists a number δ_n, $0<\delta_n<1$, such that

$$R_n(g_{\delta_n},[-1,1])_C>\tfrac{1}{4}. \tag{10}$$

For each $n\geqslant 1$ we denote g_{δ_n} briefly g_n.

Choose a sequence of indexes $1\leqslant n_0<n_1<\cdots$ such that

$$\varepsilon_{n_\nu}\leqslant\frac{1}{4\cdot 2^{\nu+2}},\quad \nu=0,1,\ldots. \tag{11}$$

Denote $x_\nu=1-1/2^\nu$ and $\Delta_\nu=[x_\nu,x_{\nu+1})$. Let λ_ν be the linear increasing function which maps Δ_ν onto $[-1,1)$. Set

$$f(x)=\begin{cases}1-\dfrac{1}{2^\nu}+\dfrac{1}{2^{\nu+1}}g_{n_\nu}(\lambda_\nu(x)), & x\in\Delta_\nu,\quad \nu=0,1,\ldots,\\ 1, & x=1.\end{cases}$$

It is readily seen that f is absolutely continuous and non-decreasing in $[0,1]$, $f(0)=0$ and $f(1)=1$. By (10) and (11) it follows that, if $n_{\nu-1}\leqslant n\leqslant n_\nu$ $(\nu\geqslant 1)$ then

$$R_n(f,[0,1])_C\geqslant R_{n_\nu}(f,\Delta_\nu)_C=\frac{1}{2^{\nu+1}}R_{n_\nu}(g_{n_\nu},[-1,1])_C>\frac{1}{4\cdot 2^{\nu+1}}\geqslant\varepsilon_{n_{\nu-1}}\geqslant\varepsilon_n$$

and therefore

$$R_n(f,[0,1])_C>\varepsilon_n,\quad n\geqslant n_0. \qquad \square$$

11.1.3 Lower bound for uniform approximation of the functions of the class V_r

Theorem 5.1 in section 5.2 established an upper bound for the rational uniform approximation of the basic class V_r. Existence of the o-effect for the rational uniform approximation for each individual function $f\in V_r$ was proved in theorem 10.4 in subsection 10.1.3. The following theorem proves the exactness of these estimates.

Theorem 11.4. (i) *For each $r\geqslant 1$ there exists a positive constant $C(r)$ such that for each $M>0$ and compact interval $[a,b]$*

$$\sup_{f\in V_r(M,[a,b])}R_n(f)_C>C(r)\frac{M(b-a)^r}{n^{r+1}},\quad n=1,2,\ldots. \tag{12}$$

(ii) *For each $r\geqslant 1$, $M>0$, compact interval $[a,b]$ and sequence $\{\varepsilon_n\}_{n=1}^\infty$, $\varepsilon_1\geqslant\varepsilon_2\geqslant\cdots>0$, $\lim_{n\to\infty}\varepsilon_n=0$, there exists a function $f\in V_r(M,[a,b])$ such that*

$$\limsup_{n\to\infty}R_n(f)_C\left(\frac{\varepsilon_n}{n^{r+1}}\right)^{-1}\geqslant 1. \tag{13}$$

Proof. Suitable change of variables shows that it is sufficient to prove the theorem only in the case $M=1$ and $[a,b]=[-1,1]$.

For each $n \geqslant 1$ set $f_n(x) = \sin(2\pi n x)/8(2\pi)^r n^{r+1}$ for $x \in [-1, 1]$. It is readily seen that $f \in V_r(1, [-1, 1])$ and $f_n((2k-1)/4n) = (-1)^{k+1} \| f_n \|_{C[-1,1]}$ for $k = -2n+1, -2n+2, \ldots, -1, 0, 1, \ldots, 2n$. Then by the Chebyshev theorem (see theorem 2.2 in section 2.2) it follows that the rational function $r \equiv 0$ is the best uniformly approximating function to f_n in R_n and

$$R_n(f_n)_C = \| f_n \|_{C[-1,1]} = \frac{1}{8(2\pi)^r n^{r+1}}$$

which implies (12).

Now we shall prove the second part of the theorem. Suppose that $\varepsilon_1 \geqslant \varepsilon_2 \geqslant \cdots > 0$ and $\lim_{n\to\infty} \varepsilon_n = 0$. Choose indices $\{n_\nu\}_{\nu=1}^{\infty}$ such that

$$n_\nu = 9^{m_\nu}, \quad 1 \leqslant m_1 < m_2 < \cdots \tag{14}$$

$$\sum_{\nu=1}^{\infty} 18 \cdot 9^r \varepsilon_{n_\nu} \leqslant 1. \tag{15}$$

Consider the function $f(x) = \sum_{\nu=1}^{\infty} g_\nu(x)$ for $x \in [-1, 1]$, where $g_\nu(x) = (\varepsilon_{n_\nu}/n_\nu^{r+1}) T_{9n_\nu}(x/2)$, $T_n(x) = \cos(n \arccos x)$. Since $g_\nu \in P_{9n_\nu}$, from Bernstein's inequality exactly as in the proof of theorem 11.1 it follows that for $k = 1, 2, \ldots$

$$\| g_\nu^{(k)} \|_{C[-1,1]} \leqslant \frac{\varepsilon_{n_\nu}}{2^k n_\nu^{r+1}} \| T_{9n_\nu}^{(k)} \|_{C[-\frac{1}{2},\frac{1}{2}]} \leqslant \frac{9^k \varepsilon_{n_\nu}}{n_\nu^{r+1-k}}. \tag{16}$$

The estimates (15) and (16) imply that the series $\sum_{\nu=1}^{\infty} \| g_\nu^{(k)} \|_{C[-1,1]}$ converges when $k = 0, 1, \ldots, r+1$ and therefore $f \in C_{[-1,1]}^{r+1}$ and $f^{(r)}(x) = \sum_{\nu=1}^{\infty} g_\nu^{(r)}(x)$ for $x \in [-1, 1]$. Using (15) we get

$$V_{-1}^{1} f^{(r)} \leqslant \sum_{\nu=1}^{\infty} V_{-1}^{1} g_\nu^{(r)} \leqslant \sum_{\nu=1}^{\infty} 18 n_\nu \| g_\nu^{(r)} \|_C \leqslant \sum_{\nu=1}^{\infty} 18 \cdot 9^r \varepsilon_{n_\nu} \leqslant 1.$$

Consequently $f \in V_r(1, [-1, 1])$.

Now consider the polynomial $P_k = \sum_{\nu=1}^{k-1} g_\nu$ of degree $9n_{k-1} \leqslant n_k$. Let $x_j = 2\cos(j\pi/9n_k)$, $3n_k \leqslant j \leqslant 6n_k$. Exactly as in the proof of theorem 11.1 we obtain by using (14)

$$f(x_j) - P_k(x_j) = \sum_{\nu=k}^{\infty} g_\nu(x_j) = (-1)^j \sum_{\nu=k}^{\infty} \frac{\varepsilon_{n_\nu}}{n_\nu^{r+1}}, \tag{17}$$

i.e. $f - P_k$ attains the maximum of its absolute value in $[-1, 1]$ with alternate signs at the points x_j. The number of these points is $3n_k + 1 \geqslant 2n_k + 2$. Then by Chebyshev's theorem it follows that P_k is the best approximating function to f in R_{n_k} and by (17)

$$R_{n_k}(f)_C = \sum_{\nu=k}^{\infty} \frac{\varepsilon_{n_\nu}}{n_\nu^{r+1}} > \frac{\varepsilon_{n_k}}{n_k^{r+1}}$$

which implies (13). □

11.1.4 Uniform approximation of convex functions

In theorem 5.11 in section 5.5 it was found an upper bound for the rational uniform approximation of the class $\mathrm{Conv}_M[a,b]$ of all functions f which are convex and continuous in $[a,b]$ and $\|f\|_{C[a,b]}\leqslant M$. Existence of the o-effect for the rational uniform approximation of each individual function $f\in\mathrm{Conv}_M[a,b]$ was established in theorem 10.9 in section 10.3. The following theorem shows that the estimates in theorem 5.11 and theorem 10.9 are exact with respect to the order.

Theorem 11.5. (i) *There exists a positive constant C such that for each $M>0$ and compact interval $[a,b]$*

$$\sup_{f\in\mathrm{Conv}_M[a,b]} R_n(f)_C > C\frac{M}{n}, \quad n=1,2,\ldots. \tag{18}$$

(ii) *For each $M>0$, compact interval $[a,b]$ and sequence $\{\varepsilon_n\}_{n=1}^{\infty}$, $\varepsilon_1\geqslant\varepsilon_2\geqslant\cdots>0$, $\lim_{n\to\infty}\varepsilon_n=0$, there exists a convex function $f\in\mathrm{Conv}_M[a,b]$ such that*

$$\limsup_{n\to\infty} R_n(f)_C\left(\frac{\varepsilon_n}{n}\right)^{-1}\geqslant 1.$$

The proof of theorem 11.5 is based on the fact that the derivative r' of each rational function r is a piecewise convex function. We shall need some auxiliary statements.

Lemma 11.1. *Let g be defined on $\Delta=[a,b]$ and*

$$g'(x)=\begin{cases} h_1, & x\in\Delta_1,\\ h_2, & x\in\Delta_2,\end{cases}$$

where $\Delta_1=(a,c),\Delta_2=(c,b)$, $a<c<b$ and $h_1<h_2$. Then for each function φ such that φ' is convex or concave in Δ the following inequality holds:

$$\|g-\varphi\|_{C(\Delta)}\geqslant\frac{(h_2-h_1)(\min\{|\Delta_1|,|\Delta_2|\})^2}{8|\Delta|}. \tag{19}$$

Proof. Denote $x_0=a$, $x_1=(a+c)/2$, $x_2=c$, $x_3=(c+b)/2$, $x_4=b$ and $d_i=[x_{i-1},x_i]$, $i=1,2,3,4$. Set $\lambda(x)=h_1+(h_2-h_1)(x-x_1)/(x_3-x_1)$. Note that $\lambda(x_1)=g'(x_1)$ and $\lambda(x_3)=g'(x_3)$.

Since φ' is convex or concave in Δ and λ is linear, then at least one of the following four inequalities holds: (i) $\varphi'(x)\leqslant\lambda(x)$ for $x\in d_1$, (ii) $\varphi'(x)\geqslant\lambda(x)$ for $x\in d_2$, (iii) $\varphi'(x)\leqslant\lambda(x)$ for $x\in d_3$, (iv) $\varphi'(x)\geqslant\lambda(x)$ for $x\in d_4$. Consider the case when the inequality (i) is valid. Then the function $g-\varphi$ is monotone on d_1

and therefore

$$\|g-\varphi\|_{C(d_1)} \geqslant \tfrac{1}{2}|g(x_0)-\varphi(x_0)-(g(x_1)-\varphi(x_1))| = \frac{1}{2}\left|\int_{d_1}(g'(x)-\varphi'(x))\mathrm{d}x\right|$$
$$\geqslant \frac{1}{2}\int_{d_1}(g'(x)-\lambda(x))\mathrm{d}x = \frac{(h_2-h_1)|\Delta_1|^2}{8|\Delta|}$$

which implies (19). The other situations are considered similarly. □

Lemma 11.2. *For each $n \geqslant 1$ there exists a function g_n such that g_n is increasing, convex and continuous in $[0, 1]$, $g_n(0) = 0$, $g_n(1) = 1$, g'_n is a step-function, $g'_n(+0) > 0$, $g'_n(1-0) < \infty$ and*

$$R_n(g_n, [0, 1])_C \geqslant \frac{1}{768n}. \tag{20}$$

Proof. Let $n \geqslant 1$ and denote $N = 16n - 1$, $h = 1/16n$ and $x_\nu = 1 - 1/2^\nu$. Set $g_n(x) = \int_0^x g'_n(t)\mathrm{d}t$ where

$$g'_n(x) = \begin{cases} 2^\nu h, & x \in (x_{\nu-1}, x_\nu), \quad \nu = 1, 2, .., N, \\ 2^N h, & x \in (x_N, 1). \end{cases}$$

It is readily seen that g_n is increasing, convex and continuous on $[0, 1]$, $g_n(0) = 0$ and $g_n(1) = \int_0^1 g'_n(x)\mathrm{d}x = 1$. Also, g'_n is a step-function, $g'_n(+0) = 2h > 0$, $g'_n(1-0) = 2^N h < \infty$.

It remains to estimate $R_n(g_n)_C$. Suppose that $r \in R_n$ and $\|g_n - r\|_{C[0,1]} = R_n(f)_C$. Clearly, there exists a division of $[0, 1]$ into at most $8n - 2$ intervals such that r' is convex or concave in each of them. From this and the fact that $N = 16n - 1$ it follows that there exists ν, $1 \leqslant \nu \leqslant N - 1$, such that r' is convex or concave in the interval $[x_{\nu-1}, x_{\nu+1}]$. Then applying lemma 11.1 to the functions g_n and r in $[x_{\nu-1}, x_{\nu+1}]$ we conclude that

$$R_n(g_n)_C = \|g_n - r\|_{C[0,1]} \geqslant \|g_n - r\|_{C[x_{\nu-1}, x_{\nu+1}]}$$
$$\geqslant \frac{(2^{\nu+1}h - 2^\nu h)(x_{\nu+1} - x_\nu)^2}{8(x_{\nu+1} - x_{\nu-1})} = \frac{1}{768n},$$

which implies (20). □

Proof of theorem 11.5. Simple change of variables shows that it is sufficient to prove the theorem only in the case $M = 1$ and $[a, b] = [0, b]$ for some $b > 0$. Then (i) follows immediately from lemma 11.2.

Now we shall prove the second part. Suppose that $\varepsilon_1 \geqslant \varepsilon_2 \geqslant \cdots > 0$ and $\lim_{n\to\infty} \varepsilon_n = 0$.

Choose indices $n_1 < n_2 < \cdots$ such that

$$\sum_{\nu=1}^{\infty} \varepsilon_{n_\nu} \leqslant 1. \tag{21}$$

Now we construct the desired function f by induction using the functions g_n from lemma 11.2. Denote $x_0 = 0$, $x_1 = 1/2$ and $\Delta_1 = [x_0, x_1]$. Let λ_1 be the increasing linear function which maps Δ_1 onto $[0, 1]$. Set $f(x) = \varepsilon_{n_1} g_{n_1}(\lambda_1(x))$ for $x \in \Delta_1$. Note that f is continuous, convex and increasing on Δ_1 and $f'(x_1 - 0) < \infty$.

Let f be defined already on $[x_0, x_\nu]$, $\nu \geqslant 1$, such that f is continuous, convex and increasing on $[x_0, x_\nu]$, $f'(x_\nu - 0) < \infty$ and $0 < x_\nu < 1$. Choose $x_{\nu+1}$ such that $x_\nu < x_{\nu+1} < 1$ and

$$f'(x_\nu - 0) \leqslant \frac{\varepsilon_{n_{\nu+1}}}{x_{\nu+1} - x_\nu} g'_{n_{\nu+1}}(+0).$$

Such a choice of $x_{\nu+1}$ is possible since $g'_{n_{\nu+1}}(+0) > 0$ by lemma 11.2. Let $\lambda_{\nu+1}$ be the increasing linear function mapping $\Delta_{\nu+1} = (x_\nu, x_{\nu+1}]$ onto $(0, 1]$. Set $f(x) = f(x_\nu) + \varepsilon_{n_{\nu+1}} g_{n_{\nu+1}}(\lambda_{\nu+1}(x))$, $x \in \Delta_{\nu+1}$. Clearly $f'(x_\nu + 0) = (\varepsilon_{n_{\nu+1}} / (x_{\nu+1} - x_\nu)) \cdot g'_{n_{\nu+1}}(+0) \geqslant f'(x_\nu - 0)$ and therefore f is continuous, increasing and convex on $[x_0, x_{\nu+1}]$. Also $f'(x_{\nu+1} - 0) < \infty$.

It is readily seen that $b = \lim_{\nu \to \infty} x_\nu$ exists and $0 < b \leqslant 1$.

Thus the function f is defined on $[0, b)$. Set $f(b) = \lim_{x \to b} f(x)$. By our construction and (21) it follows that f is continuous, increasing and convex on $[0, b]$, $f(0) = 0$ and $f(b) = \sum_{\nu=1}^{\infty} \varepsilon_{n_\nu} \leqslant 1$. In view of lemma 11.2 we conclude that for $\nu = 1, 2, \ldots$

$$R_{n_\nu}(f, [0, b])_C \geqslant R_{n_\nu}(f, \Delta_\nu)_C = \varepsilon_{n_\nu} R_{n_\nu}(g_{n_\nu}, [0,1]))_C \geqslant \frac{\varepsilon_{n_\nu}}{768 n_\nu},$$

which implies the theorem. □

11.1.5 L_p approximation of functions of bounded variation

In theorem 10.11 in section 10.4 it was proved that for each function f of bounded variation on $[0, 1]$ the following estimate holds: $R_n(f)_p = o(1/n)$ $(1 \leqslant p < \infty)$. The following theorem establishes the exactness of this estimate.

Theorem 11.6. *For each sequence* $\{\varepsilon_n\}_{n=1}^{\infty}$, $\varepsilon_1 \geqslant \varepsilon_2 \geqslant \cdots > 0$, $\lim_{n \to \infty} \varepsilon_n = 0$, *there exists a function f defined on* $[0, 1]$ *such that* $V_0^1 f \leqslant 1$ *and for* $p \geqslant 1$

$$\limsup_{n \to \infty} R_n(f)_p \left(\frac{\varepsilon_n}{n}\right)^{-1} \geqslant 1. \tag{22}$$

Proof. Set $g_n(x) = 2\pi \sin 4\pi n x$ for $x \in [0, 1]$. The function g_n vanishes at the points $x_\nu = \nu/4n$, $\nu = 0, 1, \ldots, 4n$, and has alternate signs in the consecutive intervals $\Delta_\nu = (x_{\nu-1}, x_\nu)$. Then for each rational function $r \in R_n$ there exist at least n intervals Δ_ν such that $g_n(x) r(x) \leqslant 0$ for $x \in \Delta_\nu$. This fact implies that for each $r \in R_n$ and $p \geqslant 1$

$$\| g_n - r \|_{L_p[0,1]} \geqslant \| g_n - r \|_{L_1[0,1]} \geqslant 2\pi \int_0^{1/4} |\sin 4\pi n x| \, dx = 1.$$

Consequently for $p \geqslant 1$

$$R_n(g_n)_p \geqslant 1, \quad n = 1, 2, \ldots. \tag{23}$$

Let $\varepsilon_1 \geqslant \varepsilon_2 \geqslant \cdots > 0$ and $\lim_{n\to\infty} \varepsilon_n = 0$. Choose indices $n_1 < n_2 < \cdots$ such that

$$\sum_{\nu=1}^{\infty} 16\pi \cdot 2^\nu \varepsilon_{n_\nu} \leqslant 1. \tag{24}$$

Denote $d_\nu = (1/2^\nu, 1/2^{\nu-1}]$. Let λ_ν be the increasing linear mapping of d_ν onto $(0, 1]$. Set

$$f(x) = \begin{cases} 2^\nu \dfrac{\varepsilon_{n_\nu}}{n_\nu} g_{n_\nu}(\lambda_\nu(x)), & x \in d_\nu, \quad \nu = 1, 2, \ldots, \\ 0, \quad x = 0. \end{cases}$$

Using (24) we get

$$V_0^1 f \leqslant \sum_{\nu=1}^{\infty} 2^\nu \frac{\varepsilon_{n_\nu}}{n_\nu} V_0^1 g_\nu = \sum_{\nu=1}^{\infty} 16\pi \cdot 2^\nu \varepsilon_{n_\nu} \leqslant 1.$$

In view of (23) it follows that for $\nu = 1, 2, \ldots$

$$R_{n_\nu}(f, [0,1])_p \geqslant R_{n_\nu}(f, d_\nu)_1 = \frac{\varepsilon_{n_\nu}}{n_\nu} R_{n_\nu}(g_{n_\nu}, [0,1])_1 \geqslant \frac{\varepsilon_{n_\nu}}{n_\nu},$$

which implies (22). □

11.2 Uniform approximation of functions of bounded variation and given modulus of continuity

Consider the class $V(\omega) = V(M, [a, b], \omega)$ of all functions f continuous in $[a, b]$ such that $V_a^b f \leqslant M$ and $\omega(f; \delta)_C \leqslant \omega(\delta)$ for $\delta \geqslant 0$, where $\omega(f; \delta)_C$ is the modulus of continuity of f and ω is a given modulus of continuity. In theorem 5.7 in subsection 5.3.4 we proved an estimate for the rational uniform approximation of functions $f \in V(\omega)$. The aim of this section is to prove the exactness of this estimate.

Theorem 11.7. *Let ω be a modulus of continuity such that*

$$\lim_{\delta\to 0} \frac{\omega(\delta)}{\delta} = \infty. \tag{1}$$

Then for each $M > 0$ and compact interval $[a, b]$ there exists a function $f \in V(M, [a, b], \omega)$ such that

$$\limsup_{n\to\infty} \frac{R_n(f)_C}{\inf\limits_{1 \leqslant t \leqslant n} \{M/t + \omega((b-a)/te^{n/t})\}} > 0.$$

Remark. Next we shall show that, if $\lim_{\delta\to 0}(\omega(\delta)/\delta) \neq \infty$, then $\omega(\delta)=O(\delta)$ and therefore $V(\omega)\subset \mathrm{Lip}\,1$. In this case by theorem 10.8 in section 10.2 for each $f\in V(\omega)$ we have $R_n(f)_C = o(n^{-1})$. Hence the restriction (1) in the formulation of theorem 11.7 is essential.

To prove theorem 11.7 we need some auxiliary statements. The following well-known lemma shows that for the proof of theorem 11.7 it is sufficient to consider only the case when the modulus of continuity ω is a concave function.

Lemma 11.3. *For each modulus of continuity ω*[†] *there exists a concave modulus of continuity $\tilde{\omega}$ such that*

$$\omega(\delta)\leqslant\tilde{\omega}(\delta)\leqslant 2\omega(\delta),\quad \delta\in[0,\infty). \tag{2}$$

Proof. Define $\tilde{\omega}$ as the minimal concave majorant of ω, i.e. for $\delta\geqslant 0$

$$\begin{aligned}\tilde{\omega}(\delta) &= \sup_{0\leqslant t_1\leqslant\delta\leqslant t_2}\left\{\omega(t_1)+\frac{\omega(t_2)-\omega(t_1)}{t_2-t_1}(\delta-t_1)\right\}\\ &= \sup_{0\leqslant t_1\leqslant\delta\leqslant t_2}\frac{(\delta-t_1)\omega(t_2)+(t_2-\delta)\omega(t_1)}{t_2-t_1}.\end{aligned}$$

The left-hand side inequality in (2) is obvious. Let us prove the right-hand side one. We get for $\delta>0$

$$\begin{aligned}\tilde{\omega}(\delta) &= \sup_{0<t_1<\delta<t_2}\frac{1}{t_2-t_1}((\delta-t_1)\omega(t_2)+(t_2-\delta)\omega(t_1))\\ &\leqslant \sup_{0<t_1<\delta<t_2}\frac{1}{t_2-t_1}\left((\delta-t_1)\left(\frac{t_2}{\delta}+1\right)+t_2-\delta\right)\omega(\delta)\\ &= \sup_{0<t_1<\delta<t_2}\left(1+\frac{1-t_1/\delta}{1-t_1/t_2}\right)\omega(\delta)\leqslant 2\omega(\delta)\end{aligned}$$

where we have made use of the inequality

$$\omega(t_2)=\omega\left(\frac{t_2}{\delta}\delta\right)\leqslant\left(\frac{t_2}{\delta}+1\right)\omega(\delta)\quad\text{(see section 3.1).}$$ □

Remark. The restriction that the modulus of continuity ω be concave implies that the function $\omega(\delta)/\delta$ is nonincreasing for $\delta\in(0,\infty)$. Indeed, if ω is concave and $0<\delta_1<\delta_2$, then

$$\begin{aligned}\omega(\delta_1)/\delta_1 &= \omega\left(\left(1-\frac{\delta_1}{\delta_2}\right)\cdot 0+\frac{\delta_1}{\delta_2}\delta_2\right)\Big/\delta_1\\ &\geqslant\left\{\left(1-\frac{\delta_1}{\delta_2}\right)\omega(0)+\frac{\delta_1}{\delta_2}\omega(\delta_2)\right\}\Big/\delta_1=\omega(\delta_2)/\delta_2\end{aligned}$$

[†] We remind the reader that ω is called a modulus of continuity, if ω is a continuous nondecreasing function on $[0,\infty)$ and $\omega(\delta_1+\delta_2)\leqslant\omega(\delta_1)+\omega(\delta_2)$ for $\delta_1,\delta_2\geqslant 0$, $\omega(0)=0$.

and therefore $\omega(\delta)/\delta$ is nonincreasing. From this fact and lemma 11.3 it follows that $\lim_{\delta\to 0}(\omega(\delta)/\delta)\neq\infty$ implies $\omega(\delta)=O(\delta)$, which was used in the remark after theorem 11.7.

In the proof of theorem 11.7 the main role is played by the following auxiliary theorem.

Theorem 11.8. *Let f be a function defined on $[a,b]$ and $V_a^b f\leqslant M<\infty$ and suppose that Ω is a set of m disjoint compact subintervals $\Delta_i=[a_i,b_i]$ of $[a,b]$ with $a\leqslant a_1<b_1<a_2<b_2<\cdots<a_m<b_m\leqslant b$ and $\min_{1\leqslant i\leqslant m-1}(a_{i+1}-b_i)\geqslant\varepsilon_1$, where $m\geqslant 1$ and $\varepsilon_1>0$. Suppose also that for some $\varepsilon_2>0$ it is true that for each interval $\Delta\in\Omega$, $\Delta=[u,v]$, there exist a rational function r_Δ, $\deg r_\Delta=n_\Delta$, and a set A_Δ of $k_\Delta+1$, $k_\Delta\geqslant 0$, different points $x_i\in\Delta$, arranged in increasing order, i.e. $u\leqslant x_0<x_1<\cdots<x_{k_\Delta}\leqslant v$, such that*

$$|f(x)-r_\Delta(x)|>\varepsilon_2,\quad x\in A_\Delta,$$

$f-r_\Delta$ has alternate signs at the points $x_0,x_1,\ldots,x_{k_\Delta}$ and

$$\|r_\Delta-f(u)\|_{C([a,b]\setminus\Delta)}\leqslant V_\Delta f.$$

Then $R_n(f,[a,b])_C>\frac{1}{2}\varepsilon_2$ for

$$n\leqslant\sum_{\Delta\in\Omega}k_\Delta-\sum_{\Delta\in\Omega}n_\Delta-Dm\ln\left(e+\frac{b-a}{m\varepsilon_1}\right)\ln\left(e+\frac{M}{m\varepsilon_2}\right),$$

where $D>1$ is an absolute constant.

Proof. We shall make use of some techniques from the proofs of the theorems for 'joining' of rational functions (see theorem 5.2 in section 5.2, theorem 5.3 in section 5.3). In particular, we shall apply lemma 5.1 in section 5.1 for rational approximation of a jump-function.

It is not difficult to see that, if theorem 11.8 holds in the special case when $m=2^s$, s integer, then it is valid in the general case with another absolute constant $D>1$. Thus we shall suppose that $m=2^s$, s integer.

To avoid some more complicated indexations we shall denote

$$\Omega_\mu=\{[a_i,b_{i+2^\mu}]: i=1,2,\ldots,m-2^\mu\},\quad 0\leqslant\mu\leqslant s,\quad \Omega_{-1}=\Omega$$

and

$$\Omega_\Delta=\{\Delta^*:\Delta^*\in\Omega\text{ and }\Delta^*\subset\Delta\}\text{ for each interval }\Delta\subset[a,b].$$

Denote

$$N(\mu,M,\Delta)=\sum_{\nu=0}^{\mu}6B\cdot 2^\nu\ln\left(e+\frac{2^{-\nu}|\Delta|}{\varepsilon_1}\right)\ln\left(e+2^{\mu-\nu}+\frac{2^\mu 4^{-\nu}M}{\varepsilon_2}\right),$$

where $B>1$ is the absolute constant from lemma 5.1 in section 5.1, ε_1 and

ε_2 are those from the hypotheses of the theorem, the numbers μ and M and the interval Δ are parameters.

We shall prove by induction with respect to μ the following lemma, formulated using the assumptions and notations introduced above.

Lemma 11.4. *Let* $-1 \leqslant \mu \leqslant s$. *For each* $\Delta \in \Omega_\mu$, $\Delta = [u, v]$, *there exist a rational function* r_Δ *and a set* A_Δ *of* $l_\Delta + 1$ *different points* $x_i \in \Delta$, $u \leqslant x_0 < x_1 < \cdots < x_{l_\Delta} \leqslant v$, $l_\Delta = \sum_{\Delta^* \in \Omega_\Delta} k_{\Delta^*}$, *such that*

$$|f(x) - r_\Delta(x)| > \varepsilon_2\left(1 - \sum_{\nu=0}^{\mu} \frac{1}{2^{\nu+2}}\right), \quad x \in A_\Delta,$$

$f - r_\Delta$ *has alternate signs at the points* $x_0, x_1, \ldots, x_{l_\Delta}$,

$$\| r_\Delta - f(u) \|_{C([a,b]\setminus\Delta)} \leqslant V_\Delta f$$

and

$$\deg r_\Delta \leqslant \sum_{\Delta^* \in \Omega_\Delta} n_{\Delta^*} + N(\mu, V_\Delta f, \Delta).$$

Proof. The lemma holds for $\mu = -1$ by the hypothesis of theorem 11.8.

Suppose that the lemma holds for some μ $(-1 \leqslant \mu \leqslant s-1)$. Now we shall prove it with μ replaced by $\mu + 1$.

Let $\Delta \in \Omega_{\mu+1}$. Obviously, there exist points u_1, v_1, u_2, v_2 such that $u_1 < v_1 < u_2 < v_2$, $\Delta = [u_1, v_2]$, $\Delta_1 = [u_1, v_1] \in \Omega_\mu$, $\Delta_2 = [u_2, v_2] \in \Omega_\mu$ and $u_2 - v_1 \geqslant \varepsilon_1$. Also, by our assumptions for $i = 1, 2$ there exist a rational function r_{Δ_i} and a set A_{Δ_i} of $l_{\Delta_i} + 1 = \sum_{\Delta^* \in \Omega_{\Delta_i}} k_{\Delta^*} + 1$ different points in Δ_i such that

$$|f(x) - r_{\Delta_i}(x)| > \varepsilon_2\left(1 - \sum_{\nu=0}^{\mu} \frac{1}{2^{\nu+2}}\right), \quad x \in A_{\Delta_i}, \tag{3}$$

and $f - r_{\Delta_i}$ has alternate signs at the consecutive points of A_{Δ_i},

$$\| r_{\Delta_i} - f(u_i) \|_{C([a,b]\setminus\Delta_i)} \leqslant V_{\Delta_i} f \tag{4}$$

and

$$\deg r_{\Delta_i} \leqslant \sum_{\Delta^* \in \Omega_{\Delta_i}} n_{\Delta^*} + N(\mu, V_{\Delta_i} f, \Delta_i). \tag{5}$$

Consider the rational function

$$r_\Delta(x) = (1 - \sigma(x - z)) r_{\Delta_1}(x) + \sigma(x - z) r_{\Delta_2}(x),$$

where $z = (v_1 + u_2)/2$, σ is the rational function from lemma 5.1 in section 5.1 with $\alpha = \varepsilon_1/2$, $\beta = |\Delta|$, $\gamma = \min\{1/2^{\mu+4}, \varepsilon_2/2^{\mu+4} V_\Delta f\}$ when $V_\Delta f > 0$ (the case $V_\Delta f = 0$ is trivial).

By lemma 5.1 it follows that

$$\deg\sigma \leqslant B\ln\left(e+\frac{\beta}{\alpha}\right)\ln\left(e+\frac{1}{\gamma}\right) \leqslant B\ln\left(e+\frac{2|\Delta|}{\varepsilon_1}\right)\ln\left(e+2^{\mu+4}+\frac{2^{\mu+4}V_\Delta f}{\varepsilon_2}\right)$$

$$\leqslant 6B\ln\left(e+\frac{|\Delta|}{\varepsilon_1}\right)\ln\left(e+2^{\mu+1}+\frac{2^{\mu+1}V_\Delta f}{\varepsilon_2}\right).$$

Consequently

$$\begin{aligned}\deg r_\Delta &\leqslant \deg r_{\Delta_1}+\deg r_{\Delta_2}+\deg\sigma\\ &\leqslant \sum_{\Delta^*\in\Omega_{\Delta_1}} n_{\Delta^*}+N(\mu, V_{\Delta_1}f,\Delta_1)+\sum_{\Delta^*\in\Omega_{\Delta_2}} n_{\Delta^*}+N(\mu, V_{\Delta_2}f,\Delta_2)\\ &\quad +6B\ln\left(e+\frac{|\Delta|}{\varepsilon_1}\right)\ln\left(e+2^{\mu+1}+\frac{2^{\mu+1}V_\Delta f}{\varepsilon_2}\right)\\ &\leqslant \sum_{\Delta^*\in\Omega_\Delta} n_{\Delta^*}+N(\mu+1, V_\Delta f,\Delta),\end{aligned}$$

i.e. r_Δ has the desired order. The last inequality follows from (5) and the following one:

$$\begin{aligned}&\ln\left(e+\frac{2^{-\nu}|\Delta_1|}{\varepsilon_1}\right)\ln\left(e+2^{\mu-\nu}+\frac{2^\mu 4^{-\nu}V_{\Delta_1}f}{\varepsilon_2}\right)\\ &\quad+\ln\left(e+\frac{2^{-\nu}|\Delta_2|}{\varepsilon_1}\right)\ln\left(e+2^{\mu-\nu}+\frac{2^\mu 4^{-\nu}V_{\Delta_2}f}{\varepsilon_2}\right)\\ &\leqslant 2\ln\left(e+\frac{2^{-\nu}(|\Delta_1|+|\Delta_2|)}{2\varepsilon_1}\right)\ln\left(e+2^{\mu-\nu}+\frac{2^\mu 4^{-\nu}(V_{\Delta_1}f+V_{\Delta_1}f)}{2\varepsilon_2}\right)\\ &\leqslant 2\ln\left(e+\frac{2^{-(\nu+1)}|\Delta|}{\varepsilon_1}\right)\ln\left(e+2^{(\mu+1)-(\nu+1)}+\frac{2^{\mu+1}4^{-(\nu+1)}V_\Delta f}{\varepsilon_2}\right),\end{aligned}$$

where we have applied the fact that the function $F(x,y)=-\ln(e+x)\ln(e+y)$ is convex on the set $D=\{(x,y): x,y\geqslant 0\}$. The function F is convex since $\partial^2F/\partial x^2$, $\partial^2F/\partial y^2$ and $\partial^2F/\partial x^2\cdot\partial^2F/\partial y^2-(\partial^2F/(\partial x\partial y))^2$ are nonnegative in D. Our arguments are similar to those from the proofs of theorem 5.2 in section 5.2.

Now we estimate $\|r_\Delta-f(u_1)\|_{C([a,b]\setminus\Delta)}$. Since $0\leqslant\sigma(x-z)\leqslant 1$ for $x\in(-\infty,\infty)$ (see lemma 5.1), we get from (4) for $x\in[a,b]\setminus\Delta$

$$\begin{aligned}|r_\Delta(x)-f(u_1)| &\leqslant (1-\sigma(x-z))|r_{\Delta_1}(x)-f(u_1)|\\ &\quad+\sigma(x-z)(|r_{\Delta_2}(x)-f(u_2)|+|f(u_1)-f(u_2)|)\\ &\leqslant (1-\sigma(x-z))V_{\Delta_1}f+\sigma(x-z)(V_{\Delta_2}f+V_{\Delta_1}f)\leqslant V_\Delta f\end{aligned}$$

and therefore $\|r_\Delta-f(u_1)\|_{C([a,b]\setminus\Delta)}\leqslant V_\Delta f$ as required.

It remains to estimate $|f(x)-r_\Delta(x)|$ at the points $x\in A_{\Delta_1}\cup A_{\Delta_2}$. To this end we shall make use of (3), (4), the fact that $u_2-v_1\geqslant\varepsilon_1$, the choice of α, β, γ and the properties of σ by lemma 5.1. We get for $x\in A_{\Delta_1}\subset\Delta_1$

$$\begin{aligned}|f(x)-r_\Delta(x)| &= |(1-\sigma(x-z))(f(x)-r_{\Delta_1}(x))+\sigma(x-z)(f(x)-r_{\Delta_2}(x))|\\ &\geqslant (1-\sigma(x-z))|f(x)-r_{\Delta_1}(x)|-\sigma(x-z)\\ &\quad\cdot(\|r_{\Delta_2}-f(u_2)\|_{C([a,b]\setminus\Delta_2)}+|f(u_2)-f(x)|)\\ &> \varepsilon_2\left(1-\sum_{\nu=0}^{\mu}\frac{1}{2^{\nu+2}}\right)\left(1-\frac{1}{2^{\mu+4}}\right)-\frac{\varepsilon_2}{2^{\mu+4}V_\Delta f}V_\Delta f\\ &> \varepsilon_2\left(1-\sum_{\nu=0}^{\mu+1}\frac{1}{2^{\nu+2}}\right).\end{aligned}$$

Consequently for each $x\in A_{\Delta_1}$

$$|f(x)-r_\Delta(x)|>\varepsilon_2\left(1-\sum_{\nu=0}^{\mu+1}\frac{1}{2^{\nu+2}}\right) \tag{6}$$

and $f-r_\Delta$ has alternate signs at the consecutive points of A_{Δ_1} since $f-r_{\Delta_1}$ has alternate signs there.

Similarly, one proves that for each $x\in A_{\Delta_2}$ (6) holds true and $f-r_\Delta$ has alternate signs at the consecutive points of A_{Δ_2}.

From the above arguments it follows that there exists a set A_Δ of $l_{\Delta_1}+l_{\Delta_2}+1=\sum_{\Delta^*\in\Omega_\Delta}k_{\Delta^*}+1$ different points in Δ such that

$$|f(x)-r_\Delta(x)|>\varepsilon_2\left(1-\sum_{\nu=0}^{\mu+1}\frac{1}{2^{\nu+2}}\right),\quad x\in A_\Delta,$$

and $f-r_\Delta$ has alternate signs at the consecutive points of A_Δ, as required. Thus lemma 11.4 is established with μ replaced by $\mu+1$. □

Completion of the proof of theorem 11.8. By lemma 11.4 with $\mu=s$ there exist a rational function r and a set A of $l+1=\sum_{\Delta\in\Omega}k_\Delta+1$ different points $x_i\in[a,b]$, $a\leqslant x_0<x_1<\cdots<x_l\leqslant b$, such that

$$|f(x)-r(x)|>\varepsilon_2\left(1-\sum_{\nu=0}^{s}\frac{1}{2^{\nu+2}}\right)>\frac{\varepsilon_2}{2},\quad x\in A, \tag{7}$$

and $f-r$ has alternate signs at the points $x_0, x_1,\ldots,x_l$ and

$$\deg r\leqslant\sum_{\Delta\in\Omega}n_\Delta+N(s,M,[a,b]). \tag{8}$$

From this it follows that

$$R_n(f,[a,b])_C>\frac{\varepsilon_2}{2}$$

for $n \leqslant N = \sum_{\Delta\in\Omega} k_\Delta - \sum_{\Delta\in\Omega} n_\Delta - N(s, M, [a, b]) - 1$. Indeed, suppose to the contrary that there exists a rational function q such that

$$\| f - q \|_{C[a,b]} \leqslant \varepsilon_2/2 \tag{9}$$

and

$$\deg q \leqslant N. \tag{10}$$

Then by (7) and (9) it follows that for $x \in A$

$$|r(x) - q(x)| \geqslant |f(x) - r(x)| - |f(x) - q(x)| > 0$$

and since $f - r$ has alternate signs at $x_0, x_1, \ldots, x_l$, $r - q$ has alternate signs at $x_0, x_1, \ldots, x_l$. Consequently the rational function $r - q \not\equiv 0$ has at least $l = \sum_{\Delta\in\Omega} k_\Delta$ different zeros on $[a, b]$. On the other hand by (8) and (10) we have

$$\deg(r - q) \leqslant \deg r + \deg q \leqslant \sum_{\Delta\in\Omega} n_\Delta + N(s, M, [a, b]) + N = \sum_{\Delta\in\Omega} k_\Delta - 1 = l - 1.$$

We have a contradiction. Hence

$$R_n(f, [a, b])_C > \frac{\varepsilon_2}{2}, \quad n \leqslant N.$$

Similarly as in the proof of theorem 5.2 in section 5.2 one easily verifies that

$$N(s, M, [a, b]) + 1 \leqslant Dm \ln\left(e + \frac{b - a}{m\varepsilon_1}\right) \ln\left(e + \frac{M}{m\varepsilon_2}\right)$$

and therefore

$$N \geqslant \sum_{\Delta\in\Omega} k_\Delta - \sum_{\Delta\in\Omega} n_\Delta - Dm \ln\left(e + \frac{b - a}{m\varepsilon_1}\right) \ln\left(e + \frac{M}{m\varepsilon_2}\right)$$

which establishes the theorem. □

Denote

$$\lambda(x) = \begin{cases} 0, & x \leqslant 0, \\ h, & x > 0. \end{cases}$$

Lemma 11.5. *There exist constants $B_0 > 1$ and $D_0 > 0$ such that for each α, β, $h > 0$ such that $\beta/\alpha \geqslant B_0$ there exist a rational function r^* of degree $n \geqslant D_0 \ln(\beta/\alpha)$ and $n + 1$ points $u_i \in [\alpha, \beta]$, $\alpha \leqslant u_0 < u_1 < \cdots < u_n \leqslant \beta$, such that*

$$|\lambda(x) - r^*(x)| > \frac{h}{4}, \quad x \in \{-u_i\}_{i=0}^n \cup \{u_i\}_{i=0}^n,$$

$\lambda - r^*$ has alternate signs at the points $-u_n, -u_{n-1}, \ldots, -u_0, u_0, u_1, \ldots, u_n$ and

$$\| r^* \|_{C((-\infty,\infty)\setminus[-\beta,\beta])} \leqslant h.$$

Proof. Put $B_0 = \exp\{2\pi^2/\ln 2\}$ and $D_0 = \ln 2/2\pi^2$. Let $\alpha, \beta > 0$ and $\beta/\alpha \geqslant B_0$. Consider the rational function r from lemma 9.3 in section 9.2 with $\varepsilon = \alpha/\beta$ and $n = 2[(\ln 2/2\pi^2)\ln(1/\varepsilon)]$, where $[x]$ denotes the integer part of x. Since $\beta/\alpha \geqslant B_0 = \exp\{2\pi^2/\ln 2\}$ and $\varepsilon = \alpha/\beta$,

$$\deg r = n \geqslant \frac{\ln 2}{2\pi^2}\ln\frac{\beta}{\alpha} = D_0 \ln\frac{\beta}{\alpha} \geqslant 1.$$

Put $x_i = \varepsilon^{(n-i)/n}$ for $i = 0, 1, \ldots, n$. From the choice of ε it follows that $\alpha/\beta = x_0 < x_1 < \cdots < x_n = 1$. By lemma 9.3 and the choice of ε and n we obtain

$$|\operatorname{sign} x_i - r(x_i)| > \exp\left\{-\pi^2 \frac{n}{\ln(1/\varepsilon)}\right\} \geqslant \frac{1}{2}, \quad i = 0, 1, \ldots, n,$$

and $\operatorname{sign} x - r(x)$ has alternate signs at the points $-x_n, -x_{n-1}, \ldots, -x_0, x_0, x_1, \ldots, x_n$. Finally, since n is even, by the definition of r in lemma 9.3 it follows that $|r(x)| \leqslant 1$ for $x \in (-\infty, \infty)\backslash[-1, 1]$.

It is readily seen that the rational function $r^*(x) = \frac{1}{2}h(1 + r(\beta x))$ satisfies the requirements of lemma 11.5. □

A combination of theorem 11.8 and lemma 11.5 implies the following lemma.

Lemma 11.6. *Let f be a nondecreasing function defined on $[a, b]$. Suppose that there exist intervals $d_i = [u_i, v_i]$, $i = 1, 2, \ldots, m$, such that $0 \leqslant u_1 < v_1 < u_2 < v_2 < \cdots < u_m < v_m \leqslant 1$, m is of the type $m = 4l$, l positive integer, $v_i - u_i = \eta$, $u_{2\nu} - v_{2\nu-1} = \varepsilon\ (\nu = 1, 2, \ldots, 2l)$ and $f(x) = (i-1)h$ for $x \in d_i$, $i = 1, 2, \ldots, m$, where η, ε, $h > 0$ are given numbers such that $\eta/\varepsilon \geqslant B_0$, $B_0 > 1$ is the constant from lemma 11.5.*

Then $R_n(f, [0, 1])_C > h/8$ for each

$$n \leqslant \tfrac{1}{4}D_0 m \ln\frac{\eta}{\varepsilon} - 2Dm\ln\left(e + \frac{1}{m\eta}\right),$$

where D_0 and D are the constants from lemma 11.5 and theorem 11.8 respectively.

Proof. Put $[a_i, b_i] = [u_{4i-3}, v_{4i-2}]$ for $i = 1, 2, \ldots, m/4$. Obviously $0 \leqslant a_1 < b_1 < a_2 < b_2 < \cdots < a_{m/4} < b_{m/4} \leqslant 1$ and $\min_{1 \leqslant i \leqslant m/4}(a_{i+1} - b_i) > \eta$. Denote by Ω the set of the intervals $[a_i, b_i]$.

Now we are in a position to apply lemma 11.5 with $\alpha = \frac{1}{2}\varepsilon$ and $\beta = \eta + \frac{1}{2}\varepsilon$ ($\beta/\alpha = (\eta + \frac{1}{2}\varepsilon)/\frac{1}{2}\varepsilon > 2\eta/\varepsilon > B_0$ by our assumptions). We obtain that for each interval $\Delta \in \Omega$, $\Delta = [u, v]$, there exist a rational function r_Δ of order $n_\Delta \geqslant D_0 \ln((\eta + \frac{1}{2}\varepsilon)/\frac{1}{2}\varepsilon) > D_0 \ln(\eta/\varepsilon)$ and a set A_Δ of $k_\Delta = 2n_\Delta + 2$ different points $x_i \in \Delta$, $u \leqslant x_0 < x_1 < \cdots < x_{k_\Delta} \leqslant v$, such that

$$|f(x) - r_\Delta(x)| > \frac{h}{4}, \quad x \in A_\Delta,$$

$f-r_\Delta$ has alternate signs at the points $x_0, x_1, \ldots, x_{k_\Delta}$ and

$$\|r_\Delta - f(u)\|_{C((C-\infty,\infty)\setminus\Delta)} \leqslant h = V_\Delta f.$$

Then by theorem 11.8 we conclude that

$$R_n(f,[0,1])_C > \frac{h}{8} \quad \text{for} \quad n \leqslant N = \sum_{\Delta\in\Omega} k_\Delta - \sum_{\Delta\in\Omega} n_\Delta$$
$$- D\frac{m}{4}\ln\left(e + \frac{1}{\frac{1}{4}m\eta}\right)\ln\left(e + \frac{V_0^1 f}{(\frac{1}{4}m)(\frac{1}{4}h)}\right).$$

Clearly

$$N \geqslant \tfrac{1}{4}D_0 m \ln\frac{\eta}{\varepsilon} - \tfrac{1}{4}Dm\ln\left(e + \frac{4}{m\eta}\right)\ln(e+16)$$
$$\geqslant \tfrac{1}{4}D_0 m \ln\frac{\eta}{\varepsilon} - 2Dm\ln\left(e + \frac{1}{m\eta}\right). \qquad \square$$

Proof of theorem 11.7. A simple change of variables shows that, if theorem 11.7 holds in the case $[a,b]=[0,1]$ and $M=1$, then it holds in the general case. Thus we shall suppose that $[a,b]=[0,1]$ and $M=1$. Also, in view of lemma 11.3, without loss of generality we shall suppose that the modulus of continuity ω is a concave function on $[0,\infty)$ and $\omega \not\equiv 0$, since $\lim_{\delta\to 0}(\omega(\delta)/\delta) = \infty$ from (1). Note that, because ω is concave the function $\omega(\delta)/\delta$ is nonincreasing on $(0,\infty)$.

It was proved in theorem 10.8 in section 10.2 that for the rational uniform approximation of absolutely continuous functions $f \in V(\omega)$ with 'good' modulus ω the o-effect appears. Thus to prove theorem 11.7 we shall construct a singular function like Cantor's well-known singular function.

Choose ε_0 such that $0 < \varepsilon_0 \leqslant 1$, $\omega(\varepsilon_0) \leqslant 1$ and ω is strictly increasing on $[0,\varepsilon_0]$, which is possible since ω is nondecreasing and concave on $[0,\infty)$, $\omega(0)=0$ and $\omega \not\equiv 0$. By ω^{-1} we shall denote the converse function of ω. Then ω^{-1} is strictly increasing and convex on $[0,\omega(\varepsilon_0)]$. Of course $\omega^{-1}(0)=0$. From the last facts and (1) it follows that the function $\omega^{-1}(t)/t$ is nondecreasing on $(0,\omega(\varepsilon_0)]$ and $\lim_{t\to 0}(\omega^{-1}(t)/t)=0$.

Let $\{s_\nu\}_{\nu=1}^{\infty}$ be an increasing unbounded sequence of integers such that s_ν is in the form $s_\nu = 4l_\nu$, l_ν a positive integer. Define sequences $\{m_\nu\}_{\nu=0}^{\infty}$, $\{h_\nu\}_{\nu=0}^{\infty}$, $\{\varepsilon_\nu\}_{\nu=0}^{\infty}$ and $\{\eta_\nu\}_{\nu=1}^{\infty}$ as follows: $m_0=1$, $m_\nu = s_\nu m_{\nu-1} = \prod_{j=1}^{\nu} s_j$; $h_0 = \omega(\varepsilon_0)$, $h_\nu = h_{\nu-1}/s_\nu = h_0/m_\nu$, where ε_0 is from above; $\varepsilon_\nu = \omega^{-1}(h_\nu)$; $\eta_\nu = \varepsilon_{\nu-1}/s_\nu - \varepsilon_\nu$.

We select the sequence $\{s_\nu\}_{\nu=1}^{\infty}$ such that for $\nu = 1, 2, \ldots$

(i) $\eta_\nu/\varepsilon_\nu \geqslant B_0$,

(ii) $\tfrac{1}{4}D_0 m_\nu \ln\dfrac{\eta_\nu}{\varepsilon_\nu} - 2Dm_\nu \ln\left(e + \dfrac{1}{m_\nu\eta_\nu}\right) \geqslant \tfrac{1}{5}D_0 m_\nu \ln\dfrac{\eta_\nu}{\varepsilon_\nu} \geqslant \tfrac{1}{6}D_0 m_\nu \ln\dfrac{\varepsilon_{\nu-1}}{s_\nu\varepsilon_\nu}$,

(iii) $m_{\nu-1}\omega^{-1}\left(\frac{h_0}{m_{\nu-1}}\right) \geqslant \left(m_\nu\omega^{-1}\left(\frac{h_0}{m_\nu}\right)\right)^{1/2}$,

(iv) $\frac{1}{m_\nu} \leqslant \omega(\varepsilon_0)$ and $\frac{1}{12}D_0 m_\nu \ln \frac{1}{m_\nu\omega^{-1}\left(\frac{1}{m_\nu}\right)} > 1$,

(v) $\frac{1}{24}D_0 m_\nu \geqslant 1$ and $m_\nu\omega^{-1}\left(\frac{1}{m_\nu}\right) \leqslant \frac{1}{e}$,

where B_0, D_0 and D are the positive constants from lemma 11.5 and theorem 11.8.

Such a choice of $\{s_\nu\}_{\nu=1}^{\infty}$ is possible because $\lim_{t\to 0}(\omega^{-1}(t)/t) = 0$.

Now we construct an auxiliary singular function g. Put $\Delta_1^{(1)} = (0, \varepsilon_0)$. From the choice of ε_0 we have $\Delta_1^{(1)} \subset [0, 1]$. Divide $\Delta_1^{(1)}$ into s_1 disjoint subintervals of equal length ε_0/s_1 by means of the points $u_{1,j}^{(1)} = (j-1)\varepsilon_0/s_1$, $j = 1, 2, \ldots, s_1$. Set $d_{1,j}^{(1)} = [u_{1,j}^{(1)}, v_{1,j}^{(1)}]$, $v_{1,j}^{(1)} = u_{1,j}^{(1)} + \eta_1$ for $j = 1, 2, \ldots, s_1$ and $E_1 = E_1^{(1)} = \bigcup_{j=1}^{s_1} d_{1,j}^{(1)}$. Now we define the function g on the set E_1 by $g(x) = (j-1)h_1$ for $x \in d_{1,j}$.

The set $\Delta_1^{(1)} \backslash E_1$ consists of $m_1 = s_1$ open intervals each of them having length ε_1. Number these intervals in an increasing order by $\Delta_i^{(2)}$, $i = 1, 2, \ldots, m_1$, and let $\Delta_i^{(2)} = (u_i^{(2)}, v_i^{(2)})$. We divide each interval $\Delta_i^{(2)}$ into s_2 disjoint subintervals of equal length ε_1/s_2 by means of the points $u_{i,j}^{(2)} = u_i^{(2)} + (j-1)\varepsilon_1/s_2$, $j = 1, 2, \ldots, s_2$. Set $d_{i,j}^{(2)} = [u_{i,j}^{(2)}, v_{i,j}^{(2)}]$, $v_{i,j}^{(2)} = u_{i,j}^{(2)} + \eta_2$ and $E_i^{(2)} = \bigcup_{j=1}^{s_2} d_{i,j}^{(2)}$, $E_2 = \bigcup_{i=1}^{m_1} E_i^{(2)}$.

Now we define the function g on the set E_2 by $g(x) = g(u_i^{(2)}) + (j-1)h_2$ for $x \in d_{i,j}^{(2)}$.

Similarly we do the third step. The set $\Delta_1^{(1)} \backslash (E_1 \cup E_2)$ consists of m_2 disjoint open intervals with length ε_2. Number these intervals in an increasing order by $\Delta_i^{(3)}$, $i = 1, 2, \ldots, m_2$, and let $\Delta_i^{(3)} = (u_i^{(3)}, v_i^{(3)})$. Divide each interval $\Delta_i^{(3)}$ into s_3 disjoint intervals of equal length ε_2/s_3 by means of the points $u_{i,j}^{(3)} = u_i^{(3)} + (j-1)\varepsilon_2/s_3$, $j = 1, 2, \ldots, s_3$. Set $d_{i,j}^{(3)} = [u_{i,j}^{(3)}, v_{i,j}^{(3)}]$, $v_{i,j}^{(3)} = u_{i,j}^{(3)} + \eta_3$, $E_i^{(3)} = \bigcup_{j=1}^{s_3} d_{i,j}^{(3)}$ and $E_3 = \bigcup_{i=1}^{m_2} E_i^{(3)}$. Define the function g on E_3 by $g(x) = g(u_i^{(3)}) + (j-1)h_3$ for $x \in d_{i,j}^{(3)}$.

The sets E_4, $E_5, \ldots$ are constructed similarly and the function g is defined on E_4, $E_5, \ldots$ also in a similar way.

Let g be already defined on the set $E = \bigcup_{\nu=1}^{\infty} E_\nu$ which is dense in $[0, \varepsilon_0]$. Then we set

$$g(x) = \sup\{g(t) : t \in E \text{ and } t < x\}, \quad x \in [0, 1] \backslash E.$$

Clearly, the function g is nondecreasing, continuous and singular. Also $g(0) = 0$ and $g(1) \leqslant 1$.

Now we estimate the modulus of continuity of g. It is sufficient to consider the case $0 < \delta \leqslant \varepsilon_0$. Suppose $\varepsilon_k \leqslant \delta \leqslant \varepsilon_{k-1}$, $k \geqslant 1$. Then from the definition of g it follows that

$$\omega(g, \delta)_C \leqslant \delta \frac{h_{k-1}}{\varepsilon_{k-1}} + h_k = \delta \frac{\omega(\varepsilon_{k-1})}{\varepsilon_{k-1}} + \omega(\varepsilon_k) \leqslant \omega(\delta) + \omega(\varepsilon_k) \leqslant 2\omega(\delta),$$

where we have used that $\omega(\delta)/\delta$ is nonincreasing on $(0, \infty)$. Consequently

$$\omega(g, \delta)_C \leqslant 2\omega(\delta), \quad \delta \geqslant 0. \tag{11}$$

Finally, we shall apply lemma 11.6. Let $v \geqslant 1$ and consider g over the intervals $d_{i,j}^{(v)}$, $j = 1, 2, \ldots, s_v$, $i = 1, 2, \ldots, m_{v-1}$. The function g satisfies the assumptions of lemma 11.6 with $m = m_v$, $h = h_v$, $\varepsilon = \varepsilon_v$ and $\eta = \eta_v$ ($\eta_v/\varepsilon_v \geqslant B_0$ by (i)). Then we conclude that

$$R_n(g, [0, 1])_C > \frac{h_v}{8} = \frac{h_0}{8m_v}$$

for each

$$n \leqslant N_v = \tfrac{1}{4} D_0 m_v \ln \frac{\eta_v}{\varepsilon_v} - 2Dm_v \ln\left(e + \frac{1}{m_v \eta_v}\right).$$

By the definitions of m_v, ε_v, η_v and the properties (ii)–(iv) it follows that

$$N_v \geqslant \tfrac{1}{6} D_0 m_v \ln \frac{\varepsilon_{v-1}}{s_v \varepsilon_v} = \tfrac{1}{6} D_0 m_v \ln \frac{m_{v-1}\omega^{-1}\left(\dfrac{h_0}{m_{v-1}}\right)}{m_v \omega^{-1}\left(\dfrac{h_0}{m_v}\right)}$$

$$\geqslant \tfrac{1}{12} D_0 m_v \ln \frac{1}{m_v \omega^{-1}\left(\dfrac{h_0}{m_v}\right)}$$

$$\geqslant \tfrac{1}{12} D_0 m_v \ln \frac{1}{m_v \omega^{-1}\left(\dfrac{1}{m_v}\right)}.$$

Consequently for $v = 1, 2, \ldots$

$$R_n(g, [0, 1])_C > \frac{h_0}{8m_v}, \quad n \leqslant \tfrac{1}{12} D_0 m_v \ln \frac{1}{m_v \omega^{-1}\left(\dfrac{1}{m_v}\right)}. \tag{12}$$

For each $v = 1, 2, \ldots$ choose n_v such that

$$n_v \leqslant \tfrac{1}{12} D_0 m_v \ln \frac{1}{m_v \omega^{-1}(1/m_v)} \leqslant 2n_v.$$

Set $t_\nu = \frac{1}{24} D_0 m_\nu$. From (v) it follows that $1 \leqslant t_\nu \leqslant n_\nu$. Then we get

$$\inf_{1 \leqslant t \leqslant n_\nu} \left\{ \frac{1}{t} + \omega\left(\frac{1}{t e^{n_\nu/t}} \right) \right\} \leqslant \frac{1}{t_\nu} + \omega\left(\frac{1}{t_\nu e^{n_\nu/t_\nu}} \right)$$

$$\leqslant \frac{24}{D_0 m_\nu} + \omega\left(\frac{24}{D_0 m_\nu \exp\left\{ \ln \frac{1}{m_\nu \omega^{-1}(1/m_\nu)} \right\}} \right)$$

$$\leqslant \frac{24}{D_0 m_\nu} + \omega\left(\frac{24}{D_0} \omega^{-1}\left(\frac{1}{m_\nu} \right) \right) \leqslant \frac{24}{D_0 m_\nu} + \left(\frac{24}{D_0} + 1 \right) \frac{1}{m_\nu} = \left(\frac{48}{D_0} + 1 \right) \frac{1}{m_\nu}.$$

From this and (12) it follows that

$$R_{n_\nu}(g)_C > C \inf_{1 \leqslant t \leqslant n_\nu} \left\{ \frac{1}{t} + \omega\left(\frac{1}{t e^{n_\nu/t}} \right) \right\}, \quad \nu = 1, 2, \ldots, \tag{13}$$

where $C > 0$ is a constant.

By (11) and (13) the function $f = \frac{1}{2} g$ satisfies the requirements of theorem 11.7. □

11.3 Notes

Theorems 11.1 and 11.2 in the case $k = 1$ are proved by J. Szabados (1967b); E.P. Dolženko (1967) has found a comparison between uniform rational and polynomial approximations which is closely connected to theorems 11.1 and 11.2. Theorem 11.3 is due to E.P. Dolženko (1962). Theorem 11.4 is proved by G. Freud (1970). The lower bounds in theorem 11.5 are due to A.P. Bulanov (1969). Theorem 11.6 is trivial and well-known. Theorem 11.7 (with another proof) is due to A. Pekarski (1980a). The exact lower bound for uniform rational approximation of absolutely continuous functions with given modulus of continuity is proved by A.P. Bulanov (1975b).

12

Padé approximations

One of the most popular domains in the theory of approximation of functions by means of rational functions is the theory of the Padé approximations. There exist many books and papers which consider this type of approximations. We want only to mention the excellent monograph in two volumes of Baker and Graves-Morris (1981). Here we want to consider some problems connected with the convergence of the Padé approximants, which are not entirely included in that monograph. These results are due to A.A. Gonchar and the group of mathematicians headed by him.

In section 12.1 we give the definition and some promerties of Padé approximants. In section 12.2 we have direct results for the convergence of Padé approximants – the classical theorem of Montessus de Ballore and one of its generalizations, which is due to A.A. Gonchar (1975a). In section 12.3 we give one converse theorem for the convergence of Padé approximants with fixed degree of denominator (the rows of the Padé-table) which is due to Gonchar (unpublished). In section 12.4 we give one more converse theorem of Gonchar connected with the diagonal of the Padé-table. In the notes to the chapter we give some more information about these problems.

12.1 Definition and properties of the Padé approximants

Let

$$f(z) = \sum_{\nu=0}^{\infty} f_\nu z^\nu \tag{1}$$

be a formal power series. Let n and m be two nonnegative integers.

Usually the Padé approximant $\pi_{nm} = P_{nm}/Q_{nm}$, $P_{nm} \in P_n$, $Q_{nm} \in P_m$, of order (n, m) of (1) is given by the condition

$$Q_{nm}(z)f(z) - P_{nm}(z) = O(z^{n+m+1}).^{\dagger} \tag{2}$$

† $\varphi(z) = O(z^{n+m+1}) \Leftrightarrow \limsup_{z\to 0} |\varphi(z)/z^{n+m+1}| < \infty$.

Let

$$P_{nm}(z) = a_n z^n + \cdots + a_1 z + a_0, \tag{3}$$

$$Q_{nm}(z) = b_m z^m + \cdots + b_1 z + b_0. \tag{4}$$

Let us put (1), (3) and (4) in (2):

$$(b_m z^m + \cdots + b_0)\left(\sum_{\nu=0}^{\infty} f_\nu z^\nu\right) - (a_n z^n + \cdots + a_0) = O(z^{n+m+1}).$$

If we write the conditions that the coefficients on the left side before z^k, $k = 0, \ldots, n+m$, are zero, we obtain:

$$\left.\begin{aligned} a_0 &= f_0 b_0, \\ a_1 &= f_0 b_1 + f_1 b_0, \\ a_2 &= f_0 b_2 + f_1 b_1 + f_2 b_0, \\ &\cdots \\ a_n &= \sum_{i=1}^{min(n,m)} f_{n-i} b_i + f_n b_0, \end{aligned}\right\} \tag{5}$$

$$\left.\begin{aligned} b_0 f_{n+1} + b_1 f_n + \cdots + b_m f_{n-m+1} &= 0, \\ b_0 f_{n+2} + b_1 f_{n+1} + \cdots + b_m f_{n-m} &= 0, \\ &\cdots \\ b_0 f_{n+m} + b_1 f_{n+m-1} + \cdots + b_m f_n &= 0 \end{aligned}\right\} \tag{6}$$

(we set here and in what follows $f_i = 0$ for $i < 0$).

The system (6) is a system of m linear algebraic equations with $m+1$ unknown coefficients $b_0, \ldots, b_m$, which has always a solution. If we know a solution of the system (6), we can find the coefficients of the numerator a_0, $a_1, \ldots, a_n$ from the equations (5).

One solution of the systems (5), (6) is given by

$$\left.\begin{aligned} \tilde{P}_{nm}(z) &= \begin{vmatrix} f_{n-m+1} & & \cdots & f_{n+1} \\ & & \cdots & \\ f_n & f_{n+1} & \cdots & f_{n+m} \\ \sum_{i=0}^{n-m} f_i z^{m+i} & \sum_{i=0}^{n-m+1} f_i z^{m+i-1} & \cdots & \sum_{i=0}^{n} f_i z^i \end{vmatrix}, \\ \tilde{Q}_{nm}(z) &= \begin{vmatrix} f_{n-m+1} & f_{n-m+2} & \cdots & f_n & f_{n+1} \\ f_{n-m+2} & f_{n-m+3} & \cdots & f_{n+1} & f_{n+2} \\ & & \cdots & & \\ f_n & f_{n+1} & \cdots & f_{n+m-1} & f_{n+m} \\ z^m & z^{m-1} & \cdots & z & 1 \end{vmatrix}. \end{aligned}\right\} \tag{7}$$

It is not difficult to verify that $\tilde{P}_{nm}$ and $\tilde{Q}_{nm}$ given by (7) satisfy (2).

From (7) we see that the condition $\tilde{Q}_{nm}(0) \neq 0$ is equivalent to the condition

$$C(n,m) = \begin{vmatrix} f_n & f_{n-1} & \cdots & f_{n-m+1} \\ f_{n+1} & f_n & \cdots & f_{n-m} \\ & & \cdots & \\ f_{n+m-1} & f_{n+m-2} & \cdots & f_n \end{vmatrix} \neq 0.$$

The modern definition by Baker (1973) of Padé approximant of order (n, m) is the following.

We say that the Padé approximant of order (n, m) exists if there exist two polynomials $P_{nm} \in P_n$ and $Q_{nm} \in P_m$ such that

(i) $f(z) - P_{nm}(z)/Q_{nm}(z) = O(z^{n+m+1})$,
(ii) $Q_{nm}(0) = 1$.

Then we set $\pi_{nm} = P_{nm}/Q_{nm}$ as (n, m)-th Padé approximant.

The conditions (i), (ii) are equivalent to the condition that the system (6) has a solution with $b_0 = 1$. The last condition is equivalent to the condition $C(n, m) \neq 0$.

Sometimes the problem of Padé approximation is given in the following form: find $P_{nm} \in P_n$ and $Q_{nm} \in P_m$ such that

$$f(z) - \frac{P_{nm}(z)}{Q_{nm}(z)} = O(z^{n+m+1}). \tag{8}$$

The problems (2) and (8) are equivalent if $C(n, m) \neq 0$. But if $C(n, m) = 0$, in the general case it is not so.

A solution of the problem (2) always exists (for example given by (7)), but it is possible that there does not exist a solution of the problem (8). For example it is easy to see that for $f(z) = 1 + z^2$ there does not exist a solution of (8) of order $(1, 1)$ (see Baker and Graves-Morris (1981)).

We shall not go into details when the solution of (8) exists.

In what follows Padé approximant we shall understand in the sense of Baker (i), (ii).

Usually the Padé approximants of order (n, m) are displayed in a table, called the Padé-tables, as follows:

	n			
m	0	1	2	3
0	(0, 0)	(1, 0)	(2, 0)	(3, 0)
1	(0, 1)	(1, 1)	(2, 1)	(3, 1)
2	(0, 2)	(1, 2)	(2, 2)	(3, 2)

The approximants with a given m, $(0,m)$, $(1,m)$, $(2,m)$, ..., are called a row of the Padé-table. The approximants $(0,0)$, $(1,1)$, $(2,2)$, ... are called the diagonal of the Padé-table.

As an example of a Padé approximant let us consider the (n,m)-th Padé approximant $\pi_{nm} = P_{nm}/Q_{nm}$ to e^x. The following representation was given by O. Perron:

$$\left.\begin{aligned} P_{nm}(z) &= \frac{1}{(n+m)!}\int_0^\infty t^m(t+z)^n e^{-t}\,dt, \\ Q_{nm}(z) &= \frac{1}{(n+m)!}\int_0^\infty (t-z)^m t^n e^{-t}\,dt. \end{aligned}\right\} \tag{9}$$

Obviously we have $P_{nm} \in P_n$, $Q_{nm} \in P_m$. On the other hand $Q_{nm}(0) = 1$. To prove that (9) is the (n,m)-th Padé approximant to e^x let us calculate $Q_{nm}(z)e^z - P_{nm}(z)$. We have

$$\begin{aligned} e^z Q_{nm}(z) - P_{nm}(z) &= \frac{1}{(n+m)!}\left\{\int_0^\infty (t-z)^m t^n e^{-t+z}\,dt - \int_0^\infty t^m(t+z)^n e^t\,dt\right\} \\ &= \frac{1}{(n+m)!}\int_0^z t^n(t-z)^m e^{-t}\,dt \\ &= (-1)^n \frac{z^{n+m+1}}{(n+m)!}\int_0^1 u^n(1-u)^m e^{uz}\,du. \end{aligned}$$

Since the integral is an analytic function in the neighborhood of $z = 0$, the last equality shows that $\pi_{nm} = P_{nm}/Q_{nm}$ given by (9) is the (n,m)-th Padé approximant to e^x.

We shall use in this chapter the following well-known lemma of Cauchy and Hadamard.

Lemma 12.1. *Suppose for the formal series*

$$f(z) = \sum_{\nu=0}^{\infty} f_\nu z^\nu$$

we have

$$\limsup_{n\to\infty} |f_n|^{1/n} \leqslant \rho.$$

Then f is a holomorphic function in the open disk $|z| < 1/\rho$.

12.2 Direct theorem for the rows of the Padé-table

In the theory of convergence of the Padé approximants two types of questions arise: the direct type theorems and the converse type theorems. Under direct theorem in the theory of Padé approximation we understand the following one: if we know something about the function f, for which we consider the formal power series

$$f(z)=\sum_{\nu=0}^{\infty} f_\nu z^\nu \tag{1}$$

(for example the number of the poles in some domain), what can we say about the corresponding Padé approximants, for example for their poles?

Under converse theorem in the theory of Padé approximation we understand the following one: if we know something about the Padé approximants of f, for example the number and situations of their poles, what can we say about the function f?

A typical direct theorem in Padé approximation is the classical theorem of Montessus de Ballore.

Theorem 12.1. *Let* (1) *represent the function f in a neighbourhood of $z=0$ (i.e. f is holomorphic in $z=0$). Let $D_m=\{z:|z|<R_m\}$ be the greatest disk centered at the origin inside of which f has a meromorphic continuation with no more than m poles (counting multiplicities). If D_m contains exactly k distinct poles $z_1,\dots,z_{k-1},z_k$ of f of multiplicities $p_1,\dots,p_k$ respectively and*

$$\sum_{i=1}^{k} p_i=m, \tag{2}$$

then the sequence of (n,m)-th Padé approximants, m fixed, converges uniformly to f as $n\to\infty$ on each compact K, $K\subset D'_m=D_m\backslash\{z_1,\dots,z_k\}$.

The poles of denominators Q_{nm} of the (n,m)-th Padé approximant $\pi_{nm}=P_{nm}/Q_{nm}$ tend to the poles of f in D_m.

More exactly, there exists an algebraic polynomial $Q_m\in P_m$,

$$Q_m(z)=\prod_{i=1}^{k}(z-z_i)^{p_i}\Big/(-1)^m\prod_{i=1}^{k} z_i^{p_i},$$

such that

$$\limsup_{n\to\infty}\|Q_{nm}-Q_m\|^{1/n}=\frac{\max\{|z_i|:1\leqslant i\leqslant k\}}{R_m}<1,$$

where $\|\cdot\|$ is some norm in P_m (the space of all algebraic polynomials of mth degree is a finite dimensional normed linear space and all norms in P_m are equivalent).

A.A. Gonchar (1975a) proved that even in the case when $p_1+\cdots+p_k<m$ (see the conditions of theorem 12.1) each pole of f in D_m attracts at least as many poles of π_{nm}, m fixed, as is its order of multiplicity.

We shall not prove here this direct theorem because for the converse theorem 12.3 we shall need only a weaker result (theorem 12.2). For some proofs of Montessus's theorem 12.1 see Baker and Graves-Morris (1981).

It is a well-known fact in the theory of Padé approximations that in general when we consider (n, m)-th Padé approximants to functions f with numbers of poles less than m (for example entire functions), the poles of the Padé approximants π_{nm} can form a dense set in the plane (see Perron (1957, Chapter 4), Baker and Graves-Morris (1981, Chapter 6)). We shall not consider here questions of such type. Let us mention only that usually the direct theorems for the convergence of Padé approximants are given in terms of measure or capacity (see the notes at the end of the chapter).

In what follows G is an open domain in the complex plane $\mathbb{C}$ containing the origin with boundary Γ. We set $\bar{G} = G \cup \Gamma$.

The set of all holomorphic functions in the domain G (in the closed set $\bar{G}$) we shall denote by $\mathscr{H}(G)$ ($\mathscr{H}(\bar{G})$).

We shall give first two lemmas, following Baker and Graves-Morris (1981).

Lemma 12.2. *Let $f \in \mathscr{H}(G)$ and let f be continuous in G. Then the Padé approximant of order* $(n, 0)$ (the Maclaurin polynomial of order n) *is given by*

$$\pi_{n0}(z) = \frac{1}{2\pi i} \int_\Gamma \frac{t^{n+1} - z^{n+1}}{t - z} \frac{f(t)}{t^{n+1}} \, dt. \tag{3}$$

Proof. Since

$$\frac{t^{n+1} - z^{n+1}}{t - z} = \sum_{j=0}^{n} t^{n-j} z^j$$

we obtain that for π_{n0} given by (1) we have

$$\pi_{n0} = \sum_{j=0}^{n} z^j \frac{1}{2\pi i} \int_\Gamma \frac{t^{n-j} f(t)}{t^{n+1}} \, dt = \sum_{j=0}^{n} \frac{f^{(j)}(0)}{j!} z^j. \qquad \square$$

In what follows we shall assume that appropriate (n, m)-th Padé approximants exist.

Lemma 12.3. *Let $f \in \mathscr{H}(G)$ and let f be continuous in $\bar{G}$. For every algebraic polynomial R_m of degree at most m, $R_m(0) \neq 0$, we have*

$$f(z) - \pi_{nm}(z) = \frac{z^{n+m+1}}{2\pi i Q_{nm}(z) R_m(z)} \int_\Gamma \frac{f(t) Q_{nm}(t) R_m(t)}{t^{n+m+1}(t - z)} \, dt,$$

where $\pi_{nm} = P_{nm}/Q_{nm}$ is the (n, m)-th Padé approximant to f.

Proof. Let $\pi_{nm} = P_{nm}/Q_{nm}$ be the (n, m)-th Padé approximant to f. Let us consider the Maclaurin polynomials $\pi_{(n+m)0}$ of order $n + m$ for the function $f(z) Q_{nm}(z) R_m(z)$. By lemma 12.2 we have

$$\pi_{n+m,0}(z) = \frac{1}{2\pi i} \int_\Gamma \frac{t^{n+m+1} - z^{n+m+1}}{t - z} \frac{f(t) Q_{nm}(t) R_m(t)}{t^{n+m+1}} \, dt. \tag{4}$$

Since $fQ_{nm}R_m \in \mathscr{H}(\bar{G})$, by Cauchy's theorem we get

$$f(z)Q_{nm}(z)R_m(z) = \frac{1}{2\pi i}\int_\Gamma \frac{f(t)Q_{nm}(t)R_m(t)}{t-z}\,dt. \tag{5}$$

From (4) and (5) we obtain

$$f(z)Q_{nm}(z)R_m(z) - \pi_{n+m,0}(z) = \frac{z^{n+m+1}}{2\pi i}\int_\Gamma \frac{f(t)Q_{nm}(t)R_m(t)}{(t-z)t^{n+m+1}}\,dt. \tag{6}$$

The Padé condition gives us:

$$f(z)Q_{nm}(z)R_m(z) - P_{nm}(z)R_m(z) = O(z^{n+m+1}).$$

Together with $R_m(0) = 0$, we obtain from here

$$\pi_{n+m,0}(z) = P_{nm}(z)R_m(z) + O(z^{n+m+1}) = P_{nm}(z)R_m(z)$$

and (6) gives us the statement of the lemma. □

Remark. Using lemma 12.3 Saff (1972) gives a generalization of Montessus's theorem for multipoint Padé approximation; see also Baker, Graves–Morris (1981).

Theorem 12.2. *Let $f(z)$ be an analytic function at $z = 0$ and let f be meromorphic in $\bar{D} = \{z: |z| \leqslant R\}$ with exactly s poles, counting multiplicity, $\alpha_1, \ldots, \alpha_s$, $|\alpha_i| < R$, $i = 1, \ldots, s$. Let $m \geqslant s$ and let $\pi_{nm} = P_{nm}/Q_{nm}$ be the (n, m)-th Padé approximant to f. Let there exist an algebraic polynomial $Q_m \in P_m$ such that*

$$\| Q_{nm} - Q_m \| \underset{n\to\infty}{\longrightarrow} 0, \tag{7}$$

where $\|\cdot\|$ is some norm in P_m, and the zeros of Q_m are the points $z_1, z_2, \ldots, z_m$.
Then all zeros of f are in the set $\{z_1, \ldots, z_m\}$.
On every compact set K,

$$K \subset \{z: |z| < R\} \setminus \{z_1, z_2, \ldots, z_m\},$$

we have uniform convergence of π_{nm} to f, $n \to \infty$.

Proof. Let us consider first the case when all points α_i, $i = 1, \ldots, s$, are different. Let us set

$$q_s(z) = (z - \alpha_1)\cdots(z - \alpha_s).$$

Then $fq_s \in \mathscr{H}(\bar{D})$ and lemma 12.3 with $R_m = 1$ gives us

$$f(z) - \pi_{nm}(z) = \frac{z^{n+m+1}}{2\pi i\, Q_{nm}(z)q_s(z)}\int_\Gamma \frac{f(t)q_s(t)Q_{nm}(t)}{(t-z)t^{n+m+1}}\,dt, \tag{8}$$

where $\pi_{nm} = P_{nm}/Q_{nm}$ is the (n, m)-th Padé approximant to f, and $\Gamma\{z: |z| = R\}$.

Let us set

$$D = \{z: |z| < R\}, \quad A = \{\alpha_1, \ldots, \alpha_s, z_1, \ldots, z_m\}.$$

Let K be arbitrary compact, $K \subset D \backslash A$. Then

$$\inf\{|z - \bar{z}|: z \in K, \bar{z} \in A\} = \delta > 0$$

and therefore (7) gives us that for $n > N_1$ we have

$$\inf\{|Q_{nm}(z) q_s(z)|: z \in K\} = \delta_1 > 0. \tag{9}$$

Let us set $\rho = \sup\{|z|: z \in K\}$. Since $K \subset D$ we have $0 \leqslant \rho < R$.

On the other hand the condition (5) gives us that $\|Q_{nm}\|_{C(\Gamma)} \leqslant 2 \|Q_m\|_{C(\Gamma)}$ for $n > N_2$.

Consequently from (8), (9) we obtain ($n > \max\{N_1, N_2\}$)

$$\|f - \pi_{nm}\|_{C(K)} \leqslant \left(\frac{\rho}{R}\right)^{n+m+1} \frac{2 \|f q_s\|_{C(\Gamma)} \|Q_m\|_{C(\Gamma)}}{\delta_1 (R - \rho)}. \tag{10}$$

Since $0 \leqslant \rho < R$, we obtain geometric uniform convergence of π_{nm} to f on K. Using (10) we shall prove that every α_i, $i = 1, \ldots, s$, is in the set $\{z_1, \ldots, z_m\}$. Let us assume the converse, that there exists $\alpha_{i_0} \notin \{z_1, \ldots, z_m\}$. Then there is a disk $G = \{z: |z - \alpha_{i_0}| \leqslant \theta_0\}$ with the following properties:

(a) if $n > N_3$, π_{nm} is holomorphic in G (this follows from (7)),

(b) G contains no other zeros of f except α_{i_0},

(c) $G \subset D$.

From (10) and (a)–(c) it follows that for every n, $0 < \theta < \theta_0$, for the compact $K(\theta) = \{z: \theta \leqslant |z - \alpha_{i_0}| \leqslant \theta_0\}$ we have

$$\|f - \pi_{nm}\|_{C(K(\theta))} = \mathrm{O}(q(\theta)^n), \tag{11}$$

where $0 < q(\theta) < 1$, $q(\theta)$ depends on θ.

Using the maximum principle for π_{nm} with respect to G we obtain from (11) that for $n > N_3$ we have

$$\|\pi_{nm}\|_{C(G)} \leqslant M, \tag{12}$$

where $0 < M < \infty$ (f is holomorphic in $K(\theta)$, π_{nm} is holomorphic in G for $n > N_3$).

But (12) contradicts (11) for small θ, since $|f(z)| \to \infty$, $z \to \alpha_{i_0}$.

Therefore α_{i_0} is in the set $\{z_1, \ldots, z_m\}$.

The case when α_{i_0} is a pole with multiplicity follows by continuity arguments. □

12.3 Converse theorem for the rows of the Padé-table

In this section we shall prove one theorem of A.A. Gonchar for the convergence of the rows of the Padé-table. Let us establish our notations.

Let

$$f(z)=\sum_{\nu=0}^{\infty} f_\nu z^\nu \tag{1}$$

be a formal power series, and $m \geqslant 0$ be a fixed natural number.

We denote by $\pi_{nm}=\pi_{nm}(f)$ the (n, m)-th Padé-approximant for the formal series (1), i.e. $Q_{nm}(0)=1$,

$$\pi_{nm}=P_{nm}/Q_{nm}, \quad P_{nm}\in P_n, \quad Q_{nm}\in P_m, \quad Q_{nm}(z)f(z)-P_{nm}(z)=\mathrm{O}(z^{n+m+1}).$$

We shall assume that such an approximant exists.

Let $\rho>0$. We denote by D_ρ the disk $D_\rho=\{z:|z|\leqslant\rho\}$. We set $\Gamma_\rho=\partial D_\rho=\{z:|z|=\rho\}$.

Let $\tilde{M}_m(D_\rho)$, respectively $M_m(D_\rho)$, denote the classes of functions, analytic at $z=0$, which have meromorphic continuation in D_ρ with $\leqslant m$, respectively $=m$, poles in D_ρ.

We denote by $D_m(f)$ the maximal disk in which (1) has a meromorphic continuation belonging to $\tilde{M}_m(D_m(f))$. We set $R_m(f)=R_m$ the radius of $D_m(f)$, so $R_0(f)=R_0$ is the radius of convergence of the series (1). If (1) diverges at $z=0$, we set $R_0=0$.

Theorem 12.3 (A.A. Gonchar, unpublished). *Let* (1) *represent a formal power series and let $m\geqslant 0$ be a fixed natural number. Suppose for every natural number $n\geqslant 0$ we have* $\deg Q_{nm}=m$. *Let there exist an algebraic polynomial $Q\in P_m$, $Q(0)\neq 0$, such that*

$$\limsup_{n\to\infty}\|Q_{nm}-Q\|^{1/n}\leqslant q<1 \tag{2}$$

and

$$Q(z)=\left(\prod_{k=1}^{m}(z-z_k)\right)\Big/(-1)^m\prod_{k=1}^{m}z_k, \tag{3}$$

where $\|\cdot\|$ is some norm in P_m (see 12.2).

Then

(a) $R_0>0$,
(b) $f\in M_m(D_R)$, *where* $R\geqslant\max\{|z_k|:1\leqslant k\leqslant m\}/q$,
(c) *the poles of f in D_R are the points $z_1,\dots,z_m$.*

Remark. Some of the points can coincide.

Proof. Let us have

$$Q_{nm}(z)=\left(\prod_{k=1}^{m}(z-z_{n,k})\right)\Big/(-1)^m\prod_{k=1}^{m}z_{n,k}.$$

Since

$$Q(0)\neq 0 \tag{4}$$

we have

$$0<|z_k|, \quad k=1,\dots,m.$$

Since all norms in P_m are equivalent, i.e. if $\|\cdot\|$ and $|||\cdot|||$ are two norms in P_m, then there exist constants $c_1(m)$ and $c_2(m)$, depending only on the dimension $m+1$ of P_m, such that

$$c_2(m)\|p\| \leqslant |||p||| \leqslant c_1(m)\|p\|, \quad p \in P_m,$$

and $\max\{|z_k|: 1 \leqslant k \leqslant m\} = |||Q|||$ is an equivalent norm to $\|Q\|$, from (2) it follows that

$$\limsup_{n\to\infty} |z_{n,k} - z_k|^{1/n} \leqslant q < 1. \tag{5}$$

To prove (a) let us consider the algebraic polynomial

$$Q_{nm}(z) = 1 + a_{n1}z + \cdots + a_{nm}z^m. \tag{6}$$

From (3) and (5) it follows that the sequences $\{z_{n,k}\}_{n=1}^{\infty}$, $k = 1, \ldots, m$, are convergent and therefore the sequences $\{a_{ni}\}_{n=1}^{\infty}$, $i = 1, \ldots, m$, are also convergent. Let us set $a_i = \lim_{n\to\infty} a_{ni}$, $i = 1, \ldots, m$. It follows from (2), (5), (6) that

$$Q(z) = 1 + a_1 z + \cdots + a_m z^m. \tag{7}$$

Let $M = \max_{1\leqslant k\leqslant m} |a_k|$. Let $\delta > 0$ be an arbitrary positive number. Then for every n sufficiently large, $n > N_0$, we have

$$|a_{nk}| \leqslant M + \delta, \quad k = 1, \ldots, m. \tag{8}$$

Let us set

$$\left.\begin{aligned} c &= (M+\delta)m, \\ |f_{\nu_\delta}| &= \max\{|f_\nu|: \nu \leqslant N_0\}, \end{aligned}\right\} \tag{9}$$

where f_ν are given by (1).

From the definition of the $(n-1, m)$-th Padé-approximant and (6) it follows that

$$f_n + a_{n-1,1}f_{n-1} + \cdots + a_{n-1,m}f_{n-m} = 0. \tag{10}$$

Let $n - 1 > N_0$. From (10), (8) and (9) we get

$$|f_n| \leqslant c \max_{1\leqslant k\leqslant m} |f_{n-k}|. \tag{11}$$

If we set $|f_{n-k_1}| = \max_{1\leqslant k\leqslant m} |f_{n-k}|$, then (11) gives us

$$|f_n| \leqslant c|f_{n-k_1}|, \tag{12}$$

where $1 \leqslant k_1 \leqslant m$.

If $n - k_1 > N_0$, then applying to $|f_{n-k_1}|$ estimates of the types (10), (12) we get

$$|f_n| \leqslant c^2 |f_{n-k_1-k_2}|,$$

where $|f_{n-k_1-k_2}| = \max\{|f_{n-k_1-k}|: 1 \leqslant k \leqslant m\}$.

If $n-k_1-k_2>N_0$, again using inequalities of the type (12) we obtain

$$|f_n|\leqslant c^3|f_{n-k_1-k_2-k_3}|,\quad 1\leqslant k_i\leqslant m,\quad i=1,2,3.$$

Continuing in such a way we obtain that

$$|f_n|\leqslant c^{l(n)}|f_{n-k_1-\cdots-k_{l(n)}}|, \tag{13}$$

where $n-k_1-\cdots-k_{l(n)}\leqslant N_0<n-k_1-\cdots-k_{l(n)-1}$, $1\leqslant k_i\leqslant m$, $i=1,\ldots,l(n)$. Using (19) we get

$$|f_n|\leqslant c^{l(n)}|f_{\nu_\delta}|.$$

From here we obtain

$$\limsup_{n\to\infty}|f_n|^{1/n}\leqslant\limsup_{n\to\infty}c^{l(n)/n}|f_{\nu_\delta}|^{1/n}.$$

Since evidently $l(n)\leqslant n$, from here we get

$$\limsup_{n\to\infty}|f_n|^{1/n}=A<\infty.$$

From this inequality and lemma 12.1 it follows that $R_0=1/A>0$, which proves (a).

Let us prove now (b) and (c).

From $R_0>0$ it follows that $R_m>0$.

We shall prove the following statement: in the assumption of theorem 12.3 if $f\in M_s(D_\rho)$ and $s<m$, then $R_m\geqslant\rho q^{-1}$ where q is given by (2) and (5).

It follows from the direct theorem 12.2 and the conditions of theorem 12.3 that if $f\in M_s(D_\rho)$, $s<m$, then π_{nm} converges, as $n\to\infty$, uniformly to the function f in every compact $K\subset D_\rho^*$, $D_\rho^*=D_\rho\backslash\{z:z$ is a pole of f in $D_\rho\}$, and the poles of f in D_ρ are in the set $\{z_1,\ldots,z_m\}$. Let the poles of f in D_ρ be $z'_1,\ldots,z'_s$ and let us set $q_s(z)=\prod_{k=1}^s(z-z'_k)$, $F=fq_s$. Then F is a holomorphic function in D_ρ. Since $\deg q_s<m$, the definition of the Padé-approximants gives us

$$\frac{1}{2\pi i}\int_{\Gamma_{\rho-\varepsilon}}\frac{F(z)Q_{nm}(z)}{z^{n+m+1}}\,dz=\frac{1}{2\pi i}\int_{\Gamma_{\rho-\varepsilon}}\frac{q_s(z)P_{nm}(z)+O(z^{n+m+1})}{z^{n+m+1}}\,dz=0, \tag{14}$$

where $\varepsilon>0$ is such that $|z'_i|\leqslant\rho-\varepsilon$, $i=1,\ldots,s$.

From (14) we obtain

$$\frac{1}{2\pi i}\int_{\Gamma_{\rho-\varepsilon}}\frac{F(z)Q(z)}{z^{n+m+1}}\,dz=\frac{1}{2\pi i}\int_{\Gamma_{\rho-\varepsilon}}\frac{F(z)(Q(z)-Q_{nm}(z))}{z^{n+m+1}}\,dz$$

and using (2) we get

$$\left|\frac{1}{2\pi i}\int_{\Gamma_{\rho-\varepsilon}}\frac{F(z)Q(z)}{z^{n+m+1}}\,dz\right|\leqslant c_1\left(\frac{q}{\rho-\varepsilon}\right)^{n+m+1}, \tag{15}$$

where c_1 depends on f, m, q, but not on n. Since $\varepsilon > 0$ can be arbitrary small, and

$$\frac{1}{2\pi i}\int_{\Gamma_{\rho-\varepsilon}} \frac{F(z)Q(z)}{z^{n+m+1}}\,dz = \tilde{F}_{n+m},$$

where $\tilde{F}_{n+m}$ is the $(n+m)$-th Taylor coefficient of the function FQ, we obtain from (15)

$$\limsup_{n\to\infty} |\tilde{F}_n|^{1/n} \leqslant q/\rho. \tag{16}$$

Lemma 12.1 then gives us that FQ is a holomorphic function in $D_{\rho/q}$. Since $FQ = fq_sQ$ and all zeros of q_s are in the set $\{z_1, \ldots, z_m\}$ (the zeros of Q), we conclude that the function fQ is also holomorphic in $D_{\rho/q}$. Therefore f is a meromorphic function in $D_{\rho/q}$ and has in this disk $\leqslant m$ poles, so $R_m \geqslant \rho/q$.

Let us consider now $R_m = R_m(f)$. There exist two possibilities for R_m:

(1) $R_m < \infty$;
(2) $R_m = \infty$.

In the first case the statement proved above gives us that in D_{R_m} the function f has exactly m poles. Indeed, if we assume that the number of the poles of f in D_{R_m} is less than s, then using this statement we obtain the contradiction $R_m \geqslant R_m/q$, $0 \leqslant q < 1$.

Therefore in the case (1), according to theorem 12.1, the poles of f in D_{R_m} are the points $z_1, \ldots, z_m$, which proves (c) in this case. To obtain the estimate (b) in case (1), we remark that in the disk D_ρ with $\rho = \alpha_m - \varepsilon$, $\varepsilon > 0$, $\alpha_m = \max\{|z_k| : 1 \leqslant k \leqslant m\}$, there are fewer than m poles of f. Therefore again using the statement proved above we obtain that $R_m \geqslant (\alpha_m - \varepsilon)/q$. Since $\varepsilon > 0$ can be arbitrary small, (b) is proved.

Let us consider now case (2): $R_m = \infty$. If f has exactly m poles in $\mathbb{C}$, then theorem 12.1 gives us that they are $z_1, \ldots, z_m$. We shall show that in case (2) it is impossible for f to have fewer than m poles in $\mathbb{C}$. Let us assume the contrary: $f \in M_s(\mathbb{C})$, $s < m$. Let $z'_1, \ldots, z'_s$, $s < m$, be the poles of f in $\mathbb{C}$. Let us set

$$q_s(z) = \prod_{i=1}^{s} (z - z'_i), \quad F = fq_s.$$

Let

$$F(z) = \sum_{\nu=0}^{\infty} F_\nu z^\nu.$$

Since F is an entire function, we have by lemma 12.1

$$\lim_{n\to\infty} |F_n|^{1/n} = 0. \tag{17}$$

Let us consider the denominator Q_{nm} of the (n, m)-th Padé-approximant (6):

$$Q_{nm}(z) = 1 + a_{n1}z + \cdots + a_{nm}z^m.$$

Since $\|Q_{nm} - Q\| \underset{n\to\infty}{\longrightarrow} 0$ with $\deg Q = m$, i.e. $a_m \neq 0$, we have $a_{nm} \to a_m$ (see (7)), and therefore

$$|a_{nm}| \geqslant |a_m|/2 > 0, \quad n > N_1. \tag{18}$$

We have also like (8)

$$|a_{nm}| \leqslant M + 1, \quad n > N_2. \tag{19}$$

From the definition of Padé-approximant we get

$$f(z)q_s(z) - \frac{P_{nm}(z)q_s(z)}{Q_{nm}(z)} = \mathrm{O}(z^{n+m+1}).$$

Since $\deg q_s < m$, equating the coefficient before z^{n+m} on the left side, we obtain

$$F_{n+m} + a_{n1}F_{n+m-1} + \cdots + a_{nm}F_n = 0$$

or

$$F_n = -(F_{n+m} + a_{n1}F_{n+m-1} + \cdots + a_{nm-1}F_{n+1})/a_{nm}.$$

Using (18) and (19) we obtain for $n > \max\{N_1, N_2\}$

$$|F_n| \leqslant \frac{2m(M+1)}{|a_m|} \max_{1 \leqslant k \leqslant m} |F_{n+m}|. \tag{20}$$

If we denote $c_2 = 2m(M+1)/|a_m| + 1$, we obtain from (20)

$$|F_n| \leqslant c_2 |F_{n+k_1}|, \quad 1 \leqslant k_1 \leqslant m. \tag{21}$$

If we use $Q_{n+k_1,m}$ we can obtain a similar estimate for $|F_{n+k_1}|$, continuing so we can obtain that for every nonnegative integer N we have for $n > \max\{N_1, N_2\}$

$$|F_n| \leqslant c_2^N |F_{n+k_1+\cdots+k_N}|, \quad 1 \leqslant k_i \leqslant m, \quad i = 1, \ldots, N. \tag{22}$$

From (17) it follows that for $n > N_3$ we have

$$|F_n| \leqslant \left(\frac{1}{2c_2}\right)^n. \tag{23}$$

The inequalities (22) and (23) give us for $n > \max\{N_1, N_2, N_3\}$

$$|F_n| \leqslant c_2^N \left(\frac{1}{2c_2}\right)^{n+k_1+\cdots+k_N} \leqslant c_2^N \left(\frac{1}{2c_2}\right)^{n+N} \leqslant \left(\frac{1}{2}\right)^N.$$

Since N can be arbitrary large, it follows that $F_n = 0$ for all

$n > \max\{N_1, N_2, N_3\}$, therefore F is a polynomial. But $F = fq_s$, therefore f is a rational function with denominator of degree $s < m$. This contradicts the condition of the theorem that $\deg Q_{nm} = m$ (since f is a rational function, we have $f = \pi_{nm}$ for sufficiently large n). This contradiction shows that the case $R_m = \infty$, $f \in M_s(\mathbb{C})$, $s < m$, is impossible. □

12.4 The diagonal of the Padé-table

One of the most interesting questions in the theory of Padé approximation is the asymptotic behavior of Padé approximants of order (n, n), i.e. the asymptotic behavior of the diagonal of the Padé-table of the many problems connected with this we shall consider only one converse result of A.A. Gonchar (not the stronger one; see the notes at the end of the chapter). For some direct results see the monograph of Baker and Graves-Morris (1981).

Let again

$$f(z) = \sum_{\nu=0}^{\infty} f_\nu z^\nu \tag{1}$$

be a formal power series and let us consider the (n, n)-th Padé approximant of (1)

$$f(z) - \pi_n(z) = \mathrm{O}(z^{2n+1}), \tag{2}$$

$$\pi_n(z) = p_n(z)/Q_n(z), \quad p_n \in P_n, \quad Q_n \in P_n, \quad Q_n(0) = 1. \tag{3}$$

We shall assume in this section that the Padé approximant for (1) exists for every n.

Theorem 12.4 (Gonchar, 1983). *Suppose f is represented by* (1) *and for the Padé approximant* (2), (3) *we have for every $n > N_0$ that $\deg Q_n = n$. Let π_n be holomorphic in $D = \{z: |z| \leqslant 1\}$ for $n > N_0$. Then there exists ρ_0, $0 < \rho_0 < 1$, such that $R_0 = R_0(f) \geqslant \rho_0$ and π_n converges uniformly to f in $D_{\rho_0} = \{z: |z| \leqslant \rho_0\}$.*

For the proof of this theorem we shall need two statements. The first is the well-known Hermite formula (see Walsh (1960)).

Lemma 12.4. *Let G be a domain in $\mathbb{C}$ and $g \in \mathscr{H}(\bar{G})$. Let $\alpha_k \in G$, $k = 1, \ldots, n$, and let $r \in R_{nm}$ be a rational function with poles at the points $\beta_1, \beta_2, \ldots, \beta_m$, β_i different from α_k, $i = 1, \ldots, m$, $k = 1, \ldots, n$, which interpolate g at the points α_k, $k = 1, \ldots, n$, i.e. $r(\alpha_k) = g(\alpha_k)$ for $k = 1, \ldots, n$. Then*

$$g(z) - r(z) = \frac{1}{2\pi \mathrm{i}} \int_{\partial G} \frac{\prod_{k=1}^{n}(z - \alpha_k)\prod_{k=1}^{m}(t - \beta_k)}{\prod_{k=1}^{m}(z - \beta_k)\prod_{k=1}^{n}(t - \alpha_k)} \frac{g(t)}{t - z}\, \mathrm{d}t.$$

Lemma 12.5. *Let $p \in P_n$, $\Gamma_{\rho_i} = \{z : |z| = \rho_i\}$, $i = 1, 2$, $\rho_1 < \rho_2$. Then*

$$\|p\|_{C(\Gamma_{\rho_2})} \leqslant \left(\frac{\rho_2}{\rho_1}\right)^n \|p\|_{C(\Gamma_{\rho_1})},$$

where $\|p\|_{C(\Gamma)} = \max\{|p(z)| : z \in \Gamma\}$.

Proof. Let us denote $\varphi(\rho) = \max\{|z^n p(1/z)| : |z| = 1/\rho\}$. Since $z^n p(1/z)$ is an algebraic polynomial, by the maximum principle we get $\varphi(\rho_2) \leqslant \varphi(\rho_1)$ if $\rho_2 \geqslant \rho_1$, i.e.

$$\max\left\{\frac{1}{\rho_2^n}|p(w)| : w = \rho_2\right\} \leqslant \max\left\{\frac{1}{\rho_1^n}|p(w)| : w = \rho_1\right\}. \qquad \square$$

Proof of theorem 12.4. We have

$$\pi_{n+1}(z) - \pi_n(z) = \frac{A_n z^{2n+1}}{Q_n(z)Q_{n+1}(z)}, \quad n > N_0, \tag{4}$$

since $\deg Q_n = n$, $\pi_{n+1}(z) - \pi_n(z) = \pi_{n+1}(z) - f(z) + f(z) - \pi_n(z) = O(z^{2n+3}) + O(z^{2n+1}) = O(z^{2n+1})$ and $\pi_{n+1} - \pi_n \in R_{2n+1}$.

Let us estimate A_n. Let $\xi_{n,k}$, $k = 1, \ldots, n$, be the zeros of Q_n. By the assumptions of the theorem $\xi_{n,k} > 1$ for $n > N_0$, $k = 1, \ldots, n$.

Let $\xi_{n,k}$ be an arbitrary zero of Q_n. If we multiply (4) by Q_n and set after this $z = \xi_{n,k}$ we obtain

$$A_n = -\frac{p_n(\xi_{n,k})Q_{n+1}(\xi_{n,k})}{\xi_{n,k}^{2n+1}}. \tag{5}$$

Since $Q_{n+1}(0) = 1$, we have

$$Q_{n+1}(\xi_{n,k}) = \left(\prod_{i=1}^{n+1}(\xi_{n,k} - \xi_{n+1,i})\right) \Big/ (-1)^{n+1} \prod_{i=1}^{n+1} \xi_{n+1,i}$$

$$= \xi_{n,k}^{n+1} \prod_{i=1}^{n+1}\left(\frac{1}{\xi_{n,k}} - \frac{1}{\xi_{n+1,i}}\right). \tag{6}$$

Let us remark that $Q_n(\xi_{n,k}) \neq 0$ since in the opposite case $A_n = 0$ and this contradicts $\deg Q_{n+1} = n + 1$.

From (5) and (6) we obtain

$$|A_n| \leqslant \frac{\|p_n\|_{C(\Gamma_{\rho_n})}}{\rho_n^n} \prod_{k=1}^{n+1}\left(\frac{1}{\xi_{n,k}} - \frac{1}{\xi_{n+1,k}}\right), \quad \rho_n = |\xi_{n,k}|, \tag{7}$$

since

$$|p_n(\xi_{n,k})| \leqslant \|p_n\|_{C(\Gamma_{\rho_n})}.$$

Let ρ be such that

$$2\rho(1+\rho) < (1-\rho)^2. \tag{8}$$

If we apply lemma 12.5 to $p_n, \bar{D}_\rho$ and ρ_n we obtain

$$\| p_n \|_{C(\Gamma_{\rho_n})} \leqslant \left(\frac{\rho_n}{\rho}\right)^n \| p_n \|_{C(\Gamma_\rho)}. \tag{9}$$

Using that $|1/\xi_{n,k}| \leqslant 1$, $|1/\xi_{n+1,k}| \leqslant 1$, we get from (7) and (9)

$$\begin{aligned} |A_n| &\leqslant \| p_n \|_{C(\Gamma_\rho)} \rho^{-n} \prod_{k=1}^{n+1} \left| \frac{1}{\xi_{n,k}} - \frac{1}{\xi_{n+1,k}} \right| \\ &\leqslant 2^{n+1} \| \pi_n \|_{C(\Gamma_\rho)} \frac{\| Q_n \|}{\rho^n} C(\Gamma_\rho) \leqslant 2^{n+1} \| \pi_n \|_{C(\Gamma_\rho)} \left(\frac{1+\rho}{\rho}\right)^n, \end{aligned} \tag{10}$$

since

$$|Q_n(z)| = \left| \left(\prod_{k=1}^{n} (z - \xi_{n,k}) \right) \Big/ \prod_{k=1}^{n} \xi_{n,k} \right| \leqslant (1+\rho)^n, \; |z| = \rho.$$

Let us estimate now the difference $\pi_{n+1} - \pi_n$ on the circle $\Gamma_\rho = \{z : |z| = \rho\}$. Using (4) and (10) we get

$$\| \pi_{n+1} - \pi_n \|_{C(\Gamma_\rho)} \leqslant \| \pi_n \|_{C(\Gamma_\rho)} 2^{n+1} \left(\frac{1+\rho}{\rho}\right)^n \frac{\rho^{2n+1}}{\min_{z \in \Gamma_\rho} |Q_n(z) Q_{n+1}(z)|}.$$

Since $\xi_{m,k} > 1$, $m > N_0$, we have

$$\min_{|z|=\rho} \left| \left(\prod_{k=1}^{m} (z - \xi_{m,k}) \right) \Big/ \prod_{k=1}^{m} \xi_{m,k} \right| \geqslant (1-\rho)^m.$$

From here we obtain

$$\| \pi_{n+1} - \pi_n \|_{C(\Gamma_\rho)} \leqslant \frac{2\rho}{1-\rho} \| \pi_n \|_{\Gamma_\rho} \left(\frac{2\rho(1+\rho)}{(1-\rho)^2}\right)^n. \tag{11}$$

Let us set

$$q = q(\rho) = \frac{2\rho(1+\rho)}{(1-\rho)^2} < 1 \quad \text{(see (8))}.$$

Then (11) gives us

$$\| \pi_{n+1} - \pi_n \|_{C(\Gamma_\rho)} \leqslant \| \pi_n \|_{C(\Gamma_\rho)} q^n$$

and

$$\| \pi_{n+1} \|_{C(\Gamma_\rho)} \leqslant \| \pi_n \|_{C(\Gamma_\rho)} (1 + q^n).$$

Therefore

$$\| \pi_{n+1} \|_{C(\Gamma_\rho)} \leqslant \| \pi_{N_0} \|_{C(\Gamma_\rho)} \prod_{k=N_0}^{n} (1 + q^k).$$

Since $q<1$, $\prod_{k=N_0}^{\infty}(1+q^k)$ converges and therefore we have for every n

$$\|\pi_n\|_{C(\Gamma_\rho)} \leqslant M, \tag{12}$$

where M is a constant (depending on N_0 and q, i.e. on ρ, but not on n).

Using lemma 12.4 for $g=\pi_{n+1}$, $r=\pi_n$, we obtain

$$\pi_{n+1}(z)-\pi_n(z)=\frac{1}{2\pi \mathrm{i}}\int_{\Gamma_\rho}\frac{z^{2n+1}}{t^{2n+1}}\prod_{k=1}^{n}\frac{(1-t/\xi_{n,k})}{(1-z/\xi_{n,k})}\frac{\pi_{n+1}(t)}{t-z}\mathrm{d}t, \quad |z|<\rho, \tag{13}$$

since π_n interpolates $\pi_{n+1}\, 2n+1$ times at $z=0$.

Now let $\rho'<\rho$ be such that

$$q'=\left(\frac{\rho'}{\rho}\right)^2\frac{(1+\rho)}{(1-\rho)}<1. \tag{14}$$

We get from (12), (13)

$$\|\pi_{n+1}-\pi_n\|_{C(\Gamma_{\rho'})}\leqslant\frac{M}{\rho-\rho'}\left(\frac{\rho'}{\rho}\right)^{2n+1}\left(\frac{1+\rho}{1-\rho}\right)^n$$

or

$$\limsup_{n\to\infty}\|\pi_{n+1}-\pi_n\|_{C(\Gamma_{\rho'})}^{1/n}\leqslant q'<1.$$

From here it follows that the series

$$\pi_0(z)+\sum_{n=0}^{\infty}(\pi_{n+1}(z)-\pi_n(z))$$

is convergent in $\bar{D}_{\rho'}$, i.e. $\{\pi_n\}_{n=0}^{\infty}$ converges to a function $g\in\mathscr{H}(\bar{D}_{\rho'})$ uniformly in the disk $\bar{D}_{\rho'}$, where ρ' is given by (14).

We shall show that $\{\pi_n\}_{n=0}^{\infty}$ converges in $\bar{D}_{\rho'}$ to f given by (1). Let

$$\|\pi_n-g\|_{C(\bar{D}_{\rho'})}\underset{n\to\infty}{\longrightarrow}0. \tag{15}$$

Let

$$\pi_n(z)=\sum_{\nu=0}^{\infty}\pi_{n,\nu}z^\nu, \quad n=N_0, N_0+1,\ldots,$$

and

$$g(t)=\sum_{\nu=0}^{\infty}g_\nu z^\nu.$$

From (15) it follows that

$$\pi_{n,\nu}\underset{n\to\infty}{\longrightarrow}g_\nu.$$

But on the other hand from (1)–(3) it follows that $\pi_{nk}=f_k$, $k=0,1,\ldots,2n$. Therefore $g_\nu=f_\nu$ for every $\nu=0,1,2,\ldots$ and $f\in\mathscr{H}(\bar{D}_{\rho_0})_0\rho_0=\rho'$. □

Remark. Using fine calculation it is possible to obtain for $\rho_0=\rho'$ a better estimation than obtained here. The best result is $\rho_0 = 1$ (A.A. Gonchar, 1982), which is obtained using another very complicated method.

12.5 Notes

One classical book for Padé approximation is the book of O. Perron (1957). More modern explanation see in G.D. Baker (1975) and in the encyclopedic monograph of Baker and Graves-Morris (1981).

The interest in the problem of convergence of Padé-approximants is connected with some applications in theoretical physics (see Baker (1965, 1970)).

We mentioned in section 12.2 that usually the direct theorems for convergence of Padé approximants are in terms of convergence in measure or capacity. Nuttall (1970) proved that for every meromorphic function f, analytic in $z = 0$, the sequence of the diagonal Padé approximants π_{nn} converges in measure on compact subsets of $\mathbb{C}$; see also Ch. Pommeranke (1973).

H. Wallin (1974) has shown that for an entire function the Padé approximants π_{nn} can diverge at each point $z \in \mathbb{C} \setminus 0$, i.e. the set of poles of the rational functions π_{nn} is everywhere dense in the complex plane $\mathbb{C}$.

A.A. Gonchar (1973) has shown that if f is an analytic function in $z = 0$, and a single-valued analytic function in its Weierstrass natural domain of existence W_f, then $|f - \pi_n|^{1/n} \to 0$ in capacity inside of W_f.

An extension of Montessus de Ballon's theorem for multipoint Padé approximants is given by E.B. Saff (1972).

For direct theorems concerning Padé approximants of functions which have representation by a Markov–Stieltjes integral see Gonchar (1975b), G. López (1980); for multipoint Padé approximants of such type of functions see G. López (1978a, b), A.A. Gonchar, G. Lopez (1978).

For inverse results for Padé approximants see the survey paper of G. López, V.V. Vavilov (1984). The proof of the unpublished theorems of Gonchar (12.3 and 12.4) was given to us by R. Kovacheva. For an analogous result for a more general situation concerning generalized Padé approximants see R. Kovacheva (1980). For converse problems for the rows of the Padé-table see also A.A. Gonchar (1982), Buslaev, Gonchar, Suetin (1983), R. Kovacheva (1981, 1982, 1984), E.A. Rahmanov (1980), J. Karlsson (1976), S.P. Suetin (1984), V.N. Buslaev (1982).

More for the converse results for the diagonal of the Padé-table see in J. Karlsson, Björn von Sydow (1976) (for functions with representation by integrals of Markov–Stieltjes type), A.A. Gonchar, K.N. Lungu (1981), Gonchar (1983), Gonchar and Rakhmanov (1983).

Very interesting problems exist connected with convergence of subsequences of Padé approximants, see Baker (1973), Baker, Graves-Morris (1981, p. 284).

We want to mention also a very interesting work of M.H. Gutknecht (1984), which is not directly connected with Padé approximants, but deals with the so-called Carathéodory–Fejér approximations.

APPENDIX

Some numerical results

Here we shall give some examples of numerical calculation of the best rational approximation with respect to the uniform and Hausdorff metric.

1. Uniform rational approximation of the function $|x|$ in the interval $[-1,1]$.

Let $R_{n,m}(x)$ be the rational function of order (n,m) of best uniform approximation to $|x|$ in $[-1,1]$ and let $R_{nm}(|x|)$ be the corresponding best uniform approximation. Let $T_n(x)=\cos(n\arccos x)$ be the nth Chebyshev polynomial. Then we have

$$R_{3,2}(x)=R_{2,2}(x)=\frac{0.748\,091\,27+0.728\,155\,49\,T_2(x)}{1+0.543\,689\,01\,T_2(x)},$$

$$R_{4,3}(x)=R_{4,2}(x)=\frac{0.827\,254\,25+0.963\,525\,49\,T_2(x)+0.138\,196\,6\,T_4(x)}{1+0.894\,427\,19\,T_2(x)},$$

$$R_{5,4}(x)=R_{4,4}(x)=\frac{0.867\,934\,4+1.117\,248\,5T_2(x)+0.249\,430\,3\,T_4(x)}{1+1.119\,756T_2(x)+0.134\,022\,83T_4(x)},$$

$$R_{6,5}(x)=R_{6,4}(x)=\frac{\begin{array}{c}0.893\,867\,51+1.238\,028\,3\,T_2(x)\\+0.376\,036\,71\,T_4(x)+0.031\,870\,565T_6(x)\end{array}}{1+1.263\,867\,9T_2(x)+0.265\,111\,55T_4(x)},$$

$$R_{7,6}(x)=R_{6,6}(x)$$

$$=\frac{\begin{array}{c}0.910\,565\,2+1.327\,433\,T_2(x)\\+0.485\,339\,4T_4(x)+0.068\,471\,07T_6(x)\end{array}}{1+1.369\,026\,1\,T_2(x)+0.399\,134\,4T_4(x)+0.030\,034\,3\,T_6(x)}.$$

The corresponding best uniform approximations are the following:

$$R_{3,2}(|x|)=R_{2,2}(|x|)=0.043\,689\ldots,$$
$$R_{4,3}(|x|)=R_{4,2}(|x|)=0.018\,237\ldots,$$

Table 1

	x	$\lvert x\rvert - R_{7,6}(x)$
1	−1	0.002 282 1
2	−0.837 50	−0.002 282 1
3	−0.518 50	0.002 282 1
4	−0.261 10	−0.002 282 1
5	−0.112 35	0.002 282 1
6	−0.040 15	−0.002 282 1
7	−0.010 20	0.002 282 0
8	0	−0.002 282 1
9	0.010 20	0.002 282 1
10	0.040 15	−0.002 282 1
11	0.112 35	0.002 282 1
12	0.261 10	−0.002 282 1
13	0.518 50	0.002 282 1
14	0.837 50	−0.002 282 1
15	1	0.002 282 1

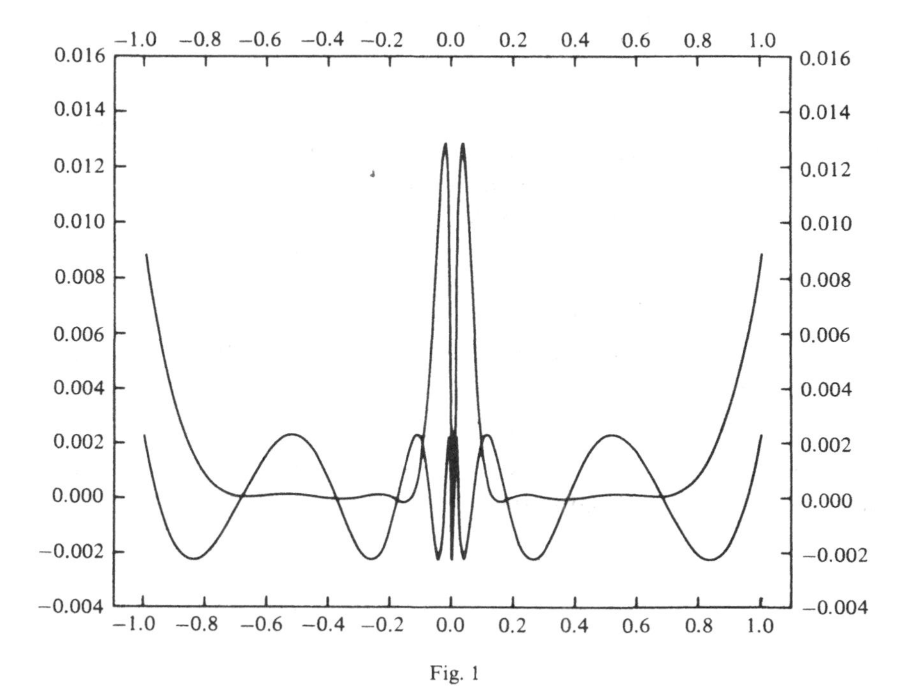

Fig. 1

Table 2

α	$R_{3,2}(x)$	$\Vert R_{3,2}-f_\alpha\Vert_{C[-1,1]}$
0.5	$\dfrac{-0.020\,942\,283-0.130\,199\,79\,T_2(x)}{1+0.893\,557\,74\,T_2(x)}$	0.079 819 1
0.4	$\dfrac{-0.048\,181\,431-0.116\,572\,35\,T_2(x)}{1+0.933\,941\,45\,T_2(x)}$	0.085 19
0.3	$\dfrac{-0.068\,560\,666-0.106\,407\,29\,T_2(x)}{1+0.963\,664\,64\,T_2(x)}$	0.089 10
0.2	$\dfrac{-0.082\,725\,071-0.099\,358\,270\,T_2(x)}{1+0.984\,095\,41\,T_2(x)}$	0.091 77
0.1	$\dfrac{-0.091\,071\,953-0.095\,202\,807\,T_2(x)}{1+0.996\,05\,T_2(x)}$	0.093 32
0.01	$\dfrac{-0.093\,808\,774-0.093\,849\,997\,T_2(x)}{1+0.999\,960\,70\,T_2(x)}$	0.093 83

$$R_{5,4}(|x|)=R_{4,4}(|x|)=0.008\,501\ldots,$$
$$R_{6,5}(|x|)=R_{6,4}(|x|)=0.004\,279\ldots,$$
$$R_{7,6}(|x|)=R_{6,6}(|x|)=0.002\,282\ldots.$$

It is interesting to compare the best uniform approximation to $|x|$ with the approximation given by Newman's rational function from section 4.1. For this purpose we give a graph of the error. On Fig. 1 the graphs of $|x|-N_{6,6}(x)$ and $|x|-R_{6,6}(x)$ are shown, where

$$N_{6,6}(x)=x\frac{p(x)-p(-x)}{p(x)+p(-x)},\quad p(x)=\prod_{k=0}^{5}(x+\xi^k),\quad \xi=e^{-1/\sqrt{6}}.$$

For the best approximation rational function $R_{7,6}(x)=R_{6,6}(x)$ we give the points of alternation and the deviation at these points (Table 1; see also Fig. 1).

Let us mention that $\Vert |x|-N_{6,6}(x)\Vert_{C[-1,1]}=0.0128\ldots$ (see Fig. 1).

2. Let us now consider the best uniform rational approximation to the function f_α which is defined in the interval $[-1,1]$ as follows:

$$f_\alpha(x)=\begin{cases}1-x/\alpha, & 0\leqslant x\leqslant\alpha,\\ 0, & \alpha\leqslant x\leqslant 1,\\ f_\alpha(-x), & -1\leqslant x<0,\end{cases}$$

$0<\alpha\leqslant 1$.

Table 3

α	$R_{5,2}(x)$	$\|R_{5,2} - f_\alpha\|_{C[-1,1]}$
0.5	$\dfrac{0.045\,749\,99 - 0.556\,257\,65\,T_2(x) + 0.092\,685\,102\,T_4(x)}{1 + 0.795\,691\,62\,T_2(x)}$	0.046 574 99
0.4	$\dfrac{0.027\,380\,810 - 0.001\,156\,141\,7\,T_2(x) + 0.079\,750\,009\,T_4(x)}{1 + 0.889\,341\,08\,T_2(x)}$	0.056 090 81
0.3	$\dfrac{0.021\,363\,073 + 0.037\,221\,628\,T_2(x) + 0.069\,125\,404\,T_4(x)}{1 + 0.946\,701\,51\,T_2(x)}$	0.065 603 33
0.2	$\dfrac{0.024\,387\,398 + 0.063\,885\,107\,T_2(x) + 0.060\,346\,917\,T_4(x)}{1 + 0.979\,512\,27\,T_2(x)}$	0.075 078 80
0.1	$\dfrac{0.033\,574\,507 + 0.081\,995\,429\,T_2(x) + 0.053\,041\,870\,T_4(x)}{1 + 0.995\,531\,10\,T_2(x)}$	0.084 494 70
0.01	$\dfrac{0.045\,440\,449 + 0.092\,883\,026\,T_2(x) + 0.047\,484\,261\,T_4(x)}{1 + 0.999\,960\,21\,T_2(x)}$	0.092 905 7

This function is an integral of an 'atom' (see the notes to Chapter 8). The Tables 2 and 3 show the best uniform approximation to the function f_α by different $\alpha \in (0, 1]$ using rational functions of order (3, 2) and (5, 2), the corresponding rational functions of best uniform approximation are denoted by $R_{3,2}(x)$ and $R_{5,2}(x)$

On Fig. 2 the graph of the functions $R_{3,2}(x)$ are given for $\alpha = 0.5$, 0.4, 0.3, 0.2, 0.1 and 0.01.

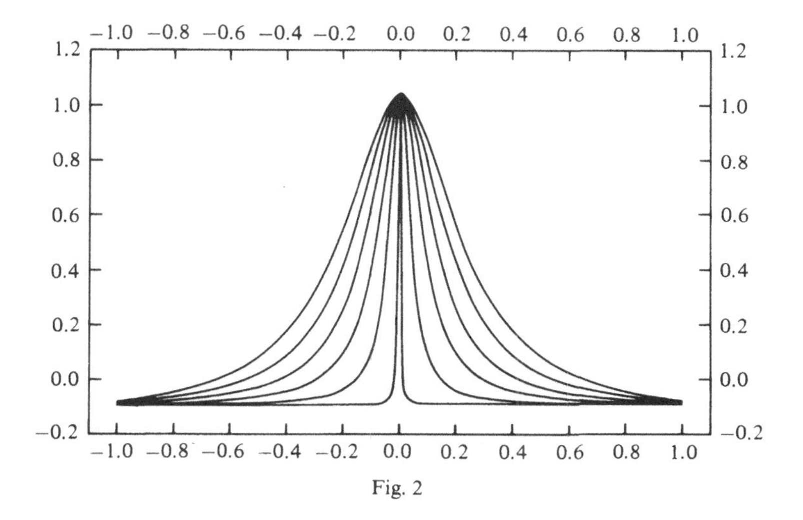

Fig. 2

The best uniform rational approximation of order (5, 6) of the function $f_{0,1}$ is the following:

$$R_{5,6}(x) = \frac{-0.000\,647\,465 - 0.000\,864\,587 T_2(x) - 0.000\,216\,969 T_4(x)}{1 + 1.501\,479\,3 T_2(x) + 0.602\,377\,42\, T_4(x) + 0.100\,897\,97\, T_6(x)}$$

(see Fig. 3).

The error is $\| R_{5,6} - f_{0,1} \|_{C[-1,1]} = 0.0224\ldots$. On Fig. 4 the graph of the error is given.

Evidently $R_{5,6}(x)$ can be used also as a good approximation of the function

$$\tilde{f}_{0.1}(x) = \begin{cases} f_{0.1}(x), & -1 \leqslant x \leqslant 1, \\ 0, & |x| > 1, \end{cases}$$

on the real line, i.e. $\| R_{5,6} - \tilde{f}_{0.1} \|_{C(-\infty,\infty)} = 0.0224\ldots$ and that is because the degree of the denominator is bigger than the degree of the numerator.

We see that the degree of the best rational uniform approximation of such 'integrals of atoms' does not depend very much on α. Let us note that the

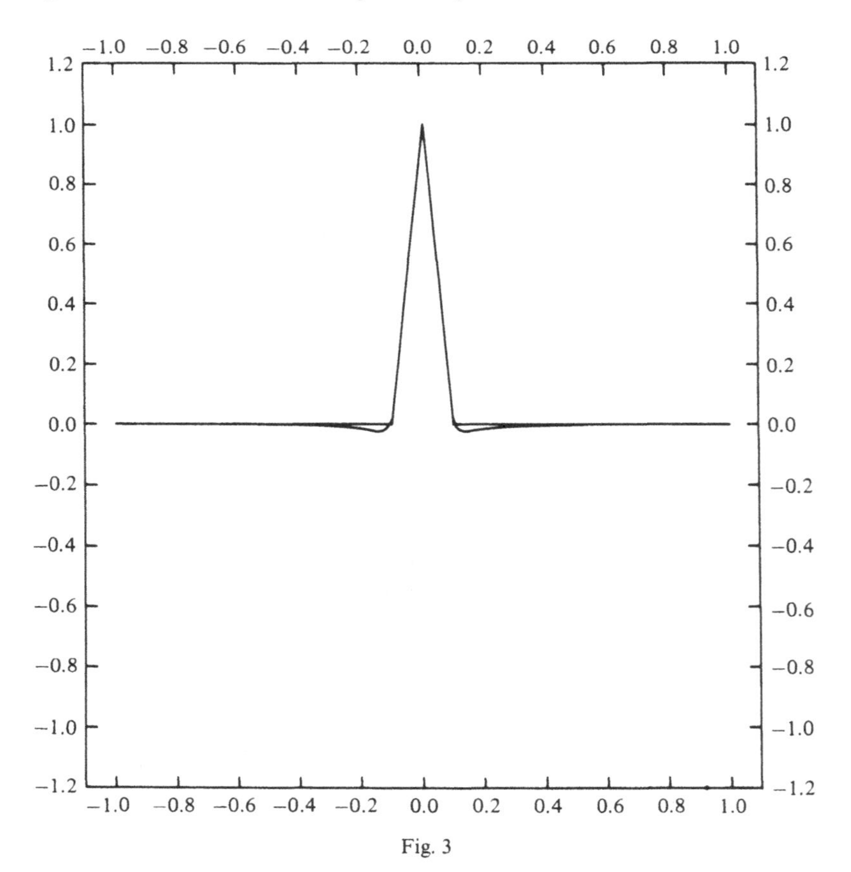

Fig. 3

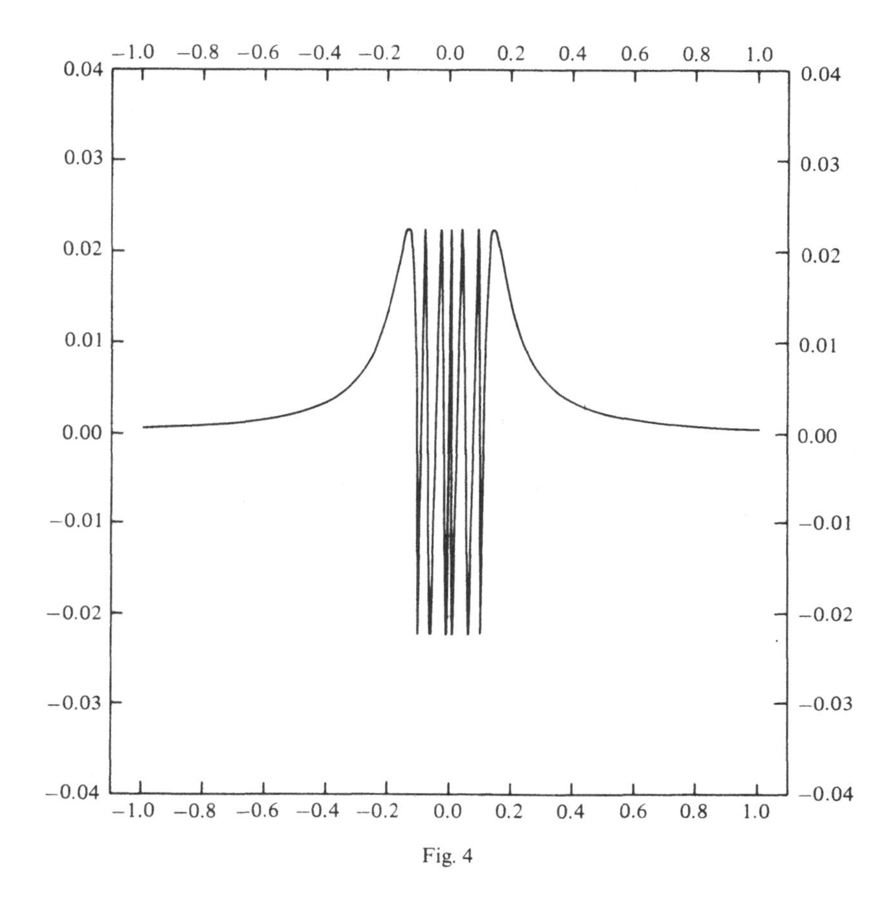

Fig. 4

corresponding polynomial approximation is bad. We have

$$\|R_{16,0}-f_{0.5}\|_C=0.0359\ldots,\quad \|R_{16,0}-f_{0.4}\|_C=0.0487\ldots,$$
$$\|R_{16,0}-f_{0.3}\|_C=0.0558\ldots,\quad \|R_{16,0}-f_{0.2}\|_C=0.0869\ldots,$$
$$\|R_{16,0}-f_{0.1}\|_C=0.2777\ldots,\quad \|R_{16,0}-f_{0.01}\|_C=0.4966\ldots,$$

and of course

$$\lim_{\alpha\to 0}\|R_{n,0}-f_\alpha\|_{C[-1,1]}=0.5.$$

3. Best Hausdorff approximation to the function sign x by rational functions in the interval $[-1,1]$.

The rational functions are a very convenient instrument for approximation of functions possessing discontinuities. We shall denote by $R^r_{n,m}(x)$ the rational function of order (n,m) of best Hausdorff approximation to sign x and by $r(R^r_{n,m}, \operatorname{sign} x)$ the Hausdorff distance with a parameter $\alpha=1$ between $R^r_{n,m}$ and sign x. Table 4 shows the difference between the best polynomial and best rational approximation:

Table 4

(n, m)	$r(R^r_{n,m}, \operatorname{sign} x)$	(n, m)	$r(R^r_{n,m}, \operatorname{sign} x)$
(2, 2)	0.171 573...	(4, 0)	0.270 8...
(3,3)	0.085 12...	(6, 0)	0.207 9...
(4, 4)	0.047 0...	(8, 0)	0.171 01...
(5, 5)	0.0278...	(10.0)	0.147 5...
(6, 6)	0.0173...	(12, 0)	0.130 11...
(7, 7)	0.011 2...	(14, 0)	0.116 96...

The graphs of the rational functions $R^r_{5,5}$ and $R^r_{10,0}$ of best Hausdorff approximation to $\operatorname{sign} x$ are given in Fig. 5.

The graphs of the rational functions $R^r_{7,7}$ and $R^r_{14,0}$ of best Hausdorff approximation to $\operatorname{sign} x$ are given in Fig. 6.

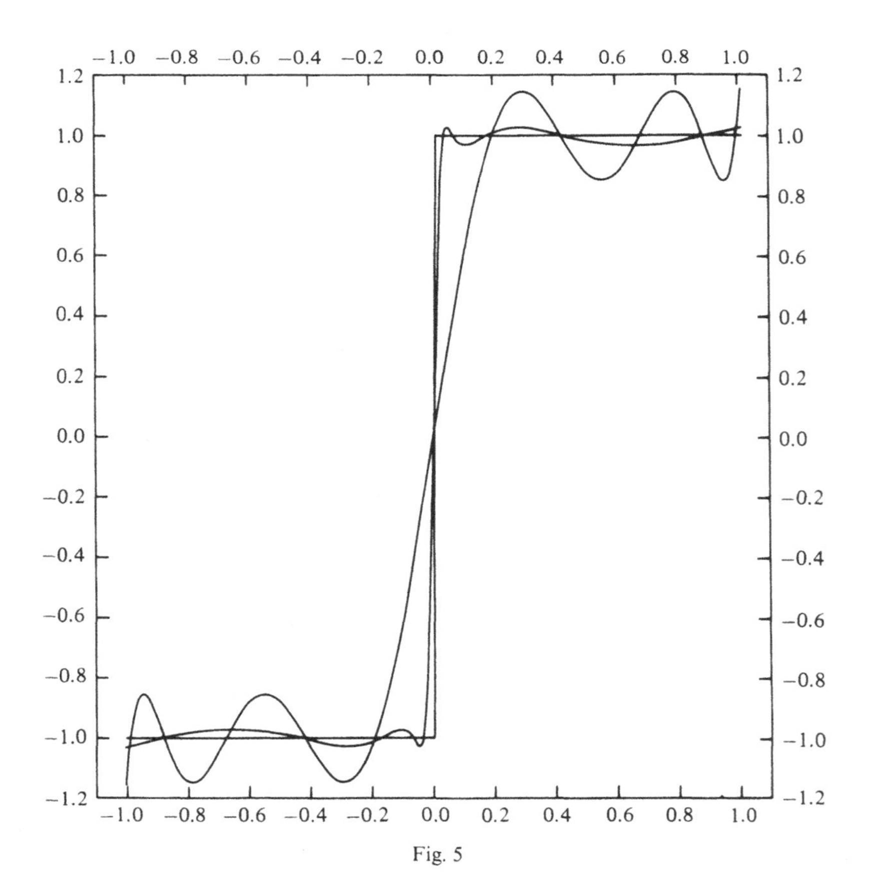

Fig. 5

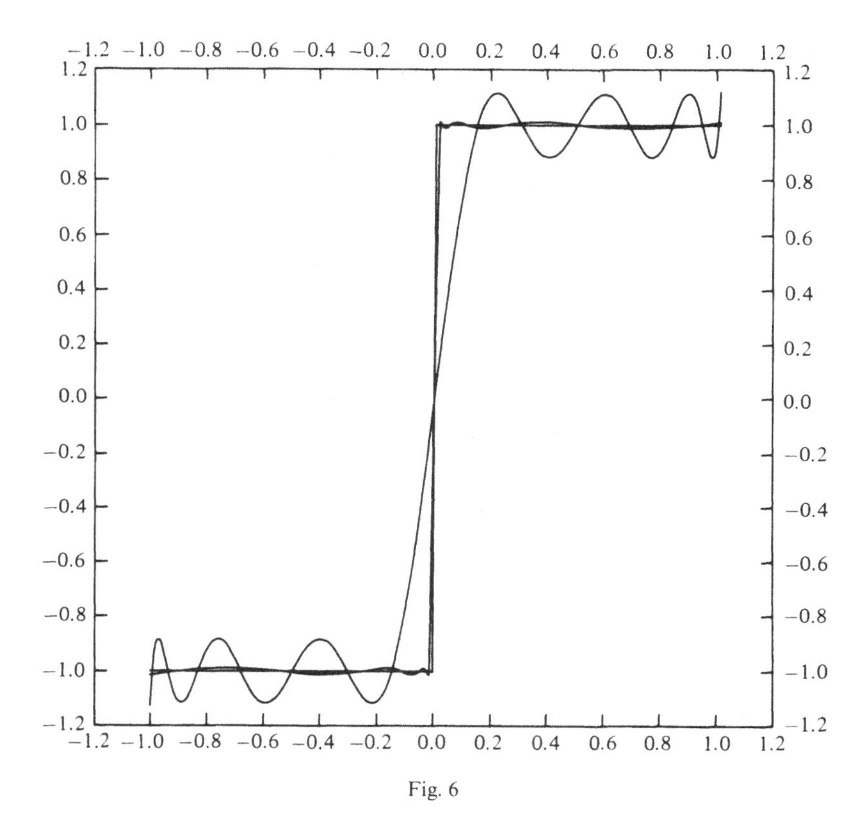

Fig. 6

Let us give the rational functions $R^r_{5,5}$ and $R^r_{7,7}$:

$$R^r_{5,5}(x) = \frac{1.818\,654 T_1(x) + 0.761\,431\,9 T_3(x) + 0.095\,369\,84 T_5(x)}{1 + 1.301\,398 T_2(x) + 0.301\,595\,6 T_4(x)},$$

$$R^r_{7,7}(x) = \frac{1.861\,95 T_1(x) + 1.004\,310 T_3(x) + 0.261\,745\,9 T_5(x) + 0.022\,530\,5 T_7(x)}{1 + 1.473\,508 T_2(x) + 0.557\,810\,6 T_4(x) + 0.084\,302\,46 T_6(x)}.$$

4. Best Hausdorff approximation of the function

$$f^*(x) = \begin{cases} 0, & -1 \leqslant x \leqslant -1/2, \\ 1, & 1/2 < x \leqslant 0, \\ -x + 1, & 0 < x \leqslant 1/2, \\ 1/2, & 1/2 < x \leqslant 1, \end{cases}$$

by polynomials and rational functions.

Let $R^r_{n,m}(x)$ be the rational function of order (n, m) of best Hausdorff approximation to f^* and let $r(R^r_{n,m}, f^*)$ be the Hausdorff distance with a parameter $\alpha = 1$ between $R^r_{n,m}$ and f^* (see Chapter 9).

We have

$$r(R^r_{7,6}, f^*) = 0.0144\ldots; r(R^r_{13,0}, f^*) = 0.0842\ldots,$$

$$R^r_{7,6}(x) = \frac{\begin{array}{l}0.498\,970\,5 + 0.903\,271\,9T_1(x) + 0.673\,600\,9T_2(x)\\ +0.414\,160\,3T_3(x) + 0.210\,746\,4T_4(x) + 0.085\,677\,84T_5(x)\\ +0.025\,629\,47T_6(x) + 0.004\,219\,460T_7(x)\end{array}}{\begin{array}{l}1 + 1.756\,451T_1(x) + 1.355\,470T_2(x)\\ +0.786\,168\,3T_3(x) + 0.404\,859\,9T_4(x)\\ +0.132\,995\,2T_5(x) + 0.038\,723\,91T_6(x),\end{array}}$$

$$\begin{aligned}R^r_{13,0}(x) = {} & 0.458\,887\,5 + 0.240\,521\,4T_1(x) - 0.348\,999\,3T_2(x)\\ & +0.073\,673\,53T_3(x) + 0.201\,407\,3T_4(x)\\ & -0.128\,726\,9T_5(x) - 0.041\,029\,47T_6(x)\\ & +0.056\,801\,36T_7(x)\\ & -0.021\,748\,364T_8(x) + 0.006\,473\,36T_9(x)\\ & +0.048\,284\,21T_{10}(x) - 0.087\,141\,37T_{11}(x)\\ & +0.037\,407\,69T_{12}(x) + 0.088\,398\,57T_{13}(x).\end{aligned}$$

In Fig. 7 the graphs of the functions f^*, $R^r_{7,6}$ and $R^r_{13,0}$ are given.

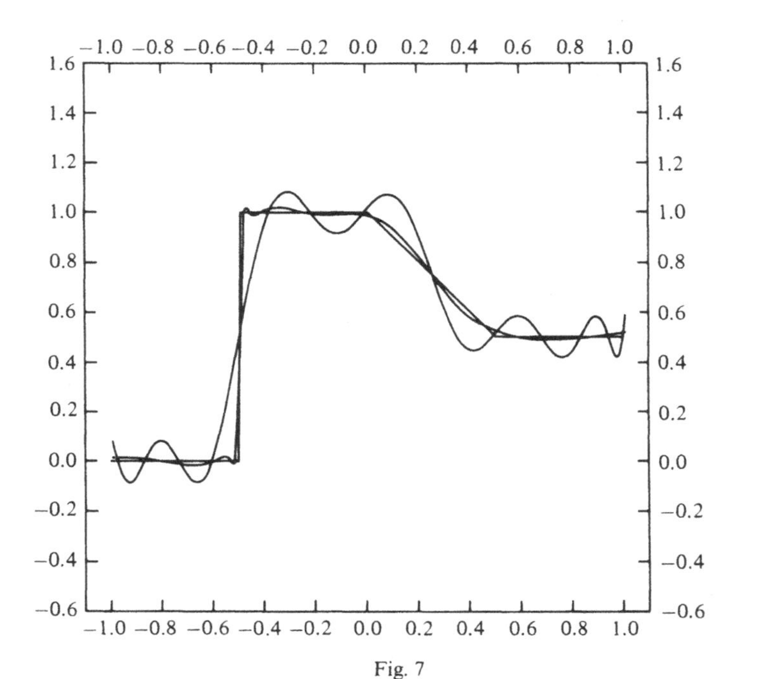

Fig. 7

REFERENCES

Abdulgaparov A.A. (1974) On the rational approximation of functions with derivative of bounded variation (Russian), *Mat. Sbornik*, **93**, 611–20.

Agahanov S.A., N.Sh. Zagirov (1978) Approximation of the function sign x in uniform and integral metric by means of rational functions (Russian), *Mat. Zametki*, **23**, 825–38.

Ahiezer N.I. (1930) On extremal properties of certain rational functions (Russian), *Dokl. Akad. Nauk. SSSR*, 495–9.

(1965) *Lectures on approximation theory* (Russian), Nauka, Moscow.

(1970) *Elements of the theory of elliptic functions* (Russian), Nauka, Moscow.

Anderson Jan E. (1980) Optimal quadrature of H^p functions, *Math. Z.*, **172**, 55–62.

Baker G.A., Jr (1965) The theory and application of the Padé approximant method, *Advances in Theoretical Physics*, 1, Academic press, New York, 1–58.

(1970) The Padé approximant method and some related generalizations, *The Padé approximant in Theoretical Physics*, Academic Press, New York, 1–39.

(1973) Existence and convergence of subsequences of Padé approximants, *J. Math. Anal. Appl.*, **43**, 498–528.

(1975) *The essentials of Padé approximants*, Academic Press, New York.

Baker G.A., Jr, P. Graves-Morris (1981) *Padé approximants*, Encyclopedia of mathematics and its applications, Part I, vol. 13, Part II, vol. 14, Cambridge Univ. P.

Barrodale I., M.J.D. Powell, F.D.K. Roberts (1972) The differential correction algorithm for rational l_∞ approximation, *SIAM J. Numer. Anal.*, **9**, 493–504.

Bergh J., J. Löfström (1976) *Interpolation spaces*, Springer-Verlag, Berlin, Heidelberg, New York.

Bergh J., J. Peetre (1974) On the spaces V_p $(0 < P \leqslant \infty)$, *Bollettino della Unione Mathematica Italiana*, **4**, 632–8.

Bernstein S.N. (1912) Sur l'ordre de la meilleure approximation des fonctions continues par les polynômes de degré donné, *Mén. Acad. royale Belg.*, **4**, 1–104.

(1952) *Collected works*, vol. I (Russian).

(1954) *Collected works*, vol. II (Russian).

Binev P., K.G. Ivanov (1985) On a representation of mixed finite differences, *Serdica*, **11**.

Boehm B. (1965) Existence of best rational Tchebycheff approximations, *Pacific J. Math.*, **15**, 19–28.

de Boor C. (1972) Good approximation by splines with variable knots, *Spline functions and approximation theory* (A. Meir and A. Sharma eds.), Birkhäuser, Basle, 57–72.

(1978) *A practical guide to splines*, Springer-Verlag, Berlin, Heidelberg, New York.

Braess D. (1984) On the conjecture of Meinardus on rational approximation of e^x, II, *J. Approx. Theory*, **40**, 375–9.

(1986) *Nonlinear approximation*, Springer-Verlag.
Brosovski B. (1965a) Über die Eindeutigkeit der rationalen Tschebyscheff Approximationen, *Numer. Math.*, **7**, 176–86.
(1965b) Über Tschebyscheffsche Approximationen mit verallgemeinerten rationalen Functionen, *Math. Zeit.*, **90**, 140–51.
(1969) Fixpunktsätze in der Approximationstheorie, *Mathematica*, **11**(54), 195–220.
Brosovski B., C. Guerreiro (1984) Conditions for the uniqueness of best generalized rational Chebyshev approximation to differentiable and analytic functions, *J. Approx. Theory*, **42**, 149–72.
Brudnyi Yu. A. (1963) A generalization of a theorem of A.F. Timan (Russian), *Dokl. Akad. Nauk SSSR*, **148**, 1237–40.
(1970) Approximation of functions of *n*th variables by quasipolynomials (Russian), *Isv. Akad. Nauk SSSR*, seria mat. **34**, 564–83.
(1971) Piecewise polynomial approximation and local approximations (Russian), *Dokl. Akad. Nauk SSSR*, 201, 1–4.
(1974) Spline approximation and functions with bounded variation (Russian), *Dokl. Akad. Nauk SSSR*, 215, 511–13.
(1979) Rational approximation and imbedding theorems (Russian), *Dokl. Akad. Nauk SSSR*, **247**, 969–72.
(1980) Rational approximation and exotic Lipschitz spaces, *Quantitative approximation*, Academic Press, New York 25–30.
Bulanov A.P. (1969) On the order of approximation of convex functions by rational functions (Russian), *Isv. Akad. Nauk SSSR*, seria mat., **33**, 1132–48.
(1975a) Asymptotics for the best rational approximation of the function sign x (Russian), *Mat. Sbornik*, **96**, 171–8.
(1975b) Rational approximation of continuous functions with bounded variation (Russian), *Isv. Akad. Nauk SSSR*, seria mat., **39**, 1142–81.
(1978) Approximation of convex functions with a given modulus of continuity by rational functions (Russian), *Mat. Sbornik*, **105**, 3–27.
Bulanov A.P., A. Hatamov (1978) On rational approximation of convex functions with a given modulus of continuity, *Analysis Mathematica*, **4**, 237–46.
Burchard H.G. (1974) Splines (with optimal knots) are better, *J. Applicable Anal.*, **3**, 309–19.
(1977) On the degree of convergence of piecewise polynomial approximation on optimal meshes, *Trans. AMS*, **234**, 531–59.
Burchard H.G., D.F. Hall (1975) Piecewise polynomial approximation on optimal meshes, *J. Approx. Theory*, **14**, 128–47.
Buslaev V.N. (1982) On the poles of *m*-th rows of the Padé-table (Russian), *Mat. Sbornik*, **117**, 435–41.
Buslaev V.N., A.A. Gonchar, S.P. Suetin (1983) On the convergence of subsequences of *m*-th rows of the Padé-table (Russian), *Mat. Sbornik*, **120**, 540–5.
Butzer P., H. Berens (1967) *Semi-groups of operators and approximation*, Springer-Verlag, Berlin.
Butzer P., H. Dyckhoff, E. Görlich, R.L. Stens (1977) Best trigonometric approximation, fractional order derivatives and Lipschitz classes, *Canad. J. Math.*, **29**, 781–93.
Butzer P., K. Scherer (1968) *Approximationsprozesse und Interpolationsmethoden*, B.F. Hochschulskripten, Mannheim.
Butzer P., R.L. Stens (1976) Chebyshev transform methods in the solution of the fundamental theorem of best algebraic approximation in the fractional case, *Colloquia Math. Soc. János Bolyai, 19, Fourier Analysis and Approximation Theory*, Budapest, 191–212.
Butzer P., R.L. Stens, M. Wehrens (1980) Higher order moduli of continuity based on the Jacobi transformation operator and best approximation, *C. R. Math. Rep. Acad. Sci. Canada*, **2**, 83–8.
Carpenter A.J., A. Ruttan, R.S. Varga (1984) Extended numerical computations on the "1/9" conjecture in rational approximation theory (to appear).
Chalmers B.L., G.D. Taylor (1983) A unified theory of strong uniqueness in uniform

approximation with constraints, *J. Approx. theory*, **37**, 29–43.
Chanturia Z.A. (1974) Modulus of variation of function and its application in the theory of Fourier series (Russian), *Dokl. Akad. Nauk SSSR*, **214**, 63–6.
Cheney E.W. (1966) *Introduction to approximation theory*, Intern. Series in Pure and Appl. Math., McGraw-Hill, New York.
Cheney E.W., H.L. Loeb (1961) Two new algorithms for rational approximation, *Numer. Math.*, **3**, 72–5.
(1962) On rational Chebyshev approximation, *Numer. math.*, **4**, 124–7.
(1964) Generalized rational approximation, *Soc. Industr. Appl. Math. J. Numer. Anal.*, **1**, 11–25.
Cody N.J., G. Meinardus, R.S. Varga (1969) Chebyshev rational approximation to e^{-x} in $[0,\infty)$ and applications to heat conduction problems, *J. Approx. Theory*, **2**, 50–65.
Coifman R.R. (1974) A real variable characterization of H^p, *Studia Math.*, **51**, 269–74.
Collatz L. (1960) Tschebyscheffsche Annäherung mit rationalen Functionen, *Abh. Math. Sem. Univ. Hamburg*, **24**, 70–8.
Collatz L., W. Krabs (1973) *Approximationstheorie: Tschebyscheffsche Approximation mit Anwendungen*, Teubner, Stuttgart.
Danchenko V.I. (1977) On the dependence of boundary properties of analytic functions of the degree of their rational approximations, *Mat. Sbornik* (Russian), **103**, 131–42.
DeVore R.A. (1983) Maximal functions and their applications to rational approximation, *Second Edmonton Conference on Approximation Theory, CMS, Conference Proceedings*, **3**, 143–57.
(1985) On the rational approximation of functions of the class $L \log L$(to appear).
DeVore R., V.A. Popov (1986) Interpolation spaces and nonlinear approximations, Proc. Inter. Conf. Lund 1986, Springer-Verlag (to appear).
DeVore R.A., Xiang-Ming Yu (1986) Multivariate rational approximation, *Trans. AMS* (to appear).
Dgafarov A.S. (1977) Averaged moduli of continuity and some connections with best approximations (Russian), *Dokl. Akad. Nauk SSSR*, **236**, 288–94.
Ditzian Z., V. Totik (1987) Moduli of smoothness, Springer-Verlag (to appear).
Dolženko E.P. (1962) The order of approximation of a function by rational functions and the properties of the function (Russian), *Mat. Sbornik*, **56**, 403–32.
(1963) Estimation for the derivatives of rational functions (Russian), *Isv. Akad. Nauk SSSR*, seria mat., **27**, 9–28.
(1966a) Uniform approximation by rational functions (algebraic and trigonometric) and the global properties of the functions (Russian) *Dokl. Akad. Nauk SSSR*, **166**, 526–9.
(1966b) Properties of the Fourier series of continuous function and the degree of rational approximation (algebraic and trigonometric) (Russian), *Mat. Sbornik*, **71**, 43–7.
(1967) Comparison of the rates of convergence of rational and spline approximation (Russian), *Mat. Zametki*, **1**, 313–20.
(1976) On the differentiability of functions with good approximation in Hausdorff and uniform metric (Russian), *Dokl. Akad. Nauk SSSR*, **230**, 765–8.
(1978) Some exact estimates for the derivatives of rational and algebraical functions (Russian), *Analysis Mathematica*, **4**, 247–68.
Dolženko E.P., V.I. Danchenko (1977) Differentiability of functions of many variables in connection with the rate of their rational approximations (Russian), *Isv. Akad. Nauk SSSR*, seria mat., **41**, 182–202.
Dolženko E.P., E.A. Sevastijanov (1974) Approximation by rational functions in integral metrics and differentiability in means (Russian), *Mat. Zametki*, **16**, 801–11.
(1976a) On approximation of functions in the Hausdorff metric (Russian), *Soviet Mat. Dokl.*, **17**, 188–91.
(1976b) On the approximation of functions in Hausdorff metric by means of piecewise monotone (in particular, rational) functions (Russian), *Mat. Sbornik*, **101**, 508–41.
Dunham G.B. (1967a) Transformed rational Chebyshev approximation, *Numer. Math.*, **10**, 147–52.

(1967b) Transformed rational Chebyshev approximation II, *Numer. Math.*, **12**, 8–10.
Dzjadik V.K. (1956) On the constructive characteristic of functions satisfying the Lipschitz condition on a finite interval of the real line (Russian), *Isv. Akad. Nauk SSSR*, seria mat., **20**, 623–42.
(1958) On the approximation of functions by polynomials on a finite interval of the real axis (Russian), *Isv. Akad. Nauk SSSR*, seria mat., **22**, 337–54.
(1966) On a new method of approximation of functions (Russian), *Ukrainskii Mat. J.*, **18**, 36–47.
(1977) *Introduction to the theory of uniform approximation of functions by polynomials* (Russian), Nauka, Moscow.
Fefferman C. (1971) Characterization of bounded mean oscillation, *Bull. Amer. Math. Soc.*, **77**, 587–8.
Fichera G. (1970) Uniform approximation of continuous functions by rational functions, *Annali di Matematica pure ed applicata*, seri IV, **84**, 375–86.
(1974) On the approximation of analytic functions by rational functions, *J. Math. phys. Sci.*, **8**, 1, 7–19.
Freud G. (1966) Über die Approximation reeler Functionen durch rationale gebrochene Functionen, *Acta Math. Acad. Sci. Hung.*, **17**, 313–24.
(1967) A contribution to the problem of rational approximation of real functions, *Studia Sci. Math. Hung.*, **2**, 419–23.
(1968) On rational approximation of absolutely continuous functions, *Studia Sci. Math. Hung.*, **3**, 383–6.
(1970) On the rational approximation of differentiable functions, *Studia Sci. Mat. Hung.*, **5**, 437–9.
Freud G., V.A. Popov (1969) On approximation by spline functions, *Proceedings of the Conference on Constructive Theory of Functions*, Budapest, 163–72.
(1970) Some questions connected with approximation by splines and polynomials (Russian), *Studia Sci. Math. Hung.*, **5**, 161–71.
Freud G., J. Szabados (1967a) Rational approximation to x^α, *Acta Math. Acad. Sci. Hung.*, **18**, 393–9.
(1967b) On rational approximation, *Studia Sci. Math. Hung.*, **2**, 215–19.
(1978) Rational approximation on the whole real axis, *Studia Sci. Math. Hung.*, **3**, 201–9.
Fuksman A.L. (1965) Structure characteristic of functions for which $e_n(f;[-1,1]) \leqslant Mn^{-(k+\alpha)}$ (Russian), *Uspehi Mat. Nauk*, **20**, 187–90.
Gaier D. (1970) Saturation bei Spline Approximation und Quadratur, *Numer. Math.*, **16**, 129–40.
Ganelius T. (1979) Rational approximation to x^α on [0,1], *Analysis Mathematica*, **5**, 19–33.
(1982) Degree of rational approximation, *Lectures on approximation and value distribution*, SMS, Montréal Univ. P., 9–78.
Garnett J.B. (1981) *Bounded analytic functions*, Academic Press, New York.
Goldstein A.A. (1963) On the stability of rational approximation, *Numer. Math.*, **5**, 431–8.
Gonchar A.A. (1955) On the best approximations by rational functions (Russian), *Dokl. Akad. Nauk SSSR*, **100**, 205–8.
(1959) Converse theorems for the best approximation on closed sets (Russian), *Dokl. Akad. Nauk SSSR*, **128**, 25–8.
(1966) Properties of functions related to their rate of approximation by rational functions, IMC Moscow, in *Amer. Math. Soc. Transl.* 4, **91**, 1970, 99–128.
(1967a) On the rate of rational approximation of continuous functions with typical singularities (Russian), *Mat. Sbornik*, **73**, 630–8.
(1967b) Estimates for the growth of rational functions and their applications (Russian), *Mat. Sbornik*, **72**, 489–503.
(1972) A local condition of single-valuedness for analytic functions (Russian), *Mat. Sbornik*, **89**, 148–64.
(1973) On the convergence of Padé approximants (Russian), *Mat. SSSR Sbornik*, **21**, 155–66.
(1974) The rate of rational approximation and the property of single-valuedness of an

analytic function in the neighborhood of an isolate singular point (Russian), *Math. SSSR Sbornik*, **23**, 254–70.

(1975a) On convergence of Padé approximants for some classes of meromorphic functions (Russian), *Math. SSSR Sbornik*, **26**, 555–75.

(1975b) On the convergence of generalized Padé approximants of meromorphic functions (Russian), *Math. SSSR Sbornik*, **27**, 503–14.

(1982) Poles of rows of the Padé-table and meromorphic continuation of functions (Russian), *Math. SSSR Sbornik*, **43**, 527–46.

(1983) On uniform convergence of diagonal Padé approximants (Russian), *Mat. SSSR Sbornik*, **46**, 539–59.

Gonchar A.A., G. L. López (1978) On Markov's theorem for multipoint Padé approximants, *Mat. SSSR Sbornik*, **34**, 449–59.

Gonchar A.A., K.N. Lungu (1981) Poles of diagonal Padé approximants and the analytic continuation of functions, *Math. SSSR Sbornik*, **39**, 255–66.

Gonchar A.A., E.A. Rahmanov (1983) On the convergence of simultaneous Padé approximants for systems of functions of Markov type, *Proceedings of the Steklov Inst. of Math.*, Issue 3, 31–50.

Gutknecht M.H. (1984) Rational Carathéodory–Fejér approximation on a disk, a circle and an interval, *J. Approx. Theory*, **41**, 257–78.

Gutknecht M.H., Lloyd N. Trefethen (1983a) Real and complex Chebyshev approximation in the unit disk and interval, *Bulletin Amer. Math. Soc.*, new series **8**, 455–8.

(1983b) Nonuniqueness of best rational Chebyshev approximation on the unit disk, *J. Approx. Theory*, **39**, 275–88.

Haar A. (1918) Die Minkowskische Geometrie und die Annäherung an stetige Funktionen, *Math. Ann.*, **78**, 294–31.

Hardy G.H., J.E. Littlewood (1928) A convergence criterion for Fourier series, *Math. Zeit*, **B. 28**, 4, 612–24.

Hardy G.H., J.E. Littlewood, G. Polya (1934) *Inequalities*, Cambridge Univ. P.

Hatamov A. (1975a) On the rational approximation of convex functions of the class Lip α (Russian), *Mat. Zametki*, **18**, 845–54.

(1975b) On the rational approximation of functions with convex derivative (Russian), *Mat. Sbornik*, **98**, 268–79.

(1977) On rational approximation of convex functions (Russian), *Mat. Zametki*, **21**, 355–70.

Ivanov K.G. (1983a) On a new characteristic of functions. II Direct and converse theorems for best algebraic approximation in $C[-1,1]$ and $L_p[-1,1]$, *Pliska*, **5**, 151–63.

(1983b) A constructive characteristic of the best algebraic approximation in $L_p[-1,1]$, $(1 \leqslant p \leqslant \infty)$, *Constructive Function Theory'81*, Sofia, 1983, 357–67.

Ivanov V.A. (1975) Direct and converse theorems of the theory of approximation in spaces L_p, $0 < p < 1$ (Russian), *Mat. Zametki*, **18**, 641–58.

Jackson D. (1911) Über die Genauigkeit der Annäherung stetiger Functionen durch ganze rationale Functionen gegebenen Grades und trigonometrische Summen gegebener Ordnung, Dis. U. Preisschr., Göttingen.

Karlsson J. (1976) Rational interpolation and best rational approximation, *J. Math. Anal. Appl.*, **53**, 38–51.

(1982) *Rational approximation of analytic functions*, Dept. of Math., Chalmers Univ. of Techn. and the Univ. of Göteburg, No. 1982–19.

Karlsson J., Björn von Sydow (1976) The convergence of Padé approximants to series of Stieltjes, *Arkiv för Matematik*, **14**, 43–53.

Kashin B.S., A.A. Saakjan (1984) *Orthogonal series* (Russian), Nauka, Moscow.

Kaufman, Jr, D.J. Leeming, G.D. Taylor (1978) A combined Rémez–differential correction algorithm for rational approximation, *Math. of Computation*, **32**, 233–42.

Kolmogorov A.N. (1948) A remark concerning the polynomials of P.L. Tchebysheff which deviate the least from a given function, *Uspehi Mat. Nauk*, **3**, 216–21.

Koosis P. (1980) *Introduction to H_p spaces*, Cambridge Univ. P.

Kovacheva R.K. (1980) Generalized Padé approximants and meromorphic continuation of functions, *Mat. SSSR Sbornik*, **37**, 337–48.
(1981) Best rational approximation in the space L_2 (Russian), *Pliska*, **4**, 15–22.
(1982) Convergence of generalized Padé approximants for holomorphic functions (Russian), *Serdica*, **8**, 82–90.
(1984) Uniform convergence of Padé approximants. General case, *Serdica*, **10**, 19–27.
Krasnoselskii M.A., Yu. B. Rutitskii (1958) *Convex functions and Orlicz spaces* (Russian), GITTL, Moscow.
Krotov V.G. (1982) On differentiability of the functions in L_p, $0 < p < 1$ (Russian), *Mat. Sbornik*, **117**, 85–113.
Lagrange R. (1965) Sur oscillations d'ordre supérieur d'une fonction numérique. *Ann. scient. Ecole norm. supér.*, **82**, 101–30.
Latter R.H. (1978) A decomposition of $H^p(R^n)$ in terms of atoms, *Studia Math.*, **62**, 92–101.
Loeb H.L. (1964) Rational Chebyshev approximation, *Notices AMS*, **11**, 335.
López G. (1978a) On the convergence of multi-point Padé approximation of Stieltjes type functions (Russian), *Dokl. Akad. Nauk SSSR*, **239**, 793–6.
(1978b) Conditions for the convergence of multi-point Padé approximation of Stieltjes type functions (Russian), *Mat. Sbornik*, **107**(149), 69–83.
(1980) On the convergence of Padé approximation of Stieltjes type meromorphic functions (Russian), *Mat. Sbornik*, **111**(153), 308–16.
López G., V.V. Vavilov (1984) Survey on recent advances in inverse problems of Padé approximation theory, *Rational approximation and interpolation*, Springer Lecture Notes in Mathematics, vol. 1105, Berlin.
Lorentz G.G. (1966) *Approximation of functions*, Holt, Rinehart and Winston, New York.
Lungu K.N. (1971) Best approximation by rational functions (Russian), *Mat. Sbornik*, **10**, 11–15.
Maehly J. (1963) Methods for fitting rational approximation; Part I, *J. Assoc. Comput. Math.*, **7**(1960), 150–62; Parts II and III, *J. Assoc. Comput. Math.*, **10**, 257–77.
Maehly H., Ch. Witzgall (1960) Tschebyscheff-Approximationen in kleinen Intervallen, I, Approximation durch Polynome, *Numer. Math.*, **2**, 142–50.
Meinardus G. (1967) *Approximation of functions; theory and numerical methods*, Springer Tracts in Natural Philosophy vol. 13, Berlin.
Meinardus G., D. Schwedt (1964) Nicht-lineare Approximationen, *Arch. Rational Mech. Anal.*, **17**, 297–326.
Motornii V.P. (1971) Approximation by algebraic polynomials in L_p metric (Russian), *Izv. Akad. Nauk SSSR*, ser. mat., **35**, 874–99.
Natanson I.P. (1949) *Constructive function theory* (Russian), Gostehizdat (English translation: Ungar, New York, 1964).
Newman D. (1964a) Rational approximation to $|x|$, *Michigan Math. J.*, **11**, 11–14.
(1964b) *On approximation theory*, Intern. Series of Numer. Math., vol. 5, Birkhäuser, Basle.
(1979a) *Approximation with rational functions*, Amer. Math. Soc., N. 41.
(1979b) Rational approximation to e^x, *J. App. Theory*, **27**, 234–5.
Newman D., H. Shapiro (1964) *Approximation by generalized rational functions*, Intern. Series of Numer. Math., vol. 5, Birkhäuser, Basle, 245–51.
Nikol'skij S.M. (1946) On the best polynomial approximation of functions which satisfy the Lipschitz condition, *Izv. Akad. Nauk SSSR*, seria mat., **10**, 295–318.
(1969) *Approximation of functions of several variables and imbedding theorems*, (Russian), Nauka, Moscow, 1st edn 1969; 2nd edn, 1977 (English translation of 1st edn, Springer-Verlag, Berlin, Heidelberg, New York, 1975).
Nitsche J. (1969a) Sätze von Jackson–Bernstein Typ für die Approximation mit Spline-Functionen, *Math. Z.*, **109**, 97–100.
(1969b) Umkehr sätze für Spline Approximation, *Compositio Math.*, **21**, 400–16.
Nuttall J. (1970) The convergence of Padé approximants of meromorphic functions, *J. Math. Anal. Appl.*, **31**, 147–53.
Opitz H.U., K. Scherer (1984a) A generalization of the Padé approximation to e^{-x} on $[0,\infty)$,

Constructive theory of functions '84, (Proceedings of the conference in Varna, 1984), Sofia, 1984, 649–57.
(1984b) On the rational approximation of e^{-x} on $[0,\infty)$, *Constructive approximation* (to appear).
Oswald P. (1980) Spline approximation in L_p $(0 < p < 1)$ metric, *Math. Nachr. Bd.*, **94**, 69–96.
Peetre J. (1963) *A theory of interpolation of normed spaces*, Lecture notes, Brasilia.
(1968) A theory of interpolation of normed spaces, *Notes de matematica*, **39**, 1–86.
(1976) *New thoughts on Besov spaces*, Duke University, Mathematics series I.
(1983) Hankel operators, rational approximation and applied questions of analysis, *Second Edmonton Conference on Approximation Theory, CMS Conference Proceedings*, vol. 3, 287–332.
Peetre J., G. Spann (1972) Interpolation of normed Abelian groups, *Ann. Mat. Pura Appl.*, **92** 217–62.
Pekarskii A.A. (1977) The successive averaging method in the theory of rational approximation (Russian), *Dokl. Akad. Nauk BSSR*, **21**, 876–8.
(1978a) Rational approximation of absolutely continuous functions (Russian), *Izv. Akad. Nauk BSSR*, ser. math., **6**, 22–6.
(1978b) Rational approximation of continuous functions with given moduli of continuity and variation (Russian), *Vesci Akad. Nauk BSSR*, seria fiz.-mat. nauk, **5**, 34–9.
(1980a) Rational approximation of singular functions (Russian), *Izv. Akad. Nauk BSSR*, seria fiz.-mat. nauk, **3**, 32–40.
(1980b) Estimates of higher derivatives of rational functions and their applications, *Izv. Akad. Nauk BSSR*, seria fiz.-mat. nauk, **5**, 21–8.
(1982) Rational approximation of absolutely continuous functions with derivatives in an Orlicz space, (Russian), *Mat. Sbornik*, **117**(159), 114–30.
(1984) Bernstein type inequalities for the derivatives of rational functions and converse theorems for rational approximation (Russian), *Mat. Sbornik*, **124**(166), 571–88.
(1985) Classes of analytic functions, determined by the best rational approximations in H_p (Russian), *Mat. Sbornik*, **127**, 1, 3–20.
(1986) Estimate for the derivatives of rational functions in $L_p[-1,1]$ (Russian), *Math. Notes*, **39**, 3, 388–94.
Peller V.V. (1980) Hankel operators of the class G_p and their applications (Rational approximation, Gaussian processes, majorant problem for operators) (Russian), *Mat. Sbornik*, **113**(155), 538–81.
(1983) Description of Hankel operators of the class G_p with $p > 0$, investigation of the rate of the rational approximation and other applications (Russian), *Mat. Sbornik*, **122**(164), 481–510.
Perron O. (1957) *Die Lehre von den Kettenbrüchen, II*, 3rd edn, Teubner, Stuttgart.
Petrushev P. (1976a) On the rational approximation of functions with convex r-th derivative, *Acta Math. Acad. Sci. Hung.*, **28**, 315–20.
(1976b) On the rational approximation of functions with convex derivative (Russian), *C.R. Acad. bulg. Sci.*, **29**, 1249–52.
(1976c) The exact order of the best uniform rational approximation of some function classes, *Colloquia Mathematica Societatis János Bolyai*, **19**. *Fourier analysis and approximation theory*, Budapest.
(1977) Uniform rational approximation of functions of bounded variation (Russian), *Pliska, Studia Math. Bulgarica*, **1**, 145–55.
(1979) Uniform rational approximation of functions in the class V_r (Russian), *Mat. Sbornik*, **108**(150), 418–32.
(1980a) Rational approximation of functions in the class V_r (Russian), *C. R. Acad. bulg. Sci.*, **33**, 1607–10.
(1980b) Best rational approximation in Hausdorff metric (Russian), *Serdica*, **6**, 29–41.
(1980c) Lower bound for the best rational approximation in Hausdorff metric (Russian), *Serdica*, **6**, 120–7.
(1980d) Rational approximation of functions of bounded variation in Hausdorff and integral

metric (Russian), *Serdica*, **6**, 202–10.

(1981) Rational and piecewise polynomial approximations (Russian), *C. R. Acad. bulg. Sci.*, **34**, 7–10.

(1983a) Connections between the best rational and spline approximations in L_p metric (Russian), *Pliska*, **5**, 68–83.

(1983b) Some new characteristics in the theory of rational approximation (Russian), *Constructive function theory '81* (Proceedings of the Conference in Varna, 1981), Sofia, 1983, 121–4.

(1984a) Relations between rational and spline approximations, *Acta Math. Sci. Hung.*, **44**, 61–83.

(1984b) On the relations between rational and spline approximations, *Constructive theory of functions '84* (Proceedings of the Conference in Varna, 1984), Sofia, 1984, 672–4.

(1985) Direct and converse theorems for spline approximation and Besov spaces, *C. R. Acad. bulg. Sci.*, **39**, 25–8.

(1987) Relations between rational and spline approximations in L_p metric, *J. App. Theory*, **50**, 141–59.

Pommeranke C. (1983) Padé approximants and convergence in capacity, *J. Math. Anal. Appl.*, **41**, 775–80.

Popov V.A. (1973) Direct and converse theorems for spline approximation with free knots, *C. R. Acad. bulg. Sci.*, **26**, 1297–9.

(1974a) On the rational approximation of functions of the class V_r, *Acta Math. Sci. Hung.*, **25**, 61–5.

(1974b) On the connection between rational and spline approximation, *C. R. Acad. bulg. Sci.*, **27**, 623–6.

(1975) Direct and converse theorems for spline approximation with free knots, *Serdica*, **1**, 218–24.

(1976a) Uniform approximation of functions with derivatives of bounded variation and its applications, *Colloquia Mathematica Societatis János Bolyai*, **19**. *Fourier analysis and Approximation theory*, Budapest, 639–47.

(1976b) Direct and converse theorems for spline approximation with free knots in L_p, *Rev. Anal. Numér. Théorie Approx.*, **5** (1976), 69–78.

(1977) Uniform rational approximation of the class V_r and its applications, *Acta Math. Acad. Sci. Hung.*, **29**, 119–29.

(1980) On the connection between rational uniform approximation and polynomial L_p approximation of functions, *Quantitative approximation*, Academic Press, New York, 267–77.

Popov V.A., P. Petrushev (1977) The exact order of the best rational approximation of convex functions by rational functions, *Mat. Sbornik*, **103**(145), 285–92.

Popov V.A., J. Szabados (1974) On a general localization theorem and some applications in the theory of rational approximation, *Acta Math. Acad. Sci. Hung.*, **25**, 165–70.

(1975) A remark on the rational approximation of functions, *C. R. Acad. bulg. Sci.*, **28**, 1303–6.

Potapov M.K. (1975) On the structure characteristics of classes of functions with given order of best approximation (Russian), *Trudy Mat. Inst. Akad. Nauk SSSR*, **134**, 260–77.

(1977) On the approximation by Jacobian polynomials (Russian), *Vestnik Moskov. Univ. Seria Mat. Meh.*, **5**, 70–82.

(1981) On the conditions for coincidence of some function classes (Russian), *Trudy Sem. I.G. Petrovskij*, **6**, 223–38.

(1983) On the approximation by algebraic polynomials in integral metric with Jacobian weight (Russian), *Vestnik Moskov. Univ. Seria Mat. Meh.*, **4**, 43–52.

Quade E.S. (1937) Trigonometrical approximation in mean, *Duke J.*, **3**, 529–43.

Rahmanov E.A. (1980) On the convergence of Padé approximation in classes of holomorphic functions (Russian), *Mat. Sbornik*, **112**, 162–9.

Ralston, A. (1965) Rational Chebyshev approximation by Remes algorithms, *Numer. Math.*, **7**, 322–30.

Remez E.Ya. (1934a) Sur un procédé convergent d'approximation successives pour déterminer les polynômes d'approximation, *C. R.*, **198**, 2063–5.
(1934b) Sur le calcul effectif des polynômes d'approximation de Tschebyscheff, *C. R.*, **199**, 337–40.
(1969) *Basis of numerical methods of Chebyshev approximation* (Russian), Kiev.
Rice J.R. (1964) *The approximation of functions, Vol. I, linear theory*, Addison-Wesley, Reading, Mass.
(1969) *The approximation of functions, Vol. II, Nonlinear and multivariate theory*, Addison-Wesley, Reading, Mass.
Russak V.N. (1973) Estimates for derivative of the rational function (Russian), *Mat. Zametki*, **13**, 493–8.
(1974) On the rational approximation on the real line (Russian), *Izv. Akad. Nauk BSSR*, seria fiz.-mat. nauk, **1**, 22–8.
(1977) A method for rational approximation on the real line (Russian), *Mat. Zametki*, **22**, 375–80.
(1979) *Rational functions as a tool for approximation* (Russian), Izdatelstvo BGU "V.I. Lenin", Minsk.
(1984) The exact orders of rational approximation of classes of functions represented as a convolution (Russian), *Dokl. Akad. Nauk SSSR*, **279**, 810–12.
Ruttan A. (1977) On the cardinality of a set of best complex rational approximation to real functions, *Padé and rational approximation* (E.B. Saff and R.S. Varga eds.), Academic Press, New York, 303–19.
Saff E.B. (1972) An extension of Montessus de Ballore's theorem on the convergence of interpolating rational functions, *J. App. Theory*, **6**, 63–7.
Saff E.B., R.S. Varga (1977) Nonuniqueness of best approximating complex rational functions, *Bull. Amer. Math. Soc.*, **83**, 375–7.
(1978) Nonuniqueness of best complex rational approximation to real functions on real intervals, *J. App. Theory*, **23**, 78–85.
Saks S. (1937) *Theory of the integral*, Warsaw.
Salem R. (1940) *Essais sur les séries trigonométriques: Actualités scientifiques et industrielles*, N° 862, Herman, Paris.
Schoenberg I.J. (1946) Contributions to problem of approximation of equidistant data by analytic functions, *Quart. Appl. Math.*, **4**, 45–99.
Schönhage A. (1973) Zur rationalen Approximerbarkeit von e^{-x} über $[0, \infty)$, *J. App. Theory*, **7**, 395–8.
(1982) Rational approximation to e^{-x} and related L_2-problems, *SIAM J. Numer. Anal.*, **19**, 1067–82.
Schumaker L. (1981) *Spline functions: basic theory*, Wiley, New York, Chichester, Brisbane, Toronto.
Semmes, S. (1982) Trace ideal criteria for Hankel operators and commutators, preprint.
Sendov, Bl. (1962) Approximation of functions by algebraic polynomials with respect to a Hausdorff type metric (Russian), *God. Sofia Univ., Fiz. -Mat. Fak.*, **55**, 1–39.
(1969) Some problems of theory of approximation of functions and sets in Hausdorff metric, *Uspehi Mat. Nauk*, **27**, 141–78.
(1979) *Hausdorff approximations*, Bulgarian Academy of Science.
(1985) The constants of H. Whitney are bounded, *C. R. Acad. bulg. Sci.*, **8**, 1299–1302.
(1987) On the theorem and constants of H. Whitney, *Constructive approximation*, **3**, 1–11.
Sendov Bl., B. Penkov (1962) ε-entropy and ε-capacity of the set of all continuous functions (Russian), *Vestnik Moskov. Univ., Mat. Meh.*, **3**, 15–19.
Sendov Bl., V. Popov (1972) The exact asymptotics of the best approximation by algebraic and trigonometrical polynomials in Hausdorff metric (Russian), *Mat. Sbornik*, **82**, 138–47.
Sevastijanov E.A. (1973) Some estimates for derivative of the rational functions in integral metrics (Russian), *Mat. Zametki*, **13**, 499–510.
(1974a) Uniform approximation by piecewise monotone functions and some applications to ϕ-variations and Fourier series (Russian), *Dokl. Akad. Nauk SSSR*, **217**, 27–30.

(1974b) On the dependence of differential properties of functions on the rate of their rational approximation in L_p metric (Russian), *Mat. Zametki*, **15**, 79–90.

(1975) Piecewise monotone and rational approximation and uniform convergence of Fourier series (Russian), *Analysis Mathematica*, **1**, 183–295.

(1978) Rational approximation and absolute convergence of Fourier series (Russian), *Mat. Sbornik*, **107**(149), 227–44.

(1980) Rate of the rational approximation of functions and their differentiability (Russian), *Izv. Akad. Nauk SSSR*, seria mat., **44**, 1410–16.

(1985) An estimate of the size of the set of those points where a given function is not differentiable by the rate of its rational approximation (Russian), *Izv. Akad. Nauk SSSR*, seria mat. (to appear).

Singer I. (1970) *Best approximation in normed linear spaces by elements of linear subspaces*, Springer-Verlag.

Somorjai G. (1976) A Müntz type problem for rational approximation, *Acta Math. Acad. Sci. Hug.*, **27**, 197–9.

Stechkin S.B. (1951) On the order of the best approximations of continuous functions (Russian), *Izv. Akad. Nauk SSSR*, seria mat., **15**, 219–41.

Stein E.M. (1970) *Singular integrals and differentiability properties of functions*, Princeton Univ., N.J.

Stein E.M., G. Weiss (1971) *Introduction to Fourier analysis on Euclidean spaces*, Princeton Univ. P., N.J.

Stens R.L. (1977) Gewichtete beste Approximation stetiger Functionen durch algebraische Polynome, *Linear spaces and approximation, Proceedings of the Conference, Oberwolfach, 1977*, Intern. Series of Numer. math. vol. 40, Birkhäuser, Basle, 107–421.

Stens R.L., M. Wehrens (1979) Legendre transform methods and best algebraic approximation, *Annales Societatis Mathematicae Poloniae Series I: Commentationes Mathematicae*, **21**, 351–80.

Storoženko E.A. (1975) Imbedding theorems and best approximations (Russian), *Mat. Sbornik*, **97**, 230–41.

(1977) On the approximation by algebraic polynomials of functions in the class L_p, $0 < p < 1$ (Russian), *Izv. Akad. Nauk SSSR*, seria mat., **41**, 652–62.

(1980) On the Jackson type theorems in H^p, $0 < p < 1$ (Russian), *Izv. Akad. Nauk SSSR*, seria mat., **44**, 946–62.

Storoženko E.A., V.G. Krotov, P. Osvald (1975) Jackson type direct and converse theorems in L_p, $0 < p < 1$, spaces (Russian), *Mat. Sbornik*, **98**(140), 395–415.

Subbotin Yu. N., N.I. Chernyh (1970) The order of the best spline approximations to some function classes (Russian), *Mat. Zametki*, **7**, 31–42.

Suetin S.P. (1984) On a converse problem for m-tuples of the Padé-table (Russian), *Mat. Sbornik*, **124** (166), 234–50.

Szabados J. (1967a) Generalization of two theorems of G. Freud concerning rational approximation, *Studia Sci. Math. Hung.*, **2**, 73–80.

(1967b) Negative results in the theory of rational approximation, *Studia Sci. Math. Hung.*, **2**, 385–90.

Szűsz P., P. Turan (1966) On the constructive theory of functions, *MTA* III, **16**, 33–45; II *Studia Sci. Math. Hung.*, **1**, 65–9; III *Studia Sci. Math. Hung.*, **1**, 315–22.

Tchebycheff P.L. (1899) Sur les questions de minima qui se rattachent à la représentation approximative des fonctions, *Oeuvres*, Vol. 1, St Petersburg, 1899, 273–378.

Tihomirov V.M. (1976) *Some problems of approximation theory* (Russian), Moskow Univ. P.

Timan A.F. (1951) An extension of the Jackson theorem on the best polynomial approximation to continuous functions on finite interval on the real line (Russian), *Dokl. Akad. Nauk SSSR*, **78**, 17–20.

(1960) Theory of approximation of functions of a real variable (Russian), GIFML, Moscow (English translation: Macmillan, New York, 1963).

Timan A.F., M.F. Timan (1950) Generalized moduli of continuity and the best approximation in mean (Russian), *Dokl. Akad. Nauk SSSR*, **71**, 17–20.

Timan M.F. (1958) Converse theorems of constructive theory of functions in L_p spaces (Russian), *Mat. Sbornik*, **46**, 125–32.

Trefethen L.M., M.H. Gutknecht (1983a) Real vs. complex rational Chebyshev approximation on an interval, *Trans. Amer. Math. Soc.*, **280**, 555–61.

(1983b) The Carathéodory–Fejér method for real rational approximation, *SIAM J. Numer. Anal.*, **20**, 420–36.

Triebel H. (1978) *Interpolation theory * Function spaces * Differential operators*, VEB Deutscher Verlag der Wissenschaften, Berlin.

Vallée-Poussin Ch. de la (1910) Sur les polynômes d'approximation et la représentation approchée d'un angle, *Bull. acad. Belgique*, 808–44.

Veidinger L. (1960) On the numerical determination of the best approximation in the Chebyshev sense, *Numer. Math.*, **2**, 99–105.

Vjacheslavov N.S. (1975) On the uniform approximation of $|x|$ by rational functions (Russian), *Dokl. Akad. Nauk SSSR*, **220**, 512–15.

Wallin H. (1974) The convergence of Padé approximants and the size of the power series coefficients, *Appl. Anal.*, **4**, 235–51.

Walsh J.L. (1934) On approximation to an analytic function by rational functions of best approximation, *Math. Z.*, **38**, 163–76.

(1949) Critical points of the polynomials and rational functions, *Amer. Math. Soc. Colloq. Bull.*, **34**, 99.

(1960) *Interpolation and approximation by rational functions in the complex domain*, AMS, Colloquium Publications, vol. XX.

Werner H. (1962) Ein Satz über diskrete Tschebyscheff-Approximation bei gebrochen linearen Funktionen, *Numer. Math.*, **4**, 154–7.

(1963) Rationale Tschebyscheff Approximation, Eigenwertteil und Differenrechnung, *Arch. Rational Mech. Anal.*, **13**, 330–47.

(1964) On the local behaviour of the rational Tchebysheff operator, *Bull. Amer. Math. Soc.*, **70**, 554–5.

Wetterling W. (1963) Ein Interpolationsverfaren zur Lösung der linearer Gleichungssysteme, die der rationalen Tschebyscheff-Approximation auftreten, *Arch. Rational Mech. Anal.*, **12**, 403–8.

Whitney H. (1957) On functions with bounded n-th differences, *J. Math. Pures Appl.*, **36**, 67–95.

(1959) On bounded functions with bounded n-th differences, *Proc. Amer. Math. Soc.*, **10**, 480–1.

Zolotarjov E.I. (1877) Application of the elliptic functions to the problems on the functions of the least and most deviation from zero (Russian), *Zapiskah Rossijskoi Akad. Nauk.*

Zygmund A. (1945) Smooth functions, *Duke Math. J.*, **12**, 47–76.

(1959) *Trigonometric series*, Vol. I, II, Cambridge Univ. P.

AUTHOR INDEX

NOTATION AND SUBJECT INDEX